Plant-Microbe Interaction—Recent Advances in Molecular and Biochemical Approaches

Volume 2: Agricultural Aspects of Microbiome Leading to Plant Defence

D1809694

Plant-Microbe Interaction—Recent Advances in Molecular and Biochemical Approaches

Volume 2: Agricultural Aspects of Microbiome Leading to Plant Defence

Prashant Swapnil
Department of Botany, Acharya Narendra Dev College, Delhi University, New Delhi, India

Mukesh Meena
Department of Botany, Mohanlal Sukhadia University, Udaipur, Rajasthan, India

Harish
Department of Botany, Mohanlal Sukhadia University, Udaipur, Rajasthan, India

Avinash Marwal
Department of Biotechnology, Vigyan Bhawan - Block B, New Campus, University College of Science, Mohanlal Sukhadia University, Udaipur, Rajasthan, India

Selvakumar Vijayalakshmi
School of Food Science and Biotechnology, Kyungpook National University, Daegu, Republic of Korea

Andleeb Zehra
Department of Botany, Institute of Science, Banaras Hindu University, Varanasi, Uttar Pradesh, India

ACADEMIC PRESS
An imprint of Elsevier

ELSEVIER

Academic Press is an imprint of Elsevier
125 London Wall, London EC2Y 5AS, United Kingdom
525 B Street, Suite 1650, San Diego, CA 92101, United States
50 Hampshire Street, 5th Floor, Cambridge, MA 02139, United States
The Boulevard, Langford Lane, Kidlington, Oxford OX5 1GB, United Kingdom

Notices
Knowledge and best practice in this field are constantly changing. As new research and experience broaden our understanding, changes in research methods, professional practices, or medical treatment may become necessary.

Practitioners and researchers must always rely on their own experience and knowledge in evaluating and using any information, methods, compounds, or experiments described herein. In using such information or methods they should be mindful of their own safety and the safety of others, including parties for whom they have a professional responsibility.

To the fullest extent of the law, neither the Publisher nor the authors, contributors, or editors, assume any liability for any injury and/or damage to persons or property as a matter of products liability, negligence or otherwise, or from any use or operation of any methods, products, instructions, or ideas contained in the material herein.

ISBN 978-0-323-91876-3

For information on all Academic Press publications
visit our website at https://www.elsevier.com/books-and-journals

Publisher: Nikki P. Levy
Acquisitions Editor: Nancy J. Maragioglio
Editorial Project Manager: Kyle Gravel
Production Project Manager: Nadhiya Sekar
Cover Designer: Christian Bilbow

Typeset by STRAIVE, India

Working together
to grow libraries in
developing countries

www.elsevier.com • www.bookaid.org

Contents

3. **Endurance of microbes against nitrogen starvation by altering the biochemical and physiological activities of plants**

Reshma Devi Ramesh, Kalaiselvi Selvaraj,
Karnan Muthusamy, Leelavathi Lakshmanan,
Steffi Pulikondan Francis, Siva Vijayakumar
Tharumasivam, and Vijayalakshmi Selvakumar

7. Ameliorative characteristics of plant growth-enhancing microbes to revamp plant growth in an intricate environment

T. Savitha and A. Sankaranarayanan

8.   Immune signaling networks in plant-pathogen interactions

Andleeb Zehra, Mukesh Meena, and Prashant Swapnil

9.   Quorum quenching strategies of endophytic bacteria: Role in plant protection

Etisha Paul, Parikshana Mathur, Charu Sharma, and Payal Chaturvedi

10. Peeking into plant-microbe interactions during plant defense

Shriniketan Puranik, Vindhya Bundela, Amanda Shylla, M. Elakkya, Livleen Shukla, and Sandeep Kumar Singh

11. Understanding plant-plant growth-promoting rhizobacteria (PGPR) interactions for inducing plant defense

Kunal Seth, Pallavi Vyas, Sandhya Deora,
Amit Kumar Gupta, Mukesh Meena, Prashant Swapnil,
and Harish

12. Revitalization of PGPR through integrating nanotechnology for sustainable development in agriculture

Gunja Vasant, Shweta Bhatt, Ragini Raghav, and
Preetam Joshi

15. Plant-microbe interactions to reduce salinity stress in plants for the improvement of the agricultural system

Yashika Maheshwari, Shalini Tailor, Avinash Marwal, and Anita Mishra

16. Metabolomic studies of medicinal plant-fungi interaction

Mahinder Partap, Abhishek Kumar, Pankaj Kumar, Shiv Shanker Pandey, and Ashish R. Warghat

17. Sustainable agricultural approach to study interaction of plants and microbes

Parul Tyagi, Ayushi Singh, Pooja Saraswat,
Ambika Chaturvedi, and Rajiv Ranjan

18. Plant-microbe interactions: Role in sustainable agriculture and food security in a changing climate

Diksha Tokas, Siril Singh, Rajni Yadav, and
Anand Narain Singh

19. Microbial interventions for improving agricultural performance under salt stress

Anisha Shashidharan and Lhea Blue

Contributors

Numbers in parentheses indicate the pages on which the authors' contributions begin.

V. Ambikapathy (79), Department of Botany, A.V.V.M. Sri Pushpam College (Autonomous), (Affiliated to Bharathidasan University, Tiruchirappalli), Thanjavur, Tamil Nadu, India

Shweta Bhatt (227), Department of Biotechnology, Atmiya University, Rajkot, India

Lhea Blue (393), Avila College of Education, Ernakulam, Kerala, India

Vindhya Bundela (167), Department of Microbiology, College of Basic Science and Humanities, Govind Ballabh Pant University of Agriculture and Technology, Udham Singh Nagar, Uttarakhand, India

Ambika Chaturvedi (331), Department of Botany, Dayalbagh Educational Institute, Dayalbagh Agra, India

Payal Chaturvedi (149), Department of Biotechnology, IIS (Deemed to be University), Jaipur, India

Sandhya Deora (201), Laboratory of Phytopathology and Microbial Biotechnology, Department of Botany, Mohanlal Sukhadia University, Udaipur, Rajasthan, India

M. Elakkya (167), Division of Microbiology, ICAR-Indian Agricultural Research Institute, New Delhi, India

Steffi Pulikondan Francis (1,33), Department of Microbiology, Cauvery College for Women (Autonomous), Tiruchirappalli, Tamil Nadu, India

S. Gomathi (79), Department of Botany, A.V.V.M. Sri Pushpam College (Autonomous), (Affiliated to Bharathidasan University, Tiruchirappalli), Thanjavur, Tamil Nadu, India

Amit Kumar Gupta (201), Laboratory of Phytopathology and Microbial Biotechnology, Department of Botany, Mohanlal Sukhadia University, Udaipur, Rajasthan, India

Harish (201), Laboratory of Phytopathology and Microbial Biotechnology, Department of Botany, Mohanlal Sukhadia University, Udaipur, Rajasthan, India

Khushboo Jain (95), Department of Biotechnology, Mohanlal Sukhadia University, Udaipur, Rajasthan, India

Subhesh Saurabh Jha (65), Department of Botany, Institute of Sciences, Banaras Hindu University, Varanasi, Uttar Pradesh, India

Preetam Joshi (227,249), Department of Biotechnology, Atmiya University, Rajkot, India

Abhishek Kumar (311), Biotechnology Division, CSIR-Institute of Himalayan Bioresource Technology, Palampur, Himachal Pradesh, India

Arun Kumar (21), Swami Keshwanand Rajasthan Agricultural University, Bikaner, Rajasthan, India

Pankaj Kumar (21,311), Department of Botany, Purnea University, Purnia; Department of Botany, KKM College, Jamui, Munger University, Munger, Bihar; Department of Biotechnology, Dr. Y.S. Parmar University of Horticulture and Forestry, Solan, Himachal Pradesh, India

Satish Kumar (21), Department of Botany, Purnea University, Purnia, India

Rima Kumari (21), Department of Botany, Purnea University, Purnia, India

Leelavathi Lakshmanan (33), Askoscen Probionics (R&D), Tiruchirappalli, Tamil Nadu, India

Yashika Maheshwari (297), Department of Biotechnology, School of Science, GSFC University, Vadodara, Gujarat, India

Avinash Marwal (95,279,297), Department of Biotechnology, Mohanlal Sukhadia University, Udaipur, Rajasthan, India

Parikshana Mathur (149), Department of Biotechnology, Central University of Rajasthan, Ajmer, India

Mukesh Meena (95,137,201,279), Laboratory of Phytopathology and Microbial Biotechnology, Department of Botany, Mohanlal Sukhadia University, Udaipur, Rajasthan, India

Tushar Mehta (279), Laboratory of Phytopathology and Microbial Biotechnology, Department of Botany, Mohanlal Sukhadia University, Udaipur, Rajasthan, India

P.F. Mishel (1), Department of Botany, Bharathidasan University, Trichy, India

Anita Mishra (95,297), Department of Science (Biotechnology), Biyani Girls College, University of Rajasthan, Jaipur, Rajasthan, India

Karnan Muthusamy (33), Askoscen Probionics (R&D), Tiruchirappalli, Tamil Nadu, India; Grassland and Forage Division, National Institute of Animal Science, Cheonan-si, South Korea

Adhishree Nagda (279), Laboratory of Phytopathology and Microbial Biotechnology, Department of Botany, Mohanlal Sukhadia University, Udaipur, Rajasthan, India

Dhaval Nirmal (249), Department of Biotechnology, Atmiya University, Rajkot, India

Shiv Shanker Pandey (311), Biotechnology Division, CSIR-Institute of Himalayan Bioresource Technology, Palampur, Himachal Pradesh; Academy of Scientific and Innovative Research (AcSIR), Ghaziabad, Uttar Pradesh, India

A. Panneerselvam (79), Department of Botany, A.V.V.M. Sri Pushpam College (Autonomous), (Affiliated to Bharathidasan University, Tiruchirappalli), Thanjavur, Tamil Nadu, India

Mahinder Partap (311), Biotechnology Division, CSIR-Institute of Himalayan Bioresource Technology, Palampur, Himachal Pradesh; Academy of Scientific and Innovative Research (AcSIR), Ghaziabad, Uttar Pradesh, India

Etisha Paul (149), Department of Biotechnology, IIS (Deemed to be University), Jaipur, India

Shriniketan Puranik (167), Division of Microbiology, ICAR-Indian Agricultural Research Institute, New Delhi, India

Ragini Raghav (227), Department of Biotechnology, Atmiya University, Rajkot, India

Reshma Devi Ramesh (33), Askoscen Probionics (R&D); Srimad Andavan Arts and Science College, Tiruchirappalli, Tamil Nadu, India

Rajiv Ranjan (331), Department of Botany, Dayalbagh Educational Institute, Dayalbagh Agra, India

L. Rene Christena (1), GENEI Laboratories Pvt. Ltd., Bengaluru, India

A. Sankaranarayanan (117), Department of Life Sciences, Sri Sathya Sai University for Human Excellence, Kalaburagi, Karnataka, India

Pooja Saraswat (331), Department of Botany, Dayalbagh Educational Institute, Dayalbagh Agra, India

T. Savitha (117), Department of Microbiology, Tiruppur Kumaran College for Women, Tiruppur, Tamil Nadu, India

Vijayalakshmi Selvakumar (33), Askoscen Probionics (R&D), Tiruchirappalli, Tamil Nadu, India; Food Science and Biotechnology, School of Agriculture and Life Sciences, Kangwon National University, Chuncheon, Republic of Korea

Kalaiselvi Selvaraj (33), Department of Microbiology, Government Arts and Science College (W), Orathanadu, Tamil Nadu, India

Kunal Seth (201), Department of Botany, Govt. Science College, Valsad, Gujarat, India

Charu Sharma (149), Department of Biotechnology, IIS (Deemed to be University), Jaipur, India

Anisha Shashidharan (393), Department of Botany, St. Albert's College (Autonomous), Ernakulam, Kerala, India

Livleen Shukla (167), Division of Microbiology, ICAR-Indian Agricultural Research Institute, New Delhi, India

Amanda Shylla (167), DBT-Institute of Bioresources and Sustainable Development, Shillong, Meghalaya, India

Anand Narain Singh (363), Soil Ecosystem and Restoration Ecology Lab, Department of Botany, Panjab University Chandigarh, Chandigarh, India

Ayushi Singh (331), Department of Botany, Dayalbagh Educational Institute, Dayalbagh Agra, India

Sandeep Kumar Singh (167), Division of Microbiology, ICAR-Indian Agricultural Research Institute, New Delhi, India

Siril Singh (363), Soil Ecosystem and Restoration Ecology Lab, Department of Botany; Department of Environment Studies, Panjab University Chandigarh, Chandigarh, India

L.S. Songachan (65), Department of Botany, Institute of Sciences, Banaras Hindu University, Varanasi, Uttar Pradesh, India

Priyankaraj Sonigra (279), Laboratory of Phytopathology and Microbial Biotechnology, Department of Botany, Mohanlal Sukhadia University, Udaipur, Rajasthan, India

Prashant Swapnil (137,201,279), School of Basic Sciences, Department of Botany, Central University of Punjab, Bathinda, Punjab, India

Shalini Tailor (95,297), Department of Biotechnology, Mohanlal Sukhadia University, Udaipur, Rajasthan, India

Sagar Teraiya (249), Department of Biotechnology, Atmiya University, Rajkot, India

Siva Vijayakumar Tharumasivam (33), Srimad Andavan Arts and Science College, Tiruchirappalli, Tamil Nadu, India

Diksha Tokas (363), Soil Ecosystem and Restoration Ecology Lab, Department of Botany, Panjab University Chandigarh, Chandigarh, India

Parul Tyagi (331), Department of Botany, Dayalbagh Educational Institute, Dayalbagh Agra, India

Gunja Vasant (227), Department of Biotechnology, Atmiya University, Rajkot, India

Pallavi Vyas (201), Laboratory of Phytopathology and Microbial Biotechnology, Department of Botany, Mohanlal Sukhadia University, Udaipur, Rajasthan, India

Ashish R. Warghat (311), Biotechnology Division, CSIR-Institute of Himalayan Bioresource Technology, Palampur, Himachal Pradesh; Academy of Scientific and Innovative Research (AcSIR), Ghaziabad, Uttar Pradesh, India

Garima Yadav (279), Laboratory of Phytopathology and Microbial Biotechnology, Department of Botany, Mohanlal Sukhadia University, Udaipur, Rajasthan, India

Rajni Yadav (363), Soil Ecosystem and Restoration Ecology Lab, Department of Botany, Panjab University Chandigarh, Chandigarh, India

Andleeb Zehra (137,279), Laboratory of Mycopathology and Microbial Technology, Department of Botany, Centre of Advanced Study in Botany, Institute of Science, Banaras Hindu University, Varanasi, Uttar Pradesh, India

About the editors

Dr. Prashant Swapnil graduated from Veer Bahadur Singh Purvanchal University, Jaunpur, India. He obtained a PhD from the Department of Botany, Banaras Hindu University, Varanasi, India. Dr. Swapnil is currently an assistant professor in the Department of Botany, Central University of Punjab, India. He has more than 25 peer reviewed international publications and more than 20 book chapters to his credit. He has participated in several international and national conferences. Dr. Swapnil has more than 8 years of teaching experience and 2 years of postdoctoral experience at the International Centre for Genetic Engineering and Biotechnology, New Delhi, India. He has expertise in cyanobacterial stress physiology as well as plant-pathogen interaction.

Dr. Mukesh Meena obtained MSc and PhD from the Department of Botany, Banaras Hindu University, Varanasi, India. He is currently an assistant professor in the Department of Botany, Mohanlal Sukhadia University, Udaipur, Rajasthan, India. Previously, he worked as a postdoctoral research fellow at the Department of Biotechnology, Faculty of Science, Jamia Hamdard, New Delhi, India. Dr. Meena is an active member of various scientific societies, including Mycological Society of India (MSI), Indian Science Congress Association (ISCA), Association of Microbiologists of India (AMI), and Asian PGPR Society of Sustainable Agriculture. He received a best paper presentation award at the Fourth International Conference on Advances in Applied Science and Environmental Technology (ASET), Bangkok, Thailand, in 2016. He has published more than 75 peer reviewed international publications and more than 40 book chapters. He has participated in several international and national conferences. Dr. Meena has expertise using various analytical tools relevant to mycopathological research.

Dr. Harish has been a faculty member at Mohanlal Sukhadia University, Udaipur, Rajasthan, India, since 2012. His lab has received funding from various agencies like University Grants Commission (UGC), Department of Science and Technology (DST), Indo-French Centre for the Promotion of Advanced Research (CEFIPRA), Rashtriya Uchchatar Shiksha Abhiyan (RUSA), and others. His research group has published more than 70 papers.

Dr. Selvakumar Vijayalakshmi is a senior researcher in the Department of Food Science and Biotechnology, Kangwon National University, South Korea. She has more than 9 years of teaching and research experience. She has published 35 papers, 7 book chapters, 11 workshops, and 19 conference presentations.

Dr. Avinash Marwal is an assistant professor in the Department of Biotechnology, Mohanlal Sukhadia University, Udaipur, Rajasthan, India. He has 13 years of research experience in molecular virology, nanobiotechnology, and bioinformatics.

Dr. Andleeb Zehra obtained a PhD from the Centre of Advanced Study in Botany, Banaras Hindu University, Varanasi, India. She is a visiting lecturer at Mianz International College, Male, Maldives. She has extensive research experience in plant-microbe interaction, fungal biology, biological control, and plant growth-promoting microbes. Dr. Zehra is a recipient of the D.S. Kothari Post Doctoral Fellowship from the University Grants Commission, New Delhi, India. She has published more than 40 articles and delivered numerous oral and poster presentations at national and international conferences.

Preface

Climate change is negatively affecting crop production and thus creating long-term food security challenges worldwide. Microbes utilize some specific metabolic capabilities from the host plant they are associated with, resulting in interactions that can either be positive, neutral, or negative. The interaction between plants and microbes alters plant health. Positively, these interactions can enhance tolerance of plants against various abiotic stresses and diseases, activate the plant immune system, help the plant adapt to environment changes, develop mycorrhizal association, and induce systematic acquired resistance. Plants have broad immune receptor arsenals that can detect and deal with all types of pathogens to understand their mechanisms to cause disease in particular plant species. Plant growth-promoting microbes can live freely or be symbiotic. These microbes can be rhizospheric, endophytic, or mycorrhizal and can adapt to stressful conditions in the environment. In challenging conditions, microbes show endurance against nitrogen starvation and exhibit biodegradation (via enzymes such as laccase, hydrolyse, peroxidase, esterase, dehydrogenase, manganese peroxidase, and lignin peroxidase) and bioaugmentation against pesticides. During interaction, plant roots release a wide range of small molecular-weight compounds in the rhizosphere, which might be helpful or even possibly harmful in nature.

Plant-Microbe Interaction—Recent Advances in Molecular and Biochemical Approaches: Volume 2 discusses maintenance of soil microflora, the ameliorative characteristics of plant growth-enhancing microbes, immune signaling networks, quorum quenching strategies, plant growth-promoting rhizobacteria (PGPR) interactions for inducing plant defense, revitalization of PGPR through nanotechnology, plant-microbe interactions under abiotic stress, metabolomic studies, and sustainable agricultural. It highlights the important aspects of plant-microbe interactions, with special emphasis on the biosynthesis of bioactive molecules.

Acknowledgments

The editors of this book are eternally grateful to all the contributing authors for their valuable chapters on the biochemical, physiological, and molecular aspects of plant-microbe interaction. We would like to extend our sincere gratitude to all the reviewers who provided their insightful suggestions to improve the quality, consistency, and content of the chapters. We would also like to express our special thanks to all the members of the editorial advisory board for their guidance and support in the compilation of this book. The editors are extremely grateful to the vice chancellor, director, dean, and head of the associated departments for their constant support and encouragement.

Prashant Swapnil
Mukesh Meena
Harish
Avinash Marwal
Selvakumar Vijayalakshmi
Andleeb Zehra

Chapter 1

Pathogen effectors: Biochemical and structural targets during plant-microbe interactions

Steffi Pulikondan Francis[a], L. Rene Christena[b], and P.F. Mishel[c]
[a]Department of Microbiology, Cauvery College for Women (Autonomous), Tiruchirappalli, Tamil Nadu, India, [b]GENEI Laboratories Pvt. Ltd., Bengaluru, India, [c]Department of Botany, Bharathidasan University, Trichy, India

1 Introduction: An overview of the plant immune system

Plant pathogens use different environmental tactics to proliferate into a plant's intercellular spaces or acquire access through wounds by gas or water pores (stomata and hydathodes). Nematodes and puffins feed directly into a plant cell by inserting a stylet. Fungi may invade or expand hyphae directly on, between, or through plant cells in the epidermal cells. Infection features (haustoria) into the plasma membrane of a host genome may be invaginated by fungi and oomycetes that are harmful and symbiotic [1]. The intracellular matrix, haustorial cytosol, and host soluble proteins are membranes with a close edge. These different pathogen groups supply effectors to the plant body to improve the fitness of the microorganism [2].

Many natural microbes are pathogenic agents that interfere with plant growth and reproduction. An intricate signaling network controls plant immune systems. The network's complexity affects visions and approaches in plant immune network research. The mode of plant immunity is defined mainly by how the common signaling network is used instead of the signaling machinery for each mode, which correlates the reliability of the immunity with the negative effects of immunity on plant fitness [3]. The plant responds to invasion by leveraging an innate immune system with two branches. The first branch notices microbes common to several groups, including pathogens, and reactivates them. The second reacts by creating an impact on host targets to pathogen virulence factors. These plant body systems and parasite compounds provide exceptional insight into biomolecular identification, tissue engineering, and phylogeny realms.

Plant-Microbe Interaction—Recent Advances in Molecular and Biochemical Approaches
https://doi.org/10.1016/B978-0-323-91876-3.00001-4

A thorough understanding of the immune function of plants will support the enhancement of crop production for food, fiber, and biofuels [4].

Plants, unlike vertebrates, feature migratory defense cells and a visceral effector. As opposed to mammals, plants depend on cells' innate immunity and periodic signals from infection sites. There are two parts to the plant immunity system. First, transmembrane pattern recognition receptors (PRRs) are utilized to react instead of microbial- or pathogen-associated molecular patterns (MAMPS or PAMPs) that slowly develop, like flagellin. Second, most resistance (R) genes that encode polymorphism nucleotide binding site-leucine rich repeat (NB-LRR) protein products operate inside the cell [5]. They are named following their typical NB and LRR domains. CATERPILLER/NOD/NLR proteins and the STAND ATPases are similar to NB-LRR proteins. NB-LRR proteins identify pathogenic effectors from various kingdoms and cause related defense reactions. Opposition to NB-LRR-mediated conditions is efficacious over illnesses that can only be developed in living host tissue (obligatory biotrophic) or hemibiotrophic pathogens (necrotrophs). [6].

The four-phase zigzag model shown in Fig. 1 describes a plant's immune system.

Phase 1 recognizes PAMPs (or MAMPs) by PRRs, which results in PAMP-triggered immunity (PTI) to prevent further colonization. Pathogen virulence-enhancing effectors are used in Phase 2. PTI can interfere with effectors. This leads to effector-triggered susceptibility (ETS).

Phase 3 is specifically recognized by a given effector created by the NB-LRR proteins, thus resulting in effector triggered immunity (ETI). Awareness is either implicit or explicit; an effector recognizes it by NB-LRR. ETI is a heightened and enhanced PTI expression that results in pest resistance. In general, hypersensitivity just at the infection site causes cell death or a hypersensitive response (HR).

In Period 4, naturally occurring infections to inhibit ETI by abolishing the ETI-suppressing effectors. New resistance specifications can be achieved with natural selection to allow ETI to reactivate. Every stage, in turn, updates the "guard hypothesis" experimental validation, and we take potential challenges into account in recognizing and controlling the plant's immune system.

This design is proportionate to [PTI—ETS+ETI] for their disease's maximal power tolerance. Step 1: Plants detect the use of PRR to cause PAMP-led immunity to micropathogens (MAMPs/PAMPs, red diamonds). Pathogens that are effective and they provide a pathogen-induced susceptibility to PTI or enable pathogens to be nutritious or dispersed (ETS) [7]. In Phase 3, an NB-LRR protein, an amplified version of PTI (Extension ETI) is recognized as one effector (represented in red) that frequently exceeds a hypersensitive cell death threshold (HR). Pathogen samples harboring the red effector selected in Phase 4 may be acquired (blue) by horizontal gene flow; this can be used to suppress ETI by pathogens. Selection favors new NB-LRR plant alleles capable of recognizing one newly acquired ETI [8].

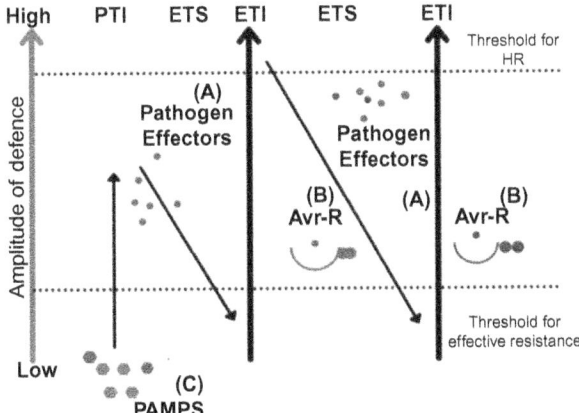

- **(A)- Pathogen effectors**-Effectors are plant pathogens that express proteins for helping the infection of certain species of plants.

- **(B)- (Avr)Avirulence** genes are characterised as the pathogen genes which regulate their specific identification by various plant genotypes.

- **(C)- PAMPs** are active clever immunological responses that safeguard the host against infection by detecting certain conserved non-self molecules.

FIG. 1   A zigzag model showing a plant immune system's quantitative performance.

2   Cell biology of effectors

Plants are sessile species; their climate is continuously threatened, and biotic stresses compound their condition. Several pathogens have severe implications for plant wellbeing. The first line of plant immunity, considered to be the oldest, relies on the identification of pathogenic molecules known as PAMPs/MAMPs [9]. PRRs identify PAMPs and activate plant immune machinery. Several obligate biotrophic phytopathogens, namely oomycetes and fungi, invade and feed on living plant cells through specialized structures known as haustoria. Deploying an arsenal of secreted proteins called effectors, these pathogens balance their parasitic propagation by subverting plant immunity without sacrificing host cells. Such secreted proteins, which are thought to be delivered by haustoria, conceivably reprogram host cells and instigate structural modifications, in addition to modulate various cellular processes. As effectors represent tools to assist disease resistance breeding, this short review provides an introduction to the relationship between the virulence function of effectors and their subcellular localization in host cells.

Like their animal counterparts, pathogenic bacteria use an array of advanced molecules known as effectors to know greater entry into plant resources for eventual colonization. These molecules significantly enhance the ability of the pathogen to spread on its host by interacting with different cell processes. R proteins, the second line of defense, are utilized by plants to track specific effectors (ETI). ETI typically results in a strong, HR that is marked by cell death and shares molecular similarities with animal apoptosis [10].

The precise transmission of pathogens is vital to their success. *Pseudomonas syringae* secretion systems, or bacterial type 3 secretion (T3SS), have been explored. Major pathogens can be injected into host cells by bacteria using syringe-style T3SS, which has a strong, mechanical structure [11]. These secretion mechanisms are absent in many fungi and oomycetes, which are biotrophic pathogens. Instead, they generate specific infections by invading into host cells, known as haustoria. The haustoria force host cells to dramatically stretch their plasma membrane, and pathogens are responsible for this mechanism.

The precise process governing effector translocalization from the outermost cells to the host cells, however, remains a mystery. For this analysis, we have divided efficient conductors into three groups depending on the subcellular regions they belong to: apoplastic, cytoplasmic, and nuclear effectors.

Plant proteases are hampered by inhibitory proteins. By blocking particular host proteases, the Avr2 effector of the biotrophic fungal pathogen *Cladosporium fulvum* weakens the fundamental defense. In contrast to the other two types of effectors, Plant protease directs cytoplasmic factors to other organelles as they move from the cytosol to the cytosol. This offers a look back at the molecular relationships between necessary biotrophic components and their hosts, clarifying this significant molecular connection and concentrating on potentially cellular components.

3 Terminology of effectors: Pathogenicity factors vs. effectors

It is critical to clear up the terminological ambiguity surrounding the cell effector because, until recently, its naming was based on host reactions. A chemical is referred to as a virulence factor when it alters the host's protective layer to benefit the pathogen [12]. Yet, the molecule is referred to as an avirulence factor when host immunoreceptors perceive the same material and do not increase the pathogenicity but rather trigger a defensive reaction. Heterogeneity in pathogenicity is ubiquitous. One effector might have both virulence and avirulence properties depending on the host. Also in a single plant species, where pathogenicity factors vs. effectors interactions between many races have been observed. This paradox leads to limitations because the plant system depends on the specific host system that they were detected in.

In the medical sector, avirulence signifies the loss of a pathogen's virulence component [13]. Because "effector" refers to all chemicals released by a pathogen during infection that alter the structure or function of a host cell, it is an inclusive and neutral term that is favored.

4   Effector form, site, and operation

Based on their localization and place of activity, effectors are grouped into three primary classes: apoplastic, cytoplasmic, and nuclear/nucleolar [14].

As the name suggests, apoplastic effectors are present outside of cells. The - cysteine-rich proteins that largely work by obstructing host proteases, hydrolases, glucanases, and other lytic enzymes are included in this family of factors, albeit they are not only ones. According to recent models, they might be the first to initiate PTI. A signal peptide and a cysteine-rich C-terminus are frequently present, representing the prototypical structure that a protein must possess to survive throughout the transition into apoplastic space [15]. Apoplastic factors might serve a much more precise purpose and act over an extended period of time to protect the pathogenic cell walls or produce antimicrobial substances out from organism in either a chelating/neutralizing approach.

Cytoplasmic effectors have a much more complex duty to deal with host cells. When the plant cytoplasm is reached, cytoplasmic effectors are involved in and appear to target components in plant defense signals. The antipathogenic vesical kinase-based presenter recognition and recruitment, an important defense factor, has been shown to target effectors from *P. syringae* [16]. Some effectors often travel to their final destination through the cytoplasm.

Nuclear effectors appear to be the ultimate weapon in pathogen arsenals, as they are designed to block the upstream immune response. Nuclear factors have the capacity to disable immune machinery master switches or reprogram host transcriptions to benefit infections. An investigation of 49 possible *Hyaloperonospora arabidopsidis* factors revealed that the nucleus was rigidly positioned 33% of the time, with the nucleocytoplasmic complex accounting for the remaining 33%. As some effectors migrate to the nucleus, some R proteins are involved in the nucleus, such as SNCI (SUPPRESSOR OF npr1-1, CONSTITUTIVE1) and N (Nucleocapsid protein). [17]. At higher temperatures, however, nuclear accumulation of SNCI and N is decreased. ETI is more vulnerable to low temperatures (10–23°C), while PTI is more active at higher temperatures (23–32°C). Bacterial pathogens have also been shown to struggle and grow at higher temperatures and more actively secrete their effectors at lower temperatures. These findings indicate a pathogenic pathology for the immune system of plants. However, several pathogens prefer temperatures for optimum growth (around 18°C).

5 Nucleolar-localized effectors

The subcellular distribution in proteins can be predicted with computers such as NOD (nodulation factors), PSORT (different organisms PSORT), and WoLF PSORT (bioinformatics tool used for the prediction of protein localization sites in cells), but very few applicant effectors have been experimentally tested in terms of their wealth of plant pathogens. Although there have been many plant pathogen-secreted effector proteins identified in the nucleus, most localization investigations used green fluorescent protein (GFP)-tagged testing. Moreover, most of these studies are quick tests and do not look at the location during infection. While GFP is a powerful tool to identify the location of subcellular effectors, results obtained should be carefully analyzed. However, as the GFP does not spread to the nucleolus, the nucleolar location is reasonable to presume. The nucleolus of plant cells is located by RXLR effectors, including HyRxLL3b and HaRxL44 from *H. arabidopsidis*. Crinkler (CRN) effectors are all in the nucleus in *Phytophtora capsici*, with at least two found to be in the nucleolus, which suggests they may be in the nucleus.

The nucleolus is a multifunctional subcellular organelle that is crucial for the biogenesis and synthesis of protein. Several DNAs are known to target the nucleolus and retroviruses. The ORF3 umbravirus, capsid virus, and nucleoprotein virus influenza are nucleoli-localized viral proteins. Since viruses depend entirely on the host machine in which their genome is translated into protein, they should target nucleolus [18]. However, one may ask why biotrophic filamentous pathogens target this subnuclear section. A newly shown target nucleolar (and nuclear) mediator subunit 19a was effector HRxL44 from the mandatory biotrophic disease *H. arabidopsidis* (MED19a).

6 Haustorial accommodation: Reprogramming cellular rearrangement

What happens when a microbe gains access to its host? How does the host satisfy the pathogen's requirements? And what are the cellular system's overall dynamics? Obligatory biotrophic pathogens, therefore, need to be subtle after invasion when interacting with their host. First and foremost, removing PTI must keep host immunity under control [19]. Second, pathogens must feed on plant cells continuously. They must spread and multiply constantly; after germination, mushroom spores emerge. The rust fungus *Uromyces appendiculatus* has been shown to use topographical indications to orient and shape infection structures. After *U. appendiculatus* senses a 0.5-μm ridge, which it considers a stomatal lip for entry into the tissue, the structure of the infection begins to develop. If the pathogen has gained footage into plant tissue, it is mainly through haustoria that nutrient acquisition and defense suppression occur. A profound sequence of *Colletotrichum gloeosporioides* growth process supports such a mechanism during *A. thaliana* infection. Effector genes are expressed in successive waves in this pathosystem associated with pathogenic transitions, while others are expressed at the oppressor stage before host invasion [20]. In reality, *Melampsora larici-populina*, the causative agent of popular leaf rust (binding biotrophic), was analyzed on a multi-stage basis to reveal that a number of small, secreted proteins are also expressed in resting urediniospores. Therefore, it can be concluded that removing immunity to plants begins before haustorial tissue structures are established.

Massive cellular host reprogramming should not be difficult to conceptualize in response to haustoria. It has also been found that there is a large amount of tonoplast around these complexes. Cells must extend their plasma membrane enormously to accommodate such vital appendages. Haustoria is isolated from the host cytoplasm (EHM). The EHM is speculated to be mostly host, sealed by a haustorial neckband from the haustoria [8]. However, both cytological and biochemical properties are different from the plasma membrane. The structure of the extrahaustorial membrane (EHM) also seems to change over time. Some plasma membrane-resident proteins have recently been identified to reside during infection in the EHM. In *H. arabidopsidis* or *Phytophthora infestans*, the PEN1 syntax (low penetration 1), the SyT1 synaptotagmin, and StREM1.3, for instance, were found in the EHMs around *P. infestans* haustoria, whereas aquaporin PIP14 and calcium APPase ACA8 remain at the plasma membrane.

Interestingly, it appears that this relocalization is pathogen dependent, as the PRR FLS2 in the EHM of *P. infestans* remains in the plasma membrane, and in *H. arabidopsidis* it is absent. However, the location of the nucleus is the most striking characteristic of this mobile rearrangement. Studies show that the nucleus of *Arabidopsis* stays close to *H. arabidopsidis* haustoria, and probably the actin cytoskeleton is guided by that. Pathogens can deliver their effectors to the nucleus faster for cell rescheduling by the proximity of the haustoria to the nucleus [21]. The pathogens would, therefore, drive the nucleus's proximity to the intruder.

6.1   Possible pathogens target: Vesicular trafficking

It is recognized that pathogens are a key element of plant defense, that of vesicular trafficking. In *H. arabidopsidis*, 26% of the effectors studied were located at membranes, most of which (18%) were associated with the endoplasmic reticulum. Due to the pathogen's presence, *Arabidopsis* cells that harbor *H. arabidopsidis* haustoria grow bulging vesicular structures relative to non-infected cells. The formation of these vesicles can be motivated by a specific vesicular movement or effectors, which can interrupt any coordinated defense response. They also may be pathogens and fast-spreading haustoria in the plasma membrane [22]. However, the support provided by observations of very similar structures of transgenic *Arabidopsis* μ-TIP-GFP plants is based on the fact that these are vacuolar structures. Other differentially localizing forms of membrane structures around haustoria produced by *H. arabidopsidis* and *P. infestans* have been demonstrated.

During infection with *H. arabidopsidis*, HaRxL17 is located at the EHM. In the absence of the pathogen, however, its ability to improve the resistance of plants is related to the role of trafficking in the cellulite membranes. Since the tonoplast is near the EHM along with HaRxL17 in case of infection, it interferes with the trafficking of plant cell membranes, which, interestingly, indicates that tonoplast plays an important role in EHM creation. However, the vesicular structures have not been confirmed to trigger a single effector, and it is not clear whether this is an answer to plant defense. Surprisingly, we are still constrained in our understanding of the extensive vacuolar biogenesis process, which justifies further research into the particular vesicular structures [23]. The possible mechanisms to block vesicular trafficking by pathogens are difficult to elucidate and these bulbous structures are eventually developed. In a point mutation in *A. thaliana*, cells cannot form central lytic vacuoles in the deubiquitinating enzyme AMSH3. Amsh3 mutant cells additionally accumulate and incorrectly load their vacuolar protein. Vacuoles are critical in various plant protection mechanisms and are proposed in impacted programmed cell death. Vacuolar treatment enzymes they mediate vacuolar membrane disruption and release vacuolar material into the cytoplasm of the cells (demonstrated for viral infection). Vacuole fusion with the plasma membrane allows the extracellular release of vacuolar material in the second proposed mechanism. Phenotypic similarities between amsh3 mutant vesicular structures and haustoria are observed [24]. The competing signature indicates that AMSH3 (or other related components) may be targeted by pathogenic substances to change the vesicular pathway.

Pathogens can also target octameric-exocyst complexes, as the exocyst architecture plays a vital role in tethering the blood vesicles and redefining cells that are integral to the plant defense. During infection, targeted exocytosis happens, and defense-related freshly synthesized compounds are given to infections, which eventually leads to the formation of the asymmetric plasma membrane. In this mechanism, it includes delivering, securing, and incorporating by secretary veins into small GTPases from the Rab and Rho families are known to be important in the plasma membrane while the exocyst complex acts as a scaffold in tethering operations [25]. The final attachment process is mediated by the integrated membrane proteins v-SNARE and t-SNARE, which fuse the plasma membrane and the bilayers to complete the process. Two exocyst sub-units, Exo70B and Exo70H1 from *Arabidopsis* plants, have already been demonstrated in mutating to be more vulnerable, their (Exo70B and Exo70H1) has their unique significance in plant immunity.

Habitual biotrophic phytopathogens have developed robust and elaborate offensive approaches, using various effector proteins to invade their host. Pathogen inventory of effectors is organized by various molecules with specific roles and capabilities. Consequently, most of the so-called effectors are cell pharmacist candidates. A harsh way of imagining the deployment of the performer is to see apoplastic effectors at the start of an assault, execute the whole task, and create the stage for more advanced weapons. At the intermediate point, true cytoplasmic effectors may be used to deactivate local surveillance, control the entire defense system, and stop the entire immune system. There is a growing report of nucleolar effectors from different pathogens, and a significant pathogenesis mechanism is likely to occur [26]. Many cellular processes, including plant protection, rely on new protein formation. Therefore, further analysis should be carried out to clarify the cell biology role of plant effectors. Certain actors also interfere with the trafficking of vesicles and thus can jeopardize vacuolar integrity, which plays a key role in plant safety.

7   Cell wall—A dispensable armor in plant immunity

Plants are a part of the natural environment and are exposed to several microorganisms that can challenge their growth and survival. Fortunately, they have a protective cell wall (CW), an essential asset to the plant because of its versatile role in maintaining integrity, communication, and defense mechanism. Nevertheless, cell wall degrading (CWDE) enzymes' secretion and stomatal entry (wounds are the common modes employed by the pathogens to enter the plant) cause leakage of nutrients and systemic infections. As reviewed by Bacete et al., the plant CW has different perception systems linked to osmolarity, mechanical stress, CW defense, and wall-derived ligand-receptor recognition, which induces the synthesis of various phytohormones and modulates the downstream genes related to immune and defense responses [27].

A typical plant CW is a heterogeneous mixture of microfibrils of cellulose, hemicellulose polymers, or the heteropolysaccharide pectin susceptible to pathogen attack. Pathogens can depolymerize and degrade the CW by secreting CWDE such as cellulases, pectinase, and hemicellulases. These enzymes are coded by multi-gene families and grouped into exo- and endopolygalacturonases, pectin methylases, xylanases, acetyl esterases, pectate/pectin lyases, and endoglucanases that have varied bond specificities within the CW [28]. To counteract such enzymes, plants secrete various inhibitors of CWDEs, the best example being polygalacturonase-inhibiting proteins (PGIPs). The others discovered in the past decade are named according to the enzymes they inhibit, such as the pectin methylesterases inhibitor, *Triticum aestivum* xylanase inhibitor, and xyloglucan endoglucanase inhibiting protein.

PGIPs are the foremost secreted during a fungal invasion, where they work to inhibit CW breakdown, fungal development, and colonization. The genes encoding PGIPs have been reported in *A. thaliana*, *Brassica campestris*, *Brassica napus*, and *Oryza sativa*, and have differential specificities despite structural homology [29]. *Botrytis cinerea* is one such fungal pathogen and secretes endopolyglacturonases (PGs) to degrade the homogalacturonans [30]. One of the PGs, BcPG1, is inhibited by PvPGIP isolated from *Phaseolus vulgaris*, as proven through inhibition assays and molecular docking assays.

Pectin lyase inhibitors (PNLIs) and pectin methylesterases inhibitors (PMEIs) are two families of proteins that can prevent the action of the pectin lyases and methylesterases secreted by *Erwinia chrysanthemi, Xanthomonas campestris, Abdopus aculeatus, X. oryzae*, and *Aspergillus niger* [31,32]. The PMEIs are predominantly found in kiwi, orange, carrot, potato, and banana, where they are highly specific. Although these enzymes are involved in pollen tube growth, senescence, and fruit maturation, their extensive role in preventing pathogen invasion is interesting. The PMEIs are known to form an elevated complex with the PME and hence aid in preventing the systemic dissemination of the tobacco mosaic virus in tomato plants [33]. In addition, since PMEs are involved in the methylesterification of the pectins, the PMEIs play an indirect role in preventing this process, thus limiting pathogen attack. A similar strategy has been adopted by *Arabidopsis* and wheat plants that demonstrated greater resistance to necrotrophic pathogens due to high methylesterification and CW stiffness in the presence of PMEIs. PNLs are another group of CWDEs secreted by bacteria and fungi that promote the cleavage of methylesterified pectin. PNLIs have been reported in sugar beet that inhibits PNLs from the fungal pathogens *Rhizoctonia solani, Phoma betae*, and *A. japonicus* [28].

Endoxylanases are another group of hydrolytic enzymes that can degrade xylan subunits in the hemicellulose, forming an integral part of the CW. These enzymes are secreted by many phytopathogenic microbes like *Fusarium graminearum, Streptomyces turgidiscabies*, and *B. cinerea* [34,35]. Based on the structural characteristics, endoxylanase inhibitors are categorized into *Triticum aestivum* xylanase inhibitors and xylanase inhibitor proteins, but they do not have any sequence homology [36]. Although these widely occur in cereals like rye, maize, and rice, they are best purified from wheat and are known as XIP-I, TAXI-I, and TAXI-II. It has been shown that increased expression of TAXI-I in wheat and *Arabidopsis* plants elevated their resistance to *B. cinerea*, a fungal pathogen. Also, the deletion of endoxylanase gene in *B. cinerea* delayed the occurrence of secondary lesions and decreased the size of the lesions in tomato leaves and grape berries. All the Xip-type and Taxi-type genes are induced by abiotic and biotic signals of phytohormones such as jasmonic acid [37]. The recombinant form of this protein is 21 kDa and inhibits only G11 xylanases of both bacterial and fungal origin. TXLIs are very stable proteins compared to counterparts with maximum activity up to 120 min in a pH of 1–12 at 100°C. This property is attributed to the high number of disulfide bonds, a typical characteristic of thaumatin-like proteins [38].

Xyloglucan endoglucanases (XEGs) are another novel class of specific enzymes that act on xyloglucan and carboxymethyl cellulose within the CW proteins and are classified into GH5, GH12, GH16, GH44, and GH74 [28]. Their corresponding inhibitor protein was first identified in tomato cells, and the homologous genes are present in maize, lotus, carrot, sorghum, and so on. A perfect example of such an inhibitor is the GmGIP1 from *P. sojae* that binds to xyloglucanase endoglucanase PsXEG1 and prevents hydrolytic activity [37]. An article by Choi et al. describes the functional characterization of a novel xyloglucan-specific *endo*-b-1,4-glucanase inhibitor1 gene, CaXEGIP1, from pepper plants. It caused significant inhibition of endo-b-1,4-glucanase from *Clostridium thermocellum*, and its overexpression in the transgenic leaves of *Arabidopsis* enhanced the resistance to *H. arabidopsidis* infection by inducing spontaneous cell death [39].

8   DAMPs, MAMPs, and NAMPs—A necessary evil

Responding to external stimuli is an essential aspect of immunological response in any living system. Hence, plants have several receptors that respond to a foreign attack or cellular damage and stimulate downstream signaling. It is critical to understand the different roles of damage-associated molecular patterns (DAMPs), MAMPs, and nematode-associated molecular patterns (NAMPs). DAMPs are byproducts of mechanical or cellular damage that produce endogenous molecules, including protein fragments, peptides, nucleotides, and amino acids, which activate the plant immune system [40]. For example, the CWDEs release fragments of homogalacturonan called oligogalacturonides (OGs) identified by the PRR–CW-associated kinase 1 in *Arabidopsis* and triggers an immune response [27]. Cellobiose and cellotriose are similar molecules released from the CW polysaccharide that has been shown to rapidly increase cytosolic Ca^{2+}, activate mitogen-activated protein kinase (MAPK), and bring about metabolic changes and synergistic immune responses [37]. Recently, extracellular ATP (eATP) has been implicated as a DAMP based on the discovery of its receptor, DORN1, where the *dorn1* mutant shows suppressed transcriptional response to wounding [41]. In addition, eATP treatment induced innate immune responses like cytosolic Ca^{2+} influx and production of nitric oxide and ROS. Plant elicitor peptide *At*Pep1, a 23-aa long peptide, was the first elicitor discovered in *A. thaliana* [40]. It is one of the derivatives of a PROPEP1, a 100-aa precursor protein, released into the apoplast upon cell injury. These peptides confer resistance to various pathogens, including *B. cinerea* and *P. infestans*. Similarly, other apoplastic peptides such as CAP-derived peptide 1, Grim Reaper peptide, *Zea mays* peptide 1, rapid alkalinization factors (RALFs), and phytosulfokine have been reported as potential DAMPs that trigger the immune pathways [40].

Some of the plasma membrane-associated PRRs in plants detect microbial attack through MAMPs or PAMPs and elicit pattern-triggered immunity (PTI) [42]. Bacterial peptidoglycan, flagellin, DNA, lipoproteins, chitin, and bacterial elongation factor Tu are the common MAMPs recognized by PRRs [41]. Such recognition with flagellin (flg22) leads to a cascade of events like the production of ROS and activation of MAPKs. Similarly, lipid A has been reported to induce plant defense responses in *Arabidopsis* and is important for symbiotic signaling. Likewise, bacterial elongation factor Tu and bacterial cold shock proteins are abundant in bacteria that are known to be perceived by the PRRs of *Brassicaceae* and *Solanaceae*, respectively [43]. An interesting study by Kim et al. shows that *Brassica rapa* leaves when treated with elf18, flg22, and chitin, induced the differential expression of genes associated with hormones and defense compounds like cinnamic acid, chorismite, L-phenylalanine metabolism, and sulfur compound biosynthesis. The same MAMPs were effective in curtailing the growth of *X. campestris* in *Brassica*-infected leaves by twofold (elf18, chitin) and eightfold (flg22), respectively.

Nematode colonization is quite challenging during plant expansion; however, plants have evolved to perceive nematode-derived signals. NAMPs associated with parasites include enzymes that degrade CWs, virulence proteins, and different transcription factors that lead to ROS bursts, activation of MAPKs, and JA/Sa signaling. Ascarosides are such small-molecule pheromones recognized through root-knot infections in several dicot and monocot plant species [41]. Interestingly the nematode secretions through the stylet also contribute significantly to modulating the expression of genes by binding to plant cell receptors. They can either act as CWDEs to cause localized CW degradation for the movement of the nematodes or act as effector proteins to modify the metabolic pathways and virulence factors that interact with plant resistance enzymes or function as regulatory proteins by binding to plant promoter genes to favor nematode feeding sites.

8.1 The dynamics of cytoskeleton in plant-microbe interactions

Microtubules (MTs) and actin microfilaments (AFs) form the basic units of cytoskeletal organization in plants that help plant growth, cell division, cell expansion, and intracellular organization. Pathogen effectors have an easier time attacking MTs and AFs and may induce PTI through RLK (Receptor-like kinases) receptors. Numerous reports highlight the physiological participation of MTs and AFs that are essential for plant immune response. For example, pathogen effector HopW1 from *P. syringae* disrupts the AF network by actin-dependent protein targeting and endocytosis [44]. HopG1 is another such effector that facilitates reorganizing the AF network that can enhance the susceptibility towards the pathogen [44]. Both AFs and MTs might be involved in the development or dynamics of PM nanodomains that induce PTI via RLK receptors. Because of their function in building immunity, both MTs and AFs are frequently targets of pathogen effectors (Table 1).

Similarly, Henty-Ridilla report that treating Tomato DC3000 with a pathogenic and nonpathogenic strain of *P. syringae* showed a rapid abundance of actin filament in the epidermal cells by recruiting FLS2, BAK1, and BIK1 proteins [45]. Also, in a separate study using epidermal leaf cells of barley, it was shown that the protein RACB controlled the focal reorganization of AFs during powdery mildew attack [46]. Additionally, myosin XI motors enhanced the density of AFs to aid the higher movement of cellular components around the body sites where the mildew fungus penetrates [44].

However, it has been seen that MT MAP65-3 (associated protein 65-3) modulates the depolymerization of MT and promotes defense against powdery and

TABLE 1 Different bacterial components that regulate the plant cytoskeleton.

Microbial component	PAMP/ effector	Affiliated cytoskeletal component	Effects
Chitin	PAMP	Actin	PTI Triggers
EF-Tu	PAMP	Actin	PTI quick response
Flagellin	PAMP	Actin	PTI reaction time is slow
AvrBST	Effector	Microtubule	Localization Distortion
HOpE1	Effector	Microtubule	MT network is distorted
HOpZ1	Effector	Microtubule	Disconnection as from MT network
HOpW1	Effector	Actin	AF network gets broken

downy mildew through the SA and JA signaling pathways [47]. A similar effect was validated using actin-depolymerizing agents, and cytochalasin E in leaves *A. thaliana* and *Brassica napus,* which were then infected with *P. syringae* and *Leptoshaeria maculans,* respectively [48]. It was observed that actin depolymerization caused a rapid increase of SA levels by isochorismate synthase that indirectly increased the plant's resistance to the respective pathogen attack. Hence, the preceding examples provide strong evidence of how the cytoskeletal system of plants brings about both physical and chemical changes during a pathogen attack. It plays a vital role in averting pathogens, inducing processes, and modifying different signal transduction pathways that ultimately trigger the immune response.

8.2   The functional importance of organelles in plant immunity

Organelles are the sites of major metabolic processes occurring within a plant cell, whereby they also actively participate in innate immune responses through interorganellar communication. Among them, chloroplasts are reported to be a source of defense signals during ETI that include ROS, nitric oxide, calcium ions, and hormones like SA, ABA, and JA. As reported by Caplan et al., chloroplasts are known to send out stromules, or dynamic tubular extensions, towards the nucleus that lead to the chloroplast-localized defensive protein buildup, NRIP1 H_2O_2 in the nucleus [49]. Using bacterial and viral effectors on *Arabidopsis* and *Nicotiana* plants, the authors proved that the constitutive expression of stromules increased HR-programmed cell death to prevent the spread of infection. The HR is also dependent on light quantity and quality, which influences the production of SA and ROS in the chloroplast and inhibits pathogen growth [50]. A similar effect has been reported in lesion-simulating disease mutant 1 of *Arabidopsis*, where a day-length-dependent increase of SA leads to uncoupled photosynthesis-related nuclear and plastid gene expression, leading to an oxidative burst [51]. Such oxidative burst can be beneficial for preventing future infections due to the generation of systemic acquired resistance (SAR) known to enhance the yield of defense molecules through the SA pathway, ethylene pathway, and other cascades [52]. The chloroplast-induced ROS have also been proved crucial against infection caused by the tobacco mosaic virus and *Xanthomonas campestris* pv. *vesicatoria*, in which transgenic tobacco plants with lower ROS production indicated a lower rate of cell death after infection [49].

Vacuoles are another set of plant organelles that have a crucial significance in immunity through their cysteine proteases, also known as vacuolar processing enzymes (VPEs). During plant immunity, vacuolar collapse occurs, culminating in HR-programmed cell death at the site of infection. Other pathogenesis-related (PR) proteins, lectins, and myrosinases also form an effective defense line during pathogen attacks. The association of PR proteins has been found in tobacco regarding resistance against *Peronospora tabacina* and *P. syringae* pv. *tabaci* [53,54]. Myrosinases are released during tissue damage from collapsed vacuoles that induce the production of isothiocyanates that are highly toxic to bacteria. Vacuoles also accumulate various secondary metabolites such as benzoxazinoids, cyanogenic glycosides, and flavonoids that partake in conferring immunity during pathogen invasion [54]. As an illustration, inactive cyanogenic glycosides are glycosides kept in the plant vacuole that can form toxic hydrocyanic acid during tissue damage by invading pathogens [55]. Likewise, in maize, benzoxazinoids are in the vacuole and inactive glucosides are preserved, which undergo degradation to form benzoxazinoids that can induce callose formation following fungal infection [56] (Fig. 2).

8.3 Pathogen effectors as immune boosters in plant immunity

Phytopathogens like bacteria, fungi, viruses, and nematodes exert their pathogenesis through effector molecules that cripple plant defense systems and initiate ETI. These proteinaceous molecules operate either in the apoplast or in the cytoplasm and bind to PTI receptors such as receptor-like proteins, receptor-like kinases, and PRRs to induce the signaling cascade downstream [57]. Most pathogen effectors function to suppress plant immunity; however, depending on the host genotype, they might have both beneficial and negative effects on the pest. Bacterial effectors are transported through various secretion systems, such as the type-III secretion system in Gram −ve's that effectively deliver effectors

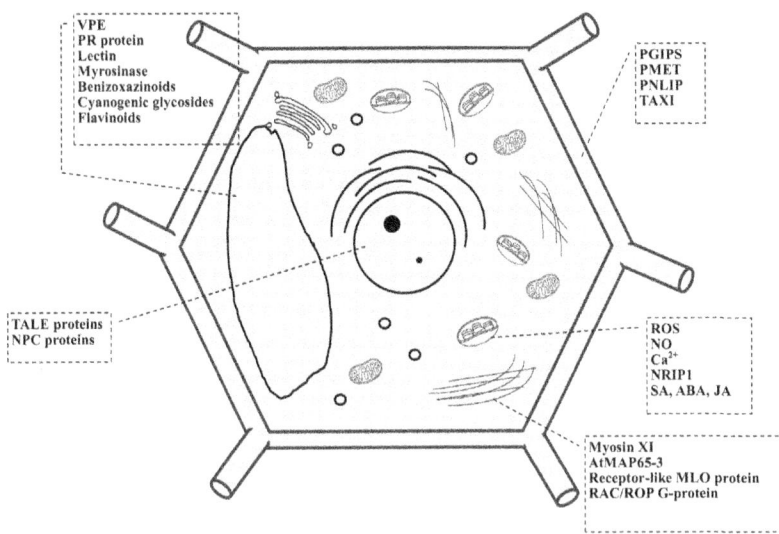

FIG. 2 Schematic diagram represents the role of different plant cell components in imparting defense against invading pathogens. Abbreviations clockwise from top: *VPE*, vacuolar processing enzymes; *PR*, pathogenesis-related proteins; *PGIPS*, polygalacturonase inhibiting proteins; *PMEI*, pectin methyl esterases inhibitors; *PNLIP*, pectin lyase inhibitor protein; *TAXI*, *Triticum aestivum* xylanase inhibitor; *ROS*, reactive oxygen species; *NO*, nitric oxide; Ca^{2+}, calcium ions; *NRIP1*, chloroplast localized defense protein; *SA*, salicylic acid; *ABA*, abscisic acid; *JA*, jasmonic acid; *AtMAP*, microtubule-associated protein; *MLO*, mildew resistance locus; *TALE*, transcription activator-like effector; *NPC*, nuclear pore complex.

into the host cells. For example, T3SS in *P. syringae* delivers effectors like AvrPtoB, HopAO1, and HopU1. AvrPtoB promotes the degradation of several PRRs through its ubiquitin ligase activity and inhibits the kinase activities of flagellin-sensing 2 protein and elongation factor Tu receptor [58]. HopAO1 is a protein tyrosine phosphatase that dephosphorylates the activated elongation factor Tu receptor [57]. On the other hand, effector HopU1, a mono-ADP-ribosyltransferase, targets translation by inhibiting RNA-binding proteins like GRP7, thus decreasing the number of PRRs at the plasma membrane to suppress immune recognition [59].

In the case of fungi and oomycetes, they either use hyphal extensions like haustoria to invade and deliver the effectors within the plant cells or colonize the intercellular spaces in the plant cells and secrete effectors to target plant defenses. For example, leaf mold-causing fungus *Cladosporium fulvum* secretes effectors Avr2, Avr4, and ECP6, which inhibit cysteine proteases and tomato plant immune receptors [60]. Some necrotrophic fungi and hemibiotrophs induce necrosis through effectors like ethylene-inducing peptide 1 (NEP)-like proteins (NLPs) that elicit immune responses to cell death. Similarly, the CRN protein effectors in *P. infestans*, such as CRN8, possess kinase activity and localize the nuclei of the host cell to interfere with the signaling pathway and induce necrosis [57].

The plant hormone secretion pathways, including SA (Salicylic acid), JA (Jasmonic acid), and ET (Ethylene), are essential targets for pathogens to manipulate signaling and influence the PTI/ETI. It was discovered that effectors from various pathogens target the precursors of SA biosynthesis. For example, Cmu 1 from *Ustilago maydis* is an effector of chorismatemutase that manipulates the shikimate pathway to thwart immunological responses and increase virulence [61]. Similarly, the filamentous phytopathogens *P. sojae* and *Verticillium dahlia* secrete isochorismatases, Pslsc1, and Vdlscl, which hydrolyze isochorismate and disrupt the metabolism of SA [62]. In *A. thaliana*, the downy mildew pathogen *H. arabidopsidis* secretes an effector HaRxL44 that degrades a mediator subunit 19a to cause elevated JA levels and diminished levels of SA [63]. Thus, growing research on effector biology provides valuable insights into immune modulation in plants and paves the way for understanding hormone dynamics during infection.

References

[1] A.M. Eid, A. Fouda, M.A. Abdel-Rahman, S.S. Salem, A. Elsaied, R. Oelmüller, S. El-Din Hassan, Harnessing bacterial endophytes for promotion of plant growth and biotechnological applications: An overview, Plan. Theory 10 (5) (2021) 935.

[2] M.R. Fishman, K. Shirasu, How to resist parasitic plants: pre- and post-attachment strategies, Curr. Opin. Plant Biol. 62 (2021), 102004.

[3] D.R. Holmes, M. Bredow, K. Thor, S.A. Pascetta, I. Sementchoukova, K.R. Siegel, J. Monaghan, A novel allele in the *Arabidopsis thaliana* MACPF protein CAD1 results in deregulated immune signaling, Genetics 217 (4) (2021).

[4] K.-W. Ma, Y. Niu, Y. Jia, J. Ordon, C. Copeland, A. Emonet, P. Schulze-Lefert, Coordination of microbe–host homeostasis by crosstalk with plant innate immunity, Nat. Plants (2021) 1–12.

[5] P. Mazumdar, P. Singh, D. Kethiravan, I. Ramathani, N. Ramakrishnan, Late blight in tomato: insights into the pathogenesis of the aggressive pathogen *Phytophthora infestans* and future research priorities, Planta 253 (2021) 1–24.

[6] C. Yan, H. Muhammad Rizwan, D. Liang, M. Reichelt, A. Mithöfer, S.S. Scholz, F. Chen, The effect of the root-colonizing *Piriformospora indica* on passion fruit (*Passiflora edulis*) development: initial defense shifts to fitness benefits and higher fruit quality, Food Chem. 359 (2021), 129671.

[7] Y.S. Rizzi, P. Happel, S. Lenz, M.J. Urs, M. Bonin, S. Cord-Landwehr, R. Kahmann, Chitosan and chitin deacetylase activity are necessary for development and virulence of *Ustilago maydis*, MBio 12 (2) (2021) 1–18.

[8] G.D. Arena, P.L. Ramos-González, L.A. Rogerio, M. Ribeiro-Alves, C.L. Casteel, J. Freitas-Astúa, M.A. Machado, Making a better home: modulation of plant defensive response by brevipalpus mites, Front. Plant Sci. 9 (2018).

[9] E.J. Andersen, S. Ali, E. Byamukama, Y. Yen, M.P. Nepal, Disease resistance mechanisms in plants, Genes 9 (2018).

[10] D.A. Bastias, M.A. Martínez-Ghersa, C.L. Ballaré, P.E. Gundel, Epichloë fungal endophytes and plant defenses: not just alkaloids, Trends Plant Sci. 22 (2017) 939–948.

[11] A.R. Bentham, J.C. de la Concepcion, N. Mukhi, R. Zdrzałek, M. Draeger, D. Gorenkin, M.J. Banfield, A molecular roadmap to the plant immune system, J. Biol. Chem. 295 (44) (2020) 14916–14935.

[12] J. Vicente, G.M. Mendiondo, J. Pauwels, V. Pastor, Y. Izquierdo, C. Naumann, M.J. Holdsworth, Distinct branches of the N-end rule pathway modulate the plant immune response, New Phytol. 221 (2) (2019) 988–1000.

[13] W. Wang, F. Jiao, Effectors of Phytophthora pathogens are powerful weapons for manipulating host immunity, Planta 250 (2019) 413–425.

[14] H.L. Wei, A. Collmer, Defining essential processes in plant pathogenesis with *Pseudomonas syringae* pv. tomato DC3000 disarmed polymutants and a subset of key type III effectors, Mol. Plant Pathol. 19 (2018) 1779–1794.

[15] Y. Zhang, Y. Liang, Y. Dong, Y. Gao, X. Yang, J. Yuan, D. Qiu, The *Magnaporthe oryzae* Alt A 1-like protein MoHrip1 binds to the plant plasma membrane, Biochem. Biophys. Res. Commun. 492 (1) (2017) 55–60.

[16] F. Zheng, L. Chen, J. Gao, F. Niu, X. Duan, L. Yin, W. Tian, Identification of autotoxic compounds from Atractylodes macrocephala Koidz and preliminary investigations of their influences on immune system, J. Plant Physiol. 230 (2018) 33–39.

[17] W. Zhu, M. Zaidem, A.L. Van de Weyer, R.M. Gutaker, E. Chae, S.T. Kim, D. Weigel, Modulation of ACD6 dependent hyperimmunity by natural alleles of an *Arabidopsis thaliana* NLR resistance gene, PLoS Genet. 14 (9) (2018).

[18] A.I. González-Hernández, E. Llorens, C. Agustí-Brisach, B. Vicedo, T. Yuste, A. Cerveró, L. Lapeña, Elucidating the mechanism of action of copper heptagluconate on the plant immune system against *Pseudomonas syringae* in tomato (*Solanum lycopersicum* L), Pest Manag. Sci. 74 (11) (2018) 2601–2607.

[19] J. Tamborski, K.V. Krasileva, Evolution of plant NLRs: from natural history to precise modifications, Annu. Rev. Plant Biol. 71 (2020) 355–378.

[20] J. Fernandez, V. Lopez, L. Kinch, M.A. Pfeifer, H. Gray, N. Garcia, K. Orth, Role of two metacaspases in development and pathogenicity of the rice blast fungus *Magnaporthe oryzae*, MBio 12 (1) (2021) 1–15.

[21] V. Calvo-Baltanás, J. Wang, E. Chae, Hybrid incompatibility of the plant immune system: an opposite force to heterosis equilibrating hybrid performances, Front. Plant Sci. 11 (2021).

[22] F.H. Correr, G.K. Hosaka, S.G.P. Gómez, M.C. Cia, C.B.M. Vitorello, L.E.A. Camargo, G.R. A. Margarido, Time-series expression profiling of sugarcane leaves infected with *Puccinia kuehnii* reveals an ineffective defense system leading to susceptibility, Plant Cell Rep. 39 (7) (2020) 873–889.

[23] S.S. Habash, P.P. Könen, A. Loeschcke, M. Wüst, K.E. Jaeger, T. Drepper, A.S.S. Schleker, The plant sesquiterpene nootkatone efficiently reduces *Heterodera schachtii* parasitism by activating plant defense, Int. J. Mol. Sci. 21 (24) (2020) 1–17.

[24] J.H. Kim, R. Hilleary, A. Seroka, S.Y. He, Crops of the future: building a climate-resilient plant immune system, Curr. Opin. Plant Biol. 60 (2021), 101997.

[25] H. Martin-Rivilla, A. Garcia-Villaraco, B. Ramos-Solano, F.J. Gutierrez-Mañero, J.A. Lucas, Bioeffectors as biotechnological tools to boost plant innate immunity: signal transduction pathways involved, Plan. Theory 9 (12) (2020) 1–25.

[26] K. Gruden, J. Lidoy, M. Petek, V. Podpečan, V. Flors, K.K. Papadopoulou, M.J. Pozo, Ménage à trois: unraveling the mechanisms regulating plant–microbe–arthropod interactions, Trends Plant Sci. 25 (2020) 1215–1226.

[27] L. Bacete, H. Mélida, E. Miedes, A. Molina, Plant cell wall-mediated immunity: cell wall changes trigger disease resistance responses, Plant J. 93 (4) (2018) 614–636.

[28] N. Juge, Plant protein inhibitors of cell wall degrading enzymes, Trends Plant Sci. 11 (7) (2006) 359–367.

[29] L. Tayi, R.V. Maku, H.K. Patel, R.V. Sonti, Identification of pectin degrading enzymes secreted by *Xanthomonas oryzae* pv. *Oryzae* and determination of their role in virulence on rice, PLoS One 11 (12) (2016) 1–15.

[30] F. Sicilia, J. Fernandez-Recio, C. Caprari, G. De Lorenzo, D. Tsernoglou, F. Cervone, L. Federici, The polygalacturonase-inhibiting protein PGIP2 of *Phaseolus vulgaris* has evolved a mixed mode of inhibition of endopolygalacturonase PG1 of *Botrytis cinerea*, Plant Physiol. 139 (3) (2005) 1380–1388.

[31] S.H. An, K.H. Sohn, H.W. Choi, I.S. Hwang, S.C. Lee, B.K. Hwang, Pepper pectin methylesterase inhibitor protein CaPMEI1 is required for antifungal activity, basal disease resistance and abiotic stress tolerance, Planta 228 (1) (2008) 61–78.

[32] V. Lionetti, A. Raiola, L. Camardella, A. Giovane, N. Obel, M. Pauly, D. Bellincampi, Overexpression of pectin methylesterase inhibitors in arabidopsis restricts fungal infection by *Botrytis cinerea*, Plant Physiol. 143 (4) (2007) 1871–1880.

[33] M.H. Chen, V. Citovsky, Systemic movement of a tobamovirus requires host cell pectin methylesterase, Plant J. 35 (3) (2003) 386–392.

[34] T. Beliën, S. Van Campenhout, M. Van Acker, J. Robben, C.M. Courtin, J.A. Delcour, G. Volckaert, Mutational analysis of endoxylanases XylA and XylB from the phytopathogen *Fusarium graminearum* reveals comprehensive insights into their inhibitor insensitivity, Appl. Env. Microbiol. 73 (14) (2007) 4602–4608.

[35] T. Maehara, H. Yagi, T. Sato, M. Ohnishi-Kameyama, Z. Fujimoto, K. Kamino, Y. Kitamura, F. St. John, S.K. Katsuro Yaoi, GH30 Glucuronoxylan-specific xylanase from *Streptomyces turgidiscabies* C56, Appl. Env. Microbiol. (2018) 1–13.

[36] D. Chmelová, D. Škulcová, M. Ondrejovic, Microbial xylanases and their inhibition by specific proteins in cereals, Kvasny Prumysl 65 (4) (2019) 127–133.

[37] R. Lorrai, S. Ferrari, Host cell wall damage during pathogen infection: mechanisms of perception and role in plant-pathogen interactions, Plan. Theory 10 (2) (2021) 1–21.

[38] E. Fierens, K. Gebruers, A.R.D. Voet, M. De Maeyer, C.M. Courtin, J.A. Delcour, Biochemical and structural characterization of TLXI, the *Triticum aestivum* L. thaumatin-like xylanase inhibitor, J. Enzyme Inhib. Med. Chem. 24 (3) (2009) 646–654.

[39] H.W. Choi, N.H. Kim, Y.K. Lee, B.K. Hwang, The pepper extracellular xyloglucan-specific endo-β-1,4-glucanase inhibitor protein gene, CaXEGIP1, is required for plant cell death and defense responses, Plant Physiol. 161 (1) (2013) 384–396.

[40] S. Hou, Z. Liu, H. Shen, D. Wu, Damage-associated molecular pattern-triggered immunity in plants, Front. Plant Sci. 10 (2019).

[41] H.W. Choi, D.F. Klessig, DAMPs, MAMPs, and NAMPs in plant innate immunity, BMC Plant Biol. 16 (1) (2016) 1–10.

[42] A. Zehra, N.A. Raytekar, M. Meena, P. Swapnil, Efficiency of microbial bio-agents as elicitors in plant defense mechanism under biotic stress: a review, Curr. Res. Microb. Sci. 2 (2021), 100054, https://doi.org/10.1016/j.crmicr.2021.100054.

[43] P. Bittel, S. Robatzek, Microbe-associated molecular patterns (MAMPs) probe plant immunity, Curr. Opin. Plant Biol. 10 (4) (2007) 335–341.

[44] K. Porter, B. Day, From filaments to function: the role of the plant actin cytoskeleton in pathogen perception, signaling and immunity, J. Integr. Plant Biol. 58 (4) (2016) 299–311.

[45] J.L. Henty-Ridilla, M. Shimono, J. Li, J.H. Chang, B. Day, C.J. Staiger, The plant actin cytoskeleton responds to signals from microbe-associated molecular patterns, PLoS Pathog. 9 (4) (2013).

[46] S.M. Schmidt, R. Panstruga, Cytoskeleton functions in plant-microbe interactions, Physiol. Mol. Plant Pathol. 71 (4–6) (2007) 135–148.

[47] E. Park, A. Nedo, J.L. Caplan, S.P. Dinesh-Kumar, Plant–microbe interactions: organelles and the cytoskeleton in action, New Phytol. 217 (3) (2018) 1012–1028.

[48] H. Leontovyčová, T. Kalachova, L. Trdá, R. Pospíchalová, L. Lamparová, P.I. Dobrev, M. Janda, Actin depolymerization is able to increase plant resistance against pathogens via activation of salicylic acid signalling pathway, Sci. Rep. 9 (1) (2019).

[49] J.L. Caplan, A.S. Kumar, E. Park, M.S. Padmanabhan, K. Hoban, S. Modla, S.P. Dinesh-Kumar, Chloroplast stromules function during innate immunity, Dev. Cell 34 (1) (2015) 45–57.

[50] S. Stael, P. Kmiecik, P. Willems, K. Van Der Kelen, Europe PMC funders group plant innate immunity – sunny side up? Trends Plant Sci. 20 (1) (2016) 3–11.

[51] R. Lv, Z. Li, M. Li, V. Dogra, S. Lv, R. Liu, C. Kim, Uncoupled expression of nuclear and plastid photosynthesis-associated genes contributes to cell death in a lesion mimic mutant, Plant Cell 31 (1) (2019) 210–230.

[52] M. Meena, P. Swapnil, K. Divyanshu, S. Kumar, T. Harish, Y. N., Zehra, A., Marwal, A., Upadhyay, R.S., PGPR-mediated induction of systemic resistance and physiochemical alterations in plants against the pathogens: current perspectives, J. Basic Microbiol. 60 (10) (2020) 828–861, https://doi.org/10.1002/jobm.202000370.

[53] P. Kumari, M. Meena, P. Gupta, M.K. Dubey, G. Nath, R.S. Upadhyay, Plant growth promoting rhizobacteria and their biopriming for growth promotion in mung bean (Vigna radiata (L.) R. Wilczek), Biocatal. Agric. Biotechnol. 16 (2018) 163–171.

[54] M.H. Madina, M.S. Rahman, H. Zheng, H. Germain, Vacuolar membrane structures and their roles in plant–pathogen interactions, Plant Mol. Biol. 101 (4–5) (2019) 343–354.

[55] J. Vetter, Plant cyanogenic glycosides, Toxicon 38 (2000) 11–36.

[56] S. Ahmad, N. Veyrat, R. Gordon-Weeks, Y. Zhang, J. Martin, L. Smart, J. Ton, Benzoxazinoid metabolites regulate innate immunity against aphids and fungi in maize, Plant Physiol. 157 (1) (2011) 317–327.

[57] T.Y. Toruño, I. Stergiopoulos, G. Coaker, Plant-pathogen effectors: cellular probes interfering with plant defenses in spatial and temporal manners, Annu. Rev. Phytopathol. 54 (6) (2016) 419–441.

[58] R.B. Abramovitch, R. Janjusevic, C.E. Stebbins, G.B. Martin, Type III effector AvrPtoB requires intrinsic E3 ubiquitin ligase activity to suppress plant cell death and immunity, PNAS USA 103 (8) (2006) 2851–2856.

[59] Z.Q. Fu, M. Guo, B.R. Jeong, F. Tian, T.E. Elthon, R.L. Cerny, J.R. Alfano, A type III effector ADP-ribosylates RNA-binding proteins and quells plant immunity, Nature 447 (7142) (2007) 284–288.

[60] J. Win, A. Chaparro-Garcia, K. Belhaj, D.G.O. Saunders, K. Yoshida, S. Dong, S. Kamoun, Effector biology of plant-associated organisms: concepts and perspectives, Cold Spring Harb. Symp. Quant. Biol. 77 (2012) 235–247.

[61] A. Djamei, K. Schipper, F. Rabe, A. Ghosh, V. Vincon, J. Kahnt, R. Kahmann, Metabolic priming by a secreted fungal effector, Nature 478 (7369) (2011) 395–398.

[62] T. Liu, T. Song, X. Zhang, H. Yuan, L. Su, W. Li, D. Dou, Unconventionally secreted effectors of two filamentous pathogens target plant salicylate biosynthesis, Nat. Comm. 5 (2014).

[63] M.C. Caillaud, S. Asai, G. Rallapalli, S. Piquerez, G. Fabro, J.D.G. Jones, A downy mildew effector attenuates salicylic acid-triggered immunity in arabidopsis by interacting with the host mediator complex, PLoS Biol. 11 (12) (2013).

Chapter 2

PGPMs-mediated improvement of crops under abiotic stress

Pankaj Kumar[a,b], Rima Kumari[a], Satish Kumar[a], and Arun Kumar[c]
[a]*Department of Botany, Purnea University, Purnia, India,* [b]*Department of Botany, KKM College, Jamui, Munger University, Munger, Bihar, India,* [c]*Swami Keshwanand Rajasthan Agricultural University, Bikaner, Rajasthan, India*

1 Introduction

Among the variety of possible plant growth-promoting microbes (PGPMs), there are free-living PGPMs, which are found in the rhizosphere and help colonize plant roots, as well as symbiotic PGPMs, which are usually comprised of rhizospheres microbes, endophytes, or mycorrhizal fungi. Root growth and root metabolism are provided by the rhizosphere, a community of microbes with the highest density of organisms responsible for both growth and metabolism. Usually, enhancement of symbiotic PGPMs generally accomplished by Fungi, protozoa, and algae together while its retarded by bacteria or where bacterial abundances is high. Although bacteria have exceptional control over plant physiology, they have a greater level of influence over plant structure than other microbes [1–4]. The several strategies through which root-colonizing bacteria might benefit plants include direct influence and indirect influence. Plant growth-promoting rhizobacteria (PGPR) are bacteria found in the rhizosphere that help plants grow [5–7]. A plant microbe that resides in plant tissues without causing illness is known as an endophyte. Bacteria, fungi, and actinomycetes can all exist together in the same location. Pathogens that have been rendered nonpathogenic but still capable of endophytic colonization by techniques such as selection or genetic alteration fall into three categories: (1) nonpathogenic, (2) pathogenic in the host, but nonpathogenic when found in endophytic colonization, and (3) pathogenic [8–10]. These organisms help plants resist physical and physiological stress as well as chemicals and bioactive substances of biotechnology [11,12].

A mycorrhizal association is the union of fungi with plant roots, as when fungi create a symbiotic association with the roots of higher plants. Many agricultural soils feature extremely high concentrations of arbuscular mycorrhizae

Plant-Microbe Interaction—Recent Advances in Molecular and Biochemical Approaches.
https://doi.org/10.1016/B978-0-323-91876-3.00007-5

21

(AM) and ectomycorrhizae (ECM) [13–15]. Most roots in the world have mycorrhizal associations [16]. Fungal nutrition is fulfilled by developing an arbuscule-like structure in the root system to access nutrients, and in return, fungi grow their hyphal network into the soil [17]. Also, the mycorrhizal connection has been proven to have growth-promoting properties, as it helps to alleviate stress on plants. Beneficial soil microbes, such as bacteria and fungi (e.g., arbuscular mycorrhizal fungus [AMF]), help develop plant tolerance to stressful situations, and microbes that inhabit specific locations (e.g., endophytes such as bacteria) can help plants adapt to various environmental conditions.

2 Plant stress due to abiotic factors

In agricultural soil, both abiotic and biotic stresses have an impact on plant growth. The key factors of light, water, carbon, and mineral nutrients control plant growth, development, and reproduction [10] (Fig. 1). Whenever these environmental conditions become extreme, plants undergo physiological and morphological modifications to adapt to these abrupt changes. Heat stress, drought, cold stress, salinity, waterlogging, and many other severe abiotic stressors significantly affect plant development and agricultural productivity. Particulate erosion occurs due to the accumulation of heavy metals and nutritional depletion. The rate of growth and yield of plants is also affected by fluctuations in temperature. The productivity of crops decreases up to 20% to 50% depending on the crop type and environmental factors where it is grown. For example, in arid or semiarid regions, plant growth is hindered due to reduced water uptake, leading to a decrease in the plant's photosynthetic efficacy.

2.1 Drought

Drought stress affects plant structure as well as human health on a global scale [18,19]. Drought may afflict 0.5% of arable land by the year 2050 [20]. Moreover, frequent and long-lasting droughts are expected to occur with global climate change [21]. Naveed et al. [22] observed how drought adversely influences the physiological, biochemical, and growth parameters of wheat

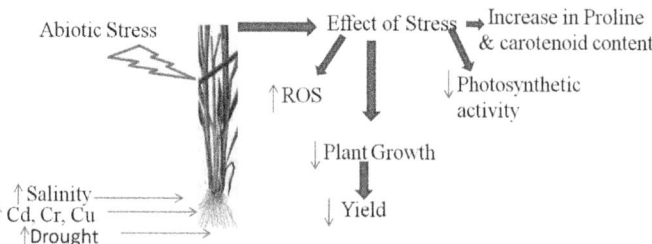

FIG. 1 A simplified concept of mechanism for abiotic stress in plants. Abiotic stress is implicated in the overproduction of reactive oxygen species (ROS), which causes cell death.

seedlings through reduced carbon dioxide assimilation, stomatal conductance (SC), relative water content (RWC), transpiration rate, and chlorophyll content. Drought hinders the supply of nutrients, which causes nutrients to be transported down to the roots by water [23]. Nitrate, sulfate, calcium (Ca), magnesium (Mg), and silicon (Si) are inaccessible because nutrient diffusion is obstructed, and the bulk flow of nutrients is blocked [24,25]. Decreases the activity of nitrate reductase as a result of reduced uptake of nitrate from soil [26]. Generation of reactive oxygen species (ROS) and free radicals such as superoxide radicals, hydrogen peroxide, and hydroxyl radicals result in oxidative stress. ROS most likely promotes lipid peroxidation and damages membrane integrity in plants, which may lead to damage to proteins, lipids, and nucleic acids [27,28]. When plants are in drought, they tend to synthesize more ethylene, which influences the quantity and quality of the plants' production [29]. The main reason for the balance disruption in plant cells due to increased salt concentration and paucity of water is the effect it has on osmotic maintenance; this eventually leads to growth problems.

Osmotic imbalance has a major impact on plant growth and development. Due to biotic or abiotic stresses, such as an increase in the development of the belowground system, reduction in the growth of the aboveground system, changes in the transport of ions (uptake, extrusion, and compartmentalization), and changes in metabolic activities, numerous responses are triggered. These responses can include increased hydration of the root zone, reduction in the growth of the canopy, changes in the movement of ions (uptake, extrusion, and compartmentalization), and altered metabolic activities. In reaction to stress, certain physiological, chemical, or cellular processes occur. In addition to direct signals such as osmotic imbalance, secondary messengers like ROS, abscisic acid (ABA), ethylene, and phospholipid operate as indirect signals. Under drought conditions, root-derived ABA can spread upwards to modulate stomatal apertures in leaves, not being confined to the primary stress points. To obtain effective plant growth and development in a water poor environment, optimal rhizosphere water content is required.

2.2   Salinity

Salinity is the main abiotic factor that negatively affects agricultural productivity. Soil salinity threatens 7% of the Earth's total land. Around the world, crops are mostly grown in places plagued by salinity, with the negative consequences being that inadequate irrigation management causes secondary salinization that affects 20% of irrigated land [30]. The main downsides of salinity include ion toxicity, nutrient deficiencies (nitrogen, calcium, potassium, phosphorous, iron, and zinc), and oxidative stress on plants [31]. Soil salinity greatly decreases the amount of phosphorus (P) that may be absorbed by plants because Ca ions precipitate with phosphate ions [32]. Some elements, such as sodium, chlorine, and boron, have specific toxic effects on plants. Osmotic stress and cell death can swiftly occur when the concentration of salt in cell walls increases excessively [33]. Saltiness has a significant impact on the initiation, growth, and photosynthetic efficiency of plants [34]. When salinity limits embryo growth, this can cause reduced microspore formation and parenchymal cell death. This can result in a failure to elongate filaments in the style and therefore a risk of early fertilization failure. Saline growth medium has a low osmotic potential of soil solution, a tendency to create specific ion effects, nutritional imbalances, or a combination of these elements, which may have various deleterious consequences on plant growth. Salinity may harm plant growth due to reduced number of phytohormones and photosynthetic products being transported to the developing plant tissue [35].

2.3 Heavy metal deposition in soil

Out of 90 metals, 53 are reported as heavy metals and not all are biologically important. Based on their solubility under physiological conditions, only 17 heavy metals are available and important to living cells. While iron (Fe), molybdenum (Mo), and manganese (Mn) are important as micronutrients, zinc (Zn), nickel (Ni), copper (Cu), vanadium (V), cobalt (Co), tungsten (W), and chromium (Cr) have been reported to be toxic elements with high or low importance as trace elements, and mercury (Hg), silver (Ag), antimony (Sb), cadmium (Cd), lead (Pb), and uranium (U) are toxic to plants and microbes [36–38]. Naturally heavy metals occur in the underlying material and the atmosphere. Anthropogenic activities like mining, combustion of fossil fuels, metal-working industries, phosphate fertilizers, and so on add to the emission and accumulation of heavy metals in ecosystems [39,40]. Toxic heavy metals generally cause inhibition of cytoplasmic enzymes and damage cell membranes, thus reducing plant growth and development [41].

2.4 Fluctuations in temperature

Temperature fluctuation affects membrane fluidity, nucleic acid, and protein structures as well as metabolite and osmolyte concentrations. Increased levels of ROS cause oxidative damage and cell death as a result of exposure to high temperatures [42]. Exposure to low temperatures decreases the functioning of plant physiological and biochemical systems, resulting in visible symptoms such as wilting, necrosis, or chlorosis [43]. This may cause plant cell membrane and lipid composition to change [44]. If these alterations are made, the cytoplasm could be compromised, and electron flow will be able to proceed through optional channels to regulate it. Concentration of calcium ions in the cell content fluctuates along with temperature, according to a study by Changes in the structure of cell organelles (thylakoid membrane and plastids), phosphorylation of the proteins found in mitochondria and thylakoid unit, as well as the activity of proteins, have all been linked to changes in temperature.

3   With growth-promoting microbes, plants can better withstand environmental challenges

Many microbes are recognized to play a significant role in plant growth promotion, nutrient management, and disease control. Alfalfa, alfalfa powder, alfalfa meal, and other alfalfa byproducts help provide nutrients to crops, promote plant growth (production of phytohormones), exert biocontrol, improve soil properties, and help with microbial leaching of inorganics. Recent uses of bacteria in soil include bioremediation of contaminated soils [45]. Burd et al. [45] noted the employment of bacteria in bioremediation of organic contaminants. Some of the most commonly reported genera of plant growth-promoting bacteria are *Rhizobium, Bacillus, Pseudomonas, Pantoea, Paenibacillus, Burkholderia, Achromobacter, Azospirillum, Microbacterium, Methylobacterium, Variovorax,* and *Enterobacter* [46]. According to Egamberdieva [47], bacteria that improve the root system by increasing the number, length, and surface area of plant roots produce phytohormones (gibberellic acid and indole acetic acid) that assist in plant growth by enhancing nutrient intake. Using bacteria from the rhizosphere, researchers from Switzerland have discovered that these microbes can make antioxidants and cytokinin, both of which promote increased concentrations of ABA and thereby inhibit the formation of free radicals (ROS). Research indicates that greater levels of enzymes involved in antioxidant activity are associated with improved resistance to oxidative stress in cells.

3.1   Resistance to drought

Studies have been conducted on the role of PGPMs in abiotic stress tolerance (drought), specifically with regards to plant adaptation [48]. To better understand the potential to ameliorate the impact of drought stress on wheat (*Triticum aestivum* L.) development, physiology, and yield, researchers investigated how endophytic bacteria (*Burkholderia phytofirmans* PsJN) can help plants thrive in the field. Plants that had their water content and carbon dioxide assimilation affected by drought showed a considerable improvement in photosynthetic rate, water-use efficiency, and chlorophyll content when infected with PsJN. Additionally, plants exposed to drought stress during the tillering and flowering stages saw a reduction in grain yield, but inoculation yielded better results, with up to 21% and 18% greater yields [22]. In the *Arabidopsis* study "induced systemic tolerance (IST) to salinity and drought was provided by *A. piechaudii* ARV8, which produced 1-aminocyclopropane-1-carboxylate (ACC) deaminase" [49]. Waqas et al. [50] found that *Phoma glomerata* and *Penicillium* sp. increased plant biomass, assimilated key nutrients, and decreased salt toxicity in cucumber. The plants treated with sodium chloride and polyethylene glycol (PEG) showed an enhanced level of salinity and drought resistance compared to untreated plants.

According to Nadeem et al. [51], using *P. fluorescens* on its own or with biochar and/or compost enhanced water availability, improved leaf water content, and decreased electrolyte leakage. Aslam et al. [52] found that drought-tolerant carbonic anhydrase-producing bacteria, which produce carbonic anhydrase, helped the growth of wheat seedlings during drought by increasing chlorophyll content, photosynthetic rate, and relative water content. AM symbiosis frequently causes changes in the transport of water into, through, and out of host plants, resulting in changes in tissue hydration and plant physiology. Using *Spinacia oleracea* L. in Zuccarini and Save [53] resulted in higher yield. Water stress tolerance has been augmented in three different AM *Glomus* species.

3.2 Sequestration of heavy metals

It is essential that the beneficial PGPMs that thrive on soils contaminated with trace metals are present for phytoremediation to succeed [4]. To deal with the toxic effects of heavy metals, they secrete acids, proteins, phytoantibiotics, and other chemicals [54]. A cadmium-resistant rhizobacteria probably assists in increasing *Brassica napus* growth by accumulating 1-aminocyclopropane-1-carboxylic acid (ACC) in roots, according to research by Dell'Amico et al. [55]. Nickel toxicity was found to be lower in canola seedlings treated with *Kluyvera ascorbata* SUCD165 [45]. The bacteria *Streptomyces acidiscabies* E13 aids in cowpea growth by producing hydroxamate-type siderophores capable of depleting nickel from the soil [56]. Bacterial inoculation on the uptake of heavy metals from PbZn mine tailings was investigated in *Brassica juncea* by Wu et al. [57]. Despite the presence of these beneficial bacteria, the levels of metals in plant tissues were unaffected. This led to a significantly larger above-ground biomass, as well as alteration in soil metal bioavailability. Trees often associate with mycorrhizal fungus in the wild. Breinigerberg soil, which has a high Zn and other heavy metal content, was used for growing tomatoes in an experiment conducted by Ouziad et al. [58]. The soil was contaminated with up to 1 mM $CdCl_2$. Compared to nonmycorrhizal controls, the plants that were colonized with the AMF *Glomus intraradices* exhibited significantly better growth. When AMF are present, plant genes that code for substances that could aid in heavy metal tolerance are deregulated.

3.3 Salinity

Ions, redox reactions, osmolytes, and polyamines are also synthesized to help plants survive salinity [59]. From 6% NaCl concentration after which they demonstrated phosphate solubilization ability and produced phytohormones, siderophores, and ACC deaminase enzyme in pot-grown tomato plants, Tank and Saraf [60] isolated several PGPR strains from tomato fields and adapted their cultures to 6% NaCl concentration. In soil where the nutrient supply was insufficient, three strains of *Pseudomonas alcaligenes* PsA15, *Bacillus polymyxa*

Bcp26, and *Mycobacterium phlei* MbP18 imparted resistance to high temperature and salt stress on maize [61]. Nonsymbiotic plants *Leymus mollis* (dunegrass) were significantly wilted and parched after 7 days and were dead after 14 days under constant exposure to 500 mM NaCl (seawater levels to simulate the plants' natural beach habitat). When challenged with 500 mM NaCl solution for 14 days, *Fusarium culmorum*-infected plants did not exhibit wilting signs until they were subjected to the high salt solution [10,62]. Inoculation of tomato plants with *Glomus mosseae* (AMF) increased plant growth under salt stress, according to a study by Al-Karaki [63]. Microaggregates and macroaggregates are formed by PGPM polysaccharides that bind soil particles (Feng et al., 2002). Bacteria secrete exopolysaccharides that improve soil structure by enhancing tolerance to water and salinity stress [64].

References

[1] H. Antoun, J. Kloepper, Plant growth promoting rhizobacteria (PGPR), in: S. Brenner, J.H. Miller (Eds.), Encyclopedia of Genetics, Academic, New York, 2001, p. 14771480.

[2] J. Barriuso, B.R. Solano, J.A. Lucas, A.P. Lobo, A. García-Villaraco, F.J.G. Manero, Ecology, genetic diversity and screening strategies of plant growth promoting rhizobacteria (PGPR), J. Plant Nutr. 1 (2008) 17.

[3] C.B. Patel, V.K. Singh, A.P. Singh, M. Meena, R.S. Upadhyay, Microbial genes involved in interaction with plants, in: H.B. Singh, V.K. Gupta, S. Jogaiah (Eds.), New and Future Developments in Microbial Biotechnology and Bioengineering, Elsevier, Singapore, 2019, pp. 171–180, https://doi.org/10.1016/B978-0-444-63503-7.00010-3.

[4] B. Saharan, V. Nehra, Plant growth promoting rhizobacteria: a critical review, Life Sci. Med. Res. 21 (2011) 1–30.

[5] P. Kumari, M. Meena, P. Gupta, M.K. Dubey, G. Nath, R.S. Upadhyay, Plant growth promoting rhizobacteria and their biopriming for growth promotion in mung bean (*Vigna radiata* (L.) R. Wilczek), Biocatal. Agric. Biotechnol. 16 (2018) 163–171.

[6] P. Kumari, M. Meena, R.S. Upadhyay, Characterization of plant growth promoting rhizobacteria (PGPR) isolated from the rhizosphere of *Vigna radiata* (mung bean), Biocatal. Agric. Biotechnol. 16 (2018) 155–162.

[7] J. Kloepper, J. Leong, M. Teintze, et al., Enhanced plant growth by siderophores produced by plant growth-promoting rhizobacteria, Nature 286 (1980) 885–886, https://doi.org/10.1038/286885a0.

[8] P.A. Backman, R.A. Sikora, Endophytes: an emerging tool for biological control, Biol. Control 46 (1) (2008) 13.

[9] M. Meena, P. Swapnil, K. Divyanshu, S. Kumar, Harish, Y.N. Tripathi, A. Zehra, A. Marwal, R.S. Upadhyay, PGPR-mediated induction of systemic resistance and physiochemical alterations in plants against the pathogens: current perspectives, J. Basic Microbiol. 60 (10) (2020) 828–861, https://doi.org/10.1002/jobm.202000370.

[10] R. Lata, S. Chowdhury, S.K. Gond, J.F. White Jr., Induction of abiotic stress tolerance in plants by endophytic microbes, Lett. Appl. Microbiol. 66 (2018) 268–276, https://doi.org/10.1111/lam.12855.

[11] A. Zehra, M. Meena, M.K. Dubey, M. Aamir, R.S. Upadhyay, Synergistic effects of plant defense elicitors and Trichoderma harzianum on enhanced induction of antioxidant defense

system in tomato against Fusarium wilt disease, Bot. Stud. 58 (2017) 44, https://doi.org/ 10.1186/s40529-017-0198-2.

[12] A. Zehra, M. Meena, M.K. Dubey, M. Aamir, R.S. Upadhyay, Activation of defense response in tomato against Fusarium wilt disease triggered by Trichoderma harzianum supplemented with exogenous chemical inducers (SA and MeJA), Braz. J. Bot. 21 (2017) 1–14, https:// doi.org/10.1007/s40415-017-0382-3.

[13] P. Christie, X. Li, B. Chen, Arbuscular mycorrhiza can depress translocation of zinc to shoots of host plants in soils moderately polluted with zinc, Plant Soil 261 (2004) 209–217, https:// doi.org/10.1023/B:PLSO.0000035542.79345.1b.

[14] A.G. Khan, M. Belik, Occurrence and ecological significance of mycorrhizal symbiosis in aquatic plants, in: A. Verma, B. Hock (Eds.), Mycorrhiza-Structure, Function, Molecular Biology and Biotechnology, Springer, Heidelberg, 1995, pp. 627–665. https://doi.org/10.1093/jxb/ern059.

[15] A. Willis, B.F. Rodriguesb, P.J.C. Harrisa, The ecology of arbuscular mycorrhizal fungi, Crit. Rev. Plant Sci. 32 (2013) 1–20, https://doi.org/10.1080/07352689.2012.683375.

[16] M. Giovannetti, L. Avio, P. Fortuna, E. Pellegrino, C. Sbrana, P. Strani, At the root of the wood wide web: self recognition and non-self incompatibility in mycorrhizal networks, Plant Signal. Behav. 1 (2006) 1–5.

[17] G.J. Bethlenfalvay, R.G. Linderman, Mycorrhizae in Sustainable Agriculture, ASA Special Publication No. 54, Madison: WI, 1992, p. 813.

[18] D. Choluj, R. Karwowska, M. Jasinska, G. Haber, Growth and dry matter partitioning in sugar beet plants (*Beta vulgaris* L.) under moderate drought, Plant Soil Environ. 50 (6) (2004), 265272.

[19] P. Rahdari, S.M. Hoseini, Drought stress: a review, Int. J. Agron. Plant Prod. 3 (2012) 443–446.

[20] B. Vinocur, A. Altman, Recent advances in engineering plant tolerance to abiotic stress: achievements and limitations, Curr. Opin. Biotechnol. 16 (2005) 123–132.

[21] J.T. Overpeck, J.E. Cole, Abrupt change in Earth's climate system, Annu. Rev. Environ. Resour. 31 (1) (2006) 1–31, https://doi.org/10.1146/annurev.energy.30.050504.144308.

[22] M. Naveed, B. Mitter, T.G. Reichenauer, K. Wieczorek, A. Sessitsch, Increased drought stress resilience of maize through endophytic colonization by *Burkholderia phytofirmans* PsJN and *Enterobacter* sp. FD17, Environ. Exp. Bot. 97 (2014) 30–39.

[23] S.S.K. Vurukonda, S. Vardharajula, A.S. Shrivastava, Enhancement of drought stress tolerance in crops by plant growth promoting rhizobacteria, Microbiol. Res. 184 (2016) 13–24.

[24] S.A. Barber, Soil Nutrient Bioavailability: A Mechanistic Approach, John Wiley & Sons, 1995.

[25] G. Selvakumar, P. Panneerselvam, A. Ganeshamurthy, Diversity utility and potential of Actinobacteria in the agro-ecosystem, in: D. Maheshwari (Ed.), Bacterial Diversity in Sustainable Agriculture. Sustainable Development and Biodiversity, vol. 1, Springer, Cham, 2014. https:// doi.org/10.1007/978-3-319-05936-5_2.

[26] F. Caravaca, M.M. Alguacil, J.A. Hernandez, A. Roldan, Involvement of antioxidant enzyme and nitrate reductase activities during water stress and recovery of mycorrhizal *Myrtus communis* and *Phillyrea angustifolia* plants, Plant Sci. 169 (1) (2005), 191197.

[27] G.A. Hendry, Oxygen free radical process and seed longevity. Seed Sci, J. 3 (2005) 141–147.

[28] C.L.M. Sgherri, P. Salvateci, M. Menconi, A. Raschi, F. Navari-Izzo, Interaction between drought and elevated CO_2 in the response of alfalfa plants to oxidative stress, J. Plant Physiol. 156 (2000) 360–366.

[29] S.Z. Ali, V. Sandhya, M. Grover, N. Kishore, L.V. Rao, B. Venkateswarlu, *Pseudomonas* sp. strain AKM-P6 enhances tolerance of sorghum seedlings to elevated temperatures, Biol. Fertil. Soils 46 (1) (2009) 4555.

[30] A.H.M.E.D. Al-Maskri, L. Al-Kharusi, H. Al-Miqbali, M.M. Khan, Effects of salinity stress on growth of lettuce (*Lactuca sativa*) under closed-recycle nutrient film technique, Int. J. Agric. Biol. 12 (3) (2010), 377380.

[31] P. Shrivastava, R. Kumar, Soil salinity: a serious environmental issue and plant growth promoting bacteria as one of the tools for its alleviation, Saudi J. Biol. Sci. 22 (2) (2015) 123–131.

[32] A. Bano, M. Fatima, Salt tolerance in *Zea mays* (L). following inoculation with *Rhizobium* and *Pseudomonas*, Biol. Fertil. Soils 45 (4) (2009), 405413.

[33] R. Munns, Comparative physiology of salt and water stress, Plant Cell Environ. 25 (2002) 239–250.

[34] G.W. Netondo, J.C. Onyango, E. Beck, Crop physiology and metabolism Sorghum and salinity II – gas exchange and chlorophyll fluorescence of sorghum under salt stress, Crop Sci. 44 (3) (2004) 806–811.

[35] M. Ashraf, Some important physiological selection criteria for salt tolerance in plants, Flora Morphol. Distrib. Funct. Ecol. Plants 199 (5) (2004), 361376.

[36] S.W. Breckle, H. Kahle, Ecological geobotany/autecology and ecotoxicology, in: Progress in Botany, Springer, Berlin, 1991, p. 391406.

[37] D.L. Godbold, A. Hüttermann, Effect of zinc, cadmium and mercury on root elongation of *Picea abies* (Karst.) seedlings, and the significance of these metals to forest die-back, Environ. Pollut. 38 (1985) 375–381.

[38] D.H. Nies, Microbial heavy-metal resistance, Appl. Microbiol. Biotechnol. 51 (1999) 730–750.

[39] M. Angelone, Trace elements concentrations in soils and plants of Western Europe, Biogeochem. Trace Met. (1992) 1960.

[40] A. Schutzendubel, A. Polle, Plant responses to abiotic stresses: heavy metal-induced oxidative stress and protection by mycorrhization, J. Exp. Bot. 53 (2002) 1351–1365.

[41] G.U. Chibuike, S.C. Obiora, Heavy metal polluted soils: effect on plants and bioremediation methods, Appl. Environ. Soil Sci. 2014 (2014).

[42] K.E. Zinn, M. Tunc-Ozdemir, J.F. Harper, Temperature stress and plant sexual reproduction: uncovering the weakest links, J. Exp. Bot. 61 (7) (2010) 1959–1968, https://doi.org/10.1093/jxb/erq053.

[43] E. Ruelland, A. Zachowski, How plants sense temperature, Environ. Exp. Bot. 69 (2010) 225–232.

[44] M. Uemura, P.L. Steponkus, Cold acclimation in plants: relationship between the lipid composition and the cryostability of the plasma membrane, J. Plant Res. 112 (1999) 245–254.

[45] G.I. Burd, D.G. Dixon, B.R. Glick, A plant growth-promoting bacterium that decreases nickel toxicity in seedlings, Appl. Environ. Microbiol. 64 (10) (1998) 36633668.

[46] M. Grover, S.k.Z. Ali, V. Sandhya, A. Rasul, B. Venkateswarlu, Role of microorganisms in adaptation of agriculture crops to abiotic stresses, World J. Microbiol. Biotechnol. 27 (2011) 1231–1240, https://doi.org/10.1007/s11274-010-0572-7.

[47] D. Egamberdieva, Alleviation of salt stress by plant growth regulators and IAA producing bacteria in wheat, Acta Physiol. Plant. 31 (2009) 861–864, https://doi.org/10.1007/s11738-009-0297-0.

[48] A. Marulanda, R. Azcón, F. Chaumont, J.M. Ruiz-Lozano, R. Aroca, Regulation of plasma membrane aquaporins by inoculation with a *Bacillus megaterium* strain in maize (*Zea mays* L.) plants under unstressed and salt-stressed conditions, Planta 232 (2010) 533–543.

[49] S. Mayak, T. Tirosh, B.R. Glick, Plant growth-promoting bacteria that confer resistance to water stress in tomato and pepper, Plant Sci. 166 (2004) 525–530, https://doi.org/10.1016/j.plantsci.2003.10.025.

[50] M. Waqas, A.L. Khan, M. Kamran, et al., Endophytic fungi produce gibberellins and indoleacetic acid and promotes host-plant growth during stress, Molecules 17 (2012) 10754–10773, https://doi.org/10.3390/molecules170910754.

[51] M. Nadeem, I. Imran, I. Taj, M. Ajmal, M. Junaid, Omega-3 fatty acids, phenolic compounds and antioxidant characteristics of chia oil supplemented margarine, Lipids Health Dis. 16 (2017) 102.

[52] A. Aslam, Z. Ahmad Zahir, H.N. Asghar, M. Shahid, Effect of carbonic anhydrasecontaining endophytic bacteria on growth and physiological attributes of wheat under water-deficit conditions, Plant Prod. Sci. 21 (3) (2018), 244255.

[53] P. Zuccarini, R. Save, Three species of arbuscular mycorrhizal fungi confer different levels of resistance to water stress in *Spinacia oleracea* L, Plant Biosyst. 150 (2016) 851–854.

[54] B. Denton, Advances in phytoremediation of heavy metals using plant growth promoting bacteria and fungi, Basic Biotechnol. 3 (2007) 1–5.

[55] E. Dell'Amico, L. Cavalca, V. Andreoni, Improvement of *Brassica napus* growth under cadmium stress by cadmium resistant rhizobacteria, Soil Biol. Biochem. 40 (2008) 74–84.

[56] C.O. Dimkpa, A. Svatos, D. Merten, G. Buchel, E. Kothe, Hydroxamate siderophores produced by *Streptomyces acidiscabies* E13 bind nickel and promote growth in cowpea (*Vigna unguiculata* L.) under nickel stress, Can. J. Microbiol. 54 (2008) 163–172.

[57] S.C. Wu, K.C. Cheung, Y.M. Luo, M.H. Wong, Effects of inoculation of plant growth-promoting rhizobacteria on metal uptake by *Brassica juncea*, Environ. Pollut. 140 (2006) 124–135.

[58] F. Ouziad, U. Hildebrandt, E. Schmelzer, H. Bothe, Differential gene expressions in arbuscular mycorrhizal-colonized tomato grown under heavy metal stress, J. Plant Physiol. 162 (2005) 634–649, https://doi.org/10.1016/j.jplph.2004.09.014.

[59] R.A. Gaxiola, R. Rao, A. Sherman, P. Grisafi, S.L. Alper, G.R. Fink, The *Arabidopsis thaliana* proton transporters, AtNhx1 and Avp1, can function in cation detoxification in yeast, Proc. Natl. Acad. Sci. U. S. A. 96 (1999) 1480–1485.

[60] N. Tank, M. Saraf, Salinity-resistant plant growth promoting rhizobacteria ameliorates sodium chloride stress on tomato plants, J. Plant Interact. 5 (2010) 51–58, https://doi.org/10.1080/17429140903125848.

[61] D. Egamberdiyeva, The effect of plant growth promoting bacteria on growth and nutrient uptake of maize in two different soils, Appl. Soil Ecol. 36 (2007) 184–189, https://doi.org/10.1016/j. apsoil.2007.02.005.

[62] R.J. Rodriguez, J. Henson, E. Van Volkenburgh, M. Hoy, L. Wright, F. Beckwith, Y. Kim, R.S. Redman, Stress tolerance in plants via habitat-adapted symbiosis, Int. Soc. Microb. Ecol. 2 (2008) 404–416.

[63] G.N. Al-Karaki, Nursery inoculation of tomato with arbuscular mycorrhizal fungi and subsequent performance under irrigation with saline water, Sci. Horticult. 109 (1) (2006) 17.

[64] V. Sandhya, Sk.Z. Ali, M. Grover, G. Reddy, B. Venkateswarlu, Alleviation of drought stress effects in sunflower seedlings by the exopolysaccharides producing *Pseudomonas putida* strain GAP-P45, Biol. Fertil. Soil 46 (2009) 17–26.

Further reading

V.A.J.M. Abadi, M. Sepehri, Effect of *Piriformospora indica* and *Azotobacter chroococcum* on mitigation of zinc deficiency stress in wheat (*Triticum aestivum* L.), Symbiosis 69 (1) (2016) 919.

A.G. Babu, P.J. Shea, D. Sudhakar, I.B. Jung, B.T. Oh, Potential use of *Pseudomonas koreensis* AGB-1 in association with *Miscanthus sinensis* to remediate heavy metal (loid)-contaminated mining site soil, J. Environ. Manag. 151 (2015), 160166.

H. Bae, R.C. Sicher, M.S. Kim, S.H. Kim, M.D. Strem, R.L. Melnick, et al., The beneficial endophyte *Trichoderma hamatum* isolate DIS 219b promotes growth and delays the onset of the drought response in *Theobroma cacao*, J. Exp. Bot. 60 (11) (2009) 32793295.

J. Bresson, F. Varoquaux, T. Bontpart, B. Touraine, D. Vile, The PGPR strain *Phyllobacterium brassicacearum* STM196 induces a reproductive delay and physiological changes that result in improved drought tolerance in *Arabidopsis*, New Phytol. 200 (2) (2013), 558569.

C. Caris, W. Hordt, H.J. Hawkins, V. Romheld, E. George, Studies of iron transport by arbuscular mycorrhizal hyphae from soil to peanut and sorghum plants, Mycorrhiza 8 (1) (1998) 3539.

E.K. Cheruiyot, L.M. Mumera, W.K. Ngetich, A. Hassanali, F. Wachira, Polyphenols as potential indicators for drought tolerance in tea (*Camellia sinensis* L.), Biosci. Biotechnol. Biochem. 71 (9) (2007) 21902197.

O. Choi, J. Kim, J.G. Kim, Y. Jeong, J.S. Moon, C.S. Park, et al., Pyrroloquinoline quinone is a plant growth promotion factor produced by *Pseudomonas fluorescens* B16, Plant Physiol. 146 (2) (2008), 657668.

K. Chookietwattana, K. Maneewan, Selection of efficient salt-tolerant bacteria containing ACC deaminase for promotion of tomato growth under salinity stress, Soil Environ. 31 (1) (2012).

Chapter 3

Endurance of microbes against nitrogen starvation by altering the biochemical and physiological activities of plants

Reshma Devi Ramesh[a,b], Kalaiselvi Selvaraj[c], Karnan Muthusamy[a,d], Leelavathi Lakshmanan[a], Steffi Pulikondan Francis[e], Siva Vijayakumar Tharumasivam[b], and Vijayalakshmi Selvakumar[a,f]

[a]Askoscen Probionics (R&D), Tiruchirappalli, Tamil Nadu, India, [b]Srimad Andavan Arts and Science College, Tiruchirappalli, Tamil Nadu, India, [c]Department of Microbiology, Government Arts and Science College (W), Orathanadu, Tamil Nadu, India, [d]Grassland and Forage Division, National Institute of Animal Science, Cheonan-si, South Korea, [e]Department of Microbiology, Cauvery College for Women (Autonomous), Tiruchirappalli, Tamil Nadu, India, [f]Food Science and Biotechnology, School of Agriculture and Life Sciences, Kangwon National University, Chuncheon, Republic of Korea

1 Introduction

When a plant grows in intimate interaction with microbial populations, the plant's nutrient intake, development, defenses, and responses to abiotic stress are affected. Microorganisms gain sustenance from plants, both circuitously by exudates or dead tissues in the soil, and without delay when microorganisms invade the internal structure of plant roots or different organs. It makes sense that microbes would have evolved special communication strategies with plants to gain access to the nutrients those plants furnish. The investigation of mechanisms, mainly the identification of the microorganisms accountable for such interactions, is still in its infancy.

Nitrogen is a crucial macroelement that all living things need. Different species can make use of a wide range of nitrogenous compounds. Most bacteria may use several organic and inorganic nitrogenous substances, and some prokaryotic organisms can use N_2, employing organic N_2 fixation. The root-associated microbiota of plant life has been investigated and proven to be

Plant-Microbe Interaction—Recent Advances in Molecular and Biochemical Approaches
https://doi.org/10.1016/B978-0-323-91876-3.00014-2
33

extensively much less complex than the microbiota of the surrounding soil, with Proteobacteria, Bacteroidetes, and Actinobacteria being the most ordinary bacteria. Metabolic signals from host cells and plant cell wall habitats help identify these bacteria. The most common nitrogen sources for microorganisms and plants are nitrate and ammonium, which are also crucial signaling molecules that influence many different aspects of metabolism and growth. As a result, it is a crucial issue for all organisms, limiting their growth and development. Most soils include predominantly inorganic nitrogen, indicating the presence of massive amounts of natural types in particular ecosystems (e.g., forests) (nitrate and ammonium). Availability of N_2 limits plant biomass and productivity in many ecosystems. Although molecular nitrogen (N_2) accounts for 78% of atmospheric gases, it cannot be used by plants. Nitrogen must be chemically or biologically changed into a plant-accessible form (nitrogen fixation).

Fixed nitrogen is a limiting nutrient in most ecosystems, with molecular nitrogen from the atmosphere serving as the biosphere's fundamental source of nitrogen. Although plants cannot directly absorb molecular nitrogen, they can produce it through the process of biological nitrogen fixation, which is only seen in prokaryotic cells. Near the end of the 19th century, about the same time that nitrogen-fixing was discovered, the rise of bacteria sticking to the root surface in soil was identified. For a long time, it was believed that only a small number of bacterial species could fix nitrogen. However, during the past 30 years, it has been found that almost all phyla of bacteria, as well as methanogenic archaea, participate in nitrogen fixation. The ability to fix nitrogen symbiotically within vascular plant nodules is disclosed by rhizobia (Alphaproteobacteria), which associates primarily with leguminous plants of one angiosperm superfamily (Fabaceae), and Frankia (in Actinobacteria), which associates with a wider variety of plants from eight families. Cyanobacteria are a huge group of nitrogen-fixing bacteria that interact with a wide range of higher and lower plants, fungi, and algae. The term "associative symbiosis" describes a sort of nitrogen-fixing bacterial colonization of the nonleguminous flora's root floor that occurs in addition to the development of distinct structures. A new bacterial class known as nitrogen-fixing endophytes was discovered as a result of the ongoing isolation of microorganisms from surface-sterilized roots.

2 Nitrogen-the primary element

Nitrogen is integral for plant growth and improvement. It is not only crucial for cell molecules such as amino acids, nucleic acids, chlorophyll, adenosine triphosphate (ATP), and several plant hormones, but it is also a predominant regulator in a number of organic strategies such as carbon metabolism, amino acid metabolism, and protein synthesis. Nitrogen is an integral macronutrient that impacts crop productivity by way of regulating growth and development. It can be observed in soils in a range of forms, such as inorganic types like nitrate and ammonium, as well as organic forms like amino acids, peptides, and lipids. Organic nitrogen can be determined in a vast range of environments, including boreal and tropical ecosystems. Nitrate and ammonium are the most regular forms. Nitrate has been tested for use by vegetation as both a nutrient and a signal metabolite. Results show that it has a great impact on plant metabolism and growth. Plants can soak up nitrogen ions as nitrate (NO_3^-) and ammonium (NH_4^+) ions. Nitrogen is essential in plant metabolic systems. As a result, nitrogen utility is indispensable and necessary to increase crop yields and output. Nitrogen is an indispensable component of protein (a protein composed of amino acids that catalyze chemical reactions and transport electrons) and chlorophyll (which also aids in photosynthesis) discovered in many essential aspects of the plant body.

3   Nitrogen's influence

Nitrogen is required for a vast variety of physiological processes. It imparts a dark green color to plant life and aids in the increase and development of plant leaves, stems, and other vegetative parts. It also encourages root development. Nitrogen boosts early plant growth, improves fruit quality, promotes the boom of leafy vegetables, and increases the protein content of fodder crops; it additionally supports the accumulation of different vitamins such as potassium and phosphorus and regulates usual plant growth. It is additionally fundamental for the characteristics of different integral biochemical compounds such as chlorophyll (for photosynthesis), more than a few enzymes (for enabling organisms to take part in biochemical approaches and nutrient absorption), and nucleic acids such as DNA and RNA (for reproduction). When nitrogen is biologically linked with carbon, hydrogen, oxygen, and sulfur, amino acids, which are the building blocks of proteins, are created. Amino acids are required for the introduction of protoplasm, which serves as a site for cell division and thus plant growth and development. Because all enzymes are made up of proteins, nitrogen is imperative for all enzymatic metabolism in plants. Nitrogen is integral for photosynthesis when you consider that it is a crucial component of the chlorophyll molecule. It not only improves the density of greens and grain crops but also increases their protein content. Because all plant enzymes are proteins, enzymatic reactions require nitrogen to proceed. It is required for numerous vitamins, such as biotin, thiamine, niacin, and riboflavin.

4   The result of nitrogen deficiency

A nitrogen deficit in the soil is one of the main causes of low crop productivity and poor crop health [1]. Growth may be hindered if cell division declines. Chlorosis is a pale green to light yellow coloration that emerges first on older leaves, normally at the tips. When nitrogen tiers are reduced, the protein content of seeds and vegetative sections decreases. In harsh conditions, flowering is extensively hampered. Nitrogen deficiency causes early maturity in some crops, resulting in a loss in manufacturing and quality. Nitrogen shortage reduces development, increases chlorosis (green to yellow leaf shade shift), and causes red and pink blotches on leaves, all of which restrict lateral bud initiation (from which leaves, stems, and branches develop).

5 Nitrogen deficiency

Several stressors triggered by complicated environmental conditions, such as vivid light, UV, extreme temperature fluctuations, drought, salinity, heavy metals, and hypoxia, have resulted in huge crop losses worldwide. Drought, high soil salinity, heat, cold, oxidative stress, and heavy metal toxicity are all examples of abiotic stress that influence and restrict agriculture worldwide. Nitrogen depletion is one of the abiotic stressors. When high-carbon natural matter, such as sawdust, is added to soil, it may lead to plant nitrogen deficiency. Soil organisms use all available nitrogen to destroy carbon sources, making nitrogen inaccessible to plants. Nitrogen stress is brought about with the aid of intense fluctuations in soil nitrogen. Protein recycling is a quintessential defense mechanism for cells, as well as a superb technique for gaining strength and ensuring survival by using recycling amino acids for protein synthesis. Nitrogen deficiency in cereal crops reduces grain yield by affecting nutrient intake, photosynthetic rate, respiratory efficiency, and enzyme activity. Early symptoms of nitrogen-stressed crop plants include chlorotic leaves, less fertile tillers, shorter plant height, and poor growth. Nitrogen deficiency causes molecular and developmental adaptation in all organisms. In higher plants, nitrogen shortage produces full-size variations in plant growth and development, such as root branching, leaf chlorosis, and decreased seed yield. The complicated and diversified physiological and biochemical modifications generated by nitrogen limit the several genes and metabolic and regulatory pathways required to generate plant adaptive responses to nitrogen constraint. Nitrogen deficit slows the development and accumulation of nonnitrogen metabolites while increasing the availability of photoassimilates for the synthesis of secondary metabolism molecules such as ascorbic acid and different natural acids.

6 Nitrogen's importance in agriculture

Nitrogen is a necessary macronutrient for plant growth. However, plants cannot utilize atmospheric nitrogen in the structure of urea, nitrate, ammonium, amino acids, and different molecules. Legumes rely on symbiotic nitrogen-fixing microorganisms to convert N_2 into ammonium ions to meet their nitrogen requirements. Food supply in emerging countries is dwindling because of climate variability. Rising international temperatures are causing variable and inappropriate rainfall events, unstable winter/summer seasons, occurrence of multiplied disorder, and crop failures. According to research, increased temperatures and a prolonged, quicker vegetative growth can create new agricultural opportunities across a wide range of agroecological zones. Climate change influences on agricultural outputs are anticipated to vary between continents, necessitating specialized adaptation strategies.

Microbial communities related to plant boom in a variety of conditions, along with acidity, alkalinity, salinity, temperature, and water scarcity, have been identified and characterized for biotechnological functions in agriculture, medicine, industry, and the environment. Crop-associated microbial diversity is indispensable for the long-term survival of agricultural manufacturing systems. Plant growth, yield, and adaptation to adversarial settings are all aided by microorganisms. There are three types of plant microbiomes: rhizospheric, phyllospheric, and endophytic. The rhizosphere is the section of soil in which plant roots alter microbial function and stability by releasing distinct substrates. The plant rhizosphere has been associated with several microbial species from the genera *Acinetobacter, Alcaligenes, Arthrobacter, Aspergillus, Azospirillum, Bacillus, Burkholderia, Enterobacter, Flavobacterium, Haloarcula, Halobacterium, Halococcus, Haloferax, Methylobacterium, Paenibacillus, Penicillium,* and *Piriform.* Endophytic microorganism are germs that colonize the inner parts of plant such as the root, stem, and seeds, inflicting damage to the host plant. *Achromobacter, Aspergillus, Azoarcus, Burkholderia, Enterobacter, Gluconacetobacter, Herbaspirillum, Klebsiella, Microbispora, Micromonospora, Nocardioides, Pantoea, Planomonospora, Pseudomonas, Penicillium, Piriformospora, Serratia, Streptomyces* and *Thermomonospor.* These microorganisms had been observed in wheat, rice, maize, soybeans, peas, and chickpeas, among others.

Microorganisms and plants frequently interact in the phyllosphere. Leaf base microbiomes are excellent candidates since they can withstand high temperatures (40–55°C) and UV exposure. Several bacteria, including *Agrobacterium, Methylobacterium, Pantoea, Penicillium,* and *Pseudomonas,* have been found in the phyllosphere of much vegetation. Crop-associated microorganisms can enhance plant life in a variety of abiotic stress situations. A variety of microbes have been found to promote plant growth, both without delay through fixing nitrogen (N_2), solubilizing minerals such as phosphorus (P), potassium (K), and zinc (Zn), and secreting siderophores, indoleacetic acids, gibberellic acids, and cytokinin, and indirectly by producing antagonistic substances, antibiotics, and lytic enzymes.

7 Microbes influence agricultural productivity

The contents of plant root exudates have an essential function in the selection and enrichment of microorganisms. Beneficial interactions between plants and microbes have developed in the form of epiphytic/endophytic/rhizospheric interactions, depending on the type and quantity of natural aspects in exudates. Crop microbiomes are necessary for agriculture because they can enhance plant growth and nutrition via organic N_2-fixation and other techniques. Microbes can expand crop yields, dispose of contaminants, forestall illness, and produce fixed nitrogen or novel chemicals. As a result of biological N_2 fixation, microbes may promote plant growth.

8 Nitrogen fixation via residing organisms

Utilizing N_2-fixing microorganisms as biofertilizers has emerged as one of the most environmentally and ecologically safe methods of increasing crop plant growth and output because nitrogen is a primary limiting element for plant growth. Biological nitrogen fixation (BNF) is a potential substitute for organic nitrogen fertilizer that may lead to increased production in agriculture. Several associative and endophytic microorganisms have been recognized that restore atmospheric nitrogen and supply it to their host plants. Many nitrogen-fixing microorganisms, such as *Arthrobacter*, *Azoarcus*, *Azospirillum*, *Azotobacter*, *Bacillus*, *Enterobacter*, *Gluconacetobacter*, *Herbaspirillum*, *Klebsiella*, *Pseudomonas*, and *Serratia*, have been isolated from the rhizosphere of some vegetation and provide constant nitrogen to related plants.

9 The nitrogen-fixation process

Nitrogen fixation also refers to organic nitrogen changes, such as the generation of nitrogen dioxide. Nitrogen fixation is the process by which nitrogen (N_2) in the environment is transformed into ammonia (NH_3). The majority of the nitrogen in the atmosphere, also known as elemental nitrogen (N_2), is inert, which means that it does not readily combine with other molecules to form new compounds. Dinitrogen is enormously inert due to the electricity of its NN triple bond. To remove one nitrogen atom from another, all three chemical bonds must be broken. Nitrogen atoms can be freed from their diatomic form (N_2) and used in a range of approaches thanks to fixation mechanisms. All varieties of existence require nitrogen fixation, each natural and manufactured, because nitrogen is indispensable to biosynthesizing basic building blocks of plants, animals, and other living forms, such as nucleotides for DNA and RNA and amino acids for proteins. As a result, nitrogen fixation is integral for agriculture and the manufacturing of fertilizer. Nitrogen-fixing microorganisms are diazotrophic bacteria.

Higher flowers and animals (e.g., termites) have developed symbiotic partnerships with diazotrophs. All other organisms depend on diazotrophs for nitrogenous chemicals, which are then utilized to construct nucleic acids, proteins, and different macromolecules through a community of metabolic processes. Each year, diazotrophs supply around 60% of the Earth's nitrogen, whereas industrially generated nitrogen accounts for roughly 25% of the total, with the remaining 15% coming from lightning, UV radiation, and other sources. As a result, diazotrophs are crucial to the continuation of the biosphere's nitrogen cycle [2]. Microbiologists have spent a lot of time investigating diazotrophs. German agronomist Hermann Hellriegel and Dutch microbiologist Martinus Beijerinck were the first describe symbiotic nitrogen fixation. BNF occurs when atmospheric nitrogen is converted to ammonia through nitrogenase, a tricky and oxygen-sensitive metalloenzyme. Nitrogenases are enzymes that involves in the fixation of nitrogen from the air with the aid of some organisms. This process can only be enhanced by a well-known group of enzymes. All nitrogenases have an iron- and sulfur-containing cofactor that forms a heterometal complex at the active site (e.g., FeMoCo).

On the other hand, some species have vanadium and iron atoms. The enzymes that catalyze nitrogenase reactions are particularly oxygen sensitive. Many microorganisms stop producing the enzyme in the presence of oxygen. Many nitrogen-fixing organisms can continue to exist in anaerobic settings, either through respiring to limit oxygen ranges or by binding oxygen to proteins.

According to Roman literature, crop rotation was once employed to obtain high yields of nonlegume vegetation when grown with legume vegetation like soybeans and peas. Nitrogen fixation, developed by Hellriegel, is a mechanism by which legumes soak up atmospheric nitrogen and replace ammonium in the soil. He found that nitrogen fixation occurs in legume nodules on the roots. Later, Beijerinck researched the mechanisms of nitrogen fixation and determined that they were microorganisms and bacteria, which he labeled rhizobia. $N_2 + 8H^+ + 8e^- \rightarrow 2NH_3 + H_2$ is the reaction for BNF. Because it affects N_2-collecting electrons, the response is recognized as a discount reaction. The specific mechanism of catalysis is unknown due to the technological obstacles biochemists confront when studying this technique in vitro. As a result, the precise order of the phases in this reaction is uncertain. While the equilibrium synthesis of ammonia from molecular hydrogen and nitrogen has a poor enthalpy of response (i.e., it releases energy), the energy barrier to activation, besides catalysis, which is executed by using nitrogenases, is extraordinarily high. The enzymatic conversion of N_2 to ammonia requires an input of chemical electricity furnished through ATP hydrolysis to overcome the activation electricity barrier.

10   Nitrogenase is a nitrogen-fixing enzyme

Nitrogenase is composed of two soluble proteins: an component I and component II. Component I protein, commonly known as MoFe protein or nitrogenase, is a 22-tetramer with a total molecular mass of 240 kDa. The nifD and nifK genes encode similar-sized subunits; for example, the isolated subunits of *Azotobacter vinelandii* MoFe-protein have 491 and 522 amino acids, respectively. The MoFe protein, also regarded as an iron-molybdenum cofactor (FeMoco), is made up of two Mo atoms, 28–34 iron (Fe) atoms, and 26–28 acid-labile sulfides. The MoFe protein consists of two types of metallic centers: the FeMo cofactor and the P-cluster pair. According to one source, the FeMo cofactor, regularly recognized as the "M-center" or "cofactor," is a secure metallocluster generated from acid-denatured MoFe protein.

The FeMo cofactor has attracted a lot of research attention because it incorporates molybdenum in a physiologically distinct structure and is assumed to be the center of substrate reduction. The P-cluster, or simply the P-cluster, is known to have a role in electron transfer between the Fe protein and the FeMo cofactor. The polypeptide folds of the subunits are similar, containing three helical/sheet domains. The three domains of every subunit are separated by a wide, shallow gap, with the FeMo cofactor at the backside of this cleft in the subunit. The P-cluster pairs are located at the interface between a pair of - and - subunits. The agency of these subunits into a dimer seems to represent the basic practical unit of the MoFe protein. The tetramer interface is stabilized with the aid of a cation binding site, most probably for calcium, produced by using ligands from each subunit. Surprisingly, the center of the six-helical barrel around the tetramer twofold axis is not filled with side chains; instead, an open channel 8–10 A wide and 35 A lengthy extends via the tetramer.

Component II, recognized as Fe protein or nitrogenase reductase, is composed of two copies of a single component with a total molecular mass of 60 kDa expressed through the *nifH* gene. Each subunit folds into a single helical/sheet area that is securely bonded at one end of the dimer by the 4Fe-4S cluster. Each subunit is composed of an eight-stranded, ordinarily parallel sheet surrounded by nine α-helices. The two subunits are linked via a molecular twofold rotation axis that runs throughout the cluster. The cluster is symmetrically coordinated through 2 cysteines, 97 and 132, supplied by each subunit, as located through biochemical and mutagenesis studies. A notable characteristic of the shape is the solvent dissemination of the 4Fe-4S cluster, which has been discovered in spectroscopic studies. Each subunit's binding sites for the terminal phosphates of bound nucleotides (as precise with the aid of the molybdate sites) are separated through 20°A from the 4Fe-4S cluster and each other. This distance between the cluster and the molybdate suggests that MgATP does not bind immediately to the cluster, as spectroscopic investigations have suggested.

This protein is made up of four nonheme Fe atoms and four acid-labile sulfides (4Fe-4S). Component I is engaged in substrate binding and discounting with the aid of binding to ATP and ferredoxin or flavodoxin proteins (Fdx or Fld). The power for the response is supplied by ATP hydrolysis, while the electrons are furnished by Fdx/Fld proteins. It is worth noting that this is a discount reaction, which means that electrons ought to be supplied to N_2 for it to be reduced to NH_4. As a result, the task of factor II is to supply electrons to component I one at a time. Before ATP is hydrolyzed to ADP, component II must provide an electron to component I. For every fixed N_2, 21–25 ATPs are required. The contact and subsequent dissociation of nitrogenase components I and II occur in numerous instances to the restoration of one N_2 molecule. Nitrogenase creates ammonia by attaching every nitrogen atom to three hydrogen atoms (NH_3). Molecular hydrogen is produced as a byproduct of the nitrogenase reaction, which is useful for those looking to synthesize H_2 as a choice for fossil fuels.

Nitrogenase reduces the substrate via three predominant types of electron transfer processes:

(1) Fe protein elimination in vivo and in vitro by using electron carriers such as flavodoxin or ferredoxin and dithionite.
(2) Single-electron switch from Fe protein to MoFe protein in a MgATP-dependent mechanism with a stoichiometry of two MgATP hydrolyzed per electron exchanged.
(3) Electron delivery to a substrate that is most likely connected to the energetic site of the MoFe protein. Under optimal conditions, the entire stoichiometry of dinitrogen discount with the aid of nitrogenase has been established.

$$N_2 + 8H^+ + 8e^- + 16MgATP \rightarrow 2NH_3 + H_2 + 16MgADP + 16Pi$$

Even at a strain of 50 atm (5065 kPa), nitrogen fixation is usually accompanied by a varying amount of proton reduction and H_2 production. This H_2 technology catalyzed with the aid of nitrogenase represents a loss of power and reductant, which is partially recovered by the motion of an absorption hydrogenase found in many nitrogen-fixing species. In an oxy-hydrogen or Knallgas reaction, it catalyzes the oxidation of H_2. Under iron deficiency, the physiological electron donor for nitrogenase in most N_2-fixing organisms is ferredoxin or flavodoxin; the donor, in turn, obtains electrons from reductants created in intermediary metabolism.

11 Ammonia is produced as a result of nitrogen fixation

Ammonia, a nitrogen fixation product, is usually assimilated via the glutamine synthetase-glutamate synthase route, just like exogenously given ammonia. Other enzymes that may also play a role in ammonia incorporation are alanine dehydrogenase and glutamate dehydrogenase. All nitrogen-fixing organisms or diazotrophs preferentially soak up ammonia or other types of fixed nitrogen (nitrate, urea, amino acids, etc.) and may also produce nitrogenase solely when such sources of mixed nitrogen are unavailable in the medium. Nitrogenase production in some nitrogen-fixing organisms may be hastily inhibited by using ammonia. This ammonia-induced reversible inhibition is related to the exchange in one subunit of nitrogenase reductase. Ammonia is generally involved in the transcriptional law of nitrogenase manufacturing in all nitrogen-fixing organisms by glutamine synthetase. Nitrogenase is irreversibly inactivated through oxygen, which additionally inhibits the synthesis of its proteins. The law of nitrogenase production by using ammonia and oxygen reduces energy and reductant waste and may additionally have advanced as an evolutionary response to the high expenses of nitrogen-fixing.

12 Nitrogen fixation and anaerobiosis

Nitrogen-fixing bacteria use a variety of approaches to minimize oxygen levels, which interfere with nitrogenase activity. Nitrogenases are the enzymes that restore nitrogen (N_2 to NH_3) and are at the heart of nitrogen fixation (N_2 to NH_3). The oxygen that dissolves the Fe-S cofactors irreversibly blocks most nitrogenases, which are key large discount complexes. O_2 binds to the Fe in nitrogenases, stopping them from binding to N_2. Nitrogen fixers have mechanisms in the region to shield nitrogenases from oxygen in vivo. One regularly occurring instance is *Streptomyces thermoautotrophicus* nitrogenase, which is unaffected by the presence of oxygen. A slime layer is a layer of water retained by a proteoglycan-rich extracellular matrix in some bacteria. This slime layer features an oxygen barrier. Some nitrogen fixers, such as Azotobacteraceae, have been related to an excessive metabolic rate, which permits oxygen reduction at the mobile membrane; however, the efficacy of this mechanism is controversial. Some nitrogen fixers, such as Azotobacteraceae, can utilize an oxygen-amendable nitrogenase under aerobic conditions, which allows for oxygen reduction at the cell membrane; however, the efficacy of this procedure is debatable. Leghemoglobin is a protein generated by using plant roots in certain prerequisites (also leghemoglobin or legoglobin). By deflecting the impact of free oxygen in the cytoplasm of infected plant cells, leghemoglobin supports the proper operation of root nodules. Leghemoglobin is an oxygen or nitrogen transporter that naturally interacts with oxygen and nitrogen in the same way. It shares chemical and structural similarities with hemoglobin. Leghemoglobin has nearly 10 times the affinity for oxygen as human hemoglobin. As a result, the oxygen awareness in the microorganism is low enough for nitrogenase activation but not high enough to bind all of the oxygen in the bacteria, permitting the bacteria to breathe.

Leghemoglobin is produced through legumes in response to rhizobia infection of the roots, as part of a symbiotic connection between the plant and nitrogen-fixing bacteria. Leghemoglobin is assumed to be formed by both bacteria and plants, with the bacterium releasing heme while the plant produces a protein precursor (an iron atom bonded in a porphyrin ring that binds oxygen). The protein and heme combine to enable the bacteria to restore nitrogen, imparting usable nitrogen to the plant and a home for the rhizobia.

13 Nitrogen-fixation genetics and regulation

Although there are many similarities between N_2-fixing organisms in terms of genetic control of nitrogen fixation and the arrangement of genes coding for the synthesis of the N_2-fixing enzyme complex, there are also some differences that may also be related to the unique requirements of corporations or character strains.

Nitrogen-fixing bacteria may also coordinate gene expression to switch on and off the proteins critical for nitrogen fixation. Fixing atmospheric nitrogen (N_2) is a time-consuming and energy-intensive process. If N_2 fixation is no longer necessary, the production of proteins required for fixation is tightly managed. The *nif* genes encode proteins that are involved in the fixation of atmospheric nitrogen into a structure that plants can use. Nitrogen-fixing microorganisms and cyanobacteria each have these genes. The *nif* genes are observed in each free-living nitrogen-fixing microorganism and plant-symbiotic bacteria. The *nif* genes encode enzymes that restore nitrogen in the atmosphere. The nitrogenase complex is the most important enzyme encoded by *nif* genes, and it is successful in converting atmospheric nitrogen to different nitrogen types like ammonia, which flora can use for a variety of reasons. In addition to the nitrogenase enzyme, the *nif* genes encode a range of regulatory proteins concerned with nitrogen fixation. In response to low fixed nitrogen and oxygen concentrations, the *nif* genes are upregulated (the low oxygen concentrations are actively maintained in the root environment). The nif regulon, which is made up of 7 operons and 17 *nif* genes, regulates nitrogen fixation. *Nif* genes have both good and bad regulators. Some of the *nif* genes are Nif A, D, L, K, F, H, S, U, Y, W, and Z. The nitrogen-sensitive NifA protein stimulates *nif* gene transcription. When there is insufficient fixed nitrogen on hand for the plant to ingest, NtrC, an RNA polymerase, initiates NifA expression. NifA then initiates the rest of the transcription for the *nif* genes. As a result, NifL inhibits NifA activity, blocking nitrogenase synthesis. NifL is regulated with the aid of different proteins that act as sensors for the concentrations of O_2 and ammonium in the surrounding environment. The *nif* genes can be found on bacteria chromosomes, however, they are more regularly located on plasmids alongside different genes involved in nitrogen fixation, such as those needed for microorganisms to communicate with their plant hosts.

14   Nitrogenase defense toward oxygen

Although the biochemical properties and exceptionally oxygen-sensitive nature of nitrogenase have proven to be identical in all nitrogen-fixing species investigated, the mechanisms that safeguard the enzyme component from the damaging results of oxygen are quite diverse. Many diazotrophs may have more than one mechanism, and in cyanobacteria, a slew of components appear to work in live performance to shield nitrogenase from both ambient and inner oxygen sources. The major topic of this assessment is the protective mechanisms in cyanobacteria. A short overview of the numerous adaptation mechanisms at work in different diazotrophs give a useful groundwork for comparison. Obligate anaerobes, such as *Clostridium pasteurianum* and

Desulfovibrio desulfunctans, lack any mechanism to guard their nitrogenase, or any other cell ingredient, from the damaging results of oxygen. As a result, they can solely exist and fix nitrogen in the whole absence of oxygen, and their natural distribution is limited to oxygen-free conditions.

Facultative bacteria, such as *Klebsiella pneumoniae*, *Bacillus polymyxa*, and *Rhodospirillum rubrum*, can thrive on blended nitrogen in the presence and absence of oxygen, however, they can only anaerobically restore nitrogen. Microaerophilic bacteria, such as *Azospirillum* species, decide on subatmospheric oxygen tiers for fixing nitrogen. They are unable to fix nitrogen in anaerobic or excessive oxygen tension conditions. Finally, aerobic bacteria, such as the *Azotobacter* species, can grow in the air on dinitrogen. Certain strains, however, may exhibit oxygen sensitivity at some stage in nitrogenase synthesis induction. The final three instructions for nitrogen-fixing microorganisms have been verified to have protective mechanisms in place.

15 Root nodule formation

15.1 I. Nod factor shape and synthesis

Lipo-chitooligosaccharides, also known as Nod factors, are Rhizobium signal molecules that play an important function in the early degrees of nodulation. The nod (nodulation) genes are bacterial genes that are involved in the synthesis of Nod factors.

These genes, with the exception of nodD, are no longer expressed in free-living bacteria. NodD binds to positive flavonoids generated via the root of the host plant; as soon as it is bound, NodD works as a transcriptional activator of the different nod genes, which encode enzymes involved in the production of Nod components. According to genetic and molecular studies, the products of nodA, nodB, and nodC loci speed up the development of the Nod backbone. Rhizobia can solely interact with a restricted number of host plants. Some rhizobia are more unrestrained than others, such as *Rhizobium* NGR234. This *Rhizobium* can nodulate a variety of tropical legumes and generates 18 exclusive Nod factors (Fig. 1).

15.2 II. Interaction with root hairs

Legume roots release growth factors, which aid in the fast multiplication of microorganisms. When rhizobia infect legume roots, they produce root hair deformation and curling, as nicely as the expression of several plant genes in the epidermis. Several plant genes have been created, and their expression is activated in the dermis all through nodulation. ENODIP and ENODS are additionally created during infection and nodule growth.

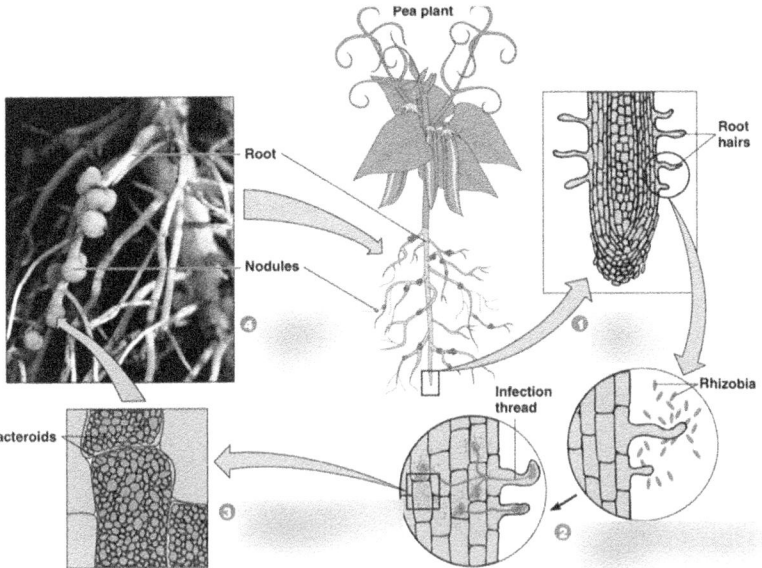

FIG. 1 (1) Rhizobia attach to root hair. (2) An infection thread is formed, through which bacteria enter root cells. (3) Bacteria change into bacteroides; packed root cells enlarge. (4) Enlarged root cells form a root nodule. *(Picture Courtesy https://quizlet.com/; Color Image from internet.)*

15.3   III. Root nodule evolution

Bacteria enter the roots via contaminated and damaged root hairs. Rhizobia adhere to root hair tips, causing them to curl tightly, trapping microorganisms inside the curls. The cytoplasmic membrane invaginates and new plant cell wall material is deposited as a result of localized hydrolysis of the plant cell wall in the coiled region. As a result, an infection thread, a tubular structure that lets a microorganism enter the plant, arises. The ultrastructure of the contamination thread's wall is comparable to that of an ordinary plant cell wall, yet the presence of positive nodulins may also give it unique characteristics. Because the required genes are expressed through cortical cells carrying an infection thread, the proline-rich early nodulins ENOD5 and ENOD12 are possibilities for contamination of thread wall components. The bacteria in the contamination thread are wrapped in a matrix that appears to be composed of plant and bacterial secretions. A 95-kD glycoprotein, for example, is determined in the contamination thread matrix and is oftentimes detected in the root cortex's intercellular gaps. Thus, it appears that interaction with bacterial floor compounds is critical in the institution of contamination threads. Hyperplasia, or fast cell division, commences quickly after launch into the cytoplasm of host cortical cells. Bacteroides are generated when bacteria shift morphologically inside the cortex. The root cells are activated, resulting in the formation of a tumor-like nodule of bacteroid-packed cells. The quantity of chromosomes in the area's host cells doubles. Chromosome doubling takes place in each polyploid and diploid legume nodule.

16   Root nodule morphology and function

The root nodule develops because of tissue proliferation prompted by way of rhizobial growth promoters, most possibly cytokinins. A mature nodule's center is the bacteroid zone, which is surrounded by numerous layers of cortical cells. Bacteroides can be discovered in the cytoplasm of plant cells by itself or in groups with peribacteroid members. Because of the presence of leghemoglobin, the energetic nodules are massive and pink, with well-developed and organized tissue. When the nodule dies, stationary-phase rhizobia are launched into the soil. The leghemoglobin pink pigment protects the N_2-fixing enzyme from oxygen damage.

17   Nitrogen fixation with the aid of free-living organisms *Azospirillum*

Azospirillum is a Gram-negative, microaerophilic, nonfermentative, free-living nitrogen-fixing bacterial genus. It has been one of the most studied plant growth-promoting bacteria (PGPB). Two key characteristics are used to describe this bacterial genus: its ability to restore atmospheric nitrogen and create a range of phytohormones such as auxins, cytokinins, and gibberellins. These two features have been recognized as the key drivers of this genus' influence on plant growth and crops. *Azospirillum* is one of the most studied PGPBs in the world, and it has been commercialized in various South American countries, including Argentina, Brazil, Uruguay, and Paraguay. Data demonstrates several aspects of the plant-bacteria interplay in both in planta and in vitro conditions. *Azospirillum* spp. have been found in saline soil, oil-contaminated soil, fermented products, fermentation tanks, sulfide springs, and microbial gas cells, albeit they are rare. The first mechanism discovered that verified *Azospirillum*'s favorable impacts on plant growth was nitrogen fixation. This mechanism is being highlighted due to the significant extent of whole nitrogen in shoots and grains observed after *Azospirillum* inoculation in wheat, sorghum, and panicum, among other cereal and grass species. In more than one greenhouse and area trial, lowering the doses of nitrogen fertilizers has proven the contribution of fixed nitrogen by way of bacteria on crops. The acetylene reduction assay (ARA) and subsequent isotopic $15N_2$ and $15N$-dilution tests have been used to investigate the incorporation of ambient nitrogen into the host plant after *Azospirillum* inoculation. While ARA has aided in the perception of *A. gramineae* associations, it has many limits when used for definitive quantification of BNF, chiefly because it is a temporary assay of enzyme activity, which is considerably reduced when plant life is disturbed. The finding that increased nitrogenase activity exists inside of contaminated roots that is sufficient to increase the inoculated plants' basic nitrogen output provides persuasive evidence that nitrogen fixing contributes the nitrogen balance of plants. Meanwhile, countless research has indicated that *Azospirillum* had a minimal effect on plant increase (an increase of 5%–18% in whole nitrogen in inoculated plants). Except in pure genetic and molecular studies, these findings almost resulted in the abandonment of *Azospirillum*'s nitrogen-fixation features. In recent years, much research has focused on nitrogen metabolism within bacterial cells, with many precise molecular pathways investigated in *Azospirillum*, which is employed as a bacterial model for nonsymbiotic nitrogen-fixing exploration. In this sense, the Sp245 stress of *A. brasilense* has been used as a model to study nitrogen metabolic pathways for the duration of the genomic age because its genome was once totally sequenced, and this stress was once physiologically described. The *nif* gene cluster was once located at two locations on the genome, one of which used to be most possibly codified for a distinct iron or vanadium nitrogenase.

Ammonia is metabolized in *Azospirillum* via two pathways: one involving glutamate dehydrogenase (gdhA) at excessive NH_4^+ concentrations and the involving glutamine synthetase (glnA) and glutamate synthetase (GOGAT) at low NH_4^+ values. The genes involved in both strategies have been discovered in every *Azospirillum* species examined to date. In recent decades, unique strategies for nitrogen-fixation research have been developed: (a) developing spontaneous ammonium-excreting mutants of *A. brasilense*; and (b) producing spontaneous ammonium-excreting mutants of *A. brasilense*. (c) the formation of paranodules, which are N-fixing sites on the roots of legume plants. On the outside, paranodules resemble legume nodules, and they can be fashioned in grasses with exogenous auxin treatment. Based on the thinking that *Azospirillum* does not release huge amounts of ammonium received from BNF on plant tissues, *A. brasilense* cells have been injected into rice and evaluated for their potential to colonize root paranodules created via treating the roots with auxins. The colonization of paranodules by way of microorganisms in handled plant life was related to great increases in plant biomass in contrast to noninoculated plants. Furthermore, nitrogenase exercise was appreciably higher in the *Azospirillum*-inoculated paranodules of the rice plants' roots than in the manipulated plants.

This mutant was able to excrete ammonium and fix nitrogen in the presence of good-sized amounts of NH_4^+, making it a good candidate for use as a biofertilizer to furnish nitrogen to graminaceous plants. This explains how more ammonium produced during nitrogen fixation is expelled. *Setaria viridis* inoculated with the HM053 strain integrates a large quantity of nitrogen via BNF, which may also be sufficient to cover the plant's everyday nitrogen demand. Furthermore, HM053 boosted wheat and barley growth as well as nif expression in planta in the course of wheat root colonization (a 300-fold increase compared to that of the wild-type strain). In discipline trials, the identical stress outperformed the original strain, resulting in a 28% increase in maize yield.

Ammonium-excreting mutants of *A. brasilense* have been shown to grow bigger plants. Furthermore, some of the mutations were tested with the usage of the paranodules colonization approach. *Azospirillum* normally impacts plant organs, specifically roots. Increased water and mineral uptake by roots were once often noted as a purpose for *Azospirillum* inoculation's useful benefits in the 1990s. Changes in root development, architecture, and size have been attributed to increased mineral intake and water absorption, rather than any particular metabolic improvement mechanism. *Azospirillum* promotes plant root growth, which enhances water absorption and nutrient adsorption quotes (including nitrogen), resulting in superior nitrogen assimilation in biomass and, more broadly, plant growth. This capacity would be mediated with the aid of bacterial colonization of the roots and/or their capability to produce a variety of phytohormones, which would manifest in most cases during the early tiers of plant development. More root exudates would penetrate the rhizosphere as a result of accelerated root biomass, increasing the bacterial population related to the roots.

18 Nitrogen-fixation cyanobacteria

Cyanobacteria, sometimes known as blue-green algae, are photosynthetic Gram-negative prokaryotes with various functions. Simple unicellular organisms (with an aggregation and colony formation tendency), unbranched filamentous forms, and lines with extra problematic branched filamentous buildings are all included. In a foremost revision of cyanobacterial taxonomy, five subgroups or sections were recognized that roughly correspond to the former orders or families of Chroococcales, Chamaesiphonales, Pleurocapsales, Nostocales (including Oscillatonaceae, Nostocaceae, and Rivulanaceae), and Stigonematales. Cyanobacteria are a type of prokaryotic, oxygen-evolving, photosynthetic Gram-negative microorganism that can live in a variety of tough environmental conditions such as nutrient deprivation, pesticides, pollution, drought, salinity, temperature, pH, mild intensity and quality, and so on. While cyanobacteria reside inside the cells of plants and are used by hosts to produce food, they serve an essential role in the ecosystem. In the late Proterozoic or early Cambrian Periods, cyanobacteria began to colonize positive eukaryote cells, imparting nourishment for the eukaryote host in exchange for an area to dwell. Cyanobacteria are usually cardio photoautotrophs, which potentially only need water, carbon dioxide, inorganic substances, and light to survive. Photosynthesis is their foremost mode of energy metabolism. In the natural world, however, it is known that some creatures can tolerate long periods of complete darkness. Furthermore, some cyanobacteria are uncommon in their capability to feed on heterotrophic substances. Cyanobacteria may also have been the first plant life to colonize exposed rock and soil surfaces. UV-absorbing sheath pigments, for example, enhance their fitness in the greater uncovered terrestrial environment. Many species can stay in the soil and other terrestrial habitats, the places in which they contribute considerably to ecosystem functioning and nutrient cycling. These bacteria comprise chlorophyll *a* and photosystems I and II, permitting them to function in oxygenic photosynthesis. *Bergey's Manual* of Systematics of Archaea and Bacteria divides the organism into following subsections. Subsection I (Chroococcales) includes unicellular cyanobacteria that reproduce through binary fission; Subsection II (Pleurocapsales) includes unicellular cyanobacteria that produce daughter cells that are smaller than the parent; and Subsection III (Oscillatoriales) includes cyanobacteria that produce trichome filaments. The era of heterocysts is a necessary section of nitrogen fixation. Heterocysts develop when filamentous cells are deprived of dissolved inorganic nitrogen. A heterocyst has a thick cell wall and just one photosystem (photosystem I), which is accountable for ATP synthesis. Photosystem II is impaired to keep away from O_2 production. O_2 inhibits nitrogenase, the enzyme responsible for N_2 fixation. The function of cyanobacterial nitrogen-fixing organisms in ecosystems and their potential, are follows

I. Aerobic N_2-fixation capable cyanobacteria
   - Space-dwelling cyanobacteria that separate N_2 fixation from oxygenic photosynthesis. Anabaena is an example of a genus with heterocystous cells.
   - Cyanobacteria are successful in differentiating between N_2 fixation and oxygenic photosynthesis. Nonheterocystous genera encompass *Gloeothece*, *Cyanothece*, and *Lyngbya*.
   - Cyanobacteria that are separated from N_2 fixation and oxygenic photosynthesis both geographically and temporally. Nonheterocystous genera consist of *Trichodesmium* and *Katagnymene*.
II. Cyanobacteria are the sole bacteria capable of fixing N_2 in anaerobic or microaerobic settings.
   - *Plectonemaboryanum* is a nonheterocystous cyanobacterium. In cyanobacterial thylakoid membranes, protein turnover has been identified as a protein that is sensitive to environmental stress conditions. Drought, food deficiency, heat, chemical stress, ozone fumigation and UV-B and visible light exposure can all have an impact on protein turnover. Numerous species can continue to exist and thrive in conditions previously considered to be inhospitable, such as desiccation, high temperatures, excessive pH, immoderate salt, and pesticides, indicating their capability to adapt to detrimental environments. Increased stages of natural and inorganic vitamins additionally make contributions to the growth and activity of terrestrial algae. Indeed, moisture can be excessive in many situations, resulting in anaerobic conditions that promote the growth of some cyanobacterial species. In these cases, temperature appears to be the most indispensable aspect of controlling algal increase and activity. *Oscillatoria* species have been recognized as the dominant algae in Bermuda grass, whereas *Anacystis* species have been found to be

the predominant algae in bentgrass. This distribution can also fluctuate depending on season and region. Pesticides are some other elements that result in the distribution and exercise of cyanobacteria. In general, most herbicides, fungicides, and soil fumigants are toxic, whereas insecticides are not. High temperatures help both the growth of blue-green algae and the productivity of phytoplankton. In rice fields with a lot of natural reliance on the soil and water, as well as characteristics like pH, temperature, and organic matter, cyanobacteria swiftly grew.

19   Nitrogen-fixing BGA

Cyanobacteria are the predominant suppliers of newly fixed nitrogen in the Arctic. They create a wide range of relationships with vegetation, including epiphytic and endophytic facultative interactions with bryophytes, lichen symbioses, and soil surface colonies that include biological soil crusts. In many terrestrial ecosystems, bryophyte-associated cyanobacteria constitute a key source of N_2; for example, in northern boreal forests, an excessive abundance of feather moss-cyanobacterial relationships make contributions of $1.5-2\,kg\,N\,ha^{-1}$ $year^{-1}$. Cyanobacteria-bryophyte interactions typically result in the highest rates of N2 fixation in polar environments due to the diversity of bryophyte species. Because their N_2 fixation costs often exceed these of other cyanobacterial symbioses, lichen-cyanobacterial symbioses are a sizeable source of fixed N_2. Furthermore, due to the fact that biological soil crusts are regularly found in many arctic ecosystems, the cyanobacteria related to them make contributions significantly to arctic N_2 imports. Certain free-living blue-green algae (cyanobacteria) restore nitrogen from the atmosphere and because they are photosynthetic, they do not compete with crop plant life or heterotrophic soil microorganisms for carbon and energy. Some nonheterocystous unicellular BGA are (*Gloeocapsa, Aphanothece, Gloeothece*, etc.) and filamentous (*Oscillatoria, Plectonema*, etc.) cyanobacteria (*Nostoc, Anabaena, Aulosira*, etc.) cyanobacteria (*Nostoc, Anabaena, Aulosira*, etc.).

In nonheterocystous species, oxygenic photosynthesis was found to be distinct from nitrogen fixation either temporally or geographically. In terms of energy, anaerobic dark conditions are no longer ideal for nitrogen fixation in these forms. As a result of geographical separation, the center nonphotosynthetic cells end up concerned with nitrogen fixation, and the outer green cells are photosynthetically active. Biofertilizer plausibly exists in the heterocystous, filamentous species of the orders Nostocales and Stigonematales, in which nitrogenase production and oxygenic photosynthesis are spatially separated and nitrogenase production is frequently light dependent. *Nostoc, Anabaena, Tolypothrix, Aulosira, Cylindrospermum, Scytonema*, and other genera abound in rice fields, all of which contribute considerably to rice fertility. Cyanobacteria can furnish the soil with 20–30 kg of nitrogen per season, which is essential for economically disadvantaged farmers who do not have enough money to buy highly priced chemical nitrogen fertilizer. Blooms of free-living cyanobacteria are generally chosen in tropical regions, and most Asian locations have rice fields infected with cyanobacteria. Vegetables, wheat, sorghum, maize, cotton, sugarcane, and other crop flowers respond to cyanobacterial biofertilizers. The traditional biofertilizer in subtropical climates is *Azolla*, a fern that harbors the heterocystous cyanobacterium *Anabaena azollae* inside its leaf cavity. In BNF, the nitrogenase enzyme is renowned for its sensitivity to molecular oxygen. Furthermore, excessive tiers of oxygen stress speed up nitrogenase subunit proteolysis, restrict nitrogenase synthesis, and result in a shortage of respiratory substrates and reductants needed for nitrogen fixation and absorption. In vivo, it is possible to undo the inhibitory effects of low oxygen levels or long exposure times. This results in higher nitrogen fixation rates and, in some diazotrophs, a posttranslational change of the Fe protein from an inactive to an active form. Diazotrophic cyanobacteria, which supply most of the constant nitrogen to the surface oceans, are the sole diazotrophs that actively generate oxygen through photosynthesis and therefore suffer from extra nitrogen restrictions. Due to this, nitrogenase only carries out a small portion of the activities that it could carry out in the actual world. Yet, it imposes a large elemental rate on diazotrophic cyanobacteria, each in terms of limited trace elements like iron and protein synthesis rates. Globally, the quantity of fixed nitrogen in the oceans has been constrained. In terms of nitrogen, cyanobacteria benefit crop flowers by producing gibberellins, auxins such as indole-3-acetic acid and indole-3-propionic acid, B12, free amino acids such as serine, arginine, glycine, aspartic acid, threonine, glutamic acid, and others, as well as extra- and intracellular polysaccharides such as xylose and galactose. These chemical compounds impart several benefits, including elevated soil structure, multiplied agricultural plant development, helpful microbes, and heavy metallic chelation. Cyanobacteria are the most common colonizers and many of them have been shown to break down tricalcium phosphate. Because rock phosphate is insoluble, it occurs in full-size portions unavailable to agricultural plants. Rock phosphate has been observed to be solubilized by using *Tolypothrix*, *Scytonema*, *Hapalosiphon*, and different cyanobacteria.

20   Cyanobacteria with a symbiotic relationship

Symbiotic cyanobacteria offer a number of benefits that make them useful in any effort to develop the number of N2-fixing symbioses include commercially significant flowers like cereals (wheat). Unlike rhizobia, most symbiotic cyanobacteria have a capacity for defending nitrogenase from oxygen inactivation (heterocysts). Cyanobacteria have an unrivaled host range (fungi, sponges, protists, and angiosperms) and are not limited to roots; however, they can create symbiotic associations with a variety of plant components and are no deeply participate to be intracellularly inside the host plant. Cyanobionts regularly provide fixed nitrogen to their hosts, but they can also supply constant carbon to nonphotosynthetic hosts. The most frequent plant hosts are bryophytes, cycads, the angiosperm *Gunnera*, the water-fern *Azolla*, and fungi (to structure lichens). Because all cyanobacteria are photoautotrophs, the majority are facultative heterotrophs, which can be found in roots, stems, leaves, and thalli and are not restrained to light-receptive components of the plant.

Hormogonia are extremely contagious agents in the majority of plant symbioses, and some, if not all, flowers launch chemical signals that stimulate development as chemoattractants that lead them into plant tissue. Plant cyanobionts are contributors to the genus *Nostoc*, which is regularly considered in nature as a free-living organism. In the laboratory, numerous cyanobacterial species that produce hormogonium, such as *Calothrix* and *Chlorogloeopsis*, have been shown to infect liverworts. Once the cyanobacterium has infiltrated the host plant, it undergoes a number of morphological, developmental, and physiological changes. Hormogonia manufacturing is slowed, however, heterocyst formation is substantially accelerated. The cyanobiont's mobile cycle charge is reduced, stopping it from outgrowing the host. CO_2 fixation is substantially inhibited, whereas nitrogen fixation is increased and ammonium absorption is reduced. The nitrogen-fixation percentages of cyanobacteria related to bryophytes are various orders of magnitude greater than those of free-living cyanobacteria. This extent is associated with an appreciably greater heterocyst frequency than in the free-living stage, which can be 6–10 times greater (only about 20% of the nitrogen constant is retained by the cyanobiont, with the remainder being communicated to the host as ammonia). Despite the speedy growth of eukaryotic oxygenic photoautotrophs over the Phanerozoic Eon, diazotrophic cyanobacteria have grown to be the "gatekeepers" of maritime production, with marine cyanobacteria acting as living fossils.

21 Arbuscular mycorrhizal fungi as nitrogen fixers

Arbuscular mycorrhizal fungi (AMF) are soil-borne fungi that can enhance plant nutrient uptake and increase plant tolerance to abiotic stress conditions. There are four AMF orders: Glomerales, Archaeosporales, Paraglomerales, and Diversisporales. They are obligate biotrophs that complete their life cycle by eating plant photosynthetic materials and lipids. The relationship of AMF with vegetation discovered 400 million years ago. The symbiotic interaction of AMF is a well-known instance of a mutualistic relationship that can influence plant growth and development. The mycelial community of the fungi goes beneath the plant's roots and promotes nutrient uptake that would in any other case be unreachable. Fungi mycelium colonizes the roots of many plants, even those of unique species, producing a common mycorrhizal community (CMN). AMF have a mutualistic interaction with around 80% of land plants. Through their extensive hyphal network, they can take mineral nutrients and water from the soil and make them accessible to the plant symbiont. The plant responds by imparting photosynthates to the AMF.

Nitrogen is a macronutrient required with the aid of flowers for growth and development. It is a necessary nutrient in plants because it is found in a variety of biomolecules such as amino acids, proteins, chlorophylls, phytohormones, and nucleic acids. Crop productivity, specifically for woody crops like the fast-growing *Populus* species, is dependent on excessive nitrogen supply. *Rhizophagus irregularis* (previously *Glomus intraradices*) is an AMF used as a soil inoculant in agriculture and horticulture. It is also one of the most necessary mycorrhizal fungi for mycoforestry; however, because it does not produce fruiting bodies, it has "absolutely no market practicable as an suitable for eating or therapeutic mushroom." *R. irregularis* is normally used in scientific studies of the effect of AMF on plant and soil improvement. The colonization of *R. irregularis* peaked earlier than that of many distinct *Glomus fungus*. There seems to be a good-sized hyphal network and sturdy intraradical spores in older host plant roots. *R. irregularis* has been located in tiny quantities in all soils, mainly those with frequent host plants, such as in woodlands and grasslands. AM symbiosis is unusual for its capability to transport nutrients. The transfer of essential mineral vitamins from AMF to the plant boosts the plant's fitness and productivity, while the AMF receives carbohydrates from the plant as an electricity source in exchange. A wide variety of genes are involved in cell division, the production of membranes, and the structural elements of cells. Surprisingly, a lack of nitrogen results in overexpression of fungal transporters, indicating an increase in dietary demand.

It was once assumed that nitrogen depletion around plant roots had decreased due to the notably fast mobility of nitrate and ammonium in soil. Depending on the plant-fungal mix and soil nitrogen sources, plant life can receive up to 42% of its nitrogen from the AM symbiont.

Ammonium appears to be the most important source of nitrogen in AM symbiosis. Unlike nitrate, ammonium can possibly be modified directly into glutamine by using the glutamine synthetase (GS) pathway and then into glutamate via glutamine oxoglutarate aminotransferase (GOGAT). Later in the metabolic process, nitrogen is incorporated into additional amides and amino acids such as arginine, alanine, and asparagine. The most frequent nitrogen type transferred via plants and AMF is arginine. *P. trichocarpa* is a temperate angiosperm tree habitat with an excessive stage of adaptive and genetic plasticity that creates symbiotic partnerships with ectomycorrhizal and AMF. *P. trichocarpa* (black cottonwood) is a deciduous tree in the willow family (Salicaceae) that is farmed for lumber and fiber merchandize in North America. The genus *Populus* was discovered in the northern hemisphere and has a variety of economically treasured species. *P. trichocarpa* was chosen as the first tree to be sequenced due to the small size of its genome, which includes 19 haploid chromosomes and is less than 4 times the size of the *Arabidopsis* genome, which is roughly 400 Mb. The human genome has 44,000 protein-coding genes. There are around 8000 pairs of duplicated genes as a result of an early genome duplication event. The full DNA code of the Western balsam poplar (*P. trichocarpa*) was revealed via DNA sequencing in 2006.

Populus species are fast-growing, strong, woody flowers that can be used as a bioenergy crop to produce biofuels and to reduce CO_2 emissions. Poplar orchards are frequently grown on marginal soils with low nitrogen. As a result, nitrogen fertilization is vital in poplar plantings to reap excessive biomass output. Throughout the *P. trichocarpa* genome, 14 ammonium and 79 nitrate transporters have been found. *Populus* has also been examined physiologically and genetically in the presence of nitrogen and phosphorus deprivation. The intraradical mycelium (IRM), which forms branching tree-like structures (arbuscules) inside the root cortical cells, transfers nutrients taken up via the ERM (Extraradical Mycelium) to the hyphal network within the host root. The arbuscules are nevertheless surrounded via the plant cell-derived periarbuscular membrane, and the intermembranous interstice is regarded as the periarbuscular gap. Mineral vitamins are transported to the periarbuscular space by AMF and absorbed through periarbuscular membrane plant nutrient transporters. AMF have been proven to take up nitrogen in the shape of organic molecules (small peptides and amino acids) just as well as complex chemical compounds in addition to ammonium and nitrate. Although flowers can utilize both organic and inorganic nitrogen, the most general nitrogen varieties in soil are nitrate (NO_3) and ammonium (NH_4^+). These nitrogen ions are picked up by specific transporters, such as nitrate (NRTs) and ammonium (AMTs) transporters, and can be absorbed locally, saved in vacuoles, or transported to different regions of the plant. For example, roots only digest a small amount of NO_3, whereas poplar leaves soak up the vast majority of NO_3. In plastids, the enzyme nitrate reductase (NR) converts cytosolic NO_3 to NO_2, which can later be changed to $NH_4^+NH_4^+$n plastids, GS and GOGAT use NH_4^+ to produce glutamine (Gln) and glutamate (Glu) [3]. NH_4^+ used to be additionally transformed to Glu in the mitochondria by means of glutamate dehydrogenase (GDH). Glutamine and Glu can be converted into different natural nitrogen forms, presenting components for nitrogen-containing chemical biosynthesis in plants. In response to inner nitrogen uptake and exterior nitrogen availability, phytohormone-mediated signaling mechanisms tightly alter plant nitrogen absorption, assimilation, and metabolism. Phytohormones such as abscisic acid (ABA) and auxin are signaling molecules that work in tandem with nitrogen alerts to control root growth and development, adjusting nitrogen acquisition in response to demand. ABA is fundamental in modulating lateral root growth and development in *A. thaliana* and *Medicago truncatula* in response to changing nitrogen levels. In response to decreasing nitrogen availability, auxin has been shown to play a role in modulating lateral root initiation and development. In the face of an excessive nitrate input, decreased root improvement in maize is related to low auxin levels. Root size and biomass are decreased when nitrogen is abundant but accelerated when nitrogen is scarce, ensuring opposing morphological adjustments in the roots. The quantity of reachable nitrogen influences the transcript tiers of genes involved in nitrogen absorption, assimilation, and metabolism, as well as root development, in poplars. Both nitrogen deficiency and excess engage unique and similar transcriptome regulatory mechanisms in poplar roots and leaves, which support morphological and physiological adaptations. Poplars may also activate their stress pathways to regulate root and leaf increase in conditions of nitrogen deficiency and excess.

22   Prospects for the future

To be environmentally sustainable in an ultra-modern society, high output yield, greater agricultural production, and soil fertility are essential. As a result, researchers have to center on a novel thought of microbial engineering based totally on favorable partitioning of distinguished biomolecules, which creates a wonderful surrounding for plant-microbe interaction. In the last few decades, the world population has more than doubled. In the meantime, food manufacturing has soared. The use of synthetic fertilizers has greatly aided increased crop output. The depth of agricultural practices has expanded in tandem with the extended use of industrial fertilizers, and a variety of fungicides, bactericides, and pesticides are being utilized in large-scale crop production, resulting in infertile soil. Researchers are attempting to discover the mechanisms of plant growth promotion, biological control, and bioremediation with the help of microbes by analyzing species and conditions that lead to benefits in plant growth. Future microbe research will depend on the improvement of molecular and biotechnological equipment to strengthen our grasp of microorganisms and achieve built-in administration of endophytic, epiphytic, and rhizospheric microbial populations.

23   Conclusion

The rhizosphere hence offers a method for obtaining culturable specified microorganisms or genes with a wide range of biotechnological makes utilized in nutrient mobilization and bioremediation. Specific strains of the rhizospheric microbiome, also known as plant nitrogen-fixing bacteria, have been proven to enhance plant growth, fitness, extreme-condition adaptability, and soil health. Biotechnology has created new possibilities for microbiome applications in the soil, including nitrogen fixation and disease biocontrol. These microorganisms have numerous metabolic and environmental needs. Microbial inoculation has a substantially greater stimulatory effect on plant improvement in both nutrient-deficient and nutrient-rich soil. Understanding microbial range and its manageable agricultural applications is essential for developing characteristics that can also be utilized as markers of plant development and production and soil health.

References

[1] A. Alexander, V.K. Singh, A. Mishra, B. Jha, Plant growth promoting rhizobacterium Steno-trophomonas maltophilia BJ01 augments endurance against N2 starvation by modulating physiology and biochemical activities of Arachis hypogea, PLoS ONE 14 (9) (2019), e0222405.

[2] R.C. Burns, R.W. Hardy, Recognition, in: Nitrogen Fixation in Bacteria and Higher Plants, Springer, Berlin, Heidelberg, 1975, pp. 3–13.

[3] H. Rennenberg, M. Dannenmann, A. Gessler, J. Kreuzwieser, J. Simon, H. Papen, Nitrogen balance in forest soils: nutritional limitation of plants under climate change stresses, Plant Biol. 11 (2009) 4–23.

Further reading

M.H. Abd-Alla, A.A. Issa, Suitability of some local agro-industrial wastes as carrier materials for cyanobacterial inoculant, Folia Microbiol. 39 (6) (1994) 576–578.

M.H. Abd-Alla, A.L.E. Mahmoud, A.A. Issa, Cyanobacterial biofertilizer improved growth of wheat, Phyton 34 (1) (1994) 11–18.

K.E. Achyuthan, J.C. Harper, R.P. Manginell, M.W. Moorman, Volatile metabolites emission by in vivo microalgae—an overlooked opportunity? Metabolites 7 (3) (2017) 39.

D.G. Adams, B. Bergman, S.A. Nierzwicki-Bauer, A.N. Rai, A. Schüßler, Cyanobacterial-plant symbioses, in: The Prokaryotes. A Handbook on the Biology of Bacteria, vol. 1, 2006, pp. 331–363.

W.N. Adger, N.W. Arnell, E.L. Tompkins, Successful adaptation to climate change across scales, Glob. Environ. Chang. 15 (2) (2005) 77–86.

M. Alexander, Introduction to soil microbiology, Soil Sci. 125 (5) (1978) 331.

R. Anandham, J. Heo, R. Krishnamoorthy, M. SenthilKumar, N.O. Gopal, S.J. Kim, S.W. Kwon, Azospirillum ramasamyi sp. nov., a novel diazotrophic bacterium isolated from fermented bovine products, Int. J. Syst. Evol. Microbiol. 69 (5) (2019) 1369–1375.

V. Araus, E.A. Vidal, T. Puelma, S. Alamos, D. Mieulet, E. Guiderdoni, R.A. Gutiérrez, Members of BTB gene family of scaffold proteins suppress nitrate uptake and nitrogen use efficiency, Plant Physiol. 171 (2) (2016) 1523–1532.

J.I. Baldani, V.L. Baldani, History on the biological nitrogen fixation research in graminaceous plants: special emphasis on the Brazilian experience, An. Acad. Bras. Cienc. 77 (3) (2005) 549–579.

Y. Bashan, L.E. de-Bashan, Advances in Agronomy, 2010.

Y. Bashan, G. Holguin, Azospirillum–plant relationships: environmental and physiological advances (1990–1996), Can. J. Microbiol. 43 (2) (1997) 103–121.

Y. Bashan, H. Levanony, Current status of Azospirillum inoculation technology: Azospirillum as a challenge for agriculture, Can. J. Microbiol. 36 (9) (1990) 591–608.

M. Benmati, C. Le Roux, N. Belbekri, N. Ykhlef, A. Djekoun, Phenotypic and molecular characterization of plant growth promoting Rhizobacteria isolated from the rhizosphere of wheat (Triticum durum Desf.) in Algeria, Afr. J. Microbiol. Res. 7 (23) (2013) 2893–2904.

I. Berman-Frank, P. Lundgren, P. Falkowski, Nitrogen fixation and photosynthetic oxygen evolution in cyanobacteria, Res. Microbiol. 154 (3) (2003) 157–164.

R.M. Boddey, R. Knowles, Methods for quantification of nitrogen fixation associated with gramineae, Crit. Rev. Plant Sci. 6 (3) (1987) 209–266.

R.M. Boddey, S. Urquiaga, B.J. Alves, V. Reis, Endophytic nitrogen fixation in sugarcane: present knowledge and future applications, Plant Soil 252 (1) (2003) 139–149.

H. Bothe, G. Neuer, I. Kalbe, G. Eisbrenner, Electron donors and hydrogenase in nitrogen-fixing microorganisms, in: Nitrogen Fixation: Proceedings of the Phytochemical Society of Europe Symposium, 1980.

J.S. Boyer, Plant productivity and environment, Science 218 (4571) (1982) 443–448.

K.E. Brigle, W.E. Newton, D.R. Dean, Complete nucleotide sequence of the *Azotobacter vinelandii* nitrogenase structural gene cluster, Gene 37 (1–3) (1985) 37–44.

B.K. Burgess, The iron-molybdenum cofactor of nitrogenase, Chem. Rev. 90 (8) (1990) 1377–1406.

R.H. Burris, G.P. Roberts, Biological nitrogen fixation, Annu. Rev. Nutr. 13 (1) (1993) 317–335.

H. Cai, Y. Lu, W. Xie, T. Zhu, X. Lian, Transcriptome response to nitrogen starvation in rice, J. Biosci. 37 (4) (2012) 731–747.

F. Cassán, M. Diaz-Zorita, Azospirillum sp. in current agriculture: from the laboratory to the field, Soil Biol. Biochem. 103 (2016) 117–130.

R.W. Castenholz, A. Wilmotte, M. Herdman, R. Rippka, J.B. Waterbury, I. Iteman, L. Hoffmann, Phylum BX. cyanobacteria, in: Bergey's Manual® of Systematic Bacteriology, Springer, New York, NY, 2001, pp. 473–599.

R.W. Castenholz, J.B. Waterbury, in: J.T. Staley, M.P. Bryant, N. Pfennig, J.G. Holt (Eds.), Bergey's Manual of Systematic Bacteriology, vol. 3, Williams & Wilkins, Baltimore, MD, 1989, pp. 1710–1727.

I. Ceballos, M. Ruiz, C. Fernández, R. Peña, A. Rodríguez, I.R. Sanders, The in vitro mass-produced model mycorrhizal fungus, Rhizophagus irregularis, significantly increases yields of the globally important food security crop cassava, PLoS ONE 8 (8) (2013), e70633.

H. Cérémonie, F. Debellé, M.P. Fernandez, Structural and functional comparison of Frankia root hair deforming factor and rhizobia Nod factor, Can. J. Bot. 77 (9) (1999) 1293–1301.

A.T.M.A. Choudhury, I.R. Kennedy, Prospects and potentials for systems of biological nitrogen fixation in sustainable rice production, Biol. Fertil. Soils 39 (4) (2004) 219–227.

C. Christiansen-Weniger, Ammonium-excreting *Azospirillum brasilense* C3: gusA inhabiting induced tumors along stem and roots of rice, Soil Biol. Biochem. 29 (5–6) (1997) 943–950.

C. Christiansen-Weniger, J.A. Van Veen, NH4+-excreting *Azospirillum brasilense* mutants enhance the nitrogen supply of a wheat host, Appl. Environ. Microbiol. 57 (10) (1991) 3006–3012.

E. D'Amelio, D.J. Des Marais, J. Cohen, Comparative Functional Ultrastructure of Two Hypersaline Submerged Cyanobacterial Mats-Guerrero Negro, Baja California Sur, Mexico, and Solar Lake, Sinai, Egypt, 1989.

J.M. Day, J. Döbereiner, Physiological aspects of N2-fixation by a Spirillum from Digitaria roots, Soil Biol. Biochem. 8 (1) (1976) 45–50.

L.E. De-Bashan, J.P. Hernandez, K.N. Nelson, Y. Bashan, R.M. Maier, Growth of quailbush in acidic, metalliferous desert mine tailings: effect of *Azospirillum brasilense* Sp6 on biomass production and rhizosphere community structure, Microb. Ecol. 60 (4) (2010) 915–927.

J. Dobereiner, History and new perspectives of diazotrophs in association with non-leguminous plants, Symbiosis 13 (1992) 1–13.

J. Dobereiner, I.E. Marriel, M. Nery, Ecological distribution of Spirillum lipoferum Beijerinck, Can. J. Microbiol. 22 (10) (1976) 1464–1473.

W.K. Dodds, D.A. Gudder, D. Mollenhauer, The ecology of Nostoc, J. Phycol. 31 (1) (1995) 2–18.

R.R. Eady, R. Issack, C. Kennedy, J.R. Postgate, H.D. Ratcliffe, Nitrogenase synthesis in *Klebsiella pneumoniae*: comparison of ammonium and oxygen regulation, Microbiology 104 (2) (1978) 277–285.

A.E. El-Enany, A.A. Issa, Cyanobacteria as a biosorbent of heavy metals in sewage water, Environ. Toxicol. Pharmacol. 8 (2) (2000) 95–101.

C. Elmerich, W.E. Newton, Associative and Endophytic Nitrogen-Fixing Bacteria and Cyanobacterial Associations, Springer, The Netherlands, 2007.

A. Ernst, H. Böhme, Control of hydrogen-dependent nitrogenase activity by adenylates and electron flow in heterocysts of Anabaena variabilis (ATCC 29413), Biochim. Biophys. Acta Bioenerg. 767 (2) (1984) 362–368.

R.D. Evans, R. Rimer, L. Sperry, J. Belnap, Exotic plant invasion alters nitrogen dynamics in an arid grassland, Ecol. Appl. 11 (5) (2001) 1301–1310.

P.G. Falkowski, Evolution of the nitrogen cycle and its influence on the biological sequestration of CO2 in the ocean, Nature 387 (6630) (1997) 272–275.

P. Fay, Heterotrophy and nitrogen fixation in Chlorogloea fritschii, Microbiology 39 (1) (1965) 11–20.

R.F. Fisher, S.R. Long, Rhizobium–plant signal exchange, Nature 357 (6380) (1992) 655–660.

C. Franche, K. Lindström, C. Elmerich, Nitrogen-fixing bacteria associated with leguminous and non-leguminous plants, Plant Soil 321 (1) (2009) 35–59.

H. Fukaki, M. Tasaka, Hormone interactions during lateral root formation, Plant Mol. Biol. 69 (4) (2009) 437.

M.M. Georgiadis, H. Komiya, P. Chakrabarti, D. Woo, J.J. Kornuc, D.C. Rees, Crystallographic structure of the nitrogenase iron protein from Azotobacter vinelandii, Science 257 (5077) (1992) 1653–1659.

D. Giani, W.E. Krumbein, Growth characteristics of non-heterocystous cyanobacterium Plectonema boryanum with N2 as nitrogen source, Arch. Microbiol. 145 (3) (1986) 259–265.

M.T. Giardi, J. Masojídek, D. Godde, Effects of abiotic stresses on the turnover of the D1 reaction centre II protein, Physiol. Plant. 101 (3) (1997) 635–642.

B.R. Glick, G. Holguin, C.L. Patten, D.M. Penrose, Biochemical and Genetic Mechanisms Used by Plant Growth Promoting bacteria, World Scientific, 1999.

K. Goethals, M. Van Montagu, M. Holsters, Conserved motifs in a divergent nod box of Azorhizobium caulinodans ORS571 reveal a common structure in promoters regulated by LysR-type proteins, Proc. Natl. Acad. Sci. 89 (5) (1992) 1646–1650.

J.W. Gotto, D.C. Yoch, Regulation of Rhodospirillum rubrum nitrogenase activity. Properties and interconversion of active and inactive Fe protein, J. Biol. Chem. 257 (6) (1982) 2868–2873.

M. Govindarajulu, P.E. Pfeffer, H. Jin, J. Abubaker, D.D. Douds, J.W. Allen, Y. Shachar-Hill, Nitrogen transfer in the arbuscular mycorrhizal symbiosis, Nature 435 (7043) (2005) 819–823.

L. Gruffman, S. Jämtgård, T. Näsholm, Plant nitrogen status and co-occurrence of organic and inorganic nitrogen sources influence root uptake by Scots pine seedlings, Tree Physiol. 34 (2) (2014) 205–213.

J. Hallmann, A. Quadt-Hallmann, W.F. Mahaffee, J.W. Kloepper, Bacterial endophytes in agricultural crops, Can. J. Microbiol. 43 (10) (1997) 895–914.

R. Haselkorn, Organization of the genes for nitrogen fixation in photosynthetic bacteria and cyanobacteria, Annu. Rev. Microbiol. 40 (1) (1986) 525–547.

R.P. Hausinger, J.B. Howard, Thiol reactivity of the nitrogenase Fe-protein from Azotobacter vinelandii, J. Biol. Chem. 258 (22) (1983) 13486–13492.

T.R. Hawkes, P.A. McLEAN, B.E. Smith, Nitrogenase from nifV mutants of Klebsiella pneumoniae contains an altered form of the iron-molybdenum cofactor, Biochem. J. 217 (1) (1984) 317–321.

A. Herrero, A.M. Muro-Pastor, E. Flores, Nitrogen control in cyanobacteria, J. Bacteriol. 183 (2) (2001) 411–425.

S. Hobara, C. McCalley, K. Koba, A.E. Giblin, M.S. Weiss, G.M. Gettel, G.R. Shaver, Nitrogen fixation in surface soils and vegetation in an Arctic tundra watershed: a key source of atmospheric nitrogen, Arct. Antarct. Alp. Res. 38 (3) (2006) 363–372.

V. Hocher, N. Alloisio, F. Auguy, P. Fournier, P. Doumas, P. Pujic, D. Bogusz, Transcriptomics of actinorhizal symbioses reveals homologs of the whole common symbiotic signaling cascade, Plant Physiol. 156 (2) (2011) 700–711.

J.B. Howard, R. Davis, B. Moldenhauer, V.L. Cash, D. Dean, Fe:S cluster ligands are the only cysteines required for nitrogenase Fe-protein activities, J. Biol. Chem. 264 (19) (1989) 11270–11274.

T. Hurek, B. Reinhold-Hurek, Azoarcus sp. strain BH72 as a model for nitrogen-fixing grass endophytes, J. Biotechnol. 106 (2–3) (2003) 169–178.

K. Huss-Danell, Tansley review no. 93. Actinorhizal symbioses and their N₂ fixation, New Phytol. (1997) 375–405.

J. Imperial, T.R. Hoover, M.S. Madden, P.W. Ludden, V.K. Shah, Substrate reduction properties of dinitrogenase activated in vitro are dependent upon the presence of homocitrate or its analogs during iron-molybdenum cofactor synthesis, Biochemistry 28 (19) (1989) 7796–7799.

A.A. Issa, M.H. Abd-Alla, T. Ohyama, Nitrogen fixing cyanobacteria: future prospect, in: Advances in Biology and Ecology of Nitrogen Fixation, vol. 2, 2014, pp. 24–48.

A.A. Issa, A.E. El-Enany, R. Abdel-Basset, Modulation of the photosynthetic source: sink relationship in cultures of the cyanobacterium *Nostoc rivulare*, Biol. Plant. 45 (2) (2002) 221–225.

A. Jacquot, Z. Li, A. Gojon, W. Schulze, L. Lejay, Post-translational regulation of nitrogen transporters in plants and microorganisms, J. Exp. Bot. 68 (10) (2017) 2567–2580.

A.P. Jangam, N. Raghuram, Nitrogen and stress, in: Elucidation of Abiotic Stress Signaling in Plants, 2015, pp. 323–339.

A.P. Jangam, N. Raghuram, Nitrogen and stress, in: Elucidation of Abiotic Stress Signaling in Plants, 2015, pp. 323–339.

S. Jansson, C.J. Douglas, Populus: a model system for plant biology, Annu. Rev. Plant Biol. 58 (2007) 435–458.

R.H. Kanemoto, P.W. Ludden, Effect of ammonia, darkness, and phenazine methosulfate on whole-cell nitrogenase activity and Fe protein modification in *Rhodospirillum rubrum*, J. Bacteriol. 158 (2) (1984) 713–720.

Y. Kapulnik, J. Kigel, Y. Okon, I. Nur, Y. Henis, Effect of Azospirillum inoculation on some growth parameters and N-content of wheat, sorghum and panicum, Plant Soil 61 (1) (1981) 65–70.

K.S. Karthika, I. Rashmi, M.S. Parvathi, Biological functions, uptake and transport of essential nutrients in relation to plant growth, in: Plant Nutrients and Abiotic Stress Tolerance, Springer, Singapore, 2018, pp. 1–49.

S. Katupitiya, P.B. New, C. Elmerich, I.R. Kennedy, Improved N2 fixation in 2, 4-D treated wheat roots associated with *Azospirillum lipoferum*: studies of colonization using reporter genes, Soil Biol. Biochem. 27 (4–5) (1995) 447–452.

I.R. Kennedy, A.T.M.A. Choudhury, M.L. Kecskés, Non-symbiotic bacterial diazotrophs in crop-farming systems: can their potential for plant growth promotion be better exploited? Soil Biol. Biochem. 36 (8) (2004) 1229–1244.

N.W. Kerby, S.C. Musgrave, P. Rowell, S.V. Shestakov, W.D. Stewart, Photoproduction of ammonium by immobilized mutant strains of Anabaena variabilis, Appl. Microbiol. Biotechnol. 24 (1) (1986) 42–46.

J. Kim, D.C. Rees, Nitrogenase and biological nitrogen fixation, Biochemistry 33 (2) (1994) 389–397.

J. Kim, D.C. Rees, Crystallographic structure and functional implications of the nitrogenase molybdenum–iron protein from *Azotobacter vinelandii*, Nature 360 (6404) (1992) 553–560.

V. Kumar, A.N. Yadav, A. Saxena, P. Sangwan, H.S. Dhaliwal, Unravelling rhizospheric diversity and potential of phytase producing microbes, SM J. Biol. 2 (1) (2016) 1009.

S.J. Leghari, N.A. Wahocho, G.M. Laghari, A. HafeezLaghari, G. MustafaBhabhan, K. HussainTalpur, T.A. Bhutto, S.A. Wahocho, A.A. Lashari, Role of nitrogen for plant growth and development: a review, Adv. Environ. Biol. 10 (9) (2016) 209–219. Available at: https://link.gale.com/apps/doc/A472372583/AONE?u=anonä75a3820&sid=bookmark-AONE&xid=21dbd6d6.

H. Liu, A.J. Able, J.A. Able, Nitrogen starvation-responsive microRNAs are affected by transgenerational stress in durum wheat seedlings, Plan. Theory 10 (5) (2021) 826.

B. Magasanik, Regulation of bacterial nitrogen assimilation by glutamine synthetase, Trends Biochem. Sci. 2 (1) (1977) 9–12.

S. Mahajan, N. Tuteja, Cold, salinity and drought stresses: an overview, Arch. Biochem. Biophys. 444 (2) (2005) 139–158.

T. Mamashita, G.R. Larocque, A. DesRochers, J. Beaulieu, B.R. Thomas, A. Mosseler, D. Sidders, Short-term growth and morphological responses to nitrogen availability and plant density in hybrid poplars and willows, Biomass Bioenergy 81 (2015) 88–97.

P.S. Maryan, R.R. Eady, A.E. Chaplin, J.R. Gallon, Nitrogen fixation by the unicellular cyanobacterium Gloeothece. Nitrogenase synthesis is only transiently repressed by oxygen, FEMS Microbiol. Lett. 34 (3) (1986) 251–255.

J.C. Meeks, J. Elhai, Regulation of cellular differentiation in filamentous cyanobacteria in free-living and plant-associated symbiotic growth states, Microbiol. Mol. Biol. Rev. 66 (1) (2002) 94–121.

B.J. Miflin, P.J. Lea, The path of ammonia assimilation in the plant kingdom, Trends Biochem. Sci. 1 (2) (1976) 103–106.

Z.D. Miller, P.N. Peralta, P. Mitchell, V.L. Chiang, S.S. Kelley, C.W. Edmunds, I.M. Peszlen, Anatomy and chemistry of *Populus trichocarpa* with genetically modified lignin content, Bioresources 14 (3) (2019) 5729–5746.

R. Mittler, Abiotic stress, the field environment and stress combination, Trends Plant Sci. 11 (1) (2006) 15–19.

T.V. Morgan, J. McCracken, W.H. Orme-Johnson, W.B. Mims, L.E. Mortenson, J. Peisach, Pulsed electron paramagnetic resonance studies of the interaction of magnesium-ATP and deuterium oxide with the iron protein of nitrogenase, Biochemistry 29 (12) (1990) 3077–3082.

M.G. Murty, J.K. Ladha, Influence of Azospirillum inoculation on the mineral uptake and growth of rice under hydroponic conditions, Plant Soil 108 (2) (1988) 281–285.

P. Mylona, K. Pawlowski, T. Bisseling, Symbiotic nitrogen fixation, Plant Cell 7 (7) (1995) 869.

F. Nadeem, Z. Ahmad, M. Ul Hassan, W. Ruifeng, X. Diao, X. Li, Adaptation of foxtail millet (*Setaria italica* L.) to abiotic stresses: a special perspective of responses to nitrogen and phosphate limitations, Front. Plant Sci. 11 (2020) 187.

W.E. Newton, Biological Nitrogen Fixation, 1992.

M. Nilsson, J. Bhattacharya, A.N. Rai, B. Bergman, Colonization of roots of rice (*Oryza sativa*) by symbiotic Nostoc strains, New Phytol. 156 (3) (2002) 517–525.

Y. Okon, C.A. Labandera-Gonzalez, Agronomic applications of Azospirillum: an evaluation of 20 years worldwide field inoculation, Soil Biol. Biochem. 26 (12) (1994) 1591–1601.

Y. Okon, P.G. Heytler, R.W.F. Hardy, N2 fixation by *Azospirillum brasilense* and its incorporation into host *Setaria italica*, Appl. Environ. Microbiol. 46 (3) (1983) 694–697.

J. Olivares, E.J. Bedmar, J. Sanjuán, Biological nitrogen fixation in the context of global change, Mol. Plant-Microbe Interact. 26 (5) (2013) 486–494.

F.O. Pedrosa, A.L.M. Oliveira, V.F. Guimarães, R.M. Etto, E.M. Souza, F.G. Furmam, C.W. Galvão, The ammonium excreting *Azospirillum brasilense* strain HM053: a new alternative inoculant for maize, Plant Soil 451 (1) (2020) 45–56.

V.T. Pham, H. Rediers, M.G. Ghequire, H.H. Nguyen, R. De Mot, J. Vanderleyden, S. Spaepen, The plant growth-promoting effect of the nitrogen-fixing endophyte *Pseudomonas stutzeri* A15, Arch. Microbiol. 199 (3) (2017) 513–517.

J. Postgate, Microbiology of the free-living nitrogen-fixing bacteria, excluding cyanobacteria, in: 4. International Symposium on Nitrogen Fixation, Canberra, ACT (Australia), 1 Dec 1980, Australian Academy of Science, 1981.

N.P.J. Price, B. RelicA, F. Talmont, A. Lewin, D. Promé, S.G. Pueppke, W.J. Broughton, Broad-host-range Rhizobium species strain NGR234 secretes a family of carbamoylated, and fucosylated, nodulation signals that are O-acetylated or sulphated, Mol. Microbiol. 6 (23) (1992) 3575–3584.

A.N.D.R.E.A. Quadt-Hallmann, J.W. Kloepper, N. Benhamou, Bacterial endophytes in cotton: mechanisms of entering the plant, Can. J. Microbiol. 43 (6) (1997) 577–582.

M. Rahmani, A.W. Hodges, C.F. Kiker, Compost users' attitudes toward compost application in Florida, Compost. Sci. Util. 12 (1) (2004) 55–60.

S. Reich, P. Böger, Regulation of nitrogenase activity in Anabaena variabilis by modification of the Fe protein, FEMS Microbiol. Lett. 58 (1) (1989) 81–86.

V.M. Reis, V.L.D. Baldani, J.I. Baldani, Isolation, identification and biochemical characterization of *Azospirillum* spp. and other nitrogen-fixing bacteria, in: Handbook for Azospirillum, Springer, Cham, 2015, pp. 3–26.

L. Reynders, K. Vlassak, Conversion of tryptophan acid by *Azospirillum brasilense*, Soil Biol. Biochem. 11 (1979) 547–548.

P.A. Roger, P.A. Reynaud, Ecology of Blue-Green Algae in Paddy Field, 1979.

A.D. Rovira, Rhizosphere research-85 years of progress and frustration, in: The Rhizosphere and Plant Growth, 1991, pp. 3–13.

S. Ruffel, G. Krouk, D. Ristova, D. Shasha, K.D. Birnbaum, G.M. Coruzzi, Nitrogen economics of root foraging: transitive closure of the nitrate–cytokinin relay and distinct systemic signaling for N supply vs. demand, Proc. Natl. Acad. Sci. 108 (45) (2011) 18524–18529.

S. Ruffel, G. Krouk, D. Ristova, D. Shasha, K.D. Birnbaum, G.M. Coruzzi, Nitrogen economics of root foraging: transitive closure of the nitrate–cytokinin relay and distinct systemic signaling for N supply vs. demand, Proc. Natl. Acad. Sci. 108 (45) (2011) 18524–18529.

R.P. Ryan, K. Germaine, A. Franks, D.J. Ryan, D.N. Dowling, Bacterial endophytes: recent developments and applications, FEMS Microbiol. Lett. 278 (1) (2008) 1–9.

S.P. Saikia, V. Jain, S. Khetarpal, S. Aravind, Dinitrogen fixation activity of *Azospirillum brasilense* in maize (*Zea mays*), Curr. Sci. (2007) 1296–1300.

S.P. Saikia, G.C. Srivastava, V. Jain, Nodule-like structures induced on the roots of maize seedlings by the addition of synthetic auxin 2, 4-D and its effects on growth and yield, Cereal Res. Commun. 32 (1) (2004) 83–89.

T. Saino, A. Hattori, Diel variation in nitrogen fixation by a marine blue-green alga, *Trichodesmium thiebautii*, Deep-Sea Res. 25 (12) (1978) 1259–1263.

K.F.D.N. Santos, V.R. Moure, V. Hauer, A.S. Santos, L. Donatti, C.W. Galvão, M.B.R. Steffens, Wheat colonization by an *Azospirillum brasilense* ammonium-excreting strain reveals upregulation of nitrogenase and superior plant growth promotion, Plant Soil 415 (1) (2017) 245–255.

S. Scherer, H. Almon, P. Böger, Interaction of photosynthesis, respiration and nitrogen fixation in cyanobacteria, Photosynth. Res. 15 (2) (1988) 95–114.

K.T. Shanmugam, F. O'gara, K. Andersen, R.C. Valentine, Biological nitrogen fixation, Annu. Rev. Plant Physiol. 29 (1) (1978) 263–276.

F.B. Simpson, R.H. Burris, A nitrogen pressure of 50 atmospheres does not prevent evolution of hydrogen by nitrogenase, Science 224 (4653) (1984) 1095–1097.

B.E. Smith, R.R. Eady, Metalloclusters of the nitrogenases, in: EJB Reviews, 1993, pp. 79–93.

S.E. Smith, I. Jakobsen, M. Grønlund, F.A. Smith, Roles of arbuscular mycorrhizas in plant phosphorus nutrition: interactions between pathways of phosphorus uptake in arbuscular mycorrhizal roots have important implications for understanding and manipulating plant phosphorus acquisition, Plant Physiol. 156 (3) (2011) 1050–1057.

B. Solheim, M. Zielke, J.W. Bjerke, J. Rozema, Effects of enhanced UV-B radiation on nitrogen fixation in arctic ecosystems, in: Plants and Climate Change, Springer, Dordrecht, 2006, pp. 109–120.

J.I. Sprent, E.K. James, Legume evolution: where do nodules and mycorrhizas fit in? Plant Physiol. 144 (2) (2007) 575–581.

R.T. St. John, V.K. Shah, W.J. Brill, Regulation of nitrogenase synthesis by oxygen in *Klebsiella pneumoniae*, J. Bacteriol. 119 (1) (1974) 266–269.

M. Staal, S. te Lintel Hekkert, G. Jan Brummer, M. Veldhuis, C. Sikkens, S. Persijn, L.J. Stal, Nitrogen fixation along a north-south transect in the eastern Atlantic Ocean, Limnol. Oceanogr. 52 (4) (2007) 1305–1316.

L.J. Stal, W.E. Krumbein, Oxygen protection of nitrogenase in the aerobically nitrogen fixing, non-heterocystous cyanobacterium *Oscillatoria* sp, Arch. Microbiol. 143 (1) (1985) 72–76.

A. Suman, P. Verma, A.N. Yadav, R. Srinivasamurthy, A. Singh, R. Prasanna, Development of hydrogel based bio-inoculant formulations and their impact on plant biometric parameters of wheat (*Triticum aestivum* L.), Int. J. Curr. Microbiol. App. Sci. 5 (3) (2016) 890–901.

Y.T. Tchan, A.M.M. Zeman, I.R. Kennedy, Nitrogen fixation in para-nodules of wheat roots by introduced free-living diazotrophs, Plant Soil 137 (1) (1991) 43–47.

D.W. Tempest, J.L. Meers, C.M. Brown, Synthesis of glutamate in *Aerobacter aerogenes* by a hitherto unknown route, Biochem. J. 117 (2) (1970) 405–407.

T.M. Tien, M.H. Gaskins, D. Hubbell, Plant growth substances produced by *Azospirillum brasilense* and their effect on the growth of pearl millet (*Pennisetum americanum* L.), Appl. Environ. Microbiol. 37 (5) (1979) 1016–1024.

E.N. Tikhonova, D.S. Grouzdev, I.K. Kravchenko, Azospirillum palustre sp. nov., a methylotrophic nitrogen-fixing species isolated from raised bog, Int. J. Syst. Evol. Microbiol. 69 (9) (2019) 2787–2793.

K.V.B.R. Tilak, N. Ranganayaki, K.K. Pal, R. De, A.K. Saxena, C.S. Nautiyal, B.N. Johri, Diversity of plant growth and soil health supporting bacteria, Curr. Sci. (2005) 136–150.

T. Tscharntke, Y. Clough, T.C. Wanger, L. Jackson, I. Motzke, I. Perfecto, A. Whitbread, Global food security, biodiversity conservation and the future of agricultural intensification, Biol. Conserv. 151 (1) (2012) 53–59.

M.R. Turetsky, The role of bryophytes in carbon and nitrogen cycling, Bryologist 106 (3) (2003) 395–409.

B. Tyler, Regulation of the assimilation of nitrogen compounds, Annu. Rev. Biochem. 47 (1) (1978) 1127–1162.

R. Uchida, Essential nutrients for plant growth: nutrient functions and deficiency symptoms, in: Plant Nutrient Management in Hawaii's Soils, 2000, pp. 31–55.

E. Uleberg, I. Hanssen-Bauer, B. van Oort, S. Dalmannsdottir, Impact of climate change on agriculture in Northern Norway and potential strategies for adaptation, Clim. Chang. 122 (1) (2014) 27–39.

K.M. Usher, B. Bergman, J.A. Raven, Exploring cyanobacterial mutualisms, Annu. Rev. Ecol. Evol. Syst. 38 (2007) 255–273.

A. Van Dommelen, A. Croonenborghs, S. Spaepen, J. Vanderleyden, Wheat growth promotion through inoculation with an ammonium-excreting mutant of *Azospirillum brasilense*, Biol. Fertil. Soils 45 (5) (2009) 549–553.

P. Verma, A.N. Yadav, S.K. Kazy, A.K. Saxena, A. Suman, Evaluating the diversity and phylogeny of plant growth promoting bacteria associated with wheat (*Triticum aestivum*) growing in central zone of India, Int. J. Curr. Microbiol. App. Sci. 3 (5) (2014) 432–447.

P. Verma, A.N. Yadav, V. Kumar, D.P. Singh, A.K. Saxena, Beneficial plant-microbes interactions: biodiversity of microbes from diverse extreme environments and its impact for crop improvement, in: Plant-Microbe Interactions in Agro-Ecological Perspectives, Springer, Singapore, 2017, pp. 543–580.

P.L. Vlek, M.Y. Diakite, H. Mueller, The role of Azolla in curbing ammonia volatilization from flooded rice systems, in: Nitrogen Economy in Tropical Soils, Springer, Dordrecht, 1995, pp. 165–174.

F. Walder, H. Niemann, M. Natarajan, M.F. Lehmann, T. Boller, A. Wiemken, Mycorrhizal networks: common goods of plants shared under unequal terms of trade, Plant Physiol. 159 (2) (2012) 789–797.

X. Wang, X. Li, S. Zhang, H. Korpelainen, C. Li, Physiological and transcriptional responses of two contrasting Populus clones to nitrogen stress, Tree Physiol. 36 (5) (2016) 628–642.

N.J. West, D.G. Adams, Phenotypic and genotypic comparison of symbiotic and free-living cyanobacteria from a single field site, Appl. Environ. Microbiol. 63 (11) (1997) 4479–4484.

B.A. Whitton, Diversity, ecology, and taxonomy of the cyanobacteria, in: Photosynthetic Prokaryotes, Springer, Boston, MA, 1992, pp. 1–51.

S.R. Wicks, Introduction and Guide to the Marine Bluegreen Algae, John Wiley and Sons, New York, NY, 1980.

J.T. Wyatt, J.K.G. Silvey, Nitrogen fixation by Gloeocapsa, Science 165 (3896) (1969) 908–909.

Y.G. Yanni, F.B. Dazzo, Enhancement of rice production using endophytic strains of *Rhizobium leguminosarum* bv. *trifolii* in extensive field inoculation trials within the Egypt Nile delta, Plant Soil 336 (2010) 129–142, https://doi.org/10.1007/s11104-010-0454-7.

C.R. Yendrek, Y.C. Lee, V. Morris, Y. Liang, C.I. Pislariu, G. Burkart, R. Dickstein, A putative transporter is essential for integrating nutrient and hormone signaling with lateral root growth and nodule development in *Medicago truncatula*, Plant J. 62 (1) (2010) 100–112.

L.E. De-Bashan, J.P. Hernandez, K.N. Nelson, Y. Bashan, R.M. Maier, Growth of quailbush in acidic, metalliferous desert mine tailings: effect of *Azospirillum brasilense* Sp6 on biomass production and rhizosphere community structure, Microb. Ecol. 60 (4) (2010) 915–927, https://doi.org/10.1007/s00248-010-9713-7.

Chapter 4

Destructive role of chemicals secreted by plants to diminish harmful microbes

Subhesh Saurabh Jha and L.S. Songachan

Department of Botany, Institute of Sciences, Banaras Hindu University, Varanasi, Uttar Pradesh, India

1 Introduction

Plant roots extend into the rhizosphere a wide spectrum of potentially important tiny molecular substances. The roots and their surroundings (i.e., the rhizosphere) are witness to some of the most intricate chemical, physical, and biological interactions of terrestrial plants with fellow plants or microbes. Interactions with plant roots in the rhizosphere include root-root, and root-microbe interactions [1]. Over the last decade, great efforts have been made to comprehend these varied connections, and in the study of plant biology the role of root exudates has lately been identified to mediate these biological interactions. For the interaction of roots with pathogenic soil bacteria, invertebrates, and rivals' root system, the rhizosphere is a quite a dynamic front [2]. However, while plant roots are concealed underneath the earth, many of the intriguing phenomena involved remain completely overlooked. The importance of chemical signals is only just beginning to be appreciated in mediating underground connections. Signals between plant roots and other soil organisms, including nearby plant roots, are mostly based on root-based compounds [3]. Chemical signals can elicit a wide range of responses in different receivers. "Chemical components may repel one creature while attracting another, or they may attract two different creatures with distinct plant effects." The mechanisms behind roots' way for deciphering the rhizosphere's many communications with other roots, soil microbes, and invertebrates are mostly unknown but in broad terms they can be defined as either favorable, detrimental or, in rare cases, neutral (Table 1). Symbiotic combinations of microbes with roots and the colonization of roots by biocontrol agents and plant growth promoting rhizobacteria (PGPR) are a few examples of positive interactions [10,11,22,23] whereas

Plant-Microbe Interaction—Recent Advances in Molecular and Biochemical Approaches
https://doi.org/10.1016/B978-0-323-91876-3.00018-X
65

negative interactions include plant competition and parasitism, bacterial or fungal pathogens, and herbivory invertebrates [15]. The negative interactions of plant roots are discussed in depth in this chapter. The factors that influence a plant's chemical signature to be identified as harmful or beneficial requires further clarification. However, many studies have shown that rhizoexudates play a significant role in determining the nature of plant and soil dynamics [10,11]. Fig. 1 and Table 1 try to explain the complex relationship between plant roots and surrounding rhizospheric microorganisms.

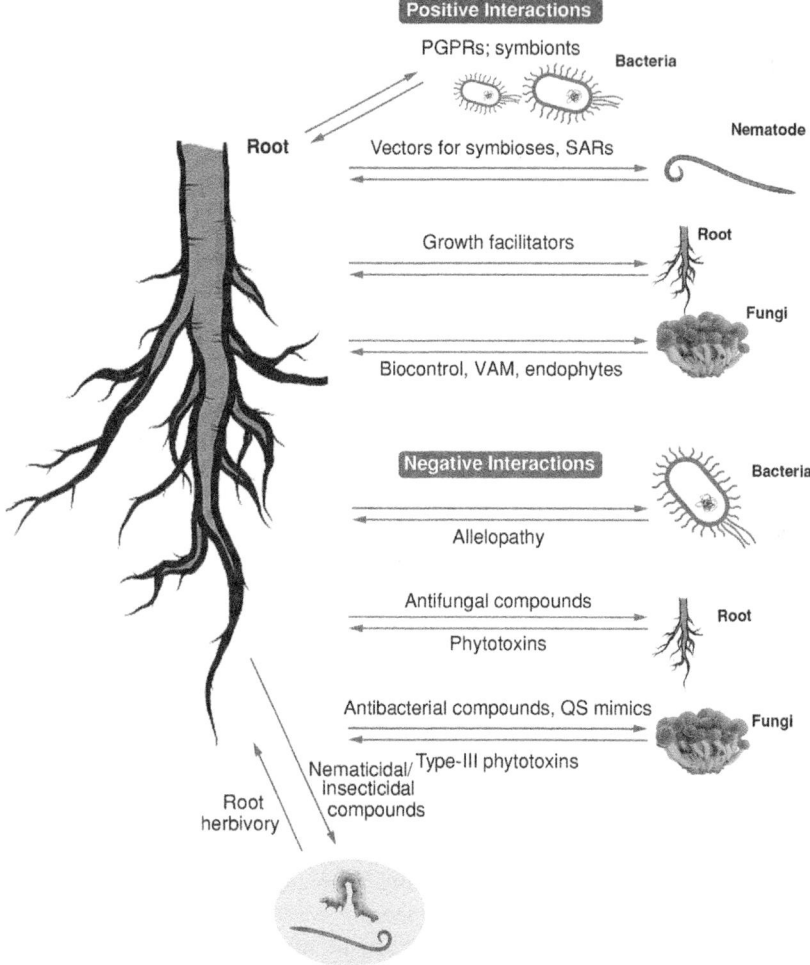

FIG. 1   Chemical rhizospheric warfare: a graphic illustration of rhizosphere interactions mediated by root exudates. *Arrow panels* represent the chemical exchange between root exudates and various microorganisms. VAM stands for vesicular arbuscular mycorrhizas, and SARs stands for systemic acquired resistance.

TABLE 1 Different types of rhizospheric interactions and their impact on plants [1,4–21].

Interaction type	Effect	Example	References
Root-nematode	Positive	Vectors for symbiosis, SARs	[12,14]
Root-root	Positive	Growth facilitators	[5,15]
Root-fungi	Positive	Biocontrol, vesicular arbuscular mycorrhizae, endophytes	[1,10,11,20]
Root-bacteria	Negative	Allelopathy	[8,9]
Root-root	Negative	Antifungal compounds, phytotoxins	[17,19]
Root-fungi	Negative	Antibacterial compounds, QS mimics, Type-III phytotoxins	[6,7]
Root-herbivory	Negative	Nematicidal/insecticidal compounds	[4,18,21]

2   Detrimental interactions between plants

2.1   Allelopathy

"Chemical interference, also known as allelopathy, is a technique employed by plants to gain an edge over their competitors. Plants that generate and release potent phytotoxins can prevent or decrease sensitive plant neighbors' establishment, growth, and survival, thus reducing the competition and increasing resources availability. Plant can release these toxins in a variety of different ways including from decomposing root and leaf tissues, live tissue leachates, green foliar volatiles, and root exudates" [8]. Plant-produced phytotoxins have a wide range of chemical structures, mode of action, and can affect various plants in different ways [24,25]. These influences on plant physiology, development, and continuance can, in turn, alter the composition and dynamics of plant and soil communities. A variety of phytotoxic chemicals have been found in plant root exudates including but not limited to (±)-catechin (*Centaurea maculosa*, spotted knapweed) [26], juglone (*Juglans nigra*, black walnut) [27], and sorgoleone (*Sorghum* spp.) [28]; only when phytotoxic root exudates are present in sufficient quantities to influence plant survival and growth can they mediate adverse plant-plant interactions. In its root exudates, *Sorghum* spp. rhizosecrete more sorgoleone than any other chemicals [29]. However, phytotoxin concentrations in the rhizosphere are determined by both chemical stability and production rates. Juglone has a minimal seasonal concentration fluctuation and is largely soil stable [27]. Sorgoleone, on the other hand, degrades rapidly in soil [30], implying that high production rates may be required to maintain phytotoxic sorgoleone concentration in the soil. The ecological significance of phytotoxic roots is also determined by the sensitivity of the plants that cohabit with allelopathic plants. In communities invaded by *Centaurea maculosa* (±)-catechin and 8-hydroxyquinoline restrict the establishment of the native North American plant population [31]. More than 20 grassland species in North America are hindered in root formation by (±)-catechin [19]. Sorgoleone, DIBOA, and 5,7,4-trihydoxy-3,5-dimethoxyflavone also inhibit the growth of weeds coexisting with *Sorghum bicolor* [16] and *Triticum aestivum* [28] in agricultural systems. However, the majority of these studies were conducted in laboratories rather than in natural settings. To quantify the function of phytotoxin formation for plant-plant interference with greater accuracy, tests with typical soil phytotoxin concentrations in realistic situations are required [32]. Nonetheless, many plants' sensitivity to a variety of phytotoxins produced by plants shows that resistance may be energy intensive and confined to a small number of species.

2.2 Biological invasions at the community level and the novel weapons' theory

Plants that are frequently exposed to allelopathic species are more prone to develop resistance to root-secreted phytotoxins over time. However, because phytotoxin resistance is likely to come at a cost in terms of energy, plants that are not exposed to the toxin on a regular basis may be less likely to develop resistance [17]. As a result, phytotoxins produced by other plants may be more susceptible to transient plant species. Similarly, phytotoxins produced by transient plants are predicted to damage a broader range of plant species than phytotoxins produced by plants that stay for lengthy periods in specific plant communities. Coevolution may lead to an increasingly complex arms race of allelochemicals with costlier resistance demands for species that are often linked, thereby, diminishing the ecological influence of direct chemical action. Biological invasions present a rare situation in which the invaders in the range have never come in contact with phytotoxins of the intruder species resulting in a significantly larger impact of these "new weapons" [33] on "naïve" native plant species in the invaded range compared to the original range of the invaders. The susceptibility of indigenous species to phytotoxins in invaders may explain some of the superior performance of exotic plants in the invaded regions [33]. Only a few species have been subjected to the new weapons' invasion hypothesis so far, but data suggests that many more exotic invaders may also be allelopathic [29]. The trials and tribulations of two invader species, *Centaurea diffusa* (diffuse knapweed) and *Centaurea maculosa* (spotted knapweed) of North Africa, are exemplary evidences of the novel weapon theory [33].

According to their results, *Centaurea diffusa* root exudates significantly hampered the growth of North American grassland species compared to the European grassland species with which *Centaurea diffusa* coexists. "Prati and Bossdorf [34] found another good example in the case of *Alliaria petiolate*, another invasive plant in North America, with considerable negative impact on North American species *Geum laciniatum* compared to its European congener *Geum urbanum*, giving strength to the new weapon theory." European *Alliaria petiolate* exudates, however, showed similar adverse effects on both congeners, showing that *Alliaria petiolate* phytotoxins may have substantial environmental impacts both in native and in invaded regions. Secondary metabolites of an invasive plant, *Carduus nutans* (musk thistle), seem to impede nodulation and nitrogen fixation in leguminous plant *Trifolium repens* (white clover). *Trifolium repens*' development and survival in *Carduus nutans*-infested agricultural regions may be drastically decreased [34]. In another example, *Empetrum hermaphroditum* (crowberry)s secondary metabolites restrict symbiotic relationships between the *Pinus sylvestris* (Scots pine) tree and mycorrhizal fungi, therefore limiting the absorption of nitrogen by *Pinus sylvestris* [35].

2.3   Associations of parasitic plants with their hosts

The establishment of connections between parasite plants and their plant host, which is detrimental for the plant and favorable for the parasite, requires the production of root exudates. Almost 4000 obligatory and nonmandatory parasitic plants have been reported by Nilsson et al. [36]. Studies on *Striga* spp. (*Striga asiatica* and *Striga hermonthica*) have helped in unraveling the applicability of root exudates in parasitic plants. *Striga* have miniscule number of seeds which survives for negligible amount of time after germination if a host connection is not established [37]. The low stocks of carbohydrates in *Striga* seeds limit the elongation of seedling roots before attachment. The closeness of an adequate host root is therefore essential to the survival of *Striga* seedlings. For germination near host roots, *Striga* seeds only germinate in the presence of the sustained (10–12h) high levels of germination inducers exuded by host roots [38]. The inductors of germination vary across various *Striga* hosts. So far, sorghum xenognosin (SXSg) has been the only *Striga* germination inducer discovered and characterized in plants. SXSg is notoriously unstable in aqueous solution [39], which is a good thing for a *Striga* germination inducer because it would not falsely implicate the presence of a host by lasting for long durations in the soil. However, because SXSg is so unstable, it was first difficult to explain how enough SXSg stayed in the soil and traveled through it to harm nearby *Striga* plants [39]. SXSg soil action was described by Fate and Lynn [40] who showed that recorcinol, a structurally identical chemical to SXSg, is generated in minute amounts alongside SXSg in *Sorghum* root exudates and stabilizes SXSg in sufficient amounts to trigger *Striga* germination. Root exudates are also important in the growth of *Striga*'s haustoria, a structure very specific in nature and are generally used by fungus to infect host roots in order to develop links with their vascular systems. The *Striga* seedling root detects the host benzoquinones, possibly through the redox activation of a receptor, and initiates haustoria formation [41]. The methods by which host benzoquinones cause haustorial development are still somewhat unknown.

3   Detrimental interactions between plants and microbes

3.1   Antimicrobial effects

Exudates from the plant roots significantly boost microbial activity in the rhizosphere [42]. To survive the constant attack from pathogenic microorganisms, plants secrete phytoalexins, defense proteins, and other unknown chemicals. There secretion is just as important as other secondary metabolites, which the plants secrete to maintain symbiotic root-microbe association result in beneficial associations for both the parties involved [1]. Flores et al. [6] discovered that soilborne bacteria and fungi were inhibited by induced, cell-specific production of pigmented naphthoquinones in *Lithospermum erythrorhizon* hairy roots. This result strongly suggests that root exudates are important in protecting the rhizosphere from dangerous microorganisms. It might be difficult to differentiate between phytoalexins, which are generated in response to pathogen infection, and phytoanticipins, which are generated constitutively and before infection, since the terms reflects in vivo antimicrobial action. In the vast majority of instances, phytoalexin concentrations in cells in direct contact with invading microbes have not been measured. One exception is a study of the quantities of different kinds of phenylpropanoids at the cellular and organ levels in *Arabidopsis thaliana* root exudates. When *A. thaliana* roots were infected with nonhost bacterial pathogens, phenylpropanoid levels were substantially greater than when roots were challenged by host bacterial pathogens. Phenylpropanoids were resistant to bacterial pathogens capable of infecting roots and producing illness, implying that these chemicals play an essential role in nonhost pathogen defense [43]. In contrast, when *A. thaliana* roots were infected with the root-pathogenic oomycete *Pythium sylvaticum*, the concentrations of indolic and phenylpropanoid secondary metabolites increased [44]. These findings show that roots and root exudates differ significantly in terms of the nature and relative quantity of main soluble phenylpropanoid elements, as well as reactions to applied biological stress. Little research has been conducted to date in order to obtain significant insight into the complex metabolic domain of antimicrobial root exudates.

3.2   Environmental associations of plants and microbes

For the process of carbon sequestration to take place in the rhizosphere, i.e., removal of carbon dioxide from the atmosphere and holding it into solid or liquid forms, plant-microbe interactions are absolutely essential with plants themselves acting as carbon sinks; it also plays a key role in nutrient cycling and the overall ecosystem functioning [45]. "The structure and strength of microbial communities in the soil influence a plant's ability to uptake minerals and nutrients. Net ecosystem changes can be impacted by plants through the release of secondary metabolites into the rhizosphere that may encourage or discourage specific bacteria's growth. This process is known as rhizodeposition, which consists of a variety of low molecular weight metabolites such as amino acids, enzymes, mucilage, and cell lysates" [46]. Since soil bacteria take advantage of this plentiful carbon supply, selective secretion of certain chemicals may promote positive symbiotic and protective partnerships, while that of others may prevent harmful connections [47]. Callaway et al. [48] hypothesized that the link between *Centaurea maculosa* and invasive weed and neighboring plant species was influenced by fungicidal treatments. These findings suggest that the mycorrhizal fungus found in these grasses aids *Centaurea maculosa*'s growth. When *Centaurea maculosa* and *Gaillardia aristata* were tested together, the opposite result was observed, with the fungus associated to *Gaillardia aristata* having a negative impact on *Centaurea maculosa*'s growth. Plant root exudates also have an impact on the amount of pollution generated in the soil and in ground water by various toxins. As a result of bacterial multiplication and survival mediated by root exudates, rhizomediation occurs, resulting in more efficient pollutant breakdown [1,2,49]. Researchers are exploring specific pairings of plant species with bacterial species or communities to maximize this process and enable even more effective and targeted degradation of environmental pollutants [49].

3.3   Root exudates' influence on surrounding nematodes

As previously stated, root exudates supply soil bacteria with a source of organic carbon, resulting in dense microbial populations in the rhizosphere [50]. These dense microbial populations acts as a food source for microbial-feeding nematodes; as a result, microbial turnover increases leading to further increase in nutrient availability to the plants. Parasitic nematodes may also prevent the buildup of root-secreted nematicidal chemicals. Yeates [21] utilized ^{14}C pulse labeling to demonstrate that *Heterodera trifolii* and other nematodes infect the roots of white clover (*Trifolium repens*), resulting in a significant increase in photosynthetically fixed ^{14}C in soil microbial biomass. Plant-parasitic nematode-infected white clover plants seem to release more organic compounds into the rhizosphere, according to these results. Microbial turnover can also be increased as a result of root feeding nematode activities via increased carbon transfers. Exudates from *Meloidogyne incognita*-infected tomato roots contain greater amounts of water-soluble ^{14}C and different metal ions than healthy roots [51]. Changes in the carbon:nitrogen (C:N) ratio of rhizospheric carbon of *Rhizoctonia solani* modifies its trophic status, making it a pathogen [4]. The effects of nematode-related changes in root exudate concentrations and feeding ratios on nematode-microbial pathogen interactions are presently being investigated. Until recently, the majority of knowledge on rhizosphere microbe-nematode interactions has come from research on rhizobia, mycorrhizal fungi, and plant pathogens [52]. Complex tri-trophic webs have been shown in this study, with nematodes and bacteria interacting with the plant host in competing, additive, or synergistic interactions. A closer look into these interactions will help us comprehend rhizosphere signaling networks mediated by root exudates.

4 Techniques used for studying various associations arbitrated by root exudates

One of the most difficult aspects of studying plant-plant and plant-microbe interactions, which are interceded by root exudates, is the hidden nature of these interactions. Understanding the structure and function of a root system, as well as a comprehensive study of the rhizosphere community, is required for root exudation research. Microbial communities' functional diversity and redundancy, as well as the quantity and distribution of plant species, must all be taken into account. Phytochemicals can be extracted from root exudates in two ways: The first approach uses mainly methanol to extract polar molecules, while the second approach uses nonpolar solvents to target nonpolar chemicals. The isolation of numerous chemical substances such as flavonoids, quinalones, carbolines, and terpenes results from the differential partitioning of root exudates. Several approaches are required for identifying plant-produced antimicrobials, characterizing rhizosphere microorganisms, and assessing microbial colonization. BIOLOG GN substrate utilization tests are frequently used to investigate microbial communities based on the diversity of metabolic functions possessed by them [53], which is based on the community's ability to use specific carbon substrates. A DNA microarray approach has also been developed for simultaneously identifying microbial communities based on their ecological function and phylogenetic affiliation [54]. Individual microbial populations within a community can be assessed for growth rate and substrate utilization using this method. Microscopy advancements have also substantially aided the investigation of root-microbe interactions. Lugtenberg et al. [55] used immunofluorescence and an rRNA-targeting probe to study the existence and metabolic activity of *Pseudomonas fluorecens* DR54 at the root tip and found out that it took the endogenous bacteria 2 days to reach the rhizosphere after inoculation. It is a difficult undertaking to screen and functionally identify the wide range of natural substances found in root exudates that affect rhizospheric microorganisms. From plant root exudates, this approach could be utilized to find antimicrobials or QS mimics. This research would shed insights on how microbial cells function physiologically in a specific environment.

5 Conclusions (Fig. 2)

We attempted to provide a comprehensive sketch of the detrimental interactions in the rhizosphere, and the benefaction of root exudates in regulating certain activities leading to these processes under controlled yet realistic conditions. However, due to the challenges of studying subsurface processes, we do not fully understand these interactions at this time. As a consequence, new methods

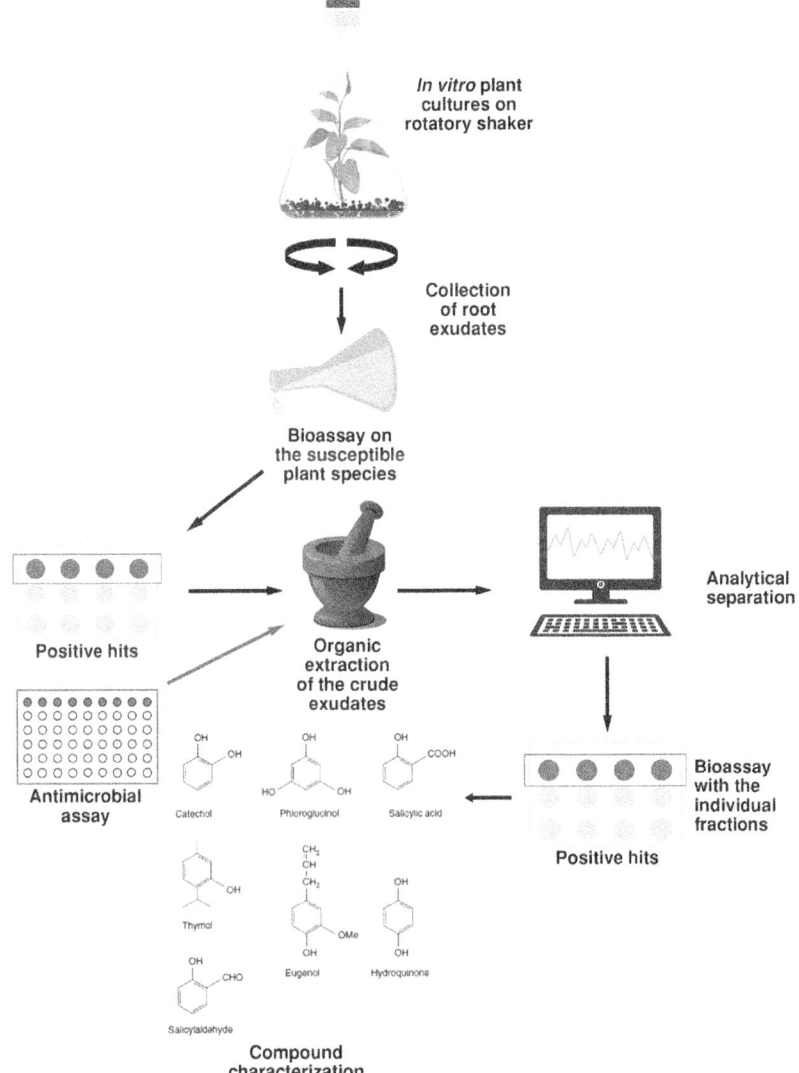

FIG. 2 A flow chart representing the assemblage, segregation, analysis, and compound classification techniques from the root exudates of plants.

for natural rhizosphere ecology studies are required, as well as cooperation among botanists, environmental engineers, and pedologists to come up with rhizotron systems that are basically laboratories constructed underground for the purpose of studying the soil and its interactions with plants and animals. Our knowledge of root-mediated activities has progressed much beyond the traditional notion that the root's sole functions are water and nutrient transfer and anchoring. Roots are increasingly becoming recognized as rhizosphere ambassadors, assisting in the communication of plants and other soil creatures. Aboveground interactions in plants may possibly translate into belowground responses in plants, according to ecological knowledge. To create molecular markers for this process, a thorough knowledge of the molecular processes involved in phytochemical secretion by roots is required. Lastly, a combination of molecular know-how of root exudation at the ecosystem scale may guide to the evolution of superior plants that absorb more nutrients, clean soils better, and keep weeds and harmful microbes at bay.

References

[1] S.S. Jha, L.S. Songachan, Mycorrhizoremediation: understanding the science behind it and it's future prospects, Mater. Today: Proc. (2021), https://doi.org/10.1016/j.matpr.2021.11.605.

[2] A. Tayang, L.S. Songachan, Microbial bioremediation of heavy metals, Curr. Sci. 120 (6) (2021) 1013, https://doi.org/10.18520/cs/v120/i6/1013-1025.

[3] D. Prasad Gond, S. Saurabh Jha, A. Kumar, S. Kumar Singh, Plant Growth Promoting Bacteria and Its Role in Green Remediation, Elsevier BV, 2021, pp. 149–163, https://doi.org/10.1016/b978-0-12-823828-8.00007-4.

[4] M. Dicke, H. Dijkman, Within-plant circulation of systemic elicitor of induced defence and release from roots of elicitor that affects neighbouring plants, Biochem. Syst. Ecol. 29 (10) (2001) 1075–1087, https://doi.org/10.1016/S0305-1978(01)00051-5.

[5] M. Faget, S. Blossfeldt, P. Von Gillhaussen, U. Schurr, V.M. Temperton, Disentangling who is who during rhizosphere acidification in root interactions: combining fluorescence with optode techniques, Front. Plant Sci. 4 (2013), https://doi.org/10.3389/fpls.2013.00392.

[6] H.E. Flores, J.M. Vicanco, V.M. Loyola-Vargas, "Radicle" biochemistry: the biology of root-specific metabolism, Trends Plant Sci. 4 (6) (1999) 220–226, https://doi.org/10.1016/S1360-1385(99)01411-9.

[7] M. Hentzer, K. Riedel, T.B. Rasmussen, A. Heydorn, J.B. Andersen, M.R. Parsek, S.A. Rice, L. Eberl, S. Molin, N. Høiby, S. Kjelleberg, M. Givskov, Inhibition of quorum sensing in Pseudomonas aeruginosa biofilm bacteria by a halogenated furanone compound, Microbiology 148 (1) (2002) 87–102, https://doi.org/10.1099/00221287-148-1-87.

[8] J.L. Hierro, R.M. Callaway, The ecological importance of allelopathy, Annu. Rev. Ecol. Evol. Syst. 52 (2021) 25–45, https://doi.org/10.1146/annurev-ecolsys-051120-030619.

[9] M.I. Hussain, S. Danish, A.M. Sánchez-Moreiras, Ó. Vicente, K. Jabran, U.K. Chaudhry, F. Branca, M.J. Reigosa, Unraveling sorghum allelopathy in agriculture: concepts and implications, Plan. Theory 10 (9) (2021), https://doi.org/10.3390/plants10091795.

[10] S.S. Jha, L.S. Songachan, Research on diversity and community composition of arbuscular mycorrhizal fungi species in India: a review, Plant Arch. 20 (2) (2020) 4201–4226.

[11] S.S. Jha, L.S. Songachan, An overview of the various methods used for mass production of arbuscular mycorrhizal fungi, IJDRBC 11 (3) (2020) 1052–1062.

[12] A. Kachroo, P. Kachroo, Mobile signals in systemic acquired resistance, Curr. Opin. Plant Biol. 58 (2020) 41–47, https://doi.org/10.1016/j.pbi.2020.10.004.

[13] S. Kalisz, S.N. Kivlin, L. Bialic-Murphy, Allelopathy is pervasive in invasive plants, Biol. Invasions 23 (2) (2021) 367–371, https://doi.org/10.1007/s10530-020-02383-6.

[14] D.F. Klessig, H.W. Choi, D.A. Dempsey, Systemic acquired resistance and salicylic acid: past, present, and future, Mol. Plant-Microbe Interact. 31 (9) (2018) 871–888, https://doi.org/10.1094/MPMI-03-18-0067-CR.

[15] L. Mommer, J. Kirkegaard, J. van Ruijven, Root-root interactions: towards a rhizosphere framework, Trends Plant Sci. 21 (3) (2016) 209–217, https://doi.org/10.1016/j.tplants.2016.01.009.

[16] L.G. Perry, C. Johnson, E.R. Alford, J.M. Vivanco, M.W. Paschke, Screening of grassland plants for restoration after spotted knapweed invasion, Restor. Ecol. 13 (4) (2005) 725–735, https://doi.org/10.1111/j.1526-100X.2005.00092.x.

[17] E.L. Rice, Allelopathy, Physiol. Ecol. (1974).

[18] A.D. Rovira, E.I. Newman, H.J. Bowen, R. Campbell, Quantitative assessment of the rhizoplane microflora by direct microscopy, Soil Biol. Biochem. 6 (4) (1974) 211–216, https://doi.org/10.1016/0038-0717(74)90053-4.

[19] J.M. Vivanco, H.P. Bais, F.R. Stermitz, G.C. Thelen, R.M. Callaway, Biogeographical variation in community response to root allelochemistry: novel weapons and exotic invasion, Ecol. Lett. 7 (4) (2004) 285–292, https://doi.org/10.1111/j.1461-0248.2004.00576.x.

[20] L.J.C. Xavier, S.M. Boyetchko, Arbuscular mycorrhizal fungi as biostimulants and bioprotectants of crops, Appl. Mycol. Biotechnol. 2 (C) (2002) 311–340, https://doi.org/10.1016/S1874-5334(02)80015-6.

[21] G.W. Yeates, Effects of plants on nematode community structure, Annu. Rev. Phytopathol. 37 (1999) 127–149, https://doi.org/10.1146/annurev.phyto.37.1.127.

[22] M. Meena, P. Swapnil, K. Divyanshu, S. Kumar, T. Harish, Y.N. Zehra, A. Marwal, R.S. Upadhyay, PGPR-mediated induction of systemic resistance and physiochemical alterations in plants against the pathogens: current perspectives, J. Basic Microbiol. 60 (10) (2020) 828–861, https://doi.org/10.1002/jobm.202000370.

[23] A. Zehra, N.A. Raytekar, M. Meena, P. Swapnil, Efficiency of microbial bio-agents as elicitors in plant defense mechanism under biotic stress: a review, Curr. Res. Microb. Sci. 2 (2021) 100054, https://doi.org/10.1016/j.crmicr.2021.100054.

[24] B.M. Chen, H.X. Liao, W.B. Chen, H.J. Wei, S.L. Peng, Role of allelopathy in plant invasion and control of invasive plants, Allelopath. J. 41 (2) (2017) 155–166, https://doi.org/10.26651/2017-41-2-1092.

[25] V.K. Singh, M. Meena, A. Zehra, A. Tiwari, M.K. Dubey, R.S. Upadhyay, Fungal toxins and their impact on living systems, in: Microbial Diversity and Biotechnology in Food Security, Springer India, 2014, pp. 513–530, https://doi.org/10.1007/978-81-322-1801-2_47.

[26] H. Pal Bais, T.S. Walker, F.R. Stermitz, R.A. Hufbauer, J.M. Vivanco, Enantiomeric-dependent phytotoxic and antimicrobial activity of (±)-catechin. A rhizosecreted racemic mixture from spotted knapweed, Plant Physiol. 128 (4) (2002) 1173–1179, https://doi.org/10.1104/pp.011019.

[27] S. Jose, A.R. Gillespie, Allelopathy in black walnut (Juglans nigra L.) alley cropping. II. Effects of juglone on hydroponically grown corn (Zea mays L.) and soybean (Glycine max L. Merr.) growth and physiology, Plant Soil 203 (2) (1998) 199–206, https://doi.org/10.1023/A:1004353326835.

[28] C.I. Nimbal, J.F. Pedersen, C.N. Yerkes, L.A. Weston, S.C. Weller, Phytotoxicity and distribution of sorgoleone in grain sorghum germplasm, J. Agric. Food Chem. 44 (5) (1996) 1343–1347, https://doi.org/10.1021/jf950561n.

[29] M.A. Czarnota, A.M. Rimando, L.A. Weston, Evaluation of root exudates of seven sorghum accessions, J. Chem. Ecol. 29 (9) (2003) 2073–2083, https://doi.org/10.1023/A:1025634402071.

[30] M.A. Czarnota, R.N. Paul, F.E. Dayan, C.I. Nimbal, L.A. Weston, Mode of action, localization of production, chemical nature, and activity of Sorgoleone: a potent PSII inhibitor in Sorghum spp. root exudates 1, Weed Technol. 15 (4) (2001) 813–825, https://doi.org/10.1614/0890-037X(2001)015[0813:MOALOP]2.0.CO;2.

[31] T.L. Weir, H.P. Bais, J.M. Vivanco, Intraspecific and interspecific interactions mediated by a phytotoxin, (−)-catechin, secreted by the roots of Centaurea maculosa (spotted knapweed), J. Chem. Ecol. 29 (11) (2003) 2397–2412, https://doi.org/10.1023/A:1026313031091.

[32] F.A. Macías, D. Marín, A. Oliveros-Bastidas, D. Castellano, A.M. Simonet, J.M.G. Molinillo, Structure-activity relationships (SAR) studies of benzoxazinones, their degradation products and analogues. Phytotoxicity on standard target species (STS), J. Agric. Food Chem. 53 (3) (2005) 538–548, https://doi.org/10.1021/jf0484071.

[33] H.P. Singh, D.R. Batish, R.K. Kohli, Autotoxicity: concept, organisms, and ecological significance, Crit. Rev. Plant Sci. 18 (6) (1999) 757–772, https://doi.org/10.1080/07352689991309478.

[34] D. Prati, O. Bossdorf, Allelopathic inhibition of germination by Alliaria petiolata (Brassicaceae), Am. J. Bot. 91 (2) (2004) 285–288, https://doi.org/10.3732/ajb.91.2.285.

[35] D.A. Wardle, K.S. Nicholson, A. Rahman, Influence of plant age on the allelopathic potential of nodding thistle (Carduus nutans L.) against pasture grasses and legumes, Weed Res. 33 (1) (1993) 69–78, https://doi.org/10.1111/j.1365-3180.1993.tb01919.x.

[36] M.-C. Nilsson, P. Högberg, O. Zackrisson, W. Fengyou, Allelopathic effects by Empetrum hermaphroditum on development and nitrogen uptake by roots and mycorrhizae of Pinus silvestris, Can. J. Bot. 71 (4) (1993) 620–628, https://doi.org/10.1139/b93-071.

[37] J.I. Yoder, Parasitic plant responses to host plant signals: a model for subterranean plant-plant interactions, Curr. Opin. Plant Biol. 2 (1) (1999) 65–70, https://doi.org/10.1016/S1369-5266(99)80013-2.

[38] A.G. Palmer, R. Gao, J. Maresh, W.K. Erbil, D.G. Lynn, Chemical biology of multi-host/pathogen interactions: chemical perception and metabolic complementation, Annu. Rev. Phytopathol. 42 (2004) 439–464, https://doi.org/10.1146/annurev.phyto.41.052002.095701.

[39] M. Chang, D.G. Lynr, D.H. Netzly, L.G. Butler, Chemical regulation of distance: characterization of the first natural host germination stimulant for Striga asiatica, J. Am. Chem. Soc. 108 (24) (1986) 7858–7860, https://doi.org/10.1021/ja00284a074.

[40] G.D. Fate, D.G. Lynn, Xenognosin methylation is critical in defining the chemical potential gradient that regulates the spatial distribution in Striga pathogenesis, J. Am. Chem. Soc. 118 (46) (1996) 11369–11376, https://doi.org/10.1021/ja961395i.

[41] W.J. Keyes, R.C. O'Malley, D. Kim, D.G. Lynn, Signaling organogenesis in parasitic angiosperms: xenognosin generation, perception, and response, J. Plant Growth Regul. 19 (2) (2000) 217–231, https://doi.org/10.1007/s003440000024.

[42] R.C. O'Malley, D.G. Lynn, Expansin message regulation in parasitic angiosperms: marking time in development, Plant Cell 12 (8) (2000) 1455–1465, https://doi.org/10.1105/tpc.12.8.1455.

[43] X. Li, X. Zhang, G. Liu, Y. Tang, C. Zhou, L. Zhang, J. Lv, The spike plays important roles in the drought tolerance as compared to the flag leaf through the phenylpropanoid pathway in

wheat, Plant Physiol. Biochem. 152 (2020) 100–111, https://doi.org/10.1016/j.plaphy.2020.05.002.

[44] H.P. Bais, B. Prithiviraj, A.K. Jha, F.M. Ausubel, J.M. Vivanco, Mediation of pathogen resistance by exudation of antimicrobials from roots, Nature 434 (7030) (2005) 217–221, https://doi.org/10.1038/nature03356.

[45] P. Kumari, M. Meena, P. Gupta, M.K. Dubey, G. Nath, R.S. Upadhyay, Plant growth promoting rhizobacteria and their biopriming for growth promotion in mung bean (Vigna radiata (L.) R. Wilczek), Biocatal. Agric. Biotechnol. 16 (2018) 163–171, https://doi.org/10.1016/j.bcab.2018.07.030.

[46] M. Hassan, J. McInroy, J. Kloepper, The interactions of rhizodeposits with plant growth-promoting Rhizobacteria in the rhizosphere: a review, Agriculture 9 (7) (2019) 142, https://doi.org/10.3390/agriculture9070142.

[47] P. Bednarek, B. Schneider, A. Svatoš, N.J. Oldham, K. Hahlbrock, Structural complexity, differential response to infection, and tissue specificity of indolic and phenylpropanoid secondary metabolism in Arabidopsis roots, Plant Physiol. 138 (2) (2005) 1058–1070, https://doi.org/10.1104/pp.104.057794.

[48] R.M. Callaway, G.C. Thelen, S. Barth, P.W. Ramsey, J.E. Gannon, Soil fungi alter interactions between the invader Centaurea maculosa and North American natives, Ecology 85 (4) (2004) 1062–1071, https://doi.org/10.1890/02-0775.

[49] I. Kuiper, E.L. Lagendijk, G.V. Bloemberg, B.J.J. Lugtenberg, Rhizoremediation: a beneficial plant-microbe interaction, Mol. Plant-Microbe Interact. 17 (1) (2004) 6–15, https://doi.org/10.1094/MPMI.2004.17.1.6.

[50] M.T. Kingsley, J.K. Fredrickson, F.B. Metting, R.J. Seidler, Environmental restoration, in: Bioremediation of Chlorinated and Polycyclic Aromatic Hydrocarbon Compounds, vol. 2, 1994.

[51] J.H. Bowers, S.T. Nameth, R.M. Riedel, R.C. Rowe, Infection and colonization of potato roots by Verticillium dahliae as affected by Pratylenchus penetrans and P. crenatus, Phytopathology 86 (6) (1996) 614–621, https://doi.org/10.1094/Phyto-86-614.

[52] S.D. Gundy, J.D. Kirkpatrick, J. Golden, The nature and role of metabolic leakage from root-knot nematode galls and infection by Rhizoctonia solani, J. Nematol. 9 (2) (1977) 113–121.

[53] R.G. Belz, K. Hurle, Differential exudation of two benzoxazinoids—one of the determining factors for seedling allelopathy of Triticeae species, J. Agric. Food Chem. 53 (2) (2005) 250–261, https://doi.org/10.1021/jf048434r.

[54] R. Dorhout, F.J. Gommers, C. Kollöffel, Phloem transport of carboxyfluorescein through tomato roots infected with meloidogyne incognita, Physiol. Mol. Plant Pathol. 43 (1) (1993) 1–10, https://doi.org/10.1006/pmpp.1993.1035.

[55] B.J.J. Lugtenberg, L. Dekkers, G.V. Bloemberg, Molecular determinants of rhizosphere colonization by Pseudomonas, Annu. Rev. Phytopathol. 39 (2001) 461–490, https://doi.org/10.1146/annurev.phyto.39.1.461.

Chapter 5

Biodegradation and bioaugmentation of pesticides using potential fungal species

S. Gomathi, V. Ambikapathy, and A. Panneerselvam
Department of Botany, A.V.V.M. Sri Pushpam College (Autonomous), (Affiliated to Bharathidasan University, Tiruchirappalli), Thanjavur, Tamil Nadu, India

1 Introduction

Around 9000 insect and mite species, 50,000 plant diseases, and 8000 weed species wreak havoc on crops all over the world. Insects and plants account for 14% and 13% of the losses, respectively. Pesticides are essential in agricultural productivity, which are utilized in nearly a third of all agricultural products. Insect damage to fruits, vegetables, and cereals would result in losses of 78%, 54%, and 32%, respectively, depending on the usage of pesticides. Crop damage from pests is reduced from 42% to 35% when pesticides are used [1].

Biodegradation involves the use of microorganisms to eliminate pollutants, which is the most promising, efficient, and cost-effective strategy. It is the complete breakdown of an organic material into its inorganic components. It results to satisfy energy requirements, the need to detoxify pollutants, and all other factors in the microbial shift [2]. Because of microorganisms' ubiquitous nature, large biomass, diversity of catalytic mechanisms, and ability to function even in the absence of oxygen and other extreme conditions, the search for pollutant-degrading microorganisms, understanding their genetics and biochemistry, and developing methods for their application in the field have become increasingly important. They have grown in importance as a human endeavor [3].

Microorganisms have been dubbed "Earth's greatest chemists" because of their practically limitless biogeochemical capabilities [4]. Microorganisms can help in pollutant transformations by generating enzymes, which reduce the activation energy needed to keep a reaction going.

Plant-Microbe Interaction—Recent Advances in Molecular and Biochemical Approaches
https://doi.org/10.1016/B978-0-323-91876-3.00013-0

Pesticides are a wide set of compounds used in agriculture, forestry, and horticulture to gain an advantage over disease-causing pest species. On the other hand, many pesticides have the potential to harm human health or the environment. Pesticides have become an indispensable feature of most intensive farming systems for food and other crops, and there is currently no viable alternate for all of their applications. As a result, pesticides will likely continue to be used indefinitely. However, more research is needed to create successful pesticide-reduction solutions through integrated pest management systems and biological control technologies. However, it is vital to keep pesticide use to a minimal level and phase out the most harmful compounds. Pesticides are naturally degraded through physical, chemical, and biological conversion processes, but their very stable and soluble nature permits them to stay in the environment for a long time. Bioremediation is one of the most successful strategies for eliminating pesticides from the environment [5]. Due to the restricted bioavailability of various pesticide-degrading bacteria in the variable subsurface environment, the bioremediation process is very uncertain [6]. The ability of microbes such as bacteria to breakdown pesticides is ascertained by a number of parameters, including pesticide transit via cell membranes, enzymatic reactions, biosurfactant synthesis, and environmental conditions such as pH, temperature, and electron acceptor availability [7,8].

Pesticides biodegrade in a number of ways depending on their nature, ambient conditions, and, most importantly, microorganism type [9]. Microorganisms play an important role in the degradation process, but fungi biotransform pesticides into benign molecules by introducing structural changes and releasing them into the soil, where bacteria can degrade them further [10]. The Phanerochaete *Chrysosporium* is one among the pesticides, and it destroys a broad spectrum of pests [11]. White-rot fungi degrade chemicals such as lindens, atrazine, diuron, terbuthylazine, metalaxyl, DDT, gamma-hexachlorocyclohexane, dieldrin, aldrin, heptachlor, chlordane, and mirex in variable degrees [12]. *Agrocybe semiorbicularis, Auricularia auricula, Coriolus versicolor, Dichomitus squalens, Flammulina velutipes, Hypholoma fasciculare, Pleurotus ostreatus*, and *Stereum hirsutum* all have the ability to breakdown pesticides like phenylamide and triazine [13].

Pesticide bioremediation is based on enzymes produced by plants and soil microbes during various metabolic activities. Enzymes are vital in the biodegradation of any xenobiotic and can aid in the future restoration of the damaged environment by rapidly refurbishing poisons. Pesticides are also degraded by enzymes in the target organism via natural detoxification mechanisms and acquired metabolic tolerance, as well as in the broader environment via soil and water microorganism biodegradation. To extract polyaromatic hydrocarbons from fresh, marine, or terrestrial water, fungal enzymes such as oxidoreductase, laccase, and peroxidases are widely used in the process of biodegradation. Organophosphorus degrading enzymes have been studied extensively over the years, and different bacteria, fungi, and cyanobacteria have been discovered and used in organophosphorus compounds as a carbon and nutrition source.

Enzymes are one of the most important bioactive molecules used for human beings and also in the sectors of industrial, environmental, and food technology when it comes to microbial sources [14]. In the current scenario, the idea of employing microorganisms as biotechnological sources of industrially important enzymes has prompted interest in extracellular biocatalytic activity in a variety of bacteria [15–17]. Because of the significant industrial value of amylase, the isolation of efficient fungal strains producing amylases well fitted to novel commercial uses is of continuing interest [18].

2   Fungi

The word "fungi" is derived from the Latin word "fungus," which means "mushroom." Fungi are nucleated, spore-bearing achlorophyllous organisms that reproduce both sexually and asexually. Their filamentous branched somatic structures are generally covered by cellular membranes which containing cellulose or chitin, or even both. In simplest terms, fungi can be described as "nongreen, nucleated thallophytes." Fungi include yeasts, molds, mushrooms, *Polyporus*, puff balls, rusts, and smuts. Mycology (Greek: mykes = mushrooms + logos = discourse) is the branch of botany that studies fungi, and a mycologist is a person who is knowledgeable in mushrooms. The founder of the study of mycology, Pier' Antonio Micheli, an Italian botanist, was the first to give maximum amount of information of fungi in his work Nova Plantarum Genera, published in 1729, and Anton De Bary (1831–88), known as the "Father of Modern Mycology," was a pioneer in the field of mycology. There are roughly 5100 genera and 50,000 species of fungi. Fungi grow in low light, damp conditions, stable temperatures, and in the presence of live or dead organic materials. They are unable to manufacture food for themselves, as a result, every fungus is both heterotrophic and holozoic like animals. The fungus is a chemoorganotroph that gets energy from the oxidation of organic substances, with an absorption-based extracellular diet. The refractory form is broken down by enzymes into a soluble form that may be absorbed. Fungi like *Beauveria bassiana*, *Cordyceps melothae*, and *Metarrihizium anisopliae* have been used to control bugs and pests. Especially, *Coelomyces*, the aquatic fungus, acts swiftly to control mosquito larvae.

3   As natural scavengers

In conjunction with saprophytic bacteria, fungi break down the rotting corpses of animals, and their waste products, and vegetation. In this way, they keep the Earth's surface clean while also allowing organisms to utilize the degraded simpler compounds. Vegetable detritus is made up of complex organic components such as cellulose, lignin, suberin, cutin, starch, glucose, pectins, and hemicellulose. Woody plants have substantial amounts of cellulose and lignin. Enzymes of both *Chaetomium globosum* (Ascomycetes) and *Merulius lacrymans* (Basidiomycetes) degrade cellulose. The enzymes, cytase and cellobiase, are released, hydrolyzing cellulose and converting it to glucose. Lignin is destroyed by the enzyme lignase, which is produced by *Polyporus adustus*, *Polyporus vesricolor*, and *Lenzites trabea* Basidiomycetes. The remaining components are degraded by fungi, which release enzymes such as hemicellulases, pectinases, and amylases (e.g., *Penicillium glaucum*, *Aspergillus oryzae*, etc.). Lipids, carbohydrates, and nitrogenous compounds are broken down by these enzymes into basic molecules like CO_2, water, ammonia, and hydrogen sulfide.

4   Humus formation

The slow breakdown of complex debris and animal carcasses in soil produces organic matter, known as humus. This treatment is known as humification. Soil fertility must be preserved at all times. It also aids in the soil's moisture retention.

5   Nitrogen fixation and biological control

Some yeasts, such as *Rhodotorula* and *Saccharomyces*, have been identified as symbiotic nitrogen fixers. The process of employing one species of a biological system to eliminate another is known as biological control. Fungi are in charge of regulating a broad range of plant illnesses and pathogens that cause disease. *Pythium* spp. affects seedlings of tobacco, tomato, mustard, chillies, and cress.

Pythium and other root rot fungi are inhibited by *Trichoderma lignorum* and *Gliocladium fimbriatum* (found in moist soils), allowing crops to grow more efficiently. Aside from that, predatory fungi can be found in the soil. They are capable of catching or killing nematodes. *Arthrobotrys oliogospora*, *Dectylell acionopaga*, *Dectylell ellipsospora*, and another nematophagous fungus *Hetero deraavenae*, a cereal cyst worm, is controlled by *Nematophthor agynophila*, an Oomycetes species.

6   Role of Mycorrhizae

Mycorrhizae are characterized as an association between fungal hyphae and roots of higher plants like vascular plants. Fungal hyphae are similar to root hairs; hence, they absorb water and minerals, and transfer them to the plant. Mycorrhizae are found in the majority of conifers, Ericaceae, and many perennial herbs. Fungi *Rhizoctonia*, *Phomci*, *Tricholoma*, *Amanda*, *Lycoperdon*, and *Scleroderma*, have shown mycorrhizal associations with a variety of plants.

7   Inorganic pesticides

Humans are always waging wars against rivals and diseases. Pesticides are one way to gain an advantage in many of these ecological interactions. Pesticides, such as insecticides, herbicides, fungicides, preservatives, and disinfectants, were used in 2.4 million tons globally in 2007. Pollution of the environment is a global issue, and the dangers and repercussions for human health are a serious concern. The current level of pollution is a personal calamity, but before filing a pollution complaint, the benefit-risk ratio must also be evaluated. Pesticides frequently damage species other than pests, including people, because their mode of action is not limited to a single species. According to the WHO, pesticide poisoning causes 3 million illnesses and up to 220,000 deaths each year, with the bulk of instances in underdeveloped countries. Pesticides produce reactive oxygen species, lowering antioxidant levels and their ability to protect cells from oxidative damage. Lipids, proteins, and nucleic acids are targeted as a result of the imbalance, and cell regulatory processes are changed. Oxidative and reactive oxygen species cause carcinogenetic, neurodegenerative, cardiovascular, pulmonary, renal, endocrine, and reproductive problems. Pesticides throw off the oxidative balance, allowing diseases and homeostasis to thrive. Some of the active ingredients are natural biochemicals extracted from specially bred plants, while others are manufactured chemicals based on dangerous metals or arsenic compounds. Most current pesticides, on the other hand, are chemist-created organic compounds. It can cost tens of millions of dollars to introduce a new pesticide and evaluate it for efficacy (activity against pests), toxicological properties, and environmental repercussions. The industry, on the other hand, is willing to pay the high development expenses if a viable pesticide against a big pest is created. Because it analyses both direct and indirect hazardous effects of chemicals, ecotoxicology has a wider reach. Indirect ecological influences include changes in habitat or food supply. Herbicide use in forest management or agriculture will change the biomass and species composition of plants on a treated area. The changes in animal habitats are profound. Even if the herbicide does not poison the animals that come into touch with it, environmental changes may have an effect on them. The ecotoxicological dangers of chemical exposure in the environment are impacted by a number of factors. The following are the most important factors to consider: (1) cellular sensitivity, (2) the chemical's intrinsic hazard, (3) exposure intensity, and (4) any potential indirect repercussions.

Hence, inorganic pesticides contain toxic metals such as arsenic, copper, lead, and mercury. Arsenic trioxide, sodium arsenite, and calcium arsenate are herbicides and soil sterilants. Paris green, lead arsenate, and calcium arsenate are insecticides. They survive for a long period in terrestrial settings, only to be flushed out and destroyed by wind or water. In the current scenario, synthetic organic insecticides have mostly supplanted inorganic pesticides. One of the most well-known examples is Bordeaux mixture, which is a mixture of copper-based chemicals used as a fungicide to preserve fruit and vegetables.

8   Toxicological classification of pesticides

Pesticides such as insecticides, herbicides, fungicides, rodenticides, wood preservatives, garden chemicals, and household disinfectants are used to kill pests. These pesticides differ in their physical, chemical, and identical properties from one class to the other. As a result, its worthwhile to categorize them based on their characteristics and study them inside their various groups. Synthetic pesticides are man-made chemicals that do not exist in nature. They are classified into a variety of classes based on the requirements. Drum suggests three potential pesticide classification techniques [19] (Table 1 and Fig. 1).

TABLE 1 Some of the microorganisms for pesticides degradation.

Types of microorganism	Name of the pesticides	Degradation
Bacterial organism	*Pseudomonas* sp.	Endosulfan, endrin, hexachlorocyclohexane, methyl parathion, monocrotophos, aldrin, chlorpyrifos, coumaphos, DDT, diazinon
	Bacillus sp.	Methyl parathion, monocrotophos, parathion, polycyclic aromatic hydrocarbons. Coumaphos, diazinon, dieldrin, DDT endosulfan, endrin, glyphosate
	Alcaligenes	Chlorpyrifos, endosulfan
	Flavobacterium	Diazinon, glyphosate, methyl parathion, parathion
Actinomycetes	*Micromonospora* sp. *Actinomyces* sp., *Nocardia* sp., and *Streptomyces*	Aldrin, carbofuran, chlorpyrifos, diuron, diazinon
Fungi	Fungi that cause white rot *Cladosporium* sp., *Rhizopus* sp., *Aspergillus fumigatus* *Penicillium* sp., *Aspergillus* sp., *Fusarium* sp., *Mucor* sp., *Trichoderma* sp., *Mortierella* sp.	DDT, diuron, endosulfan, esfenvalerate, fenitrothion, fenitrooxon, fipronil, heptachlor epoxide, lindane, malathionmetalaxyl, pentachlorophenol, terbuthylazine, 2,4-D
Algae	Algae, little green algae Chlamydomonas	Parathion, phorate, atrazine, fenvalerate, DDT, patoran

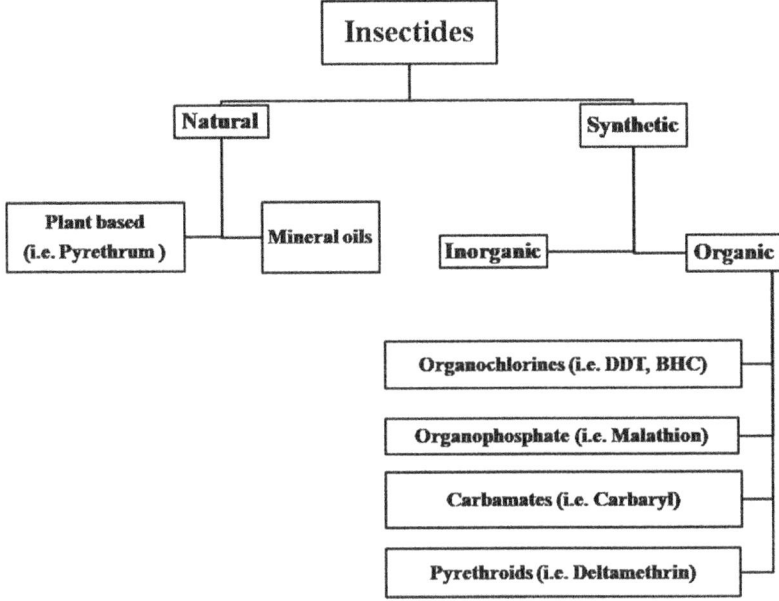

FIG. 1   Classification of insecticides.

9   Fungi in removal and degradation of pesticides

Aspergillus allhabadii, Aspergillus aultaceus, and *Aspergillus candidus, Alternaria brassicola, A. exserohilumturcicum, Drechslera australiensis, Drechslera halodes, Drechslera hawaiiensis, Drechslera* sp., *Humicola insolens, Humicola brevis,* and *Fusarium acuminum* are the isolates from those genera; 7 are identified as Ascomycotes, 2 are Deuteromycotes, and 1 is a Zygomycetes filamentous fungus. The majority of the fungal isolates are grouped into species such as *Alternaria, Aspergillus, Drechslera,* and *Fusarium;* these are considered as the most prevalent genera. *Curvularia, Exserohilum, Humicola, Rhizopus,* and *Torula* were the most frequently isolated genera [20].

10   Isolation of fungi from soil [21]

In a conical flask with a capacity of 250 mL, a 10-g soil sample was combined with 100 mL of sterile distilled water. To create a homogeneous suspension and successive dilutions of the soil sample, it was poured into the flask and was shaken in an electric shaker. The numbers 10^{-2} and 10^{-3} were calculated. PDA media was used to plate 1 mL of a 10^{-3} dilution in Petridishes. The media's pH was altered to 5.6. To prevent bacterial growth, streptomycin sulfate $(100\,mg^{-1})$ was added to the medium. The plates were incubated at

$25 \pm 2°C$ for 5 days, and the fungi that appeared on the media were recorded. Population of fungi g^{-1} is calculated as below:

$$\text{Dry wt.of the soil.} = \frac{\text{Mean number of propagules in dilution plate}}{\text{Wt.of the dry soil}}$$
$$\times \text{ dilution factor}$$

$$\text{Percentage frequency} = \frac{\text{No.of soil samples from which fungi were recorded}}{\text{No.of soil samples}}$$
$$\times 100$$

11   Observations

On PDA plates, colonies with different morphologies were enumerated individually. A section of the colony's developing edge was taken up with a pair of needles and mounted on a clean slide with lactophenol cotton blue stain. To help staining and remove any air bubbles formed, the slide was gently heated in a spirit lamp. The extra discoloration was removed using tissue paper, and the cover slip was then sealed with clear nail polish. The slides were observed by using a compound microscope. Microphotography of certain fungal species was also achieved using a Nikon Optiphot microscope (Japan).

12   Identification

Using standard manuals such as Soil fungi [22], Dematiaceous Hyphomycetes [23], More Dematiaceous Hyphomycetes [24] and Hyphomycetes, colony color and morphology, hyphal structure, spore size, shapes, and spore-bearing structures of the fungi were recorded [25].

13   TLC and HPLC analyses

TLC plates were prepared by evenly spreading fine silica gel powder with a binder starch on 15×15 cm clean glass plates, which were then dried in a 1000°C oven. The two lines on the chromate plate are designated with an appropriate standard. As much as 100 mL of the extracted samples were spotted on the base line and dried with a capillary tube. The chromate plate was then placed vertically in a container that contained a solvent system (mobile phase) of n--hexane:acetone:ethyl acetate (80:10:10). When the developing solvent approached the ending line, the plate was removed, dried, and the components were detected by iodine vapors. The separated product was subjected to HPLC analysis in the same way as the rest of the process.

14   GC-MS analysis

The final microbial degradation extracts were dissolved in acetone and GC-MS analyzed [26]. 1 mL of the extract was injected into the injection port of a gas chromatograph (Shimadzu gas chromatograph, GC-2018 Plus A equipped with Ni (550 MBq) ECD electron capture detector, Shimadzu Co., Tokyo, Japan) at 25°C with a purge flow rate of 50 mL for 1 min. The GC was performed using helium as the carrier gas and a 0.25-mm diameter column with a 0.25-mm film thickness of 5% phenyl methyl polysiloxane.

15   Biodegradation

In the pesticide degradation process, a pesticide turns into a harmless substance that is compatible with the environment and applied as biodegradation. Pesticides can be degraded in plants, animals, soil, and water, but the most common sort of degradation occurs in the soil, where microorganisms, especially fungi and bacteria, which feed on pesticides reduce their existence. The soil fumigant methyl bromide, the herbicide dalapon, and the fungicide chloroneb are all pesticides that are degraded by microorganisms.

The following factors must be considered for optimal pesticide biodegradation in the soil.

1. Organisms must have the enzymatic activity required for rapid pollutant breakdown in order to reduce the concentration of contamination.
2. Bioavailability of the pollutant must be the goal.
3. Microbial/plant development and enzymatic activity require favorable soil parameters.
4. Biological treatment must be less expensive than alternative methods of pollutant removal.

16   Strategies for biodegradation

The following strategies are required for effective biodegradation/biosorption of a specific pollutant.

1. **Intrinsic/passive bioremediation** is a natural method of contaminant removal by native microorganisms at a very slow rate.
2. **Biostimulation**: Nitrogen and phosphate must be added to the soil to enhance indigenous microorganisms.
3. **Bioventing**: A biostimulation technique in which gases such as oxygen and methane are supplied to the soil to enhance microbial activity.
4. **Bioaugmentation**: The inoculation of microorganisms into a contaminated site, to aid biodegradation, is known as bioaugmentation.

5. **Composting**: Contaminated soil piles are created and processed using aerobic thermophilic microorganisms to break down toxins. Piles are manually stirred and wet on a regular basis to enhance microbial activity.
6. **Phytoremediation**: This can be accomplished directly through the cultivation of heavy metal-accumulating plants or indirectly through the promotion of microbes in the rhizosphere by plants.
7. **Bioremediation**: This refers to the use of microorganisms to detoxify toxic substances in the soil and other environments.
8. **Mineralization**: A variety or group of microbes present in the said completely converts an organic pollutant to its inorganic ingredient.

Although there are a variety of biodegradation techniques available, the following are the most important:

1. Bacterial degradation: Pesticides are degraded by the majority of bacterial species. The majority of pesticides degrade somewhat, resulting in the generation and aggregation of metabolites.
2. Fungal degradation: Fungi degrade pesticides by instigating slight structural changes, removing toxins, and releasing into the soil, where bacteria can biodegrade them still further.
3. Microbial degradation: For example, in soil, the herbicide 4-butyric acid (2, 4-D B) and the insecticide phorate are activated microbiologically to create hazardous metabolites for plants and insects.
4. Altering the toxicity spectrum: Some fungicides like PCNB control a specific group of organisms, but they metabolize to produce chemicals that inhibit different species of organisms, which are transformed in soil to chlorinated benzoic acids, which kill pests.
5. Solubilized leaching: Leaching can be solubilized, so it is used to remove pesticides. Dichloro diphenyl tricholroethane (DDT), the well-known chlorinated pesticide, is an example of pesticide biodegradation.

Pesticide-degrading microorganisms can come from a variety of sources. Pesticides are mostly used on agricultural crops; therefore, soil, along with pesticide industrial effluent, sewage sludge, activated sludge, wastewater, natural waters, sediments, places near pesticide manufacture, and even some live species, is the primary source of these chemicals. Pesticide degrading microorganisms have been isolated from a broad range of locations that have been contaminated with pesticides in general. In many laboratories, collections of microorganisms were characterized by their identification, growth, and breakdown of pesticides. The isolation and characterization of pesticide-degrading microorganisms paves the way for novel strategies to be used to repair damaged environments [27,28].

The microbial systems that eliminate organic pollutants from the natural environment are very critical. In biodegradation biotechnology, environmental microbiology and analytical geochemistry are significant underlying branches of science. The study of microorganisms from pure-culture isolates, laboratory enrichment cultures, and polluted field sites has advanced our understanding of biodegradation in general and aromatic-hydrocarbon biodegradation in particular. New analytical and molecular tools have increased our understanding of the processes, occurrence, and identification of active actors participating in the biodegradation of organic pollutants in the environment ranging from sequencing the DNA of biodegrading bacteria [29,30]. White rot fungi have been suggested as potential bioremediation agents, especially for pollutants that are difficult to break down by bacteria. This ability shows the production of extracellular enzymes that interact with a diverse variety of chemical compounds. Extracellular enzymes involved in lignin degradation include lignin peroxidase, manganese peroxidase, laccase, and oxidases. A number of pesticide-degrading microorganisms have been discovered, and the list is increasing. The three main enzyme involved in degradation are esterases, glutathione S-transferases (GSTs), and cytochrome P450 [31] (Fig. 2).

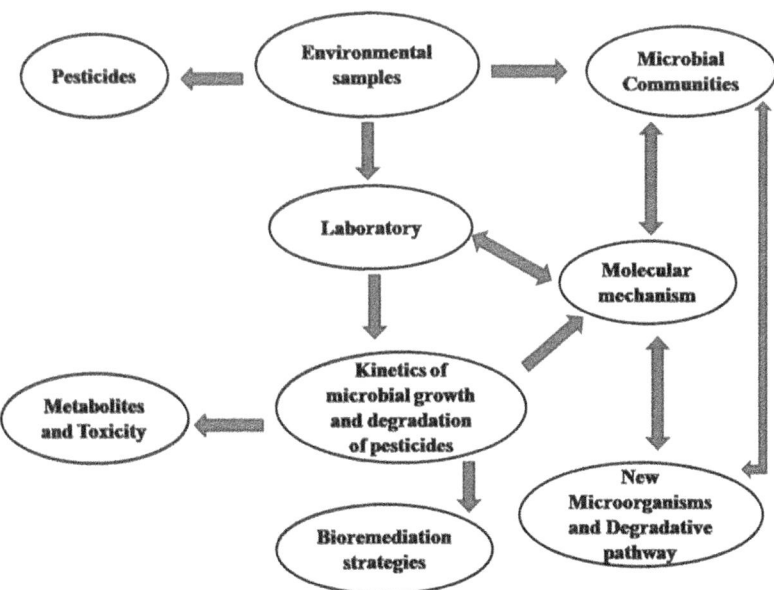

FIG. 2 Relationships between pesticides and microbial communities.

17   Enzymes involved in the biodegradation process

Many pesticides' biology is dependent on enzymes' technology [32]. Pesticides can be transformed using some enzymes, which is a novel treatment method for removing toxic compounds from polluted environments. Pesticide breakdown that is catalyzed by some enzymes could be more successful than the current chemical techniques. Many pesticides are catalyzed by some targeting specific enzymes which have importance for physiological functions. Some pesticides are activated in situ by enzymatic action, and many pesticides work by targeting specific enzymes with important physiological responsibilities. Pesticide chemicals are degraded by enzymes in the target organism via intrinsic detoxification processes and evolved metabolic resistance, as well as in the wider environment via biodegradation by soil and water microbes [33,34].

Genetically modified *Escherichia coli* enzymatically degrades methyl parathion and numerous other Ops; PNP is detected by HPLC [35]. *Micrococcus* sp. has been found to have the ability to degrade OP insecticides like cypermethrin by direct enzymatic activity [36]. The fungus *Conidiobolus* also destroys lindane by enzyme activity. The lack of a metabolite was confirmed by GC-ECD and GC-MS, demonstrating that this fungus completely destroys lindane [37]. A study employing extracellular enzyme obtained from fungi was used to investigate the degradation of atrazine (AT) and alachlor (AL) [38]. The FDS-1 strain of *Burkholderia* sp. may enzymatically degrade nitrophenyl at 30 degrees Celsius and pH 7.0 [39]. GMO bacteria strains have enzymes that can degrade a broad range of pesticides, including OPs, carbamates, and pyrethroids [40]. The most common enzymes employed to break down organochlorinated pesticides are dehydrochlorinated enzymes, hydrolytic enzymes, and dehydrogenases. Genes from the Lin family, which have traditional functional codes, are involved [41].

In the first metabolic stage, hydrolases, esterases, oxidases (MFO), and glutathione *S*-transferases (GST) are engaged; in the second metabolic stage, enzymes catalyze metabolic reactions such as hydrolysis, oxidation, addition of an oxygen to a double bound, oxidation of an amino group (NH_2) to a nitro group, addition of a hydroxyl group to a benzene ring, dehalogenation, reduction of a nitro group (NO_2) to an amino group, replacement of a sulfur with an oxygen, metabolism of side chains, and ring cleavage. Microbes' metabolic potential, which is regulated by both accessibility and bioavailability, determines their ability to detoxify pollutant compounds [9,42] (Fig. 3).

There are three stages of pesticide metabolism. Stage I showed that metabolism changes the characteristics of a parent chemical via oxidation, reduction, or hydrolysis to produce a more water soluble and usually less poisonous derivative than the parent. A pesticide metabolite is conjugated to a sugar or amino acid in the second stage, increasing water solubility and lowering toxicity when compared to the original pesticide. The third stage involves the conversion of stage II metabolites into secondary conjugates, which are also nontoxic. These

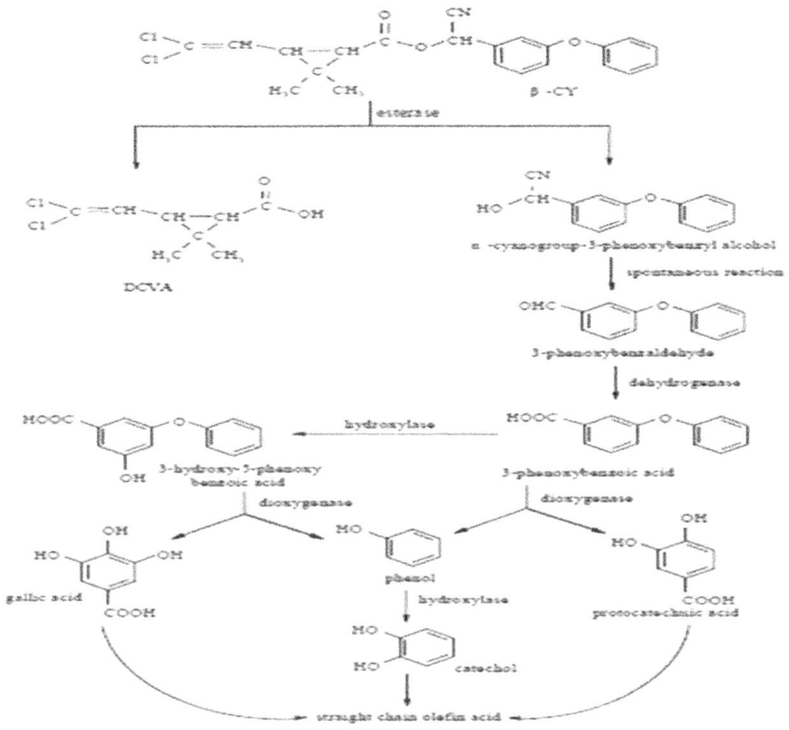

FIG. 3   Degradation pathway of beta-CY (β-CY) by *Aspergillus niger* YAT.

actions are carried out by fungi and bacteria, which produce internal and extracellular enzymes, hydrolytic enzymes, peroxidases, oxygenases, and so on [43].

18   Conclusions

Despite tremendous advances in pesticide microbial degradation and the identification of the majority of pesticide-degrading microbial strains, microbial bioremediation has been limited due to its poor degradable efficiency and the status of the environment. Mineralization and cometabolism were the main mechanisms for further degradation of pesticides and their intermediate products. While pesticides' group and molecular structure determined their degradation behavior in the microbial environment, chemical structure determined their solubility, with molecular orientation, spatial structure, chemical functional groups, intermolecular attraction, and repulsion characteristics influencing pesticide ingestion by flies. With the advancement of genetic engineering and molecular biology on the one hand, and the application of gene recombination techniques on the other, researchers have begun to transfer to the production of effective engineering bacteria in the present scenario. They employed an enzyme gene, on the other hand, to produce a vector that could successfully express pesticide degrading characteristics. The aim was to improve degradation efficiency by increasing the expression level of specific proteins or enzymes, which would alleviate the problem of some enzymes in the environment not being able to be stabilized and maintain a high level of activity. In conclusion, using microbiological agents or fertilizer preparations is an effective way to prevent pesticide pollution in the environment.

References

[1] D. Pimentel, Environmental and economic costs of the application of pesticides primarily in the United States, in: M. Pimentel, D. Pimentel (Eds.), Food, Energy, and Society, third ed., CRC Press, 2007.

[2] D. Paul, G. Pandey, J. Pandey, R.K. Jain, Accessing microbial diversity for bioremediation and environmental restoration, Trends Biotechnol. 23 (2005) 135–142.

[3] M. Megharaj, B. Ramakrishnan, K. Venkateswarlu, N. Sethunathan, R. Naidu, Bioremediation approaches for organic pollutants: a critical perspective, Int. J. Environ. Sci. 37 (2011) 1362–1375.

[4] M.T. Madigan, J.M. Martinko, P.V. Dunlap, D.P. Clark, Brock Biology of Microorganisms, Pearson Prentice Hall, Upper Saddle, 2012, p. 992.

[5] P. Satapute, M.K. Paidi, M. Kurjogi, S. Jogaiah, Physiological adaptation and spectral annotation of arsenic and cadmium heavy metal-resistant and susceptible strain *Pseudomonas taiwanensis*, Environ. Pollut. 251 (2019) 555–563.

[6] G. Odukkathil, N. Vasudevan, Toxicity and bioremediation of pesticides in agricultural soil, Rev. Environ. Sci. Biotechnol. 12 (4) (2013) 421–444.

[7] M.S. Mohamed, Degradation of methomyl by the novel bacterial strain *Stenotrophomonas maltophilia* M1, Electron. J. Biotechnol. 12 (4) (2009) 6–7.

[8] M. Narayanan, S. Kumarasamy, M. Ranganathan, Enzyme and metabolites attained in degradation of chemical pesticides ß cypermethrin by *Bacillus cereus*, Mater. Today: Proc. (2020), https://doi.org/10.1016/j.matpr.2020.05.722.

[9] M.L. Ortiz Hernandez, E. Sanchez Salinas, E. Dantan Gonzalez, M.L. Castrejon-Godinez, Pesticide biodegradation: mechanisms, genetics and strategies to enhance the process, in: Biodegradation Life Science, IntechOpen, 2013, pp. 251–287.

[10] L. Gianfreda, M.A. Rao, Potential of extra cellular enzymes in remediation of polluted soils: a review, Enzym. Microb. Technol. 35 (2004) 339–354.

[11] R.L. Singh, Principles and Applications of Environmental Biotechnology for a Sustainable Future, Springer, Singapore, 2017.

[12] V.M. Pathak, Handbook of Research on Microbial Tools for Environmental Waste Management, IGI Global, 2018.

[13] G. Bending, D.M. Friloux, A. Walker, Degradation of contrasting pesticides by white rot fungi and its relationship with ligninolytic potential, FEMS Microbiol. Lett. 212 (2002) 59–63.

[14] K. Kathiresan, S. Manivannan, Alpha-amylase production by *Penicillium fellutanum* isolated from mangrove rhizospheric soil, Afr. J. Biotechnol. 5 (2006) 829–832.

[15] E.A. Abu, S.A. Ado, D.B. James, Raw starch degrading amylase production by mixed culture of *Aspergillus niger* and *Saccharomyces cerevisae* grown on *Sorghumpomace*, Afr. J. Biotechnol. 4 (8) (2005) 785–790.

[16] I. Akpan, M.O. Bankole, A.M. Adesemowo, G.O. Latunde Dada, Production of amylase by *A. niger* in a cheap solid medium using rice band and agricultural materials, Trop. Sci. 39 (1999) 77–79.

[17] A. Pandey, P. Nigam, V.T. Soccol, D. Singh, R. Mohan, Advances in microbial amylases, Biotechnol. Appl. Biochem. 31 (2000) 135–152.

[18] V. Rekha, M.D. Saifuddin, S.R. Ahammad, P.K. Dhal, Isolation of amylase producing fungi from paddy-field soil, Ecol. Environ. Conserv. 19 (4) (2013) 1041–1044.

[19] R. Kaur, G.K. Mavi, S. Raghav, Pesticides classification and its impact on environment, Int. J. Curr. Microbiol. App. Sci. 8 (3) (2019) 1889–1897.

[20] M. Sangeetha, K. Kanimozhi, A. Panneerselvam, R. Senthil Kumar, Biodegradation of pesticide using fungi isolated from paddy fields of Thanjavur District, India, Int. J. Curr. Microbiol. Appl. Sci. 5 (10) (2016) 348–354. https://doi.org/10.20546/ijcmas.2016.510.039.

[21] J.H. Warcup, The soil plate method for isolation of fungi from soil, Nature 166 (1950) 117–118.

[22] J.C. Gillman, A Manual of Soil Fungi, Revised second ed., Oxford and IBH Publishing Company, Calcutta, Bombay, New Delhi, 1957, p. 436 (Indian Reprint).

[23] M.B. Ellis, Dematiaceous Hyphomycetes, Commonwealth Mycological Institute, Kew, Surrey, England, 1971.

[24] M.B. Ellis, More Dematiaceous Hypomycetes, Commonwealth Mycological Institute, Kew, Surrey, England, 1976.

[25] C.V. Subramanian, Hypomycetes, ICAR Publications, New Delhi, 1971.

[26] J. Liang, T. Cheng, Y. Huang, J. Liu, Petroleum degradation by *Pseudomonas* sp. ZS1 is impeded in the presence of antagonist *Alcaligenes* sp. CT10, AMB Express 8 (1) (2018) 88.

[27] M.M. Raffia, P.B.B.N. Charyulu, *Azospirillum*—biofertilizer for sustainable cereal crop production: current status, in: Recent Developments in Applied Microbiology and Biochemistry, 2021, pp. 193–209, https://doi.org/10.1016/B978-0-12-821406-0.00018-7.

[28] M.L. Ortiz Hernandez, E. Sanchez Salinas, A. Olvera Velona, J.L. Folch Mallol, Pesticides in the environment: impacts and its biodegradation as a strategy for residues treatment, in: M. Stoytcheva (Ed.), Pesticides—Formulations, Effects, Fate, InTech, 2011, https://doi.org/10.5772/13534.

[29] W. Elliot, L. Zhengfan, J. Aicha, S. Alessandro, A data driven approach to reducing household food waste, Sustain. Prod. Consum. 29 (2021) 600–613, https://doi.org/10.1016/j.spc.2021.11.004.

[30] C.O. Jeon, E.L. Madsen, In situ microbial metabolism of aromatic hydrocarbon environmental pollutants, Curr. Opin. Biotechnol. 24 (3) (2012) 474–481.

[31] C. Bass, L.M. Field, Gene amplification and insecticide resistance, Pest Manag. Sci. 67 (8) (2011) 886–890.

[32] P. Riya, T. Jagatpati, Biodegradation and bioremediation of pesticides in soil: its objectives, classification of pesticides, factors and recent developments, World J. Sci. Technol. 2 (7) (2012) 36–41.

[33] C. Scott, G. Pandey, C.J. Hartley, C.J. Jackson, M.J. Cheesman, M.C. Taylor, R. Pandey, J.L. Khurana, M. Teese, C.W. Coppin, K.M. Weir, R.K. Jain, R. Lal, R.J. Russell, J.G. Oakeshott, The enzymatic basis for pesticide bioremediation, Indian J. Microbiol. 48 (2008) 65–79.

[34] A. Trigo, A. Valencia, Cases I systemic approaches to biodegradation, FEMS Microbiol. Rev. 33 (2009) 98–108.

[35] H. Zhang, C. Yang, C. Li, L. Li, Q. Zhao, C. Qiao, Functional assembly of a microbial consortium with autofluorescent and mineralizing activity for the biodegradation of organophosphates, J. Agric. Food Chem. 56 (17) (2008) 7897–7902.

[36] P.N. Tallur, V.B. Megadi, H.Z. Ninnekar, Biodegradation of cypermethrin by *Micrococcus* sp. strain CPN 1, Biodegradation 19 (1) (2008) 77–82.

[37] V. Nagpal, M.C. Srinivasan, K.M. Paknikar, Biodegradation of hexachlorocyclohexane (Lindane) by a non-white rot fungus conidiobolus 03-1-56 isolated from litter, Indian J. Microbiol. 48 (1) (2008) 134–141.

[38] E.M. Chirnside, W.F. Ritter, M. Radosevich, Bioremediation strategies for pesticide contaminated soil: IV. Biodegradation using a selected microbial consortium following pretreatment with fungal enzymes, in: Proceedings of the 9th International In Situ and On-Site Bioremediation Symposium, Baltimore, MD, USA, 2007, pp. 1233–1240.

[39] W.S. Lan, J.D. Gu, J.L. Zhang, Coexpression of two detoxifying pesticide degrading enzymes in a genetically engineered bacterium, Int. Biodeterior. Biodegrad. 58 (2) (2006) 70–76.

[40] Z. Liu, Q. Hong, J.H. Xu, W. Jun, S.P. Li, Construction of a genetically engineered microorganism for degrading organophosphate and carbamate pesticides, Int. Biodeterior. Biodegrad. 58 (2) (2006) 65–69.

[41] M.K. Javaid, M. Ashiq, M. Tahir, Potential of biological agents in decontamination of agricultural soil, Scientifica (2016), https://doi.org/10.1155/2016/1598325. 1598325. *Hindawi Publishing Corporation.*

[42] B. Ramakrishnan, M. Megharaj, K. Venkateswarlu, N. Sethunathan, R. Naidu, Mixtures of environmental pollutants: effects on microorganisms and their activities in soils, Rev. Environ. Contam. Toxicol. 211 (2011) 63–120.

[43] L.L. Van Eerd, R.E. Hoagland, R.M. Zablotowicz, J.C. Hall, Pesticide metabolism in plants and microorganisms, Weed Sci. 51 (4) (2003) 472–495.

Chapter 6

Agricultural management by improving beneficial microflora

Shalini Tailor[a], Khushboo Jain[a], Avinash Marwal[a], Mukesh Meena[b], and Anita Mishra[c]

[a]*Department of Biotechnology, Mohanlal Sukhadia University, Udaipur, Rajasthan, India,* [b]*Laboratory of Phytopathology and Microbial Biotechnology, Department of Botany, Mohanlal Sukhadia University, Udaipur, Rajasthan, India,* [c]*Department of Science (Biotechnology), Biyani Girls College, University of Rajasthan, Jaipur, Rajasthan, India*

1 Introduction

The living beings from the primitive ones to the most developed all require food for their survival and growth. Food being the major component for living is cultivated and harvested throughout the world using various agricultural techniques. For agriculture practice, one requires land which is composed of soil. Hence, it could be said that without the soil, there is no existence of food. Soil has the greatest affinity and dwelling place for a wide range of microflora which in turn is found to be beneficial or harmful, in some cases, for the growth of the plants. The presence or the absence of these microorganisms in the soil totally depends on the soil quality, texture, region where it is present, and the various abiotic factors of that ecosystem. The soil of a specific region may grow certain sort of plant types better as compared to the other soil type since the soil varies from region to region [1]. Soil possesses several vital characteristics and functions such as nutrient recycling, growth of plants and crops, the dwelling place for living creatures, especially microbes, water-holding capacity, filtration, degradation of organic as well as inorganic matter, and many others [1]. Soil serves as the raw material for any kind of agricultural activity in order to produce food or fiber. A soil particle gets surrounded by a large population of microbes in an ecosystem. One gram of soil approximately contains 10 billion bacteria and builds the highest number as well as the biomass of the soil microflora [2].

In India, 60% of the land is under cultivation means that a large population depends majorly on the agricultural produce for their daily requirement of food and feed. Not only India but also most of the countries around the world are

Plant-Microbe Interaction—Recent Advances in Molecular and Biochemical Approaches
https://doi.org/10.1016/B978-0-323-91876-3.00004-X

dependent on agriculture and farms for the fulfillment of their food requisites. In order to feed such a large population globally, in the areas restricted under cultivation/agricultural practice, one has to increase their production [3]. In the past 60 years, the use of chemical-based fertilizers, pesticides, and insecticides has increased drastically for high-yield production. This uncontrolled use/consumption of the chemical fertilizers leads to the loss of soil fertility that affects its nutritive value, which in turn disturbs the soil microflora and ultimately deteriorates the soil quality. Once the soil quality gets affected, it badly hampers the microbes present in the soil which are crucial for enhancing the nutrient value, nitrogen and phosphorous fixation, biodegradation of complex molecules, balancing the ecosystem, etc., and makes it much difficult to bring back all those vital features to the soil for getting ample amount of crop production through agriculture.

This review mainly focuses on the issues of soil quality and how the affected quality hinders the microflora present in the soil. It also focuses on the restoration of the beneficial microbes along with the regulation and improvement in the conventional agricultural techniques that could help in the enrichment of the soil, its nutrients, fertility as well as microbiota.

2   Role of microflora in agriculture/plant growth

Microflora is the microorganisms that are found associated with the plant and its various parts. Majorly, these microbes reside in the root region forming the rhizosphere; those present in the inner tissues of the plant forms the endosphere, and those residing on the surface tissues of the plants forms the phylloplane [4]. These microbes show interaction with the plants as well as their surroundings, which may be beneficial or harmful at the same time [5]. Many microbes show symbiotic association with the plant parts and hence, they both get benefited from each other [6]. Various others serve as parasites and maybe pathogenic in nature [7]. The interaction of these microbes with the biotic (plants) and abiotic (climate, soil pH, temperature, stress, soil nutrients, water concentration, etc.) factors either enhance or decline crop production [8]. The biotic and abiotic factors are highly responsible for shaping the phenotype and the genotype of these microbes found in the soil of a specific region in an ecosystem [9]. The microflora found in the soil of a specific region is limiting and adapted to that particular environment. These microbes not only help in the growth of the plant but also help in fighting against the pathogenic strains trying to deteriorate the yield of the crop plant. The microflora present in the soil in the vicinity of plants may be either bacteria (mostly), fungus, or viruses [10,11]. Depending on the type of microbes present in the soil, different tasks are being performed by them which are depicted in Table 1 [12].

The major roles of the microflora found in the soil are as follows.

TABLE 1 Functions performed by soil microflora.

	Microorganisms		
S. no.	Bacteria	Virus	Fungi
1	Nitrogen fixation	Biopesticides	Nutrient recycling
2	Nutrient recycling		Biofertilizers
3	Biofertilizers		Biopesticides
4	Biopesticides		Phosphate mobilization
5	Phosphate mobilization		Phytoremediation
6	Phytoremediation		Decomposition

2.1   Nitrogen fixation

Nitrogen (N) is the element present in the highest concentration in the environment. Around 78% part of the air is composed only of nitrogen. In agriculture, the most crucial element required by the plants for their growth and development is nitrogen. Without nitrogen, soil fertility, as well as the growth efficiency of the plants, is highly affected. Though nitrogen is present in a huge amount in the environment, it is not readily available in the elemental form to the plants or the soil. Hence, the process of conversion of the elemental nitrogen (N_2) into the different forms that could be easily and efficiently absorbed by the soil and transferred to the plants ultimately, is known as mineralization [13]. The conversion of ammonium into nitrate is known as nitrification, which is carried out by the specific types of microbes. These microbes perform the conversion act either in the free-living state or as the symbionts. The first ever studied and the most important microbe that performs nitrogen fixation is the rhizobium. Rhizobium is found associated in the root region of the plants, especially of the family Legumenacae. These soil microbes form nodules in the root region of the plants and fix the atmospheric N_2 into ammonia (NH_3). This ammonia is then converted into various other forms that are taken up by the plants for further processing and convert finally into organic molecules. These molecules later form the vital part of the plant cells such as DNA, amino acids, other plant proteins, etc. [14]. Other free-living bacteria also have the nitrogen-fixation capacity though in a lesser amount (Fig. 1), yet they promote plant growth by contributing the element to the soil and plants. *Herbaspirillum, Clostridium, Burkholderia, Azospirillum, Azotobacter, Gluconacetobacter, Paenibacillus,* and *Methanosarcina* are the important soil microbes found associated with the rhizosphere [15].

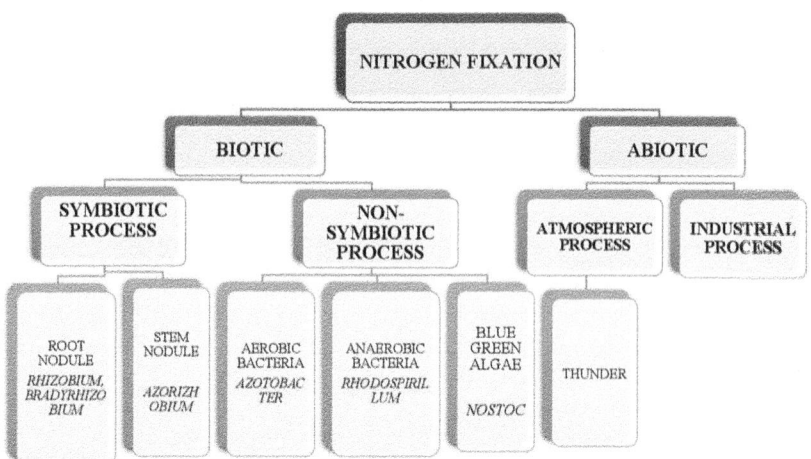

FIG. 1 Type of nitrogen fixation.

2.2 Nutrient recycling

Soil serves as the storehouse for the nutrients. These nutrients are categorized as macronutrients, micronutrients, and trace elements, depending on the amount they are utilized and needed by the plants for their growth and development. These nutrients exist in the environment in the organic form such as nucleic acids, proteins, fatty acids, carbohydrates, minerals, etc. But, plants could not absorb them in the organic form. So, these nutrients are broken down into simpler monomeric units so that they become readily available and absorbed easily by the plants. A specific set of microbes, especially saprophytes and fungi, performs these crucial phenomena of degrading the complex organic matter into the consumable simpler inorganic matter through various hydrolytic enzymes secreted by these microbes into the soil, where they are found associated and aggregated[16]. These microbes have generally are generally termed as the decomposers, which form an important class in the food web of any ecosystem. Without the decomposers, the nutrient pool will get deficient and all the dead and decayed organic matter will remain like this only. The nutrients present in an ecosystem exist in various forms like inorganic form when in soil or soil solution, organic when in soil and plant organic matter. Most elements that are absorbed by the soil and delivered to the plants are present in their cationic forms [13]. The major elements present in the soil organic matters are nitrogen (N), phosphorus (P), and potassium (K) along with sulfur (S). It has been reported that in the topsoil region, approximately 1000 pounds of nitrogen, 230 pounds of phosphorus, and 165 pounds of sulfur are found per acre of one percentage point of the organic matter [13]. Though the major component of any type of organic matter is carbon instead of nitrogen, microbes show a higher affinity toward nitrogen instead of carbon. Hence, carbon is recycled and fixed in a lesser amount as compared to nitrogen.

2.3   Biofertilizers

Due to the unstoppable increase in the population, the need for food has also increased abruptly, though globally but the developing countries are much affected. The developing countries fulfill the food requirement through agriculture, which poses a great responsibility to produce such a large amount of food to feed such a huge population which led to the tremendous use of chemical fertilizers to increase the yield of the crops. Although the yield increased with the time, the adverse effects of the chemical-based fertilizers became more severe and toxic, not only to the soil or plants but also for the living beings and to the environment. To overcome this growing issue of chemical fertilizers, scientists introduced a novel and environment-friendly approach by utilizing the living agents (microbes and their secretions) as fertilizers and named them as the biofertilizers. Biofertilizers served as an efficient component for sustainable agriculture having positive and beneficial effects on soil fertility along with good production [17,18]. Biofertilizers are those natural elements that are made from microbes; either of the same strain or having different strains; enhancing the plant growth by making nutrients available to a greater amount [19].

Biofertilizers composed of microbial formulations are capable of promoting plant growth by easily solubilizing the nutrients and supplying them to the plants[20]. These may consist of plant growth-promoting bacteria (PGPB) and plant growth-promoting rhizobacteria (PGPR) of various important classes [21]. These PGPB promote the growth of plant, whereas the PGPR evoke the root and shoot development. The PGPR are found in the rhizosphere of the plant. They not only enhance the plant growth but also help in controlling the diseases that could badly affect its growth as well as the crop yield. There are several modes through which the PGPR promote shoot and root growth. It can be mainly because of their secretions such as some phytohormones, plant secondary metabolites, or root exudates. They possess the potential to fight against the pathogenic microbes responsible for causing disease and control further invasion. PGPR and PGPB provide resistance to the growing plants from the abiotic factors such as stress [22], drought or salinity, etc. [23]; by interacting with plants physically, chemically, and biologically [24–26].

2.4   Biopesticides

Just like chemical-based fertilizers, the use of chemical pesticides was also at a huge rate over the years, approximately from the period of the green revolution [20]. These pesticides were used to control the pests that were harming the crops in the large agriculture fields and destroying the tons of the crop yield. Although the pesticides controlled the issue with time, their harmful effects begin to arise which disturbed the biotic (bioaccumulation, loss of biodiversity, invasion of secondary pests, etc.) [27]as well as the abiotic components (hampered soil quality, resistant species development, etc.), especially the nontargeted ones. This led to the demand for the development of a natural substitute for these harmful, toxic chemical pesticides. Biopesticides are the formulations of natural materials such as microbes, plants, animals, and other related compounds, which help in suppressing the growth of pests by different mechanisms known [28–31]. The biopesticides can be categorized according to their preparations or the materials from which they are made; namely, microbial biopesticides [32], biochemical biopesticides, and plant-incorporated protectants (PIPs). To prepare the natural pesticides, several bacteria, viruses, and fungi are deployed along with some specific biochemical compounds. The main motive of biopesticides is to control the pest's growth and attack on the agriculture fields, but they are not responsible for imparting any benefit in vital processes such as photosynthesis, development, growth, or other biological or physiological processes.

The microbes being utilized in the formulation of biopesticides belong to different genera, to name a few are; *Bacillus thuringiensis* (Bt) makes around 90% of the microbial biopesticides in the market; however, *Beauveria bassiana*, *Baculovirus*, *Steinernema*, *Nosema*, and *Chlorella* have shown significant roles in controlling the pest population [33–36]. The different modes of action of these biopesticides having an inhibitory impact are endotoxins, metabolic toxins, some growth regulators, and other inhibitors [37,38].

2.5 Phosphate mobilization

Phosphorus is another element that exists in nature in a bulk amount. It is present in the rocks and sediments in the inorganic state and also present in the organic state in the biotic components and soil as well. It is a major nutrient for the growth and development of plants hence provided in the form of fertilizers to the plants. But the plants cannot absorb and utilize P in this state since it is immobilized/fixed in its native state. Hence, it is needed to be solubilized and mineralized [15] for which various microbes such as PGPR and mycorrhizal fungi play a vital role in converting the inorganic phosphorus into the utilizable form like phosphates. These two processes occur by different modes, i.e., solubilization requires acid degradation; on the other hand, mineralization occurs through the action of enzyme phosphatases. The insoluble P is converted into the soluble form through the action of acids released by the bacterial and fungal groups like citric acid and gluconic acid. In mineralization, these microbial components secrete several enzymes out of which phosphatases degrade the phosphoric esters into the consumable form of phosphorus.

2.6   Phytoremediation

The continuous and abundant use of chemical-based fertilizers, insecticides, and pesticides for enhancing the growth of plants and the soil quality as well over the years had posed a threat to the soil microbiome along with the plant growth and the environment by releasing toxins and generation of contaminants and recalcitrants. These contaminants hindered the absorption of the beneficial components/nutrients from the soil because of their increased quantity that disrupted the whole of the microflora in the rhizosphere as well as phyllosphere in the agricultural fields.

To overcome this problem of increased concentration of the hazardous elements or heavy metals from the soil, plants, and whole of the ecosystem; a remediation process came into existence by using natural biological agents like microbes and plants [39]. The process in which plants perform the act of remediation or cleansing of the soil and improving the plant health and growth was termed as phytoremediation. There are selected plant species having the genetic potency to remove, degrade, metabolize, or immobilize a wide range of contaminants found in an ecosystem in association with the microbes that colonize the rhizosphere (Fig. 2).

Phytoremediation is a natural, nontoxic, environmental-friendly process but it takes a lot of time in cleaning the specific contaminant. Phytoremediation includes various steps, which are briefly explained. In phytoextraction, the contaminants are taken up by the plant roots and are transferred to the other parts of the plants from the roots through absorption, precipitation, or concentration of these contaminants. The next step is known as phytostabilization in which the contaminants are immobilized in the soil either by the process of precipitation within the roots of the plant or by accumulating and absorbing them through the roots. This immobilization helps reduce the amount as well as mobility of the contaminants into the environment. One of the crucial steps in phytoremediation is the breakdown of the contaminants through various physiological and metabolic processes by the plants that have absorbed them, which is commonly known as phytotransformation or phytodegradation. The process where the contaminants are degraded within the rhizosphere by the microbes colonizing in the roots and converting them into nontoxic compounds is known as phytostimulation. When the absorbed contaminants from the soil by the plants are transformed into volatile products and liberated into the environment through transpiration is known as phytovolatilization. This process is best suited for heavy metal and hydrocarbon contaminants. Last but not the least is the phenomenon known as rhizofilteration, which involves the prominent role of plant roots in adsorbing, precipitating, or absorbing the contaminants present in the soil or groundwater in the vicinity of the plant roots. Rhizofilteration generally performs the cleaning of groundwater, wastewater, or surface water [40].

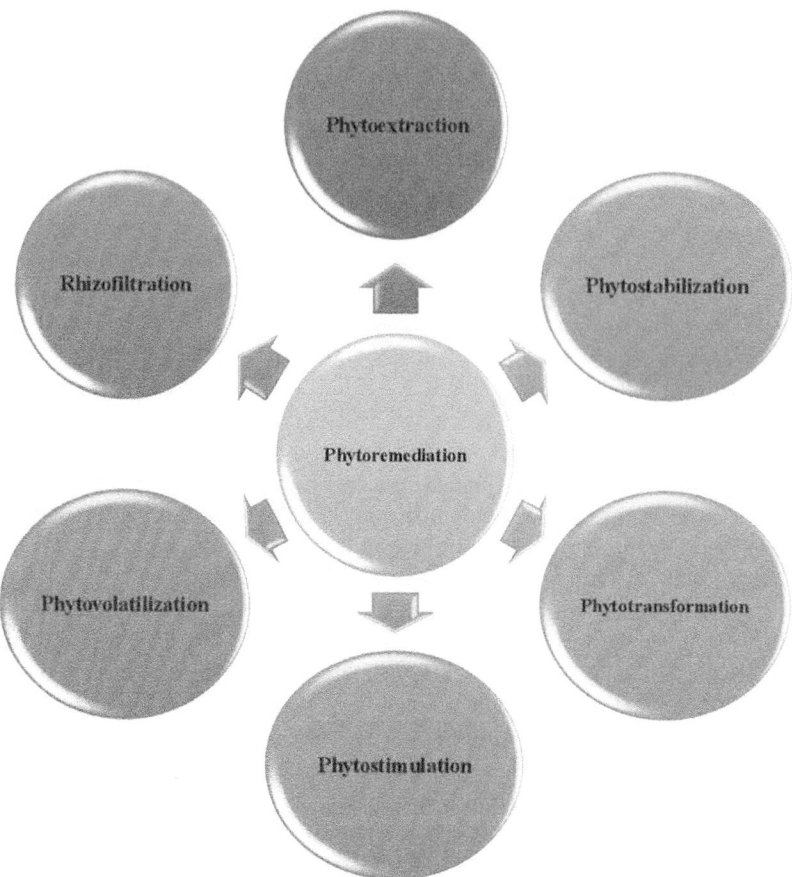

FIG. 2 Mechanism of phytoremediation.

3 Types of microflora

Soil is the dwelling place for a variety of microbes which in turn interact with the soil particles by forming soil-microbe aggregates which help in making the nutrient available to the plants in the consumable form. These beneficial microflora provide vital characteristics for the enhanced growth and development of the plants without affecting soil fertility and nutrient deficiency. A certain specific class of microbes is dedicated to continuously enriching the soil by making the atmospheric nutrients readily available by degrading them into the most consumable form such as cations, chelated ions, oxidized sulfur, and many various other micronutrients without which plant growth is hindered. Not only do they promote the growth of the plants, but they also release numerous plant secondary metabolites, phytohormones, antibiotics providing defense against phytopathogens along with tolerance toward abiotic stress like drought, salinity, etc. [41]. Table 2

TABLE 2 Group of beneficial microbes with their functions.

S. no.	Group	Example	Function	Benefit	References
1	Free living	*Azotobacter, Clostridium, Nostoc, Klebsiella* sp., *Anabaena*	Nitrogen fixation	Enhances plant growth and soil fertility	[42]
2	Symbiotic	*Rhizobium, Anabaena azollae*	Nitrogen fixation	Enhances plant growth, soil fertility, protection from biotic-abiotic stress, phytohormone secretion and carbon utilization	[43]
3	Associative symbiotic	*Azospirillum*	Nitrogen fixation	Plant growth, stress tolerance through phytohormone (IAA) signaling and defense response	[44]
4	Fungi	*Trichoderma* sp., *Penicillium* sp., *Aspergillus* sp.	Nitrogen fixation	Biocontrol agent, biofungicides	[15]
5	Arbuscular mycorrhiza	*Glomus* sp., *Sclerocystis* sp., *Rhizoctonaia solani*	Phosphate mobilization	Abiotic stress tolerance, promotes plant growth and performance, nutrient availability	[45,46]
6	Silicate and zinc solubilizers	*Bacillus* sp.	Micronutrient solubilizer	Secretes phytohormones and metabolites against pathogens, prevents from abiotic stress	[47]
7	PGPR	*Pseudomonas fluorescens*	Disease control and resistance to abiotic stress	Secretes phytohormones and plant defense metabolites, stress resistance ability	
8	Gram-negative bacteria with filamentous development	*Streptomyces*	Phosphate solubilization	Potential biofertilizer, secrete secondary metabolites possessing antibiotics	[48]

shows the important species of microbes beneficial for plant growth and development, along with other crucial characteristics.

4 Modes of microflora enrichment

4.1 Phytohormone modulation

Several microbes that are colonizing the rhizosphere and phyllosphere regions of a plant tend to produce and secrete various plant hormones such as ethylene, gibberellic acid, auxin, cytokinin, and abscisic acid promoting plant growth and development [49]. The phytohormones secreted by the microflora have varying effects when used in combination or alone on the growth and health of the plants. According to various research, it has been reported that the most potential phytohormone secreted is auxin; mainly indole acetic acid (IAA). IAA is secreted by most classes of the microflora associated with plants. Several symbiotic bacteria responsible for nitrogen fixation have been reported to secrete IAA and gibberellic acid, which is found to enhance plant growth and yield, e.g., *Gluconacetobacter diazotrophicus* and *Bacillus amyloliquefaciens*. Also, it has been reported that species such as *Burkholderia caryophylli*, *Pseudomonas* spp., and *Achromobacter piechaudii* were found to be responsible for regulating the ethylene level in plants by producing aminocyclopropane 1-carboxylic acid (ACC)-deaminase enzyme.

4.2 Organic farming and bioavailability of nutrients

Due to the deleterious and highly toxic effects of chemical-based fertilizers, pesticides, fungicides, and insecticides, the soil efficacy and health is deteriorating day by day. It also affects the plants' growth because these toxic elements get accumulated into the soil and delivered to the plants as well. This leads to the bioaccumulation and biomagnification of these toxic chemicals into the soil microflora as well as into the food chain of that ecosystem. All these hazardous effects proposed the new agriculture technique called organic farming. In the past few years, organic farming has captivated the agriculture practice, which replaced chemical-based fertilizers, pesticides, and other conventional techniques that pose hazardous impacts on both the biotic as well as abiotic components. Organic farming involves the use of various strategies to replenish soil nutrients along with better and healthier crop production [50]. Organic farming is a nontoxic, biodegradable, and environment-friendly approach that has overpowered the use of chemical components (Fig. 3).

Using the various techniques of organic farming helps in nutrient enrichment and easy availability to the plant, which eventually enhances crop production. Practicing crop rotation with legumes and the main crop is beneficial in maintaining adequate levels of nitrogen and phosphorus [13]. Usage of

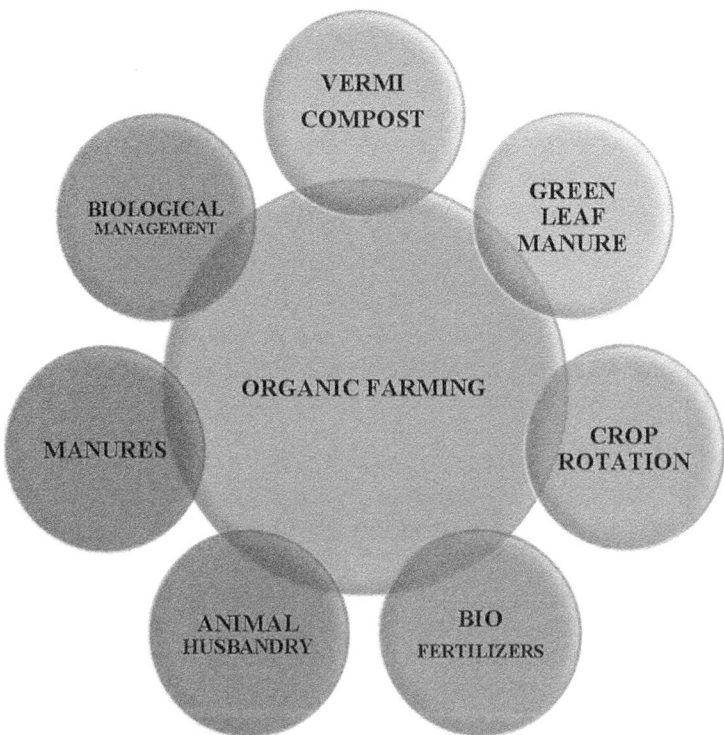

FIG. 3 Steps in organic farming.

excessive fertilizers also leads to the toxicity of chemicals in the soil, which is passed on to the plants ultimately. Hence, the green manure along with compost shows a positive result in balancing the important nutrients and degrading the harmful elements from the soil.

4.3   Improvement in soil structure

Soil is the most important part of agriculture without which no one can grow such a bulk quantity of crops and plants. So, it becomes necessary to understand and maintain the soil structure of a specific region. Soil structure deals with the different size ranges of the soil particles along with the pores existing between them. Since soil does not remain in the form of individual particles but as the soil aggregates. In the formation of the soil aggregates, the role of soil microflora plays a vital role. Soil microflora is the major colonizing creatures of the soil. While performing their several physiological and metabolic processes, these microbes secrete organic compounds like polysaccharides, organic acids along with their body structures like hyphae, which are highly responsible for generating the soil aggregates or clumps. Vermicompost involves the use of earthworms and various other soil-dwelling microbes' to help in forming and regulating the soil structure, which gets disturbed by the continuous agricultural practices. Hence, it is recommended to maintain a healthy population of soil microflora to have a good soil structure and soil organic matter that will help in enriching the soil nutrients by capturing them in the soil aggregates and will transmit directly to the plants for their enhanced growth [15].

4.4 Siderophore production

Another important function performed by the soil microflora in the production of low molecular weight organic compounds in the environment deficient in iron. The scarcity of iron elements in the soil leads to the secretion of siderophores. The most important function of siderophore is to chelate the iron in ferric iron form ($Fe3+$) present in the environment (soil or water) and make it available to the rhizospheric microbes and plants for their growth and function. The siderophores secreted by microbes vary from species to species. Siderophores are not only capable of iron chelation but also prevent phytopathogens from causing the disease to the crops. They prevent the disease-causing microbes by limiting the iron accessibility for the pathogens [51]. Also, they show higher affinity when used as the biocontrol agents in the farm fields and bioremediation process. These vital functions of siderophores make them a suitable component for maintaining the microflora and increasing plant growth.

4.5 Soil pH

The first and the foremost requisite for the microflora to establish and flourish in the soil is to check and regulate the soil pH. Soil pH plays a crucial role, not only for the growth of the crop plants but also in maintaining a healthier population of the rhizoflora and other microbes in the soil. Extensive usage of chemical fertilizers, soil erosion, bioaccumulation, and biomagnification leads to increased toxicity in the soil, which in turn disturbs the soil pH by increasing the acidity which badly affects the plant growth and the microflora population. Soil pH regulates the soil biology and helps in nutrient cycling also. The best optimum pH for the soil is between pH 6 and 7. This is the best-suited soil hydrogen ion concentration for most of the microbes as well as the plants. Though there are exceptions like alkaliphiles and acidophiles that require high alkaline and acidic pH, respectively, for their proper functioning. Micronutrients like aluminum and manganese concentration become highly toxic at low pH levels, whereas few other soil micronutrients become easily available and readily absorbable at an acidic pH like iron and zinc. Few microbes are best grown at a higher alkaline pH especially the rhizobium as they require pH 7 and above for the nitrogen fixation and make molybdenum available to the rhizobium for the same. A fungus shows better activity in acidic pH, whereas most bacteria and earthworms are best active at neutral pH. The cation exchange capacity is also affected by the soil pH and it increases with the increase in the soil pH [13].

5 Regulation of agricultural practice for sustainable soil microbial growth

The relation of soil and the microflora associated with it is the soul for the soil structure, soil health, soil fertility, crop health, and its production. But due to the old and deprived agriculture practices, the efficiency of the soil, as well as the amount of microflora residing in the soil, has been affected drastically. Once the beneficial microbiome gets disrupted, it becomes difficult to replenish and colonize them again. In the absence of these microbes whether in the rhizo-sphere, phyllosphere, or the endophytic zone, the actual performance of the soil to form soil aggregates, maintain the structure of the soil with good porosity, aeration, nutrients availability, and cycling and ultimately transferring them to the crops is hampered. One of the major reasons for this problem is the wrong agriculture techniques, which are required to be improvised and must specifi-cally be regulated at a fixed time interval so that the microflora can be incor-porated into the soil in the appropriate quantity and the nutrients can be enriched that will lead to the sustainable agriculture production. Not only the wrong farming techniques but also the limited resources act as a barrier in the path of sustainable microflora as well as sustainable crop production.

To overcome this problem, many molecular techniques have been employed for getting an ample amount of crop production without affecting the soil micro-flora. It may include changing the genotype of the microbe or pathogen, incor-poration of bioformulation containing either the individual microbe or the consortium, the presence of the phytopathogens in the soil that affects the activ-ity of the microbes present in the soil through the interactions between them [52], and the unfavorable abiotic components such as drought, salinity, pH, tem-perature, etc., which helps in developing more active strains with good defense mechanisms [53].

For these changes, the techniques like gene sequencing, microscopy, HPLC, RT-PCR, and many others have been widely used that helped in reactivating the beneficial soil microbiota along with greater production within the limited resources[54]. For example, in a finding done by Rossmann et al. [55] in the wheat crop (*Triticum aestivum*) using the sequencing method, it was found that the rhizosphere of the wheat plant was colonized majorly by the Proteobacteria, Actinobacteria, Acidobacteria, Ascomycota, Chytridiomycota, and Basidiomy-cota. Once the type of microflora is known to the scientists, it becomes easier to understand the basic requirements (nutrients, pH, and water-holding capacity) of those microbes for sustainable growth and agriculture production. The agri-culture production can be increased without affecting the microflora by

TABLE 3 Plant microbe and its effect on the plant.

S. no.	Plant	Beneficial microbe	Effects	References
1	Wheat	*Nostoc ellipsosporum* and *Nostoc punctiforme*	Physical architecture, nutritional intake, and microorganism behavior are all improved	[56]
2	Oat	*Klebsiella* sp.	Inoculated plants grew faster, had more water content, and had heavier roots	[42]
3	Stevia	*Streptomyces* species	Salt tolerance improvement	[57]
4	Wheat	*Bacillus siamensis*	Reduces cadmium accumulation and improves growth and antioxidant defense	[58]
5	Ryegrass	*Pseudomonas aeruginosa*, *Burkholderia gladioli*	Improved phytostabilization of Cu and Cd	[59]

modifying the regular agriculture practices with the following steps summarized below [13]. Table 3 represents the effect of beneficial microbes on the specific plants demonstrated by the respective researchers.

5.1 Reduced tillage and soil compaction

Tillage is a year-old agriculture technique that is found to have many harmful effects on the soil and microbes. Excessive tillage causes increased oxygen concentration in the soil, which in turn enhances the activity of the soil microbes. The increased biological activity of these microbes leads to the increased degradation of the soil organic matter thereby decreasing its amount. Due to the lack of organic matter found to exist in the soil aggregates, the soil health and nutrient availability begins to decline which in turn affects the beneficial soil microflora population.

Inversion tillage, on the other hand, leads to increased exposure of the soil because of reduced soil-crop bonding; causing heavy soil erosion. Due to tillage practice in the agriculture field, it badly affects the hyphal network created by the mycorrhizal fungi leading to its destruction and declining the fungal population. So, a specific amount of beneficial soil microflora along with the organic matter in the form of manure, compost, etc. has to be added to counteract the adverse effects of tillage.

Excessive usage of farm equipment and foot traffic in the agriculture fields is highly responsible for soil compaction, especially when the soil is moist and wet. It leads to the clogging of soil pores and makes it hard and less aerated. Clogged pores in the soil lead to the improper supply of gases, nutrients, and water as well. Hence, to prevent the soil moisture, amount of microflora, their dwelling place, and the soil organic matter, tillage must be avoided or reduced. But in organic farming, tillage helps in the removal of the weeds as no weedicides or herbicides are used.

5.2   Increase organic matter inputs

For sustainable agriculture production, the role of soil organic matter is of vital importance. The organic matter is present in the soil aggregates surrounded and generated by the soil microflora, which helps in fulfilling their nutrient requirements. Hence, it is important to decrease the loss of organic matter through microbial degradation, soil erosion, or runoff. For this, the soil must be supplied with natural/green crop residues, compost, manure, and crop rotation also.

Frequent analysis of the soil organic matter concentration could be a possible solution for keeping surveillance for its adequate amount. Agricultural labs support in this regard by checking and analyzing the soil samples at a regular time interval and providing data to add the required amount of organic matter.

5.3   Cover crop usage

In agroecosystem, the use of cover crops is practiced at a large scale to prevent and manage beneficial soil characteristics such as soil organic matter, soil erosion, soil quality, fertility, water-holding capacity, microbiota, etc. Cover crops are the fast-growing crops that are grown to cover the soil and may be grown with or without the cash crops. They also help in controlling pests and diseases as these cover crops are grown in rotation with the cash crops and other varieties.

Cover crops themselves provide the organic matter in the form of biomass produced by them. It also prevents soil compaction by increasing the soil pores if tap roots containing cover crops are grown, whereas the soil aggregation and nutrient cycling are achieved with the fibrous root cover crops. Legumes, as well as nonlegumes, could be used in the form of cover crops depending upon the need of the soil for nutrients, organic matter, quality, and health. It also prevents losses from occurring due to the leaching of the soil minerals. Buckwheat, rye, and radish are a few examples of cash crops.

5.4 Decreased use of pesticide and increased biodiversity of beneficial organisms

Excessive use of chemical pesticides and insecticides is one of the major problems that badly affect the beneficial and nontargeted insects and microbes present in the soil. For reducing its effectiveness and increasing biodiversity; a new biological approach could serve as a boon for the beneficial microflora, other organisms, and sustainable agriculture [60]. Farmscaping is defined as a biological whole-farm technique used for increasing the beneficial organisms that will control and promote pest management; an important strategy for sustainable agriculture. The pest management through Farmscaping involves the utilization of hedgerows, cover crops, insectary plants, and water bodies also for captivating and assisting beneficial organisms such as arthropods, insects, reptiles, bats, and many others. Farmscaping also plays an important role in preventing soil erosion and water runoff that will provide stability to the soil and add organic matter to it, making the soil healthier and microflora rich.

5.5 Crop rotation

Crop rotation is a technique that has been used extensively, especially in the developing countries, to date. The major motive behind rotating the crops is to prevent the population buildup of pests, weeds, and pathogens over time. This happens because the pets/insects dwelling in a specific crop type do not give the opportunity to grow and flourish themselves for a longer duration, as one crop type will be changed with the other, in a certain period. This aids in destroying the pest population and disease-causing life cycle as well as managing the health of the crops. Excess of a specific nutrient into the soil can also be minimized through crop rotation.

5.6 Adequate supply of nutrients

It is well known that nutrients are vital not only for the microflora or crops but are also important for soil health and fertility. Nutrients, either in excess or deficient amount, both badly affect the microbiota and their activity. It also hampers the soil organic matter, development of toxic compounds, growth of pests, disturbed soil pH, and many other harmful effects. Thus, it becomes a major requirement to add, maintain, and recycle the adequate amount of the nutrients into the soil at a specific time interval. The application and types of nutrients can also be a limiting factor. If the nutrients are in the volatile state then they will be lost into the atmosphere in the form of gases, if they are highly soluble, they will run off or get eroded with water and will pollute the water reservoirs and other ecosystems. Hence, it becomes very important to understand the soil type, the types of microflora residing in it, the type of crops to be grown, and the type of nutrients that will be needed by the crops to grow properly. Majorly, nitrogen and phosphorus acquire the top place for maintaining a good amount of vital nutrients and microflora activity. Organic farming techniques such as crop rotation, use of cover crops, use of compost, manure, and crop residues help in maintaining the proper nutrients and aid in sustainable agriculture production at a large scale.

6 Limitations

For achieving the sustainable development goals, one has to understand the things which are lacking for optimizing the soil quality, fertility, microflora, and eventually the agriculture production. One of the major reasons for this is the ignorance toward the soil microbiota that could serve as a promising tool for sustainable agriculture, within the limited resources [53]. Few of such limitations are listed here that affect the biotic and abiotic components, which in turn disturb the crop production sustainably.

Lack of scientific knowledge: Lack of scientific knowledge regarding the biotic (soil microflora, organic matter, crop type) and abiotic (soil type, nutrients, climatic conditions, soil pH, temperature, etc.) elements to achieve maximum crop yield using the soil microflora.

Excessive use of chemical fertilizer and pesticide to increase crop production: Though it increased the production in the long run, the soil and the beneficial microbes are badly affected. It leads to soil toxicity, increased resistance species of pathogens, pests, and insects, polluted soil, water, and food as well.

Lesser use of biofertilizers and biopesticides in the farm field: Although organic farming is being practiced globally but at a very little pace. This leads to very little use of organic compost, bioformulations that will gradually help in replenishing the soil and crop health. The soil type, as well as the environmental factors, responds differently for the biofertilizers which lead to lower production [61].

Old agriculture techniques: Crop production without proper types of equipment, wrong agriculture practices, and ignorance toward the beneficial soil microbiota are a few of the major reasons for lower production.

7 Conclusions

The world's population is growing day by day. To fulfill the food requirement of such a vast population, globally, one has to find a sustainable alternative [62], because in the coming future the natural resources will be on the verge of declination and will not be adequate for such a large population. Instead of relying on chemical-based formulations, the use of bioformulations should be maximized at a large scale. This will help in replenishing the soil nutrients along with the microflora enrichment. Without the presence of microflora, soil quality and health cannot be managed. Field trials with modified microbes generated by advanced molecular technology [63] must be opted to manage the microbial population along with more crop production.

Acknowledgment

The authors are thankful to the University Grant Commission (UGC) under Start-Up-Grant-Scheme, New Delhi, India for the financial assistance (No. F.30-510/2020 (BSR) FD Diary No. 8839).

References

[1] Soil Quality, United States Department of Agriculture, Natural Resource Conservation, 1995. https://www.nrcs.usda.gov/wps/portal/nrcs/detail/national/technical/nra/rca/?cid=nrcs143_014198.

[2] J.J. Hoorman, Role of soil bacteria, Agric. Nat. Resour. (2016). https://ohioline.osu.edu/factsheet/anr-36.

[3] S. Tailor, M. Meena, A. Marwal, Application of nanotechnology in management of various plant diseases, in: R.K. Singh, Gopala (Eds.), Innovative Approaches in Diagnosis and Management of Crop Diseases, Volume 3: Nanomolecules and Biocontrol Agents, Apple Academic Press, Taylor & Francis Group, 2021, https://doi.org/10.1201/9781003187844-1 (Chapter 1).

[4] T.R. Turner, E.K. James, P.S. Poole, The plant microbiome, Genome Biol. 14 (2013) 209, https://doi.org/10.1186/gb-2013-14-6-209.

[5] A. Marwal, R.K. Gaur, Disease-causing seed pathogenic microorganisms and their management practices, in: A.K. Tiwari (Ed.), Advances in Seed Production and Management, 2021, pp. 185–200, https://doi.org/10.1007/978-981-15-4198-8_9 (Chapter 9).

[6] N.M. Sudheep, A. Marwal, N. Lakra, K. Anwar, S. Mahmood, Fascinating fungal endophytes role and possible beneficial applications: an overview, in: D.P. Singh, H.B. Singh, R. Prabha (Eds.), Plant-Microbe Interactions in Agro-Ecological Perspectives. Volume 1: Fundamental Mechanisms, Methods and Functions, 2017, pp. 255–273, https://doi.org/10.1007/978-981-10-5813-4_13 (Chapter 13).

[7] R. Prajapat, A. Marwal, P.N. Jha, Erwinia carotovora associated with potato: a critical appraisal with respect to Indian perspective, Int. J. Curr. Microbiol. App. Sci. 2 (10) (2013) 83–89.

[8] A. Marwal, A.K. Srivastava, R.K. Gaur, Improved plant tolerance to biotic stress for agronomic management, Agrica 9 (2020) 84–100, https://doi.org/10.5958/2394-448X.2020.00013.9.

[9] A. Marwal, R. Kumar, R.K. Verma, M. Mishra, R.K. Gaur, S.M.P. Khurana, Genomics and molecular mechanisms of plant's response to abiotic and biotic stresses, in: S.M.P. Khurana, R.K. Gaur (Eds.), Plant Biotechnology: Progress in Genomic Era, 2019, pp. 131–146, https://doi.org/10.1007/978-981-13-8499-8_6 (Chapter 6).

[10] A. Marwal, A.K. Sahu, R.K. Gaur, First report of airborne begomovirus infection in Melia azedarach (Pride of India), an ornamental tree in India, Aerobiologia 30 (2) (2014) 211–215.

[11] A. Marwal, R.K. Verma, M. Mishra, R. Kumar, R.K. Gaur, Mastreviruses in the African world: harbouring both monocot and dicot species, in: R.V. Kumar (Ed.), Geminiviruses: Impact,

Challenges and Approaches, Springer, Cham, 2019, pp. 85–102, https://doi.org/10.1007/978-3-030-18248-9_5 (Chapter 5).

[12] A. Marwal, A.K. Srivastava, R.K. Gaur, Plant viruses as biopesticides, in: H. Singh, A. Vaishnav (Eds.), New and Future Developments in Microbial Biotechnology and Bioengineering. Sustainable Agriculture: Advances in Microbe-Based Biostimulants, Elsevier, 2022. ISBN:9780323855785.

[13] C. White, M. Barbercheck, Managing Soil Health: Concepts and Practices, Penn State Extension, 2017. https://extension.psu.edu/managing-soil-health-concepts-and-practices.

[14] V.C. Pankievicz, F.P. do Amaral, K.F. Santos, B. Agtuca, Y. Xu, M.J. Schueller, G. Stacey, Robust biological nitrogen fixation in a model grass-bacterial association, Plant J. 81 (2015) 907–919.

[15] G. Kumar, M.M. Rashid, S. Nanda, Beneficial microorganisms for stable and sustainable agriculture, Biopestic. Int. 17 (2021) 17–27. DocID: https://connectjournals.com/02196.2021.17.17.

[16] M.D. Toor, M. Adnan, Role of soil microbes in agriculture; a review, Open Access J. Biog. Sci. Res. 4 (2) (2020) 2020, https://doi.org/10.46718/JBGSR.2020.04.000091.

[17] A. Bargaz, K. Lyamlouli, M. Chtouki, Y. Zeroual, D. Dhiba, Soil microbial resources for improving fertilizers efficiency in an integrated plant nutrient management system, Front. Microbiol. 9 (2018) 1606, https://doi.org/10.3389/fmicb.2018.01606.

[18] M. Singh, D. Singh, A. Gupta, K.D. Pandey, P.K. Singh, A. Kumar, Plant growth promoting rhizobacteria, in: A.K. Singh, A. Kumar, P.K. Singh (Eds.), PGPR Amelioration in Sustainable Agriculture, Elsevier, Cambridge, MA, 2019, pp. 41–66, https://doi.org/10.1016/B978-0-12-815879-1.00003-3.

[19] U. Riaz, S.M. Mehdi, S. Iqbal, H.I. Khalid, A.A. Qadir, W. Anum, Bio-fertilizers: eco-friendly approach for plant and soil environment, in: K.R. Hakeem, R.A. Bhat, H. Qadri (Eds.), Bioremediation and Biotechnology: Sustainable Approaches to Pollution Degradation, Springer, Cham, 2020, pp. 188–214, https://doi.org/10.1007/978-3-030-35691-0_8.

[20] S. Mohod, G.P. Lakhawat, S.K. Deshmukh, R.P. Ugwekar, Production of liquid biofertilizers and its quality control, Int. J. Emerg. Trends Eng. Basic Sci. 2 (2) (2012) 158–165.

[21] M. Meena, P. Swapnil, K. Divyanshu, S. Kumar, Y.N. Tripathi, A. Zehra, A. Marwal, R.S. Upadhyay, PGPR-mediated induction of systemic resistance and physiochemical alterations in plants against the pathogens: current perspectives, J. Basic Microbiol. 60 (10) (2020) 828–861.

[22] P. Swapnil, M. Meena, S.K. Singh, U.P. Dhuldhaj, A. Marwal, Vital roles of carotenoids in plants and humans to deteriorate stress with its structure, biosynthesis, metabolic engineering and functional aspects, Curr. Plant Biol. 26 (2021) 100203.

[23] A. Marwal, A. Sahu, R.K. Gaur, New insights in the functional genomics of plants responding to abiotic stress, in: R.K. Gaur, P. Sharma (Eds.), Molecular Approaches in Plant Abiotic Stress, CRC Press, Science Publishers, Taylor & Francis Group, 2014, pp. 158–180, https://doi.org/10.1201/b15538. ISBN:13:978-1-4665-8894-3 (Chapter 10).

[24] C. Dimkpa, T. Weinand, F. Asch, Plant-rhizobacteria interactions alleviate abiotic stress conditions, Plant Cell Environ. 32 (2009) 1682–1694, https://doi.org/10.1111/j.13653040.2009.02028.x.

[25] J.W. Kloepper, C.-M. Ryu, S. Zhang, Induced systemic resistance and production of plant growth by Bacillus spp, Phytopathology 94 (2004) 1259–1266.

[26] V. Gupta, Beneficial microorganisms for sustainable agriculture, Microbiol. Aust. 3 (33) (2012) 113–115, https://doi.org/10.1071/MA12113.

[27] E.O. Fenibo, G.N. Ijoma, T. Matambo, Biopesticides in sustainable agriculture: a critical sustainable development driver governed by green chemistry principles, Front. Sustain. Food Syst. 5 (2021) 619058, https://doi.org/10.3389/fsufs.2021.619058.

[28] P.G. Marrone, Pesticidal natural products—status and future potential, Pest Manag. Sci. 75 (2019) 2325–2340, https://doi.org/10.1002/ps.5433.

[29] A. Marwal, R.K. Gaur, Plant viruses as an engineered nanovehicle (PVENVs), in: R.K. Gaur, S.M.P. Khurana, P. Sharma, T. Hohn (Eds.), Plant Virus-Host Interaction: Molecular Approaches and Viral Evolution, second ed., Academic Press, Elsevier, 2021, pp. 525–536. (Chapter 22). Hardcover ISBN: 9780128216293, eBook ISBN: 9780128244838.

[30] M. Nuruzzaman, Y. Liu, M.M. Rahman, R. Dharmarajan, L. Duan, A.F.M.J. Uddin, et al., Nanobiopesticides: composition and preparation methods, in: O. Koul (Ed.), Nano-Biopesticides Today and Future Perspectives, Academic Press, London, 2019, pp. 69–131.

[31] M.A.C. Wattimena, F.S. Latumahina, Effectiveness of botanical biopesticides with different concentrations of termite mortality, J. Belantara. 4 (2021) 66–74, https://doi.org/10.29303/jbl. v4i1.630.

[32] A. Marwal, R.K. Gaur, Nanophytovirology: an emerging field for disease management, in: S. Topolovec-Pintaric (Ed.), Plant Diseases—Current Threats and Management Trends, IntechOpen, 2019, ISBN: 978-1-78985-115-1, https://doi.org/10.5772/intechopen.86653. ISBN:978-1-78985-116-8.

[33] M.H. Abu-Dieyeh, A.K. Watson, Efficacy of *Sclerotinia* minor for dandelion control: effect of dandelion accession, age and grass competition, Weed Res. 47 (2007) 63–72, https://doi.org/ 10.1111/j.1365-3180.2007.00542.x.

[34] A.L. Gonçalves, The use of microalgae and cyanobacteria in the improvement of agricultural practices: a review on their biofertilising, biostimulating and biopesticide roles, Appl. Sci. 11 (2021) 871, https://doi.org/10.3390/app11020871.

[35] E.M. Radwan, M.A. El-Malla, M.A. Fouda, R.A.S. Mesbah, Appraisal of positive pesticides influence on pink bollworm larvae, *Pectinophora gossypiella* (Saunders), Egypt. Acad. J. Biol. Sci. F. Toxicol. Pest Control 10 (2018) 37–47, https://doi.org/10.21608/eajbsf.2018.17018.

[36] D.C. Steinkraus, N.P. Tugwell, *Beauveria bassiana* (Deuteromycotina: Moniliales) effects on *Lygus lineolaris* (Hemiptera: Miridae), J. Entomol. Sci. 32 (1997) 79–90, https://doi.org/ 10.18474/0749-8004-32.1.79.

[37] S.A. Dar, S.H. Wani, S.H. Mir, A. Showkat, T. Dolkar, T. Dawa, Biopesticides: mode of action, efficacy and scope in pest management, J. Adv. Res. Biochem. Pharmacol. 4 (2021) 1–8.

[38] T.C. Sparks, R. Nauen, IRAC: mode of action classification and insecticide resistance management, Pest. Biochem. Physiol. 121 (2015) 122–128, https://doi.org/10.1016/j. pestbp.2014.11.014.

[39] S. Sharma, M. Meena, P. Swapnil, A. Marwal, A.K. Gupta, Removal of volatile organic compounds and heavy metals through the biological-based process, in: M. Shah, S. Rodriguez-Couto, J. Biswas (Eds.), Innovative Role of Biofiltration in Wastewater Treatment Plants (WWTPs), Elsevier, 2022, pp. 45–64, https://doi.org/10.1016/B978-0-12-823946-9.00003-6. ISBN:9780128239469 (Chapter 3).

[40] R. Bhagea, B. Rouksaar, C. Christabelle, G. Keshavi, N. Huda,Phytoremediation ppt. slideshare, 2014.

[41] A. Marwal, S.S. Verma, S. Trivedi, R. Prajapat, Reactive oxygen species: its effects on various diseases, J. Adv. Biotechnol. 3 (1) (2014) 122–134, https://doi.org/10.24297/jbt.v3i1.1684.

[42] S. Sapre, I. Gontia-Mishra, S. Tiwari, *Klebsiella sp.* confers enhanced tolerance to salinity and plant growth promotion on oat seedlings (*Avena sativa*), Microbiol. Res. 206 (2018) 25–32.

[43] Y. Mabrouk, I. Hemissi, I.B. Salem, S. Mejri, M. Saidi, O. Belhadj, Potential of rhizobia in improving nitrogen fixation and yields of legumes, in: E.C. Rigobelo (Ed.), Symbiosis, IntechOpen, 2018, https://doi.org/10.5772/intechopen.73495.

[44] J. Fukami, P. Cerezini, M. Hungria, *Azospirillum*: benefits that go far beyond biological nitrogen fixation, AMB Express 8 (2018) 73, https://doi.org/10.1186/s13568-018-0608-1.

[45] N. Begum, C. Qin, M.A. Ahanger, S. Raza, M.I. Khan, M. Ashraf, N. Ahmed, L. Zhang, Role of arbuscular mycorrhizal fungi in plant growth regulation: implications in abiotic stress tolerance, Front. Plant Sci. 10 (2019) 1068, https://doi.org/10.3389/fpls.2019.01068.

[46] N.H. Duc, A.T. Vo, I. Haddidi, H. Daood, K. Posta, Arbuscular mycorrhizal fungi improve tolerance of the medicinal plant *Eclipta prostrata* (L.) and induce major changes in polyphenol profiles under salt stresses, Front. Plant Sci. 10 (2021), https://doi.org/10.3389/fpls.2020.612299.

[47] E.A. Kazerooni, et al., Biocontrol potential of *Bacillus amyloliquefaciens* against *Botrytis pelargonii* and *Alternaria alternata* on *Capsicum annuum*, J. Fungi (Basel) 7 (2021), https://doi.org/10.3390/jof7060472.

[48] A. Sakineh, S. Ayme, S. Akram, S. Naser, Streptomyces strains modulate dynamics of soil bacterial communities and their efficacy in disease suppression caused by *Phytophthora capsici*, Sci. Rep. 11 (2021) 1–14.

[49] L. Boiero, D. Perrig, O. Masciarelli, C. Penna, F. Cassán, V. Luna, Phytohormone production by three strains of *Bradyrhizobium japonicum* and possible physiological and technological implications, Appl. Microbiol. Biotechnol. 74 (2007) 874–880.

[50] S. Tailor, K. Jain, A. Marwal, M. Meena, K. Anbarasu, R.K. Gaur, Outlooks of nanotech in organic farming management, Def. Life Sci. J. 7 (1) (2022) 52–60, https://doi.org/10.14429/dlsj.7.16763.

[51] X. Shen, H. Hu, H. Peng, W. Wang, X. Zhang, Comparative genomic analysis of four representative plant growth-promoting rhizobacteria in *Pseudomonas*, BMC Genomics 14 (2013), https://doi.org/10.1186/1471-2164-14-271.

[52] A. Marwal, R.K. Gaur, Host plant strategies to combat against viruses effector proteins, Curr. Genomics 21 (6) (2020) 401–410.

[53] L.F. Santos, F.L. Olivares, Plant microbiome structure and benefits for sustainable agriculture, Curr. Plant Biol. 26 (2021) 2214–6628, https://doi.org/10.1016/j.cpb.2021.100198.

[54] R. Singh, Microbial biotechnology: a promising implement for sustainable agriculture, in: New and Future Developments in Microbial Biotechnology and Bioengineering, Elsevier, 2018.

[55] M. Rossmann, J.E. Pérez-Jaramillo, V.N. Kavamura, J.B. Chiaramonte, K. Dumack, A.M. Fiore-Donno, L.W. Mendes, M.M.C. Ferreira, M. Bonkowski, J.M. Raaijmakers, T.H. Mauchline, R. Mendes, Multitrophic interactions in the rhizosphere microbiome of wheat: from bacteria and fungi to protists, FEMS Microbiol. Ecol. 96 (2020), https://doi.org/10.1093/femsec/fiaa032.

[56] R. Nisha, B. Kiran, A. Kaushik, C.P. Kaushik, Bioremediation of salt affected soils using cyanobacteria in terms of physical structure, nutrient status and microbial activity, Int. J. Environ. Sci. Technol. 15 (2018) 571–580.

[57] S.T. Tolba, M. Ibrahim, E.A. Amer, D.A. Ahmed, First insights into salt tolerance improvement of Stevia by plant growth-promoting Streptomyces species, Arch. Microbiol. 201 (2019) 1295–1306.

[58] S.A. Awan, N. Ilyas, I. Khan, M.A. Raza, A.U. Rehman, M. Rizwan, A. Rastogi, R. Tariq, M. Brestic, *Bacillus siamensis* reduces cadmium accumulation and improves growth and antioxidant defense system in two wheat (*Triticum aestivum* L.) varieties, Plan. Theory 9 (2020) 878, https://doi.org/10.3390/plants9070878.

[59] T. Ke, G. Guo, J. Liu, C. Zhang, Y. Tao, P. Wang, Y. Xu, L. Chen, Improvement of the Cu and Cd phytostabilization efficiency of perennial ryegrass through the inoculation of three metal-resistant PGPR strains, Environ. Pollut. 271 (2021), https://doi.org/10.1016/j. envpol.2020. 116314.

[60] S. Mahmood, N. Lakra, A. Marwal, N.M. Sudheep, K. Anwar, Crop genetic engineering: an approach to improve fungal resistance in plant system, in: D.P. Singh, H.B. Singh, R. Prabha (Eds.), Plant-Microbe Interactions in Agro-Ecological Perspectives. Volume 2: Microbial Interactions and Agro-Ecological Impacts, Springer Nature, Singapore, 2017, pp. 581–591, https://doi.org/10.1007/978-981-10-6593-4_23 (Chapter 23).

[61] E.K. Mitter, M. Tosi, D. Obregón, K.E. Dunfield, J.J. Germida, Rethinking crop nutrition in times of modern microbiology: innovative biofertilizer technologies, Front. Sustain. Food Syst. 5 (2021) 606815, https://doi.org/10.3389/fsufs.2021.606815.

[62] B.K. Singh, P. Trivedi, Microbiome and the future for food and nutrient security, Microb. Biotechnol. 10 (2017) 50–53, https://doi.org/10.1111/1751-7915.12592.

[63] A. Marwal, R.K. Gaur, Molecular markers: tools for genetic analysis, in: A.S. Verma, A. Singh (Eds.), Animal Biotechnology 2nd Edition Models in Discovery and Translation. Section II: Animal Biotechnology: Tools and Techniques, Academic Press, Elsevier, 2020, pp. 353–372, https://doi.org/10.1016/B978-0-12-811710-1.00016-1 (Chapter 18).

Chapter 7

Ameliorative characteristics of plant growth-enhancing microbes to revamp plant growth in an intricate environment

T. Savitha[a] and A. Sankaranarayanan[b]
[a]*Department of Microbiology, Tiruppur Kumaran College for Women, Tiruppur, Tamil Nadu, India,*
[b]*Department of Life Sciences, Sri Sathya Sai University for Human Excellence, Kalaburagi, Karnataka, India*

1 Introduction

Due to an increasing population, the world's natural resources are being rapidly depleted. To combat the growing problem of global food insecurity, farming efficiency needs to be improved. Conservative agricultural methods involving the haphazard utilization of substances like fertilizers and pesticides for enhancing productivity are a threat to agroecosystems. At the same time, anthropogenic activities and worldwide changes in climate are deleterious, causing irreparable damage to the ecosystem. In this scenario, plant growth-promoting microbes (PGPMs) are attracting the attention of environmentalists. For example, rhizosphere bacteria and fungi can be used as biofertilizers and biopesticides to effectively manage a range of agricultural problems. The function of PGPMs in the remediation of ecosystems is accomplished by the elimination or alleviation of recalcitrant abiotic stresses and climate changes. They are promising tools for maintaining and preserving environmental sustainability. Advanced knowledge and explorative studies on plant-microbe interactions pave a path for food safety in terms of quality, quantity, and ecological sustainability.

Plant-Microbe Interaction—Recent Advances in Molecular and Biochemical Approaches
https://doi.org/10.1016/B978-0-323-91876-3.00006-3

2 Plant growth enhancing microbes (PGEMs)

The most part of the zone of the rhizosphere is a problem area for microbial actions contributed basically beside native microscopic organisms and parasites [1]. There are two types of PGEMs: plant growth-promoting rhizobacteria (PGPR) and plant growth-promoting fungi (PGPF). Microbial activities like solubilization of inorganic mixtures, deprivation and mineralization of organic complexes, and discharge of organically dynamic materials like phytohormones, chelators, and antiinfection agents work with a ton in plant growth augmentation [2]. A rhizobacterial strain is viewed as a putative PGPR if it has explicit plant development-advancing qualities and can upgrade plant development upon inoculation. An optimal PGPR strain should exhibit several characteristics, as identified in Fig. 1 [3].

2.1 Plant growth-promoting rhizobacteria

PGPR are soil microbes that can induce plant growth either in the root of the plants or leaves and sometimes in plant tissues. The genera of PGPR reported to be beneficial for plant growth include *Pseudomonas, Bacillus, Rhizobium* (*Mesorhizobium, Sinorhizobium, Bradyrhizobium,* and *Azorhizobium*), *Klebsiella, Arthrobacter, Azospirillum, Enterobacter, Alcaligenes, Azotobacter,* and

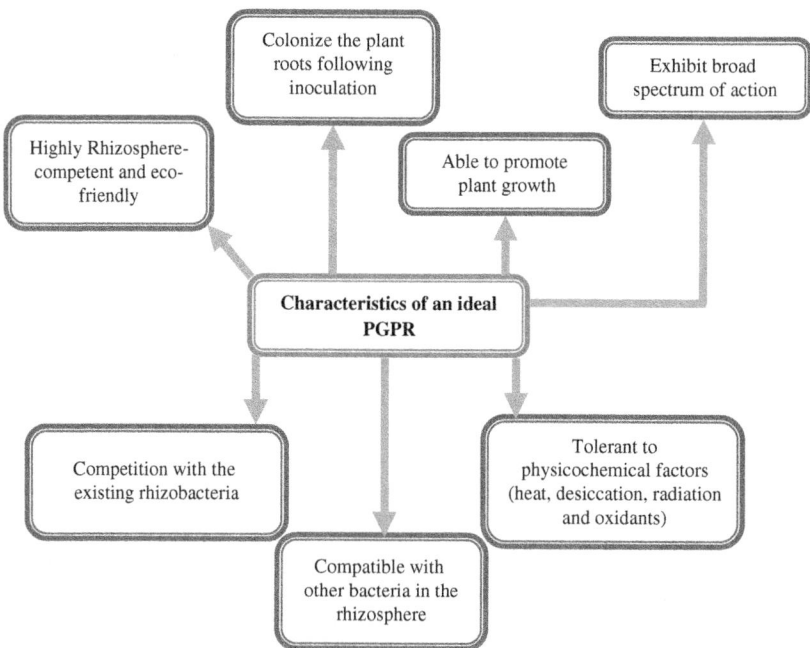

FIG. 1 Important characteristics of an ideal PGPR.

Serratia [4]. These microorganisms play a crucial role in atmospheric nitrogen fixation [5], inorganic phosphate solubilization, iron confiscation [6], and phytohormone amalgamation [7,8].

PGPR exhibits its role in plant development indirectly by preventing the detrimental effect of phytopathogens [9,10]. Mechanisms of biocontrol mediated by PGPR include production of allochemicals, induction of plant resistance to pathogens and abiotic stresses, and competition for a substrate or ecological niche [11]. Actinomycetes strains such as *Streptomyces* sp., *Streptosporangium* sp., *Micromonosopra* sp., and *Thermobifida* sp. have good biocontrol potential against root pathogenic fungi [12]. They exert beneficial activities on plants, including producing phytohormones and antibiotics and debasing enzymes on fungal cell walls.

2.2 Plant growth-promoting fungi

Rhizosphere fungi including *Penicillium*, *Phoma*, *Trichoderma*, *Fusarium*, *Aspergillus*, and arbuscular mycorrhiza fungus (AMF) have attracted attention for their plant growth-supporting actions. Significant mechanisms by which PGPF help plant growth include manufacturing plant hormones, decomposing organic matter, solubilizing inaccessible nutrient elements in soil, and defending plants from biotic and abiotic stresses. Indirect growth promotion by PGPF is carried out through niche segregation, predation, antibiosis, mycoparasitism, and induced systemic resistance (ISR). In certain conditions, these mechanisms may act in combination to enhance plant growth.

3 Mechanisms of PGPR and plant growth improvement

As shown in Fig. 2, PGPR assume a significant role, both directly and indirectly, in the enhancement of plant growth. The direct mechanisms of plant development include growth hormone production, phosphorous solubilization, atmospheric nitrogen fixation, and iron sequestration. The indirect mechanisms include fighting for nutrients, protecting against phytopathogens, and reducing the harmful effects of biotic stresses via the fabrication of low molecular weight substances such as ketones, aldehydes, alcohols, sulfides, and ammonia, secondary metabolites with antagonistic properties, and cell wall-degrading enzymes [13–15].

3.1 Direct plant growth enhancement

3.1.1 Biological nitrogen fixation

Nitrogen (N), a fundamental element for crop development, is widely accessible in the environment but cannot be utilized by plants in its vaporous structure. Biological nitrogen fixation (BNF) is a method of converting atmospheric nitrogen into an easily accessible form (e.g., ammonia), which can be utilized by

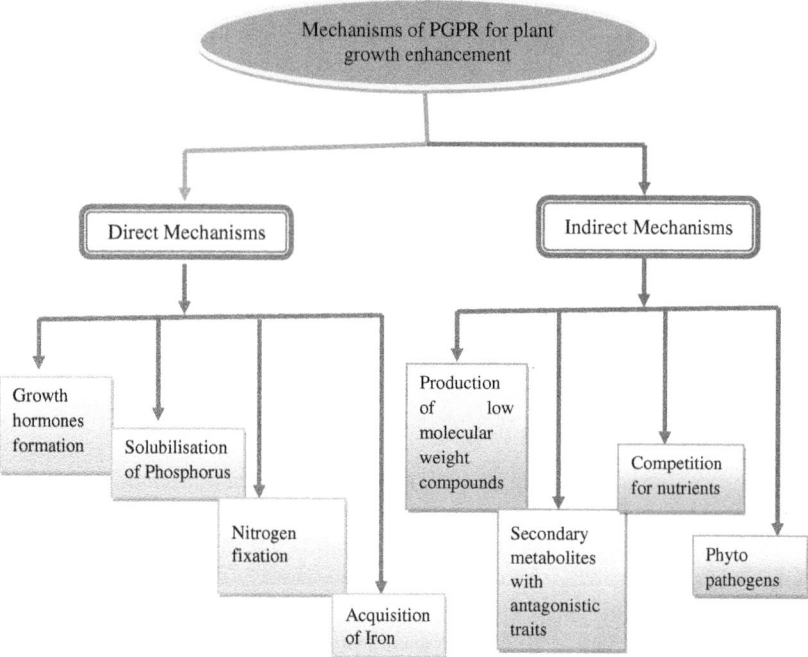

FIG. 2 Mechanisms of PGPR for plant growth enhancement.

plants through the response among rhizobia and leguminous plants such as pigeon pea, groundnut, and chickpea. A large segment of N anchored by legumes is collected in grains and deposited in roots and shoots. Crops like rice, wheat, and sugarcane can also can fix atmospheric N with free-living bacteria/diazotrophs, for example, *Azospirillum* and cyanobacteria. Rhizobia can be coated on seeds of legumes for better N fixation. They can stick within the soil intended for several years in the lack of their congregation [16]. Some actinomycetes are capable of being used in BNF, such as *Micromonospora*, *Agromyces*, and *Streptomyces* [17].

3.1.2 Phosphate solubilization

Phosphorous (P) is found in embedded form (with inorganic or organic molecules), however, plants can utilize H_2PO_4 and/or HPO_4^{2-} [18]. P is widely used as an artificial fertilizer, however, its unregulated and unmanaged use has destructive outcomes on the climate. The P-solubilizing microbes by various enzyme mediated pathways, it mineralized and consists of acid phosphatases, C-P lyase, D-alpha glycerolophosphate, phosphonoacetate hydrolase, phosphor hydrolases, and phytase which solubilizes the bound form so that they are accessible to plants [19]. Plant growth-promoting microorganisms like *Streptomyces*,

Pseudomonas, and *Bacillus* are well known for P solubilization. Actinomycetes with excessive P-solubilizing tendencies have been documented [20].

3.1.3 Phytohormone production

Plant-coupled microbes are notable for generating phytohormones (auxins) to facilitate plant growth. Phytohormones influence the morphological and physiological processes of plants. The phytohormones revolutionize the growth of prototype of plants and result in stretched and long roots among superior exterior regions, allowing the plant to contact water and nutrients from profound profundities of soil.

3.1.4 Iron sequestration

Naturally, irons subsist because of insoluble types of hydroxides and oxyhydroxides, which are inaccessible to plants. Siderophores (elevated attraction iron-chelating substances) supply iron to plants and enhance its bioavailability in the rhizosphere [21].

3.2 Indirect plant growth enhancement

Indirect plant growth mechanisms alludes to apply of PGP mediators for supervising the detrimental property of biotic stresses (like insect pests and pathogens) to perk up on the whole healthiness of the plant. Similarly, PGP microorganisms are known as biocontrol agents that combat pests and alleviate pathogenic effects.

3.2.1 Competition

Plant root exudates assume a momentous task indecisive the precise microbial group living in its locality. For example, it has been reported that flavonoids and phenolic compounds in the rhizosphere can affect the symbiosis between beneficial rhizobacteria and plants [22]. Research shows that amino acids, sugars, organic acids, and vitamins are important nutrients for microbes. Antagonism during the struggle for existing nutrients is a form of biocontrol by microbes to resist pathogens and inhibit disease in plants [22]. The assembly of hydrolases, antibiotics, volatile compounds, and siderophores are alternative mechanisms put forth by PGPMs [23].

3.2.2 Cell wall-degrading enzymes

The cell walls of plant insects and pathogenic fungi contain macromolecules like lipids, cellulose, chitin, glucans, and proteins. PGPMs are recognized for their ability to generate cell wall-mortifying enzymes. These enzymes interrupt and lyse the cell walls of insect pests and pathogens. PGPMs are extensively stated to produce these hydrolytic enzymes, for example, chitinase, peroxidase, protease, and glucanase [24].

3.2.3 Antibiosis

Antibiosis is a control system of PGPMs. The diffusible compounds formed by BCA and/or PGP bacteria have been recognized to hinder plant pathogens in the zone of the rhizosphere. A broad spectrum of antibiotics such as macrolides, nucleosides, polyenes, benzoquinones, and aminoglycosides are produced by PGPMs. Actinobacteria are the foremost manufacturer of antibiotics. For instance, the whole range of microbial bioactive molecules changed into approximately 33,500 and out of which 13,700 were produced by Actinobacteria [25].

4 Role of PGPR in alleviating plant abiotic stress

4.1 Management of abiotic stress factors by PGPR

Drought, salinity, extreme temperature changes, heavy metal contamination, flooding, and ultraviolet irradiations are abiotic stress components that have direct adverse effects on plant growth and ultimately lead to steep declines in yields. Bacterial association in escalating abiotic stress tolerance and attractive defense responses in plants exposed to diverse stress elements [26,27]. Heavy metals exert harmful impacts on biota and microbes by hindering vital functional groups and changing their dynamic conformations [28,29]. Table 1 lists various abiotic stresses and the PGPR that can manage them.

TABLE 1 Management of abiotic stress factors by PGPR.

Abiotic stress	Plant	PGPM	Effects	References
Drought	Glycine max	Klebsiella variicola	Improves plant growth by inducing adventitious root	[30]
Salinity	Oryza sativa	Bacillus amyloliquefaciens SN 13	Increases antioxidant activity	[31]
Salinity	Glycine max	Pseudomonas simiae AU	Upregulates vegetative storage proteins; increases chlorophyll content	[32]
Drought	Zea mays	Enterobacter sp. Achromobacter sp.	Increases growth and enhances drought tolerance	[33]

TABLE 1 Management of abiotic stress factors by PGPR—cont'd

Abiotic stress	Plant	PGPM	Effects	References
Cadmium	Oryza sativa	Klebsiella pneumoniae MCC 3091	Enhances cadmium tolerance; alleviates phytotoxicity	[34]
Copper, Chromium	Triticum aestivum	Bacillus cereus	Alleviates heavy metal tolerance; increases antioxidant activity	[35]

4.1.1 Salinity

Salinity is one of the dynamic factors that cripple plant growth and yield. Salinity is caused by ordinary actions that enhance the concentration of dissolved salts like sodium chloride in the soil. Salinity tolerance in plants is reliant on the plant's physiological system, period of coverage to saline conditions, the availability of salt around roots, confined soil-water associations, and microclimate conditions (temperature, humidity, etc.). Plants respond to salinity stress in diverse ways, but most plants are vulnerable to increased salt concentrations and experience stunted growth and development. Soil microbiome and particularly halo-tolerant PGPB acquire plant nutrients required for growth through a variety of mechanisms, including the dissolution of phosphate compounds and the production of siderophores and phytohormones, among others. Extraordinary consideration must be compensated to the intonation of plant ethylene intensity by utilizing the ACC as the solitary carbon source of nitrogen, owing to its whole bang on plant growth and development. Plant phytohormone signaling by PGPB is an excellent technique to improve yield. In addition, the mutual possession of diverse traits has a beneficial impact on plants during periods of salinization [36].

4.1.2 Drought

Drought is a significant abiotic stress adversely affecting agriculture worldwide. The application of PGPR during drought conditions is beneficial for crop production. The PGPR expeditiously colonize the root rhizosphere and ascertain both free-living and cozy relations with host plants. These associations lead to enhanced crop productivity and alleviation of biotic and abiotic stresses via a diversity of mechanisms [37,38]. PGPR may assume crucial roles as biocontrol agents, alleviators of abiotic stress, biofertilizers, and remediators of toxins in the soil. PGPR use various mechanisms to manage plant drought tolerance, including modifying the host root system structural design, scavenging reactive oxygen species by antioxidant production, managing oxidative stress via the biosynthesis and metabolism of phytohormones, osmoregulation, secreting extracellular polysaccharide that may assume as humectants, and transcriptional regulation of host stress reactive genes [38–40].

4.1.3 Heavy metal tolerance

Heavy metal pollution in soils is harmful to most microbes and can restrain the efficacy of inoculants. Heavy metals can diminish soil fertility, encumbrance in rhizosphere microbial inhabitants, plant photosynthetic effectiveness, reasoned nutrient inequality and diminution in yield [41]. Beneficial microbes such as *Bacillus subtilis*, *Alcaligenes faecalis*, and *Pseudomonas aeruginosa* can remediate contaminated soils, increasing plant tolerance to heavy metals [42].

4.2 Essential nutrients enhancement by PGPR

PGPR improves nitrogen bioavailability by expanding the root surface region and root morphology to effect an elevated nitrogen uptake. Other PGPR varieties influence nitrogen bioavailability by changing nitrogen forms to accessible ones or distressing the root nutrient transport system [43]. In a research study, PGP *Bacillus* sp. mixtures, unruffled of different *Bacilli* species, activated the appearance of genes influential nitrate and ammonia uptake and transfer, in addition, to increase host plant growth and development in *Arabidopsis thaliana*. Bacilli-induced *Arabidopsis thaliana* highlighted considerably higher transcript levels of nitrate transporters NRT1 (AtNRT1), NRT2 (AtNRT2), and ammonium transporters AMT1 (AtAMT1), which resulted in superior nutrient uptake and plant growth. Liu et al. [44] reported that in *Arabidopsis* when inoculating *Bacillus subtilis* strain GB03 having high efficiency. Jang et al. [45] suggested that enhanced growth of plants incited by coupled PGPR might be incompletely accomplished by enhanced convenience and acquisition of nitrogen. Enhanced nitrogen accessibility, auxins synthesis, and P-solubilization were renowned in peanut *Arachis* hypogeal inoculated with a conglomerate of diazotrophic root derivation bacteria isolated from the halophyte *Arthrocnemum indicum*. In one study, inoculation of *Pseudomonas* sp., *Agrobacterium* sp., *Klebsiella* sp., and *Ochrobacterium* sp. resulted in improved salt tolerance in peanut plants, which was associated with reduced stage of reactive oxygen species [46].

In salty soils, phosphate-solubilizing halotolerant bacteria establish plant growth and reduce the undesirable effect of salt [47]. The growth and phosphate-solubilizing capabilities of *Bacillus megaterium* were considerably increased due to its transformation to sodium chloride stress [48]. Around 30%–65% of the total P in soil is available in organic form, which is delivered from organophosphates by microbes because of mineralization processes [49]. Under stress conditions, production of siderophores is the foremost bacterial system of providing plants with accessible forms of iron [50]. It establishes a function in chlorophyll synthesis and the continuance of chloroplast structure and function; it has an effective role in DNA synthesis and respiration and proceeds as a prosthetic group component of many enzymes, as well as those implicated in redox reactions. Regardless of its vast abundance in the lithosphere, the plant accessibility of iron is very restricted due to its decreased solubility [51].

4.3 Rhizobacterial phytohormones for the alleviation of abiotic stress

Alleviation of abiotic stress in plants may be achieved by one of the mechanisms of bio phytohormone compounds having similar configuration to the plant, production by rhizobacterial strains. Major phytohormones for plant growth improvement are auxins, cytokines, gibberellins, abscisic acid, and ethylene [52].

4.3.1 Auxins

Auxins are potent molecules naturally formed by plants that play roles in cell division, differentiation, and extension as well as alleviation of the abiotic stress environment [53]. They can be secreted and excreted by nearly 80% of rhizosphere bacteria, including *Azosprillum* sp., *Azotobacter* sp., *Enterobacter* sp., and *Pseudomonas* sp. [54]. This emphasized that in the lead inoculation with the selected strain, the roots concealed elevated amounts of tryptophan and simultaneously bacteria synthesized indole acetic acid (IAA) in the rhizosphere promotes plant development [55]. At reduced levels, bacterial auxins promote elongation of major plant roots, whereas in increased concentrations, they promote the arrangement of lateral and adventitious roots [56].

4.3.2 Cytokines

Cytokines are important compounds in plant growth and development as well as the continuance of root and shoot meristem activity, lateral root and nodule arrangement, embryogenesis, vascular development, apical dominance in retort to environmental stimuli, and root elongation [57]. Cytokine-synthesizing microbes like *Azospirillum* sp., *Arthrobacter* sp., *Bacillus* sp., and *Pseudomonas* sp. have improved glycine max root and shoot biomass and the proline content in tissues in salt stress conditions [58]. *Bacillus aryabhattai* strain SRB02 produces cytokines and develops soybean growth in nitrosative, oxidative, and temperature grades [59]. The twinned role of bacterial cytokines consists of optimizing nutrient contribution and adapting to host immunity in plants contaminated with pathogens [60].

4.3.3 Gibberellins

Gibberellins are plant hormones that can combat abiotic and other physiological stresses [61]. These compounds regulate various developmental processes in plants, including regulation of seed dormancy, promotion of root growth and root hair abundance, germination, quiescence, ripening of fruits, and flowering [62]. Production of gibberellins has been established in various rhizosphere microbes such as *Herbaspirillum seroprdicae, Acetobacter diazotrophicus, Azospirillum* sp., and *Bacillus* sp. [63]. Various research studies reported that gibberellins produced by bacteria can promote plant growth and yield. Seeding of maize roots with various *Azospirillum* strains increased the levels of gibberellins in the roots and enhanced growth [64]. Moreover, gibberellins increase the thermo-tolerance of plants [61].

4.3.4 Abscisic acid

Abscisic acid (ABA) assumes a hormonal role as an inhibitor of plant growth and metabolic activities. It is a sesquiterpenoid compound that facilitates plant seed development and maturation, seed induction and dormancy of buds, synthesis of proteins and osmolytes, senescence processes, and regulation of the ability of plants to continue to exist under biotic and abiotic stress conditions [65]. Tsukanova et al. [66] reported that drought resistance is mediated by the presence of ABA in the rhizosphere region of the plant. ABA-producing strain *Bacillus aryabhattai* (SRB02), retrieved as of soybean rhizosphere, extensively enhances the host plant biomass and formation of nodules below in the drought stress environment [59].

4.3.5 Ethylene

Ethylene is a plant growth monitor that plays an important role in diverse phases of plant ontogenesis as well as plant growth, maturity, flowering, germination, and senescence. It enhances the configuration of adventitious roots, arouses germination of seeds, and breaks seed dormancy in addition to concerned in stress signaling pathways. Overproduction of ethylene can be induced by biotic and abiotic stresses like temperature gradients, flooding, pathogen interaction, salinity, drought, and metals [67].

5 Role of PGPB as biocontrol agents

Biocontrol of phytopathogens by PGPB is an emerging alternative to the use of pesticides and chemical fertilizers. It is cost effective, eco-friendly, and long lasting, making it an attractive method for inhibiting phytopathogens in agriculture [68]. Bacteria such as *Bacillus, Enterobacter, Pseudomonas*, and *Streptomyces* are being explored as biocontrol agents because of their antagonist activity.

***Bacillus* sp.** These are gram-positive, rod-shaped, endospore-producing bacteria belonging to the family *Bacillaceae* (class Bacilli, phylum Bacillota). They have proven to be effective PGPB in the agriculture sector [69] due to their three major contributions in rice crops: (1) increased yield, (2) abiotic stress tolerance, and (3) reduction during the disease incidence. The colonization of *Bacillus* sp. on crop roots increases crop yields [70]. Abiotic stresses like drought and salinity are chief threats to rice growth and yield. Several research studies show that PGPB has versatile roles in crosstalk amid stresses and phytohormones within rice, especially in osmolyte biosynthesis followed by an osmotic adjustment. Salt-tolerant PGPB (*Bacillus tequilensis* strain UPMRB9 and *Bacillus aryabhattai* strain UPMRE6) inoculated on rice plants exhibited valuable effects on transpiration, photosynthesis, and stomatal conductance [71]. In plant disease management, *Bacillus* reins in the explosion of phytopathogens by restraining plant immunity [72,73], which is called ISR, an important mechanism to protect the plant against phytopathogens. First, *Bacillus* triggers ISR by inducing agents such as antimicrobial metabolites produced by PGPB, then it establishes plant antioxidant enzymes such as polyphenol oxidase, peroxidase, chitinase, beta 1,3 glucanase, and phenylalanine ammonia-lyase. This aids plants in reducing the level of reactive oxygen species, which is a basis of oxidative stress throughout phytopathogenic infection [72,74].

***Pseudomonas* sp.** These are gram-negative, rod-shaped, polar flagellated bacteria belonging to the phylum Pseudomonadota, class Gammaproteobacteria, and family Pseudomonadaceae. This species and its products are widely applied in large-scale biotechnological functions [75]. It plays various roles as a clinically significant and opportunistic nosocomial pathogen, phytopathogen, and biocontrol agent. Specific examples in these categories include the human pathogen *Pseudomonas aeruginosa* [76], the plant pathogen *Pseudomonas syringae* [77], and the nonpathogenic biocontrol agents *Pseudomonas fluorescens* and *Pseudomonas putida* [78,79]. As a phytopathogen, *Pseudomonas syringae* colonizes plant tissue by entering through plant leaves via the stomata, proliferates in the intercellular space, and ultimately establishes necrotic lesions that are habitually bounded by chlorotic halos [80].

***Enterobacter* sp.** These are gram-negative, rod-shaped, nonspore-forming bacteria belong to the phylum Pseudomonadota, class Gammaproteobacteria, and family Enterobacteriaceae. The potency of *Enterobacter* sp. toward the improvement of sustainable farming systems observed in PGPB functions in three diverse pathways: (1) synthesizing meticulous compounds for the plants, (2) assisting the uptake of nutrients from the soil, and (3) preventing plant disease [81]. In a 2020 study, two bacterial strains (BSB1 and BCB11) secluded starting the field illustrated antagonists actions enroute for *Burkholderia glumae* were recognized as belonging to the genus *Enterobacter* [82].

***Streptomyces* sp.** These are complex gram-positive filamentous bacteria belong to the phylum Actinomycetota, class Actinomycetia, and family Streptomycetaceae. They have specific capabilities of producing an assortment of bioactive amalgams, which is valuable in both medicine and agriculture. They have attracted global attention because of their powerful production of extracellular enzymes [83] and antibiotics [84]. The characterization of this species as PGPB is related to their antagonistic and plant growth-enhancing activities [85]. The specific pathways of plant growth-promoting activities like IAA, siderophores, HCN, cellulose, and chitinase have also been decoded in the genome of *Streptomyces* strains [86].

6 Applications of PGPM for sustainable agriculture

PGPMs assume a significant function in plant growth in an extensive assortment of systems. Their mechanisms of action consists of abiotic stress tolerance in plants, production of volatile organic compounds, nutrient fixation for easy uptake by the plant, the production of siderophores as plant growth regulators, and the production of protection enzyme-like glucanase, ACC deaminase, and chitinase for the hindrance of disease [87]. However, the mechanisms of action of diverse PGPMs vary depending on the type of host plants. PGPMs can furthermore increase plant adsorption of water and nutrients, enhancing root development and increasing plant enzymatic capability. They also exhibit synergistic causes to improve their effect on plants by promoting plant growth or hindering pathogens. They are an exceptional substitute for chemical fertilizers and pesticides [88].

7 Challenges and prospects

The usage of microbial-based products as bio-inoculants faces many challenges. Development of a new PGPR strain as an efficient bio-inoculant requires preliminary screening in the laboratory before it can be used in the field. Basic initial screening of axenic cultures of PGPR strains alone exhibits effective plant growth promotion under field conditions. The large-scale production and application of PGPR requires addressing many important issues and challenges, as shown in Fig. 3.

The agriculture industry plays a pivotal role in ensuring and maintaining food security for human survival. Chemical fertilizers are hazardous to the soil and environment, whereas biofertilizers are natural products that do not threaten the ecosystem. Naturally derived fertilizers have been demonstrated to be an essential component of sustainable agriculture to maintain long-term fertility of soil and sustain crop productivity.

The triangle of communications among bio-inoculant microorganisms, residential soil microbiota, and host plants is favored for plant growth and increased productivity of crops as well as maintenance of the integrity of global health and biogeochemical cycling.

Government sectors and federal agencies should endorse the use of biofertilizers as an eco-friendly substitute for crop enhancement and entrepreneurs should invest in or establish biofertilizer industries. In addition, an awareness

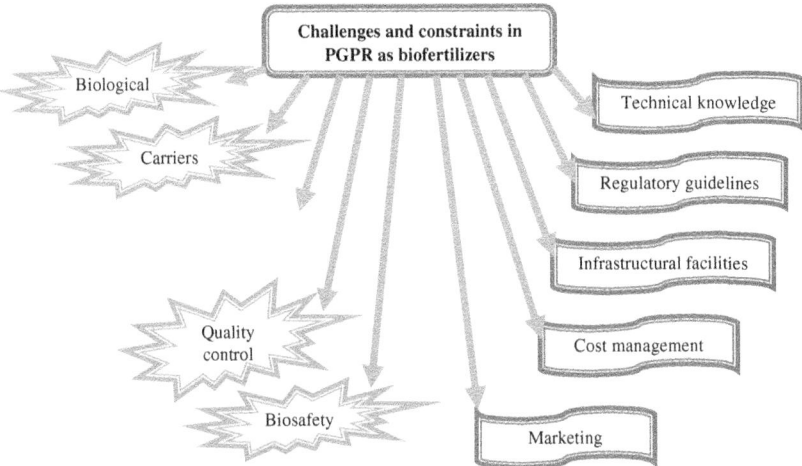

FIG. 3 Challenges and constraints in PGPR as biofertilizers.

campaign should be organized to communicate to farmers and consumers the benefits of applying microbe-based biofertilizers for a green and clean environment for future generations.

8 Conclusion

The use of PGPMs is a powerful and profitable practice for improving crop efficiency and food quality as well as creating sustainable and eco-friendly agricultural practices. These microbes have the facility to inhabit plant roots provided that reimbursement to the hosts, by adapt the assembly of phytohormones, escalating the accessibility of soil nutrients and the confrontation aligned with pathogens. Using PGPMs as biofertilizers and eco-friendly substitutes to artificial agrochemicals such as chemical fertilizers and pesticides to promote plant growth and crop yield, improve soil fertility, and control phytopathogens encourages sustainable agriculture. Useful PGPMs include *Bacillus*, *Flavobacterium*, *Azospirillum*, *Rhizobium*, *Enterobacter*, *Frankia*, *Pseudomonas*, *Klebsiella*, *Trichoderma*, *Clostridium*, *Serratia*, and *Streptomyces* species. The superior brunt of PGPR in conditions of bio fertilization, biocontrol and bioremediation, all of which put forth a constructive persuade on crop efficiency and functioning ecosystem, support should be given to its functioning in agriculture. With advanced scientific and technological development, PGPR can be successfully used to ensure the stability and productivity of agrosystems, thus paving the way for food security worldwide.

References

[1] L.M. Nelson, Plant growth promoting rhizobacteria (PGPR): prospects for new inoculants, Plant Manag Netw. (2004), https://doi.org/10.1094/CM-2004-0301-05-RV.

[2] Y. Kapulnik, Y. Okon, Plant growth promotion by rhizosphere bacteria, in: Y. Waisel, A. Eshel, U. Kafkafi (Eds.), Plant Roots: The Hidden Half, Marcel Dekker, New York, 2002, pp. 869–885.

[3] P. Vejan, R. Abdullah, T. Khadiran, S. Ismail, B.A. Nasrulhaq, Role of plant growth promoting rhizobacteria in agricultural sustainability—a review, Molecules 21 (2016) 573.

[4] M. Ahemad, M. Kibret, Mechanisms and applications of plant growth promoting rhizobacteria: current perspective, J. King Saudi Univ. Sci. 26 (2014) 1–20.

[5] B. Hirel, T. Tetu, P.J. Lea, F. Dubois, Improving nitrogen use efficiency in crops for sustainable agriculture, Sustainability 3 (12) (2011) 1452–1485.

[6] R.Z. Sayyed, S.B. Chincholkar, M.S. Reddy, N.S. Gangurde, P.R. Patel, Siderophores producing PGPR for crop nutrition and phytopathogens suppression, in: K.D. Maheshwari (Ed.), Bacteria in Agrobiology: Disease Management, Springer, Berlin, 2013, pp. 449–471.

[7] D.K. Maheshwari, S. Dheeman, M. Agarwal, Phytohormone producing PGPR for sustainable agriculture, in: D.K. Maheshwari (Ed.), Bacterial Metabolites in Sustainable Agroecosystem, Springer, Cham, 2015, pp. 159–182.

[8] M. Meena, P. Swapnil, K. Divyanshu, S. Kumar, T.Y.N. Harish, A. Zehra, A. Marwal, R.S. Upadhyay, PGPR-mediated induction of systemic resistance and physiochemical alterations in plants against the pathogens: current perspectives, J. Basic Microbiol. 60 (10) (2020) 828–861, https://doi.org/10.1002/jobm.202000370.

[9] M. Meena, P. Swapnil, A. Zehra, M.K. Dubey, M. Aamir, C.B. Patel, R.S. Upadhyay, Virulence factors and their associated genes in microbes, in: H.B. Singh, V.K. Gupta, S. Jogaiah (Eds.), New and Future Developments in Microbial Biotechnology and Bioengineering, Elsevier, 2019, pp. 181–208, https://doi.org/10.1016/B978-0-444-63503-7.00011-5.

[10] S. Mishra, N.K. Arora, Evaluation of rhizospheric *Pseudomonas* and *Bacillus* as bio control tool for *Xanthomonas campsteris* pv *campseteris*, World J. Microbiol. Biotechnol. 28 (2) (2012) 693–702.

[11] S. Compant, B. Duffy, J. Nowak, C. Clement, E.A. Barka, Use of plant growth promoting bacteria for biocontrol of plant diseases: principles, mechanisms of action and future prospects, Appl. Environ. Microbiol. 71 (2005) 4951–4959.

[12] M. Sreevidya, S. Gopalakrishnan, H. Kudapa, R.K. Varshney, Exploring plant growth-promotion actinomycetes from vermin compost and rhizosphere soil for yield enhancement in chickpea, Braz. J. Microbiol. 47 (1) (2016) 85–95.

[13] R. Dey, K.K. Pal, K.V.B.R. Tilak, Plant growth-promoting rhizobacteria in crop protection and challenges, in: A. Goyal, C. Manoharachary (Eds.), Future Challenges in Crop Protection Against Fungal Pathogens, Springer, New York, 2014, pp. 31–58.

[14] P. Kumari, M. Meena, R.S. Upadhyay, Characterization of plant growth promoting rhizobacteria (PGPR) isolated from the rhizosphere of *Vigna radiata* (mung bean), Biocatal. Agric. Biotechnol. 16 (2018) 155–162.

[15] P. Kumari, M. Meena, P. Gupta, M.K. Dubey, G. Nath, R.S. Upadhyay, Plant growth promoting rhizobacteria and their biopriming for growth promotion in mung bean (*Vigna radiata* (L.) R. Wilczek), Biocatal. Agric. Biotechnol. 16 (2018) 163–171.

[16] N. Sanginga, S.K.A. Danso, K. Mulongoy, A.A. Ojeifo, Persistence and recovery of introduced *Rhizobium* 10 years after inoculation on *Leucaena leucocephala* grown on an Alfisol in south western Nigeria, Plant Soil 159 (1994) 199–204.

[17] A. Sellstedt, K.H. Richau, Aspects of nitrogen-fixing Actinobacteria, in particular free-living and symbiotic Frankia, FEMS Microbiol. Lett. 342 (2013) 179–186.

[18] E. Smyth, Selection and Analysis of Bacteria on the Basis of Their Ability to Promote Plant Development and Growth (PhD thesis), University College Dublin, Ireland, 2011.

[19] B.R. Glick, Plant growth-promoting bacteria: mechanisms and applications, Scientifica 15 (2012) 963401, https://doi.org/10.6064/2012/963401.

[20] R. Jog, M. Pandya, G. Nareshkumar, S. Rajkumar, Mechanism of phosphate solubilisation and antifungal activity of Streptomyces spp. isolated from wheat roots and rhizosphere and their application in improving plant growth, Microbiology 160 (2014) 778–788.

[21] M. Rajkumar, N. Ae, M.N.V. Prasad, H. Freitas, Potential of siderophores-producing bacteria for improving heavy metal phytoextraction, Trends Biotechnol. 28 (2010) 142–149.

[22] S.A. Palaniyandi, S.H. Yang, L. Zhang, J.W. Suh, Effects of actinobacteria on plant disease suppression and growth-promotion, Appl. Microbiol. Biotechnol. 97 (2013) 9621–9636.

[23] M. Wan, G. Li, J. Zhang, D. Jiang, H.C. Huang, Effect of volatile substances of *Streptomyces platensis* F-1 on control of plant fungal diseases, Biol. Control 46 (2008) 552–559.

[24] K.F. Chater, S. Biro, K.J. Lee, T. Palmer, H. Schrempf, The complex extracellular biology of *Streptomyces*, FEMS Microbiol. Rev. 34 (2010) 171–198.

[25] J. Berdy, Thoughts and facts about antibiotics: where we are now and where we are heading, J Antibiot. 65 (2012) 385–395.

[26] L.B. Bruno, C. Karthik, Y. Ma, K. Kadirvelu, H. Freitas, M. Rajkumar, Amelioration of chromium and heat stresses in Sorghum bicolour by Cr_6^+ reducing-thermotolerant plant growth promoting bacteria, Chemosphere 244 (2020) 125521, https://doi.org/10.1016/j.chemosphere.2019.12552.

[27] V. Ramírez, J.-A. Munive, L. Cortes, J. Munoz-Rojas, R. Portillo, A. Baez, Long-chain hydrocarbons (C21, C24, and C31) released by *Bacillus* sp. MH778713 break dormancy of mesquite seeds subjected to chromium stress, Front. Microbiol. 11 (2020) 741, https://doi.org/10.3389/fmicb.2020.00741.

[28] L.G. Li, Y. Xia, T. Zhang, Co-occurrence of antibiotic and metal resistance genes revealed in complete genome collection, ISME J. 11 (2017) 651–662, https://doi.org/10.1038/ismej.2016.15.

[29] S.R. Manoj, C. Karthik, K. Kadirvelu, P.I. Arulselvi, T. Shanmugasundaram, B. Bruno, M. Rajkumar, Understanding the molecular mechanisms for the enhanced phytoremediation of heavy metals through plant growth promoting rhizobacteria: a review, J. Environ. Manag. 254 (2020) 109779, https://doi.org/10.1016/j.jenvman.2019.109779.

[30] A.Y. Kim, R. Shahzad, S.M. Kang, C.W. Seo, Y.G. Park, H.J. Park, I.J. Lee, IAA-producing Klebsiella variicola AY13 reprograms soybean growth during flooding stress, J. Crop. Sci. Biotechnol. 20 (2017) 235–242, https://doi.org/10.1007/s12892-017-0041-0.

[31] C.S. Nautiyal, S. Srivastava, P.S. Chauhan, K. Seem, A. Mishra, S.K. Sopory, Plant growth-promoting bacteria Bacillus amyloliquefaciens NBRISN13 modulates gene expression profile of leaf and rhizosphere community in rice during salt stress, Plant Physiol. Biochem. 66 (2013) 1–9, https://doi.org/10.1016/j.plaphy.2013.01.020.

[32] A. Vaishnav, S. Kumari, S. Jain, A. Varma, D.K. Choudhary, Putative bacterial volatile-mediated growth in soybean (*Glycine max* L. *Merrill*) and expression of induced proteins under salt stress, J. Appl. Microbiol. 119 (2015) 539–551, https://doi.org/10.1111/jam.12866.

[33] S. Danish, M.Z. Hye, S. Hussain, M. Riaz, M.F. Qayyum, Mitigation of drought stress in maize through inoculation with drought tolerant acc deaminase containing PGPR under axenic conditions, Pak. J. Bot. 52 (2020) 49–60, https://doi.org/10.30848/PJB2020-1(7).

[34] K. Pramanik, S. Mitra, A. Sarkar, T. Soren, T.K. Maiti, Characterization of cadmium-resistant *Klebsiella pneumonia* MCC 3091 promoted rice seedling growth by alleviating phytotoxicity of cadmium, Environ. Sci. Pollut. Res. 24 (2017) 24419–24437, https://doi.org/10.1007/s11356-017-0033-z.

[35] T.U. Hassan, A. Bano, I. Naz, Alleviation of heavy metals toxicity by the application of plant growth promoting rhizobacteria and effects on wheat grown in saline sodic field, Int. J. Phytoremediation 19 (2017) 522–529, https://doi.org/10.1080/15226514.2016.1267696.

[36] S. Shilev, Plant growth promoting bacteria mitigating soil salinity stress in plants, Appl. Sci. 10 (2020) 7326, https://doi.org/10.3390/app10207326.

[37] D. Barnawal, N. Bharti, S.S. Pandey, A. Pandey, C.S. Chanotiya, A. Kalra, Plant growth-promoting rhizobacteria enhance wheat salt and drought stress tolerance by altering endogenous phytohormone levels and TaCTR1/TaDREB2 expression, Physiol. Plant. 161 (2017) 502–514, https://doi.org/10.1111/ppl.12614.

[38] C. Forni, D. Duca, B.R. Glick, Mechanisms of plant response to salt and drought stress and their alteration by rhizobacteria, Plant Soil 410 (2017) 335–356, https://doi.org/10.1007/s11104-016-3007-x.

[39] M. Meena, P. Swapnil, R.S. Upadhyay, Isolation, characterization and toxicological potential of tenuazonic acid, alternariol and alternariol monomethyl ether produced by *Alternaria* species phytopathogenic on plants, Sci. Rep. 7 (2017) 8777, https://doi.org/10.1038/s41598-017-09138-9.

[40] C.B. Patel, V.K. Singh, A.P. Singh, M. Meena, R.S. Upadhyay, Microbial genes involved in interaction with plants, in: H.B. Singh, V.K. Gupta, S. Jogaiah (Eds.), New and Future Developments in Microbial Biotechnology and Bioengineering, Elsevier, Singapore, 2019, pp. 171–180, https://doi.org/10.1016/B978-0-444-63503-7.00010-3.

[41] T. Mimmo, Y. Pii, F. Valentinuzzi, S. Astolfi, N. Lehto, B. Robinson, et al., Nutrient availability in the rhizosphere: a review, Acta Hortic. 1217 (2018) 13–28, https://doi.org/10.17660/ActaHortic.2018.1217.2.

[42] R.J. Ndeddy Aka, O.O. Babalola, Effect of bacterial inoculation of strains of *Pseudomonas aeruginosa, Alcaligenes feacalis* and *Bacillus subtilis* on germination, growth and heavy metal (Cd, Cr, and Ni) uptake of *Brassica juncea*, Int. J. Phytoremed. 18 (2016) 200–209, https://doi.org/10.1080/15226514.2015.1073671.

[43] P. Calvo, S. Zebelo, D. McNear, J. Kloepper, H. Fadamiro, Plant growth-promoting rhizobacteria induce changes in *Arabidopsis thaliana* gene expression of nitrate and ammonium uptake genes, J. Plant Interact. 14 (2019) 224–231, https://doi.org/10.1080/17429145.2019.1602887.

[44] K. Liu, J.A. McInroy, C.-H. Hu, J.W. Kloepper, Mixtures of plant-growth-promoting rhizobacteria enhance biological control of multiple plant diseases and plant-growth promotion in the presence of pathogens, Plant Dis. 102 (2017) 67–72, https://doi.org/10.1094/PDIS-04-17-0478-RE.

[45] J.H. Jang, S.-H. Kim, I. Khaine, M.J. Kwak, H.K. Lee, T.Y. Lee, W.Y. Lee, S.Y. Woo, Physiological changes and growth promotion induced in poplar seedlings by the plant growth-promoting rhizobacteria *Bacillus subtilis* JS, Photosynthetica 56 (2018) 1188–1203, https://doi.org/10.1007/s11099-018-0801-0.

[46] S. Sharma, J. Kulkarni, B. Jha, Halo tolerant rhizobacteria promote growth and enhance salinity tolerance in peanut, Front. Microbiol. 7 (2016) 1600, https://doi.org/10.3389/fmicb.2016.01600.

[47] H. Etesami, G.A. Beattie, Mining halophytes for plant growth—promoting halo tolerant bacteria to enhance the salinity tolerance of non-halophytic crops, Front. Microbiol. 9 (2018) 148, https://doi.org/10.3389/fmicb.2018.00148.

[48] S. Thant, N.N. Aung, O.M. Aye, N.N. Oo, T.M.M. Htun, A.A. Mon, K.T. Mar, K. Kyaing, K.K. Oo, M. Thywe, P. Phwe, T.S.Y. Hnin, S.M. Thet, Z.K. Latt, Phosphate solubilization of *Bacillus megaterium* isolated from non-saline soils under salt stressed conditions, J. Bacteriol. Mycol. 6 (2018) 335–341, https://doi.org/10.15406/jbmoa.2018.06.00230.

[49] E.T. Alori, B.R. Glick, O.O. Babalola, Microbial phosphorus solubilization and its potential for use in sustainable agriculture, Front. Microbiol. 8 (2017) 971, https://doi.org/10.3389/fmicb.2017.00971.

[50] L. Jian, X. Bai, H. Zhang, X. Song, Z. Li, Promotion of growth and metal accumulation of alfalfa by co inoculation with *Sinorhizobium* and *Agrobacterium* under copper and zinc stress, PeerJ. 7 (2019) e6875, https://doi.org/10.7717/peerj.6875.

[51] X. Zhang, D. Zhang, W. Sun, T. Wang, The adaptive mechanism of plants to iron deficiency via iron uptake, transport, and homeostasis, Int. J. Mol. Sci. 20 (2019) 2424, https://doi.org/10.3390/ijms20102424.

[52] V. Shah, A. Daverey, Phytoremediation: a multidisciplinary approach to clean up heavy metal contaminated soil, Environ. Technol. Innov. 18 (2020) 100774, https://doi.org/10.1016/j.eti.2020.100774.

[53] S. Paque, D. Weijers, Auxin: the plant molecule that influences almost anything, BMC Biol. 14 (2016) 67, https://doi.org/10.1186/s12915-016-0291-0.

[54] S.-H. Park, M. Elhiti, H. Wang, A. Xu, D. Brown, A. Wang, Adventitious root formation of in vitro peach shoots is regulated by auxin and ethylene, Sci. Hortic. 226 (2017) 250–260, https://doi.org/10.1016/j.scienta.2017.08.053.

[55] Y. Liu, L. Chen, N. Zhang, Z. Li, G. Zhang, Y. Xu, Q. Shen, R. Zhang, Plant-microbe communication enhances auxin biosynthesis by a root-associated bacterium, *Bacillus amyloliquefaciens* SQR9, Mol. Plant Microbe Interact. 29 (2016) 324–330, https://doi.org/10.1094/MPMI-10-15-0239-R.

[56] E.H. Verbon, L.M. Liberman, Beneficial microbes affect endogenous mechanisms controlling root development, Trends Plant Sci. 21 (2016) 218–229, https://doi.org/10.1016/j.tplants.2016.01.013.

[57] A. Osugi, H. Sakakibara, How do plants respond to cytokinins and what is their importance? BMC Biol. 13 (2015) 102, https://doi.org/10.1186/s12915-015-0214-5.

[58] I. Naz, A. Bano, T. Ul-Hassan, Isolation of phytohormones producing plant growth promoting rhizobacteria from weeds growing in Khewra salt range, Pakistan and their implication in providing salt tolerance to *Glycine max* L, Afr. J. Biotechnol. 8 (2009) 5762–5766, https://doi.org/10.5897/AJB09.1176.

[59] Y.G. Park, B.G. Mun, S.M. Kang, A. Hussain, R. Shahzad, C.W. Seo, A.-Y. Kim, S.-U. Lee, K.Y. Oh, D.Y. Lee, I.-J. Lee, B.-W. Yun, *Bacillus aryabhattai* SRB02 tolerates oxidative and nitrosative stress and promotes the growth of soybean by modulating the production of phytohormones, PLoS One 12 (2017) e0173203, https://doi.org/10.1371/journal.pone.0173203.

[60] S.S. Akhtar, M.F. Mekureyaw, C. Pandey, T. Roitsch, Role of cytokinins for interactions of plants with microbial pathogens and pest insects, Front. Plant Sci. 10 (2020) 1777, https://doi.org/10.3389/fpls.2019.01777.

[61] S.-M. Kang, A.L. Khan, M. Waqas, S. Asaf, K.-E. Lee, Y.-G. Park, A.-Y. Kim, M.A. Khan, Y.-H. You, I.-J. Lee, Integrated phytohormone production by the plant growth-promoting rhizobacterium *Bacillus tequilensis* SSB07 induced thermo tolerance in soybean, J. Plant Interact. 14 (2019) 416–423, https://doi.org/10.1080/17429145.2019.1640294.

[62] J. Binenbaum, R. Weinstain, E. Shani, Gibberellins localization and transport in plants, Trends Plant Sci. 23 (5) (2018) 410–421, https://doi.org/10.1016/j.tplants.2018.02.005.

[63] R. Nagel, J.E. Bieber, M.G. Schmidt-Dannert, R.S. Nett, R.J. Peters, A third class: functional gibberellins biosynthetic operon in beta-proteobacteria, Front. Microbiol. 9 (2018) 2916, https://doi.org/10.3389/fmicb.2018.02916.

[64] L.T.M. Revolti, C.H. Caprio, F.L.C. Mingotte, G.V. Moro, *Azospirillum* spp. potential for maize growth and yield, Afr. J. Biotechnol. 17 (2018) 574–585, https://doi.org/10.5897/AJB2017.16333.

[65] K. Shu, W. Zhou, F. Chen, X. Luo, W. Yang, Abscisic acid and gibberellins antagonistically mediate plant development and abiotic stress responses, Front. Plant Sci. 9 (2018) 416, https://doi.org/10.3389/fpls.2018.00416.

[66] K.A. Tsukanova, J.J.M. Meyer, T.N. Bibikova, Effect of plant growth-promoting *Rhizobacteria* on plant hormone homeostasis, South Afr. J. Bot. 113 (2017) 91–102, https://doi.org/10.1016/j.sajb.2017.07.007.

[67] J. Vacheron, E. Combes-Meynet, V. Walker, B. Gouesnard, D. Muller, Y. Moenne-Loccoz, C. Prigent-Combaret, Expression on roots and contribution to maize phytostimulation of 1-aminocyclopropane-1-decarboxylate deaminase gene *acdS* in *Pseudomonas fluorescens* F113, Plant Soil 407 (2016) 187–202, https://doi.org/10.1007/s11104-016-2907-0.

[68] H. Etesami, Plant growth promotion and suppression of fungal pathogens in rice (Oryza sativa L.) by plant growth-promoting bacteria, in: Field Crops: Sustainable Management by PGPR, Springer, Cham, Switzerland, 2019, pp. 351–383.

[69] R. Radhakrishnan, A. Hashem, E.F. Abd Allah, Bacillus: a biological tool for crop improvement through bio-molecular changes in adverse environments, Front. Physiol. 8 (2017) 667.

[70] B.K. Kashyap, M.K. Solanki, A.K. Pandey, S. Prabha, P. Kumar, B. Kumari, Bacillus as plant growth promoting rhizobacteria (PGPR): a promising green agriculture technology, in: Plant Health Under Biotic Stress, Springer Nature Singapore, Singapore, 2019, pp. 219–236.

[71] R. Shultana, A.T. Kee Zuan, M.R. Yusop, H.M. Saud, Characterization of salt-tolerant plant growth-promoting rhizobacteria and the effect on growth and yield of saline-affected rice, PLoS One 15 (2020) e0238537.

[72] A. Rais, Z. Jabeen, F. Shair, F.Y. Hafeez, M.N. Hassan, *Bacillus* spp. a bio-control agent enhances the activity of antioxidant defense enzymes in rice against *Pyricularia oryzae*, PLoS One 12 (2017) e0187412.

[73] J. Shafi, H. Tian, M. Ji, *Bacillus* species as versatile weapons for plant pathogens: a review, Biotechnol. Biotechnol. Equip. 31 (2017) 446–459.

[74] P. Jin, Y. Wang, Z. Tan, W. Liu, W. Miao, Antibacterial activity and rice-induced resistance, mediated by C15 surfactin A, in controlling rice disease caused by *Xanthomonas oryzae* pv. oryzae, Pestic. Biochem. Physiol. 169 (2020) 104669.

[75] O.F. Anayo, E.C. Scholastica, O.C. Peter, U.G. Nneji, A. Obinna, L.O. Mistura, The beneficial roles of *Pseudomonas* in medicine, industries, and environment: a review, in: Pseudomonas aeruginosa—An Armory Within, Intech Open, London, UK, 2016.

[76] S.P. Diggle, M. Whiteley, Microbe profile: *Pseudomonas aeruginosa*: opportunistic pathogen and lab rat, Microbiology 166 (2020) 30–33.

[77] C.E. Morris, D.C. Sands, B.A. Vinatzer, C. Glaux, C. Guilbaud, A. Buffiere, S. Yan, H. Dominguez, B.M. Thompson, The life history of the plant pathogen Pseudomonas syringae is linked to the water cycle, ISME J. 2 (2008) 321–334.

[78] B.V. David, G. Chandrasehar, P.N. Selvam, Pseudomonas fluorescens: a plant-growth-promoting rhizobacterium (PGPR) with potential role in biocontrol of pests of crops, in: Crop Improvement Through Microbial Biotechnology, Elsevier, Amsterdam, The Netherlands, 2018, pp. 221–243.

[79] R. Kandaswamy, M.K. Ramasamy, R. Palanivel, U. Balasundaram, Impact of *Pseudomonas putida* RRF3 on the root transcriptome of rice plants: insights into defense response, secondary metabolism and root exudation, J. Biosci. 44 (2019) 98.

[80] S.S. Hirano, C.D. Upper, Bacteria in the leaf ecosystem with emphasis on Pseudomonas syringae—a pathogen, ice nucleus, and epiphyte, Microbiol. Mol. Biol. Rev. 64 (2000) 624–653.

[81] V. Kumar, L. Jain, S.K. Jain, S. Chaturvedi, P. Kaushal, Bacterial endophytes of rice (*Oryza sativa* L.) and their potential for plant growth promotion and antagonistic activities, S. Afr. J. Bot. 134 (2020) 50–63.

[82] P.G.C. Atuesta, W. Murillo Arango, J. Eras, D.F. Oliveros, J.J. Méndez Arteaga, Rice-associated rhizobacteria as a source of secondary metabolites against *Burkholderia glumae*, Molecules 25 (2020) 2567.

[83] S. Mukhtar, A. Zaheer, D. Aiysha, K. Abdulla Malik, S. Mehnaz, Actinomycetes: a source of industrially important enzymes, J Proteom. Bioinform. 10 (2017) 316–319.

[84] G.A. Quinn, A.M. Banat, A.M. Abdelhameed, I.M. Banat, *Streptomyces* from traditional medicine: sources of new innovations in antibiotic discovery, J. Med. Microbiol. 69 (2020) 1040–1048.

[85] Z.R. Suarez-Moreno, D.M. Vinchira-Villarraga, D.I. Vergara-Morales, L. Castellanos, F.A. Ramos, C. Guarnaccia, G. Degrassi, V. Venturi, N. Moreno-Sarmiento, Plant-growth promotion and biocontrol properties of three *Streptomyces* spp. isolates to control bacterial rice pathogens, Front. Microbiol. 10 (2019) 290.

[86] G. Subramaniam, V. Thakur, R.K. Saxena, S. Vadlamudi, S. Purohit, V. Kumar, A. Rathore, A. Chitikineni, R.K. Varshney, Complete genome sequence of sixteen plant growth promoting Streptomyces strains, Sci. Rep. 10 (2020) 10294.

[87] P. García-Fraile, E. Menendez, R. Rivas, Role of bacterial biofertilizers in agriculture and forestry, AIMS Bioeng. 2 (2015) 183–205, https://doi.org/10.3934/bioeng.2015.3.183.

[88] R.M. Dos Santos, P.A.E. Diaz, L.L.B. Lobo, E.C. Rigobelo, Use of plant growth-promoting rhizobacteria in maize and sugarcane: characteristics and applications, Front. Sustain. Food Syst. 4 (2020), https://doi.org/10.3389/fsufs.2020.00136.

Chapter 8

Immune signaling networks in plant-pathogen interactions

Andleeb Zehra[a], Mukesh Meena[b], and Prashant Swapnil[c]

[a]*Laboratory of Mycopathology and Microbial Technology, Department of Botany, Centre of Advanced Study in Botany, Institute of Science, Banaras Hindu University, Varanasi, Uttar Pradesh, India,* [b]*Laboratory of Phytopathology and Microbial Biotechnology, Department of Botany, Mohanlal Sukhadia University, Udaipur, Rajasthan, India,* [c]*School of Basic Sciences, Department of Botany, Central University of Punjab, Bathinda, Punjab, India*

1 Introduction

Immune responses in plants rely on the modulation of signaling activities in response to pathogen attack in a spatiotemporal distribution that is appropriate. Plants have two main immune systems, both of which are activated by particular receptors via recognizing microbial compounds [1]. Pathogen-associated molecular pattern (PAMP)-triggered immunity (PTI), which is mediated by cell surface-localized pattern recognition receptors (PRRs), and effector triggered immunity (ETI), which is mediated primarily by intracellular nucleotide-binding domain and leucine-rich repeat (LRR) receptors, are the two types of host receptors that activate these two layers of the immune system [2,3]. Receptor kinases (RKs) and receptor-like proteins (RLPs) are very identical to toll-like receptors (cell-surface immunological receptors) [4]. NLRs (nucleotide-binding leucine-rich repeat receptors) are nucleotide-binding intracellular immunological receptors found in plants and mammals [5]. Pattern recognition receptors (PRRs) on the cell surface of plants tend to perceive pathogen or microbe or host-derived immunogenic molecular patterns, whereas NLRs are responsible for sensing the more diverse pathogen effector proteins transported inside the plant cell [6].

Plant immunological signaling regulated by both PRRs and NLRs has made significant progress in the recent decade, particularly in *Arabidopsis*. This chapter summarizes the pathogen effectors and their roles, the nature of the effectors, their involvement in inhibiting PTI responses, pathogen recognition, and subsequent reaction perception. In addition, we highlight recent progress in

Plant-Microbe Interaction—Recent Advances in Molecular and Biochemical Approaches
https://doi.org/10.1016/B978-0-323-91876-3.00015-4

137

understanding the network complexity of regulatory networks leading to plant immunity, from pathogen sensing to signaling pathways to immune responses.

2 Immunogenic signals and microbial recognition

Plant defenses are activated by immunogenic signals in a variety of ways. Microbe-associated molecular patterns (MAMPs) are highly consistent across taxonomic groups [3,7]. During infection, plants use lytic enzymes, which are similar to peptidylglycan and epitope (bacterial flagellin) flg22, to activate the epitopes related to MAMP from affecting pathogens [8]. Pathogens damage the cell wall of the plant tissue by using certain enzymes, ultimately synthesizing extracellular nicotinamide adenine dinucleotide (NAD) and ATP [4]. As damage-associated molecular patterns (DAMPs), these plant-derived compounds can be exploited to activate immune responses. According to Bacete et al. [9], in plants, cell wall integrity helps the plants to detect pathogens. Plants produce phytocytokines (plant elicitor peptides; Peps), which is recognized by PRRs (PRR-associated receptors) and further regulate plant immunity in the same way as cytokines performs in animals [10]. Peps induce the defense process through PEPR1 that is LRR-RKs (leucine-rich repeat receptor kinases), and PEPR2 by interacting with each other [11]. Peps cleavage is regulated by various calcium-dependent cysteine proteases such as cytoplasmic type-II metacaspases [12].

Pathogens create effectors in the plant cell's intracellular or extracellular areas to enhance parasitism [13]. However, when the host plant attains homologous immune receptors, coevolution between the host and the pathogen can make disease effectors immunogenic [14]. PRRs distinguish immunogenic apoplastic effectors, DAMPs and MAMPs, while NLRs distinguish intracellular effectors. As a result, any one infection involves many immune receptors detecting multiple danger signals.

MAMPs are found in both pathogens and in commensal and helpful bacteria. According to a recent review, beneficial microbes can actively decrease immunogenic responses of the PRRs of the plants by avoiding detection [4]. Normal MAMP perception, on the other hand, actively helps in the case of microbial homeostasis due to mutations, which impair perception by MAMP while also causing an aberrant apoplast aqueous microenvironment resulting in leaf dysbiosis [15]. In *Medicago truncatula*, chitin oligomers alone can induce immunological responses, but a combination of lipochitooligosaccharide and oligomers of chitin improves symbiosis synergistically [16]. Although the exact mechanism is unknown, this discovery emphasizes the significance of integrating diverse microbial signals for optimal host plant responses.

A recent and well-designed study revealed that differentiated roots of *Arabidopsis* may combine MAMP signals with cell damage to activate defenses mechanism in response to pathogenic microbes but not commensal bacteria [4]. These differentiated roots express very few PRRs and respond poorly to

MAMPs, allowing commensal bacteria to colonize to their full potential. Pathogenic bacterium infection, on the other hand, produces cell damage in these roots and stimulates the expression of numerous PRRs in surrounding cells, allowing the latter to activate powerful defenses in response to MAMPs. Interpretation by plants of a wound signal and a MAMP by plants is established as a true danger signal. Immune signaling from each receptor is frequently studied independently, and different immune receptors are thought to be functionally redundant. However, it is unclear if a single immunogenic signal is enough to alert the host plant to the threat's nature. During contact with a single microbe, various signals such as MAMPs, DAMPs, wounding, phytocytokines, and/or symbiotic signal molecules are detected, and the integration of these signals could be crucial for plants to discriminate friends from adversaries.

3   Extracellular recognition by pattern recognition receptors

Plant PRRs are receptor-like kinases (RLKs) or receptor-like proteins (RLPs) with an extracellular domain for MAMP recognition that are found on the plasma membrane [17]. PRRs are classified as either transmembrane receptor kinases or transmembrane receptor-like proteins, with the latter lacking an evident intrinsic signaling component [18,19]. In plants, the receptor kinase gene family has expanded dramatically: the *Arabidopsis thaliana* genome contains over 610 members, many of which are susceptible to biotic stresses [20]. In *A. thaliana*, there are 57 members of the receptor-like protein family. The growth of these families contrasts with the situation in animals, which have 12 toll-like receptors that perform the same function as PRRs in plants [17].

Plants identify a wide range of PAMPs, including proteins, carbohydrates, lipids, and tiny molecules like ATP [21]. The *A. thaliana* receptor kinase FlA-GEllIn SEnSInG 2 (FlS2), which binds bacterial flagellin directly and then assembles an active signaling complex, is the best example of PAMP recognition [22]. Although the PAMP paradigm implies that all PAMPs should be recognized by all species, it has been discovered that this is not always the case. For example, recognition of the bacterial elongation factor EF-Tu appears to be limited to the Brassicaceae [23]. Similarly, the rice Xa21 receptor confers race-specific resistance to the bacterial disease caused by *Xanthomonas oryzae* and has recently been discovered to function as a PRR for a new sulfonated bacterial protein known as Ax [24].

It is generally known that PRRs require coreceptors to perceive ligands [25]. For example, somatic-embryogenesis receptor-like kinases (SERK) are required as coreceptors for flagellin sensing 2 (FLS2) and the EF-TU receptor (EFR) [26]. Similarly, another LysM-RK chitin elicitor receptor kinase 1 (CERK1) is required as a coreceptor for lysin motif (LysM)-containing receptor kinase 5 (LYK5), a chitin receptor [27]. The contact of FLS2 with BAK1 is initiated by flg22, and many accessory LRR-RKs with short LRR domains regulate it extensively. In the resting state, BAK1-interacting RLK2 (BIR2) and

BIR3 negatively control the FLS2-BAK1 relationship by sequestering BAK1 [28]. A comprehensive cell-surface interaction network was discovered in a systems investigation of 200 *Arabidopsis* LRR-RKs, in which various LRR-RKs are coupled through short LRR-RKs as hubs [29]. Importantly, APEX and NIK1 inhibit the flg22-induced FLS2-BAK1 connection, whereas FIR increases it. Disentangling how distinct accessory LRR-RKs coordinate during MAMP recognition and immune activation will be a future task.

When PRRs are activated, a number of different important signaling modules are activated [30]. RLCKs serve an important role in connecting PRRs to downstream signaling modules. Botrytis-induced kinase 1 (BIK1) and several PBS1-like (PBL) kinases of the RLCK subfamily VII, as well as brassinosteroid signaling kinase 1 (BSK1) of the RLCK subfamily II, are among them in *Arabidopsis* [26]. After being triggered by PRRs, these RLCKs phosphorylate and activate downstream components. BIK1 phosphorylates the N terminus of NADPH oxidase respiratory burst oxidase homolog D to produce ROS, for example (RbohD) [26]. The activation of calcium-dependent kinase 5 (CPK5) and direct Ca^{2+} binding to EF-hands in the RbohD N terminus is also required for the MAMP-induced ROS burst. It was also observed that a cysteine-rich receptor kinase phosphorylates the C terminus of RbohD and enhances NADPH oxidase activity [22].

In addition to the previously mentioned coreceptors and accessory RKs, PRRs can interact with RKs with distinct ectodomains. The malectin domain RK reduced the sensitivity of oomycetes interacting with FLS2, EFR, and CERK1 in a constitutive manner to positively control antifungal and antibacterial immunity by influencing the development of PRR-coreceptor complexes [31]. Furthermore, *Cantharanthus roseus* receptor-like kinase is required for ligand-induced interactions, most likely through acting as a scaffold protein [32]. As a result, different scaffold proteins could be recruited by a single PRR to both undoubtedly or negatively adjust its function.

4 Intracellular recognition

Intracellular recognition is the perception, which is recognized by effectors or intracellular receptors of PVM (pathogen virulence molecules). Intracellular recognition activates effector-triggered immunity (ETI), which is considered as the second pathogen-sensing mechanism in plants [2,33]. Recognition is mediated by nucleotide binding (NB) domains and LRR (leucine-rich repeats) domains containing proteins (Fig. 1). Plants with NB-LRR proteins cause resistance against biotic stress, and also contain toll, interleukin-1 receptor, and resistance protein (TIR) domain (N-terminal) [34]. Some other class of NB-LRR proteins is incorporated with the coiled coil (CC) domain (N-terminal), which is involved in the PAMP induction of innate immunity in animals [7,35]. NB-LRR proteins have the ability to recognize pathogen effectors either directly or indirectly. Direct recognition is a physical association and is

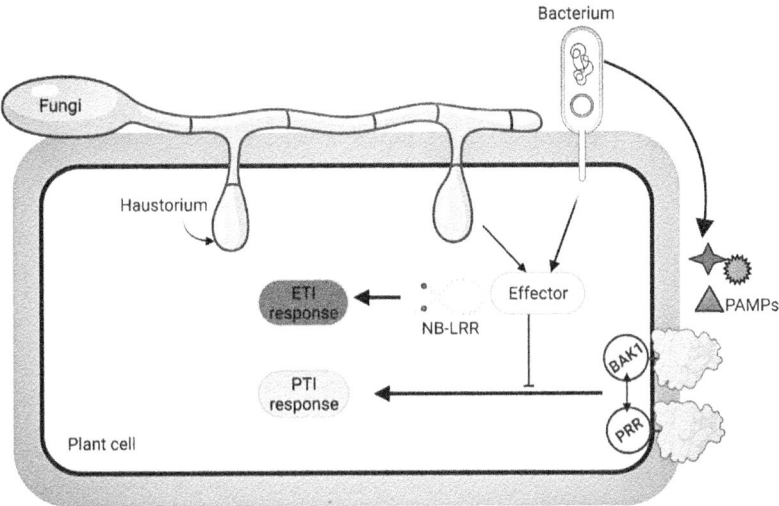

FIG. 1 Schematic representation of pathway of plant immunity. Bacterial and fungal pathogens start to propagate at the surface of the plant tissue. Most of the fungi start to extend their haustoria and penetrate the cell wall. These pathogens release PAMPs, which are recognized by PRRs. Furthermore, PRRs interact with BAK1 to stimulate PTI. The effector proteins released by fungal and bacterial pathogens suppress the PTI. Some effectors are recognized by NB-LRR, which stimulates the ETI signaling pathway. *PAMPs*, pathogen-associated molecular patterns; *PRRs*, pattern recognition receptors; *BAK1*, brassinosteroid insensitive 1-associated kinase 1; *PTI*, PAMP-triggered immunity. *(Created with BioRender.com.)*

assayed by yeast two-hybrid screening [36], while indirect recognition can be done by some accessory proteins. These accessory proteins are associated with the NB-LRR protein (Fig. 2A). These associations of effector proteins and receptors provide the signal of strong selection in between alleles in the host and pathogen populations with high levels of sequence polymorphism and different recognition specificities [37]. Indirect effector interaction is mediated by a pathogen virulence target accessory protein and these interactions cause changes in the accessory protein and make it easy to be recognized by the NB-LRR protein [38]. There are different models that have been proposed to understand the effector and accessory protein association, in which the guard model (studied in *Arabidopsis thaliana* RIN4 proteins) assumes that NB-LRR proteins protect the targeted accessory protein [39]. This RIN4 forms an association with the NB-LRR proteins RPM1 and show resistance to *Pseudomonas syringae* 2 [40,41]. The guard model hypothesizes that the effectors target RIN4, which is still unclear. According to Jones and Dangl [2], in the absence of RPM1 and RPS2, RIN4 should be modified through evolution to avoid complex association with the effector proteins, while in the presence of RPM1 and RPS2 it will favor effector binding to help in recognition. To solve this evolutionary problem, a decoy model has been proposed, which discusses

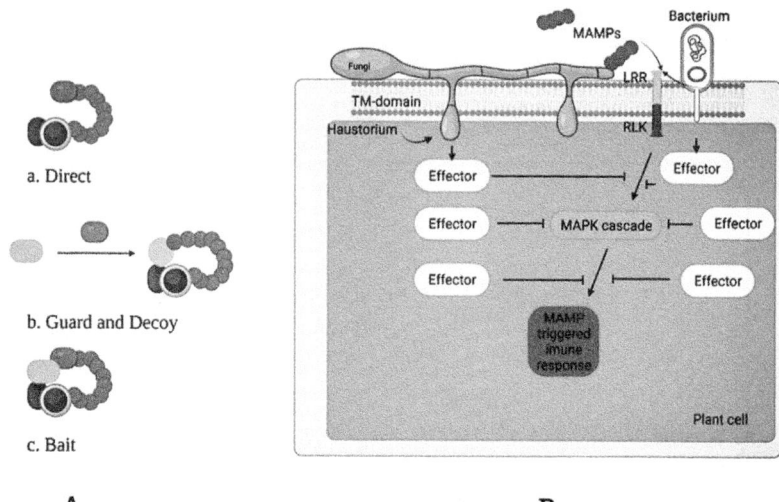

FIG. 2 (A) Diagrammatic representation shows the recognition of effectors by NB-LRR either directly or indirectly. (a) The effector *(purple)* physically binds with the receptor *(green, blue, and red)* to stimulate triggers immune signaling. (b) The effector *(green)* alters the confirmation of an accessory protein *(yellow)*, which is further recognized by NB-LRR. (c) Effector *(green)* interacts with accessory protein *(yellow)* facilitating direct binding to be recognized by NB-LRR. (B) Resistance in plants associated with MAMP recognition by RLK activates MAPK and causes MAMP-triggered immune response. *NB-LRR*, nucleotide-binding-leucine-rich repeat; *MAMP*, microbe-associated molecular patterns; *RLK*, receptor-like kinases; *MAPK*, mitogen-activated protein kinase.

the duplication of the effector target gene that could reduce evolutionary limitations and permit the accessory protein to contribute exclusively in effector sensitivity [42–44]. The decoy model only discusses the insight of effectors by the NB-LRR protein. This model fails to describe the necessity of enzymatic activity and role of proteins in response to defense [45]. Further, the decoy model has been modified as a bait and switch model in which an accessory protein (linked with NB-LRR) as "bait" interacts with an effector, which triggers signaling; the succeeding recognition result ensues between the NB-LRR protein and effector [46]. It is very necessary to know that bait and switch models are not fully understood [45].

5 Immune signaling pathway

Plants have evolved different mechanisms to block the entry of pathogens or microbes through physical barriers like cuticles (waxy) and rigid cell walls, and synthesizing different secondary metabolites as antimicrobial agents. Plants depend on innate immunity, which is acquired thorough systemic signals developing from the pathogen-attacked cells to recall previous infections [47].

Recently, it has been reported that the plants adopted a zig-zag model, which stated that plants show a defense mechanism through two molecular mechanisms [2] (Fig. 2B). Through the first mechanism plant membrane receptors, also known as PRRs, recognize microbe-associated molecular patterns (MAMP) through pattern recognition receptors (PRRs or plasma membrane bound proteins) such as receptor like proteins (RLP) and chitin elicitor receptor kinase 1 (CERK1) at the surface of plant cells to induce MAMP triggered immunity (MTI) which rapidly results in the influx of calcium, accumulation of reactive oxygen species [48,49], activation of mitogen-activated protein kinase (MAPK) phosphorylation cascades [50], alterations in the cell wall, and expression of defense genes [51–53]. Nucleotide-binding leucine-rich repeat (NB-LRR) or receptors (NLRs) are the intracellular immune receptors that induce effector-triggered immunity (ETI) [39]. There are two groups of NLR proteins based on N-terminal coiled coil-NLR (CC-NLR) and TIR (toll-interleukin-1 receptor)-NLRs. These NLR proteins stop the proliferation of pathogens by Ca^{2+} signaling, ROS and nitric oxide production, membrane alteration, reprogramming of defense genes, and several other hypersensitive responses such as programmed cell death [54,55]. The interaction of effectors and NLR triggers NLR in the active state by the conversion of ADP to ATP and initiates the signaling mechanism [56,57]. ETI induced by NLR either through direct or indirect recognition (by recognizing guarders host target protein or plant decoy protein) [58–60]. Guardee proteins have immune function like signaling, while decoy proteins do not show assessable resistant function. The NLRs of plants induce apoptosis, which is very similar to NLRs of animals except caspase, which is not identified in plants. Recently, it has been stated that the CC-NLR protein ZAR1 (HOPZ-Activated Resistance 1) form "resistosomes" (in the oligomeric state) to activate NLR functions in plants [60]. ZAR1 associates with pseudokinase RKS1 (resistance-related kinase 1) and recognizes the effector proteins (bacterial). The complex association of ZAR1-RKS1 recognizes uridylyl transferase and RLCK (receptor-like cytoplasmic kinases). Uridylylated PBL2 (PBS1-like protein 2) interacts with the RKS1-ZAR1 complex and causes conformational alteration in the NB domain and releases ADP from ZAR1, which results in a depleted ZAR1 complex. The complex of ZAR1-RKS1-PBL2UMP activated by dATP or ATP to initiate the second signaling step and leads to NLR protein oligomerization [1]. By following these two mechanisms, plants activate signaling cascade events to protect themselves from pathogen attacks.

6 Future perspectives

Plants face different biotic stresses such as oomycetes, fungi, phytoplasmas, bacteria, viruses, and nematodes during their lifetime. To protect themselves from the pathogenic microorganisms plants acquire different strategies to enhance the strength of immune systems after the induction of pathogens

[61,62]. Apart from using different chemical agents, biocontrol strategies were found more eco-friendly to induce resistance in plants against pathogens. Several practices with the natural plant defense mechanisms give novel strategies for acquiring better management to resist disease in plants. Genetic engineering of PRRs and R proteins ultimately enhances the recognition ability of plants regarding the microbes. Furthermore, there are several important signaling components of PTI and ETI that still need to be studied regarding how NB-LRR can be activated by effector recognition. To study this, we need genome sequencing of both pathogen and host to facilitate the recognition of effector proteins, through effector expression patterns. Using next-generation sequencing will open the way to study host-pathogen systems. However, substantially structural cell biology and biochemical applications will be helpful to know the molecular and biochemical events related to the activation of receptor and signaling pathways.

Acknowledgment

The author MM is thankful to Mohanlal Sukhadia University, Udaipur, for providing the necessary facilities during the course of the study.

Conflict of interest

The authors declare no conflict of interest. The funders had no role in the design of the study; in the collection, analyses, or interpretation of data; in the writing of the manuscript, or in the decision to publish the results.

References

[1] R. Nishad, T. Ahmed, V.J. Rahman, A. Kareem, Modulation of plant defense system in response to microbial interactions, Front. Microbiol. 11 (2020) 1298.

[2] J.D. Jones, J.L. Dangl, The plant immune system, Nature 444 (2006) 323–329.

[3] A. Zehra, N.A. Raytekar, M. Meena, P. Swapnil, Efficiency of microbial bio-agents as elicitors in plant defense mechanism under biotic stress: a review, Curr. Res. Microb. Sci. 2 (2021) 100054, https://doi.org/10.1016/j.crmicr.2021.100054.

[4] J.M. Zhou, Y. Zhang, Plant immunity: danger perception and signaling, Cell 181 (5) (2020) 978–989.

[5] J.D. Jones, R.E. Vance, J.L. Dangl, Intracellular innate immune surveillance devices in plants and animals, Science 354 (6316) (2016) aaf6395.

[6] R. Schellenberger, M. Touchard, C. Clément, F. Baillieul, S. Cordelier, J. Crouzet, S. Dorey, Apoplastic invasion patterns triggering plant immunity: plasma membrane sensing at the frontline, Mol. Plant Pathol. 20 (11) (2019) 1602–1616.

[7] F.M. Ausubel, Are innate immune signaling pathways in plants and animals conserved? Nat. Immunol. 6 (10) (2005) 973–979.

[8] J. Chen, Y. Qiao, G. Chen, C. Chang, H. Dong, B. Tang, X. Cheng, X. Liu, Z. Hua, Salmonella flagella confer anti-tumor immunological effect via activating flagellin/TLR5 signaling within tumor microenvironment, Acta Pharm. Sin. B 11 (10) (2021) 3165–3177.

[9] L. Bacete, H. Melida, E. Miedes, A. Molina, Plant cell wall-mediated immunity: cell wall changes trigger disease resistance responses, Plant J. 93 (4) (2018) 614–636.

[10] A.A. Gust, R. Pruitt, T. Nürnberger, Sensing danger: key to activating plant immunity, Trends Plant Sci. 22 (9) (2017) 779–791.

[11] K. Yamaguchi, T. Kawasaki, Pathogen-and plant-derived peptides trigger plant immunity, Peptides 144 (2021) 170611.

[12] S. Hou, J. Zhang, P. He, Stress-induced activation of receptor signaling by protease-mediated cleavage, Biochem. J. 478 (10) (2021) 1847–1852.

[13] T.Y. Toruño, I. Stergiopoulos, G. Coaker, Plant-pathogen effectors: cellular probes interfering with plant defenses in spatial and temporal manners, Annu. Rev. Phytopathol. 54 (2016) 419–441.

[14] A. Rana, M. Ahmed, A. Rub, Y. Akhter, A tug-of-war between the host and the pathogen generates strategic hotspots for the development of novel therapeutic interventions against infectious diseases, Virulence 6 (6) (2015) 566–580.

[15] J. Chen, M. Clinton, G. Qi, D. Wang, F. Liu, Z.Q. Fu, Reprogramming and remodeling: transcriptional and epigenetic regulation of salicylic acid-mediated plant defense, J. Exp. Bot. 71 (17) (2020) 5256–5268.

[16] F. Feng, J. Sun, G.V. Radhakrishnan, T. Lee, Z. Bozsóki, S. Fort, A. Gavrin, K. Gysel, M.B. Thygesen, K.R. Andersen, S. Radutoiu, A combination of chitooligosaccharide and lipochitooligosaccharide recognition promotes arbuscular mycorrhizal associations in Medicago truncatula, Nat. Commun. 10 (1) (2019) 1–12.

[17] P.N. Dodds, J.P. Rathjen, Plant immunity: towards an integrated view of plant–pathogen interactions, Nat. Rev. Genet. 11 (8) (2010) 539–548.

[18] G.P. Amarante-Mendes, S. Adjemian, L.M. Branco, L.C. Zanetti, R. Weinlich, K.R. Bortoluci, Pattern recognition receptors and the host cell death molecular machinery, Front. Immunol. 9 (2018) 2379.

[19] M. Meena, P. Swapnil, A. Zehra, M.K. Dubey, M. Aamir, C.B. Patel, R.S. Upadhyay, Virulence factors and their associated genes in microbes, in: H.B. Singh, V.K. Gupta, S. Jogaiah (Eds.), New and Future Developments in Microbial Biotechnology and Bioengineering, Elsevier, 2019, pp. 181–208, https://doi.org/10.1016/B978-0-444-63503-7.00011-5.

[20] A. Marshall, R.B. Aalen, D. Audenaert, T. Beeckman, M.R. Broadley, M.A. Butenko, A.I. Caño-Delgado, S. de Vries, T. Dresselhaus, G. Felix, N.S. Graham, Tackling drought stress: receptor-like kinases present new approaches, Plant Cell 24 (6) (2012) 2262–2278.

[21] N.M. Sanabria, H. van Heerden, I.A. Dubery, Molecular characterisation and regulation of a *Nicotiana tabacum* S-domain receptor-like kinase gene induced during an early rapid response to lipopolysaccharides, Gene 501 (1) (2012) 39–48.

[22] I. Albert, C. Hua, T. Nürnberger, R.N. Pruitt, L. Zhang, Surface sensor systems in plant immunity, Plant Physiol. 182 (4) (2020) 1582–1596.

[23] C. Zipfel, G. Kunze, D. Chinchilla, A. Caniard, J.D. Jones, T. Boller, G. Felix, Perception of the bacterial PAMP EF-Tu by the receptor EFR restricts Agrobacterium-mediated transformation, Cell 125 (4) (2006) 749–760.

[24] J.N. Tripathi, J. Lorenzen, O. Bahar, P. Ronald, L. Tripathi, Transgenic expression of the rice Xa21 pattern-recognition receptor in banana (Musa sp.) confers resistance to *Xanthomonas campestris* pv. *musacearum*, Plant Biotechnol. J. 12 (6) (2014) 663–673.

[25] Y. Ma, R.K. Walker, Y. Zhao, G.A. Berkowitz, Linking ligand perception by PEPR pattern recognition receptors to cytosolic Ca^{2+} elevation and downstream immune signaling in plants, Proc. Natl. Acad. Sci. 109 (48) (2012) 19852–19857.

[26] D. Tang, G. Wang, J.M. Zhou, Receptor kinases in plant-pathogen interactions: more than pattern recognition, Plant Cell 29 (4) (2017) 618–637.

[27] L. Buendia, A. Girardin, T. Wang, L. Cottret, B. Lefebvre, LysM receptor-like kinase and LysM receptor-like protein families: an update on phylogeny and functional characterization, Front. Plant Sci. 9 (2018) 1531.

[28] T. Koller, A.F. Bent, FLS2-BAK1 extracellular domain interaction sites required for defense signaling activation, PLoS One 9 (10) (2014) e111185.

[29] Y. Huang, P. Jamieson, L. Shan, The APEX approaches: a unified LRR-RK network revealed, Trends Plant Sci. 23 (5) (2018) 372–374.

[30] D. Li, M. Wu, Pattern recognition receptors in health and diseases, Curr. Signal Transduct. Ther. 6 (1) (2021) 1–24.

[31] H. Mang, B. Feng, Z. Hu, A. Boisson-Dernier, C.M. Franck, X. Meng, Y. Huang, J. Zhou, G. Xu, T. Wang, L. Shan, Differential regulation of two-tiered plant immunity and sexual reproduction by ANXUR receptor-like kinases, Plant Cell 29 (12) (2017) 3140–3156.

[32] D. Ji, T. Chen, Z. Zhang, B. Li, S. Tian, Versatile roles of the receptor-like kinase Feronia in plant growth, development and host-pathogen interaction, Int. J. Mol. Sci. 21 (21) (2020) 7881.

[33] S.T. Chisholm, G. Coaker, B. Day, B.J. Staskawicz, Host–microbe interactions: shaping the evolution of the plant immune response, Cell 124 (2006) 803–814.

[34] N.J. Gay, M. Gangloff, Structure and function of Toll receptors and their ligands, Annu. Rev. Biochem. 76 (2007) 141–165, https://doi.org/10.1146/annurev.biochem.76.060305.151318.

[35] S.E. Girardin, D.J. Philpott, B. Lemaitre, Sensing microbes by diverse hosts. Workshop on pattern recognition proteins and receptors, EMBO Rep. 4 (2003) 932–936.

[36] Y. Jia, S.A. McAdams, G.T. Bryan, H.P. Hershey, B. Valent, Direct interaction of resistance gene and avirulence gene products confers rice blast resistance, EMBO J. 19 (2000) 4004–4014.

[37] M. Ravensdale, A. Nemri, P.H. Thrall, J.G. Ellis, P.N. Dodds, Co-evolutionary interactions between host resistance and pathogen effector genes in flax rust disease, Mol. Plant Pathol. 12 (1) (2011) 93–102.

[38] R.A. van der Hoorn, S. Kamoun, From guard todecoy: a new model for perception of plant pathogen effectors, Plant Cell 20 (2008) 2009–2017.

[39] J.L. Dangl, J.D. Jones, Plant pathogens and integrated defence responses to infection, Nature 411 (2001) 826–833, https://doi.org/10.1038/35081161.

[40] D. Mackey, Y. Belkhadir, J.M. Alonso, J.R. Ecker, J. Dangl, *Arabidopsis* RIN4 is a target of the type III virulence effector AvrRpt2 and modulates RPS2-mediated resistance, Cell 112 (2003) 379–389.

[41] D. Mackey, B.F. Holt III, A. Wiig, J.L. Dangl, RIN4 interacts with Pseudomonas syringae type III effector molecules and is required for RPM1-mediated resistance in Arabidopsis, Cell 108 (6) (2002) 743–754.

[42] J.R. Gutierrez, et al., Prf immune complexes of tomato are oligomeric and contain multiple Pto-like kinases that diversify effector recognition, Plant J. 61 (2009) 507–581.

[43] T.S. Mucyn, A. Clemente, V.M. Andriotis, A.L. Balmuth, G.E. Oldroyd, B.J. Staskawicz, J.P. Rathjen, The tomato NBARC-LRR protein Prf interacts with Pto kinase in vivo to regulate specific plant immunity, Plant Cell 18 (10) (2006) 2792–2806.

[44] C. Zipfel, J.P. Rathjen, Plant immunity: AvrPto targets the frontline, Curr. Biol. 18 (5) (2008) R218–R220.

[45] T.S. Mucyn, A.J. Wu, A.L. Balmuth, J.M. Arasteh, J.P. Rathjen, Regulation of tomato Prf by Pto-like protein kinases, Mol. Plant-Microbe Interact. 22 (2009) 391–401.

[46] S.M. Collier, P. Moffett, NB-LRRs work a 'bait and switch' on pathogens, Trends Plant Sci. 14 (2009) 521–529.

[47] E. Reimer-Michalski, U. Conrath, Innate immune memory in plants, Semin. Immunol. 28 (2016) 319–327, https://doi.org/10.1016/j.smim.2016.05.006.

[48] A. Podgórska, M. Burian, B. Szal, Extra-cellular but extra-ordinarily important for cells: apoplastic reactive oxygen species metabolism, Front. Plant Sci. 8 (2017) 1353, https://doi.org/10.3389/fpls.2017.01353. R218–R220.

[49] P. Yuan, E. Jauregui, L. Du, K. Tanaka, B. Poovaiah, Calcium signatures and signaling events orchestrate plant–microbe interactions, Curr. Opin. Plant Biol. 38 (2017) 173–183, https://doi.org/10.1016/j.pbi.2017.06.003.

[50] E. Jeworutzki, M.R.G. Roelfsema, U. Anschütz, E. Krol, J.T.M. Elzenga, G. Felix, T. Boller, R. Hedrich, D. Becker, Early signaling through the Arabidopsis pattern recognition receptors FLS2 and EFR involves Ca^{2+}-associated opening of plasma membrane anion channels, Plant J. 62 (3) (2010) 367–378.

[51] G. de Lorenzo, S. Ferrari, F. Cervone, E. Okun, Extracellular DAMPs in plants and mammals: immunity, tissue damage and repair, Trends Immunol. 39 (2018) 937–950, https://doi.org/10.1016/j.it.2018.09.006.

[52] Y. Saijo, E.P. Loo, S. Yasuda, Pattern recognition receptors and signaling in plant–microbe interactions, Plant J. 93 (2018) 592–613, https://doi.org/10.1111/tpj. 13808.

[53] R. Willmann, H.M. Lajunen, G. Erbs, M.-A. Newman, D. Kolb, K. Tsuda, et al., Arabidopsis lysin-motif proteins LYM1 LYM3 CERK1 mediate bacterial peptidoglycan sensing and immunity to bacterial infection, Proc. Natl. Acad. Sci. USA 108 (2011) 19824–19829, https://doi.org/10.1073/pnas.1112862108.

[54] J.L. Caplan, A.S. Kumar, E. Park, M.S. Padmanabhan, K. Hoban, S. Modla, K. Czymmek, S.P. Dinesh-Kumar, Chloroplast stromules function during innate immunity, Dev. Cell 34 (2015) 45–57, https://doi.org/10.1016/j.devcel.2015.05.011.

[55] C.B. Patel, V.K. Singh, A.P. Singh, M. Meena, R.S. Upadhyay, Microbial genes involved in interaction with plants, in: H.B. Singh, V.K. Gupta, S. Jogaiah (Eds.), New and Future Developments in Microbial Biotechnology and Bioengineering, Elsevier, Singapore, 2019, pp. 171–180, https://doi.org/10.1016/B978-0-444-63503-7.00010-3.

[56] J. Kourelis, R.A.L. van der Hoorn, Defended to the nines: 25 years of resistance gene cloning identifies nine mechanisms for R protein function, Plant Cell 30 (2018) 285–299, https://doi.org/10.1105/tpc.17.00579.

[57] M. Ravensdale, M. Bernoux, T. Ve, B. Kobe, P.H. Thrall, J.G. Ellis, et al., Intramolecular interaction influences binding of the Flax L5 and L6 resistance proteins to their AvrL567 ligands, PLoS Pathog. 8 (2012) e1003004, https://doi.org/10.1371/journal.ppat.1003004.

[58] E.H. Chung, F. El-Kasmi, Y. He, A. Loehr, J.L. Dangl, A plant phosphoswitch platform repeatedly targeted by type III effector proteins regulates the output of both tiers of plant immune receptors, Cell Host Microbe 16 (2014) 484–494, https://doi.org/10.1016/j.chom.2014.09.004.

[59] V. Ntoukakis, I.M. Saur, B. Conlan, J.P. Rathjen, The changing of the guard: the Pto/Prf receptor complex of tomato and pathogen recognition, Curr. Opin. Plant Biol. 20 (2014) 69–74, https://doi.org/10.1016/j.pbi.2014.04.002.

[60] J. Wang, M. Hu, J. Wang, J. Qi, Z. Han, G. Wang, et al., Reconstitution and structure of a plant NLR resistosome conferring immunity, Science 364 (2019) eaav5870, https://doi.org/10.1126/science.aav5870.

[61] R. Panstruga, J.E. Parker, P. Schulze-Lefert, SnapShot: plant immune response pathways, Cell 136 (2009) 978.e1–978.e3, https://doi.org/10.1016/j.cell.2009.02.020.

[62] A. Piasecka, N. Jedrzejczak-Rey, P. Bednarek, Secondary metabolites in plant innate immunity: conserved function of divergent chemicals, New Phytol. 206 (2015) 948–964, https://doi.org/10.1111/nph.13325.

Chapter 9

Quorum quenching strategies of endophytic bacteria: Role in plant protection

Etisha Paul[a], Parikshana Mathur[b], Charu Sharma[a], and Payal Chaturvedi[a]
[a]*Department of Biotechnology, IIS (Deemed to be University), Jaipur, India,* [b]*Department of Biotechnology, Central University of Rajasthan, Ajmer, India*

1 Introduction

Endophytes are microorganisms usually bacteria and fungi that are found associated with plants. Endophytes just like other microbes exhibit the phenomenon of quorum sensing (QS) and quorum quenching (QQ). In quorum sensing, signaling compounds like *N*-acyl homoserine-lactones (AHL) and cyclic peptides are produced by microbes to communicate among themselves and to interact with plants and animals. Endophytes provide protection to plants from pathogens and thus are important for maintaining plant health. One of the strategies used by these endophytes to control infections is quorum quenching in which quorum sensing signals produced by pathogens are degraded leading to reduction in their virulence. This bioactive nature of endophytes is continuously being explored for protecting plants against potential pathogens and are, thus, considered as important biocontrol agents nowadays. This chapter focuses on quorum sensing and quorum quenching abilities of endophytes with specific reference to disease management for plant protection.

2 Endophytes

Microorganisms generally colonize healthy plant tissues and reside inside them either completely or for some time. These microbes called endophytes are non-pathogenic and have been reported from roots, cotyledons, stems, flowers, seeds, and fruits of different species of plants. They can be isolated by surface sterilizing the plant segment with different agents and inoculating the sterilized segments on suitable media [1]. Alternatively, the sterilized plant parts can be

Plant-Microbe Interaction—Recent Advances in Molecular and Biochemical Approaches
https://doi.org/10.1016/B978-0-323-91876-3.00005-1
149

crushed in sterile saline solution and the serially diluted suspension can be spread on preferred media [2]. A number of endophytic species of bacteria or fungi can be isolated from a single plant [3]. Endophytes enter plant tissue by defeating defense mechanisms of the host [4]. In one study, the mechanism of entry of an endophyte *Bacillus subtilis* BSn5 in *Arabidopsis thaliana* and *Amorphophallus konjac* was explored. It was reported that *Bacillus subtilis* produces a lantibiotic subtilomycin, which suppresses the defense mechanism of plant to inhibit the entry of microbes. This in turn favors the colonization of plants by its endophyte [5].

Endophytes help in nutrient uptake in host plants. They produce different phytohormones such as indole acetic acid (IAA), cytokinin, auxins, and gibberellic acids and thereby enhance plant growth. They can degrade organic compounds and, thus, help in nutrient cycling. Endophytes also act as biocontrol agents, as they protect plants as well as animals form pathogens [6,7]. They also enable plants to tolerate abiotic stresses like drought, salinity, heavy metals, etc. They produce a variety of secondary metabolites such as terpenoids, alkaloids, phenols, and phytohormones, which act as antibacterial, antitumor, antifungal, antiviral, antioxidant, and antiinflammatory compounds [8]. These are nowadays also used in the synthesis of nanoparticles [9]. Endophytes also show a quorum sensing (QS) phenomenon to communicate with each other and use these sensing signals for gene expression. QS is based on the production of signaling compounds (such as *N*-acyl homoserine lactones (AHL) and cyclic peptides) and is used for communication and interactions of microbes with plants, animals, and within their own community. Inhibition of the phenomenon of QS by enzymes/bioactive metabolites is known as quorum quenching (QQ). Many endophytes are capable of inhibiting QS signals produced by plant pathogens and thus protect plants against pathogen attack. They help in promoting plant growth by reducing the virulence of phytopathogens and maintaining plant's health [10]. Similarly, incorporating AHL-degrading enzymes in plants has proven beneficial for inhibiting the growth of invading pathogens. This strategy can be explored in agriculture for crop improvement and its protection from infectious diseases.

3 Quorum sensing

During high cell density, microorganisms communicate with each other through chemical signals in a QS process [11]. These QS signals consist of *N*-acyl homoserine lactones (AHL), autoinducing peptides (AIPs), and autoinducer-2 (AI-2) molecules. Gram-positive and Gram-negative bacteria use different signals to communicate with each other [12]. QS was first studied in the marine bacterial species, *Vibrio fischeri*. QS enables bacteria to coordinate their behavior and control many physiological processes [13] (Fig. 1).

FIG. 1 Physiological processes regulated by quorum sensing in bacteria.

3.1   Quorum sensing systems

In general, the QS system comprises of three key components:

- Synthase protein—produces QS signals.
- Quorum sensing signals—AHL molecules (Autoinducer-1 or AI-1), autoinducer 2 (AI-2), and cyclic peptides.
- Response regulator—binds QS signals and regulates the expression of target genes [14].

Different phytopathogens have been studied for their QS systems (Table 1). The details of such systems are explained in the subsequent sections.

3.1.1   AHL-based quorum sensing system

It is found usually in Gram-negative bacteria that produces N-acyl homoserine lactones (AHL), (sometimes referred to as AI-1) as their signaling molecules. AHL structures are made up of a homoserine lactone (HSL) ring associated with an acyl chain ranging from C4-C18 carbon atoms [19]. Many Gram-negative bacteria use the AI synthase enzyme (a member of the LuxI family) to synthesize AHL molecules from S-adenosylmethionine (SAM) and acyl carrier protein (ACP). AHL diffuses across the cell and accumulates in the environment.

TABLE 1 Phytopathogens and their QS systems.

S. no.	Bacteria	Type of QS system	Disease produced	References
1	*Erwinia* sp. Eca and Ecc	AHL and AI-2-based QS	Soft-rot disease in plants	[15]
2	*Xanthomonas* species	Diffusible signal factors (DSF)-based QS having *rpf* cluster	Blight disease in rice crops (Xoo) and black rot in crucifers, cauliflower, cabbage (Xcc)	[16]
3	*Pseudomonas syringae* pv. *tabaci* 11528	AHL-based QS system PsyI/PsyR	Wild-fire disease in agricultural crops	[17]
4	*Dickeya* species	Vfm-based QS	Soft rot disease of many fruits and vegetables	[18]
5	*Pectobacterium* spp.	AHL-based QS system ExpI and ExpR	Soft rot disease of many agriculture crops and plants	[18]

When the concentration of AHL increases, it binds to its specific receptors (LuxR-type proteins) and regulates the target gene expression [20].

The endophytic bacterium *Serratia plymuthica* G3 strain isolated from wheat stem was reported to have two types of QS systems, SplIR and SpsIR, which produced a wide variety of HSL signals. AHL synthase (*splI* and *spsI*) and response regulator genes (*splR* and *spsR*) of these two systems were found to be homologs of the LuxI/LuxR system and were involved in signal production and regulation of gene expression. QS signals produced by SplI included C6-HSL, C7-HSL, 3-oxo-HSL derivatives (3-oxo-C6-HSL, 3-oxo-C7-HSL), and a 3-hydroxy derivative (3-hydroxy-C6-HSL) whereas SpsI produced non-substituted C4-HSL and C5-HSL. It was also reported that these QS systems in *Serratia plymuthica* G3 strain were also responsible for the regulation of anti-fungal activity and exoenzyme production [21].

3.1.2 Autoinducing peptide-based quorum sensing system

It is found in Gram-positive bacteria that produce oligopeptides as autoinducers called autoinducing peptide (AIP) instead of HSL. The QS system in these bacteria is regulated by a two component system that has a sensor kinase protein to detect the signal produced and a response regulator protein to regulate the expression of the target gene. The AIPs are initially produced as precursor compounds (pro-AIP), processed, and secreted outside the cell via specialized transporter proteins. The size of the modified AIPs can vary in different bacterial species and may range from 5 to 17 amino acids, which can be either linear or cyclic [22]. As the cell density increases, a high level of AIP also accumulates in the environment. Its high concentrations are detected by a sensor kinase receptor protein, which phosphorylates a conserved histidine residue present on it after binding. The activated kinase in turn phosphorylates a conserved aspartate residue present on a response regulator, which results in the regulation of expression of QS target gene activity. In Gram-positive bacteria, all these components involved in QS, including generation and processing of pro-AIPs, are encoded by a single operon that is activated by a phosphorylated response regulator [23].

In *Bacillus subtilis*, common soil bacteria, the development of competence and the sporulation process are dependent on a QS system that produces peptides and is regulated by a two-component system. The peptide that mediates competence is called ComX which, after processing, accumulates in high concentrations in the environment on increase in cell density. This is detected by ComP (sensor kinase), which becomes phosphorylated and transfers the phosphorylation signal to ComA (response regulator protein). The phosphorylated ComA protein, in turn, enhances the expression of the *comS* gene, whose product ComS inhibits the proteolysis of ComK. ComK regulates the transcription of structural genes involved in the development of competence. Similarly, during the sporulation process in *Bacillus subtilis*, the CSF peptide (sporulation factor) is generated. During high cell density, CSF accumulates extracellularly and is transported inside the cell by an ABC-type transporter called the Opp protein. High concentrations of CSF inhibit the expression of the *comS* gene and promote sporulation. Whereas, at low concentrations, CSF inhibits RapC, which is a ComA-specific phosphatase increasing the levels of activated ComA that favors competence [24].

3.1.3   Autoinducer-2-based quorum sensing system

It is found in both Gram-negative and in Gram-positive bacteria and uses AI-2 as the QS signal. AI-2 produced by one bacterium can be sensed by other bacteria and thus it is used for interspecies communication. The production, release, and signal transfer of AI-2 is under regulation of the *lux*S operon in Gram-negative *Escherichia coli*. The synthesis of AI-2 requires the LuxS enzyme in the presence of which *S*-adenosylhomocysteine (SAH) is converted to homocysteine to form an AI-2 precursor called 4,5 dihydroxy-2,3-pentanedione (DPD). The DPD formed can then undergo different types of intramolecular cyclization producing different types of AI-2 in different bacterial species. The AI-2 system is also involved in biofilm formation by many bacteria. The detection of AI-2 in these bacteria is also made through a two-component system that involves a sensor kinase, which transfers a phosphorylation signal to a central signal relay protein that, in turn, activates the response regulator. In many other bacteria, the presence of AI-2 is detected by an ABC transporter called the Lsr-receptor complex. *Klebsiella* spp. are free-living nitrogen fixers found in soil; they are also present as endophytes of many plants. The presence of the *lsr* operon required for the AI-2 ABC transporter has been observed in these microbes. *Sinorhizobium meliloti*, a nitrogen-fixing endophyte associated with the leguminous plants of *Medicago*, *Melilotus*, and *Trigonella* genera is also reported to have specific receptors for AI-2 on its pSymB plasmid. However, the presence of the *luxS* gene has not established in this endophyte [25].

3.1.4 Virulence factor modulating (VFM)-based quorum sensing system

Besides these QS systems, *Dickeya* species that causes the soft-rot disease of many fruits and vegetable crops has a new and different type of QS system called the virulence factor modulating (vfm) QS system. This system is found only in *Dickeya* sp. and is studied in detail in *ickeya dadantii* 3937. In this phy-topathogen, the *vfm* locus regulates the production of genes involved in plant cell wall-degrading enzymes and is also necessary for establishing virulence. The signaling molecule involved in QS has not been completely identified and characterized but is reported to have modified amino acids and fatty acids in its structure. All the components involved in synthesis, production of QS sig-nals, and virulence regulation are part of the *vfm* locus. The two-component reg-ulatory system is known as VfmHI. VfmH is activated by binding to a QS signal and in turn activates VfmE, which is a transcriptional activator, activates viru-lence factors that include genes for plant cell wall degrading enzymes and the Vfm system [18].

4 Virulence and pathogenicity in plants

Virulence is the ability of pathogens to cause infection to the host. A number of virulence factors are produced by bacterial phytopathogens, which lead to var-ious diseases in plants. In many of these pathogens, the expression of genes involved in virulence is regulated by quorum sensing signals. In phytopatho-gens, the production of many virulence factors that include plant cell wall degrading enzymes, factors inducing necrosis, type III secreted systems, etc. is regulated by QS [17,26].

Two phytopathogenic species of *Erwinia*, namely, *Erwinia carotovora* ssp. *atroseptica* (*Eca*) and *Erwinia carotovora* ssp. *carotovora* (*Ecc*) causes soft-rot disease in plants. It uses AHL and AI-2 sensing signals to regulate the QS pro-cess. This phytopathogen is reported to have two types of QS systems mediated by both AHL and AI-2 that play a vital role in its virulence. Both *Ecc* and *Eca* strains produced 3-oxo-C6 HSL signals, which regulates extracellular enzyme and virulence factor production [15]. In a study, complex microbial interactions mediated by the QS systems of a plant pathogen (*Pseudomonas savastanoi* pv. *Savastanoi* (*Psv*)), an endophyte (*Erwinia toletana*) and an epiphyte (*Pantoea agglomerans*) of the olive tree were studied. *Psv* produces olive knot disease and it was reported to be aggravated by other two residents, which are usually not pathogenic. Both the resident flora and pathogen showed the presence of AHL-based QS systems leading to the development of polybacterial disease [27]. The endophytic bacteria *Ralstonia solanacearum* is a pathogen of various plants like tomato, potato, and eggplant. It causes bacterial wilt disease in egg-plants. The QS system of this pathogen is activated by the production of a volatile compound 3-hydroxy-palmitic acid methyl ester (3OH-PAME), which is the product of the *phcB* gene. This signal regulates the expression of virulence factors [28]. *Agrobacterium tumefaciens* infects many plants, causes crown gall disease inhibiting the growth and yield of the plant. Upon infection, it transfers the tumor-inducing (Ti) plasmid to plant parts that contain virulence genes for tumor production. The QS system of *Agrobacterium tumefaciens* works similarly to LuxI/R type and is involved in regulating the conjugative transfer of plasmids and their replication. TraI is an AHL synthase, homolog of LuxI (produces 3-oxo-C8 HSL) and TraR is a regulator protein, homolog of LuxR. Besides these two genes, the QS system also contains a TraM protein, which can suppress the transcriptional activity of TraR [17,29]. The phytopathogen *Pseudomonas syringae* pv. *tabaci* 11,528 causes wild-fire disease in agricultural crops. Its QS system has also been identified as a homolog of the LuxI-LuxR system and is named PsyI/PsyR. PsyI produces signal molecules 3-oxo-C6-HSL and C6-HSL, which repress the genes involved in biofilm formation, motility, and pili assembly during the log phase [17]. *Pectobacterium* spp. causes soft-rot disease of many agriculture crops and plants including potato. Virulence in it is regulated by a QS mechanism that produces AHL signals similar to that present in the LuxI/LuxR type. LuxI and LuxR homologs are known as ExpI and ExpR in many *Pectobacterium* sp. Two major QS signals, 3-oxo-C8-HSL and 3-oxo-C6-HSL, are produced by ExpI. They regulate the production of virulence factors like plant cell wall degrading enzymes (like pectinases, cellulases, proteases, xylanases, and phospholipases). Similarly, *Pectobacterium atrosepticum* produces 3-oxo-C6-HSL, C6-HSL, 3-oxoC8-HSL, and 3-oxo-C10-HSL QS signals to regulate the synthesis of virulence factors [18].

Besides the AHL and AI-2-mediated QS systems, many phytopathogens have QS systems that are based on the production of diffusible signal factors (DSFs). This system has been identified in many pathogenic species of *Xanthomonas* that infects many crops such as tomato, rice, pepper, etc. *Xanthomonas oryzae* pv. *Oryzae* (Xoo) causes bacterial blight disease in rice crops. *Xanthomonas campestris* pv. *Campestris* (Xcc) causes black rot in crucifers, cauliflower, cabbage, etc. These pathogens produce different DSF signals such as cis-9-methyl-2-decenoic acid, cis-2-undecenoic acid, DSF (cis-11-methyl-2-dodecenoic acid), BDSF (cis-2-dodecenoic acid), CDSF (cis-11-

methyldodeca-2,5-dienoic), and IDSF (cis-10-methyl-2-dodecenoic acid) through which they control the production and expression of virulence factors. The gene cluster involved in encoding the DSF system is named the *rpf* cluster in which rpfF is a DSF synthase that produces different DSF signals. rpfC and rpfG are part of two-component signaling systems having RpfC as the hybrid sensor kinase and RpfG as the response regulator that are activated by phosphorylation. At high cell density, DSF signals are produced and the QS system is activated leading to synthesis of many virulence factors like plant cell wall degrading enzymes, adhesion proteins, *exo*-polysaccharides, iron uptake factors, proteins involved in oxidative stress, and biofilm formation [16].

The abovementioned studies have shown that the regulation and production of these virulence factors in many phytopathogens are controlled by the QS system. Virulence produced in response to QS signals can be inhibited using the QQ strategy. QQ is an alternative approach to control plant infection and diseases in an ecofriendly way. It can be of wide applicability in agriculture as it can protect plants against pathogens and increase their yield.

5 Quorum quenching

The phenomenon of inhibiting quorum sensing is known as quorum quenching (QQ). It is controlled by different quorum quenching compounds. These QQ compounds are noninhibitory to bacterial growth and directly interact with quorum sensing components. They work enzymatically by the degradation of signaling molecules or can block signal generation and signal reception [30].

5.1 Strategies for quorum quenching

The inhibition in the communication of the bacteria using QQ molecule from endophytes is a useful approach to control virulence and many bacterial diseases [31]. The endophytes play an important role in plant defense using a QQ strategy [32]. On the basis of structure and functions, quorum sensing inhibitors can be classified into two groups. The first group contains a molecule that is structurally similar to QS signals like halogenated furanones and synthetic autoinducing peptides which are similar to AHL and AIP signals. The second group contains small chemicals that act as enzyme inhibitors [33]. Different strategies have been found in plants to inhibit quorum sensing processes, which can control virulence and pathogenicity. These strategies include the inhibition of detection or synthesis of QS signals and degradation of QS signals (Fig. 2).

5.1.1 Inhibition of QS signal biosynthesis

The synthesis of small AHL molecules that act as QS signals can be inhibited by various compounds, which lead to the inhibition of the QS process. AHL molecules are synthesized using substrates such as SAM and acylated ACP in the

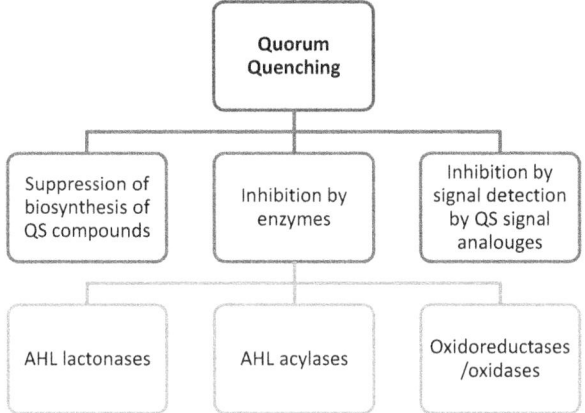

FIG. 2 Strategies for quorum quenching.

presence of AHL synthase enzyme. Inhibition of the FabI enzyme (that generates an acyl chain for AHL synthesis) will result in the inhibition of AHL production. FabI is reported to be inhibited by antibacterial agents such as diazobroines and triclosan of which the latter suppresses the synthesis of C4-HSL. Some compounds such as S-anhydroribosyl-l-homocysteine and S--homoribosyl-l-cysteine act as structural analogs of LuxS and competitively inhibit its activity. The synthesis of both AHL and AI-2 takes place in the presence of MTAN ($5'$-methylthioadenosine/S-adenosylhomocysteine nucleosidase); two classes of MTAN inhibitors are known, Immucillin A (ImmA) derivatives and DADMe-ImmA. Few reports are available on the inhibitors of autoinducing peptides. Components like ribosomes and peptidases involved in their synthesis are also used in bacterial growth and survival. Thus, their inhibition as a part of a QQ process also exerts bactericidal effects. Ambuic acid, a fungal secondary metabolite, can inhibit the synthesis of QS in some Gram-positive bacteria [19]. QS can also be inhibited by blocking the efflux pumps that help in the movement of long chain of AHL across the membrane [13].

5.1.2 Inhibition by enzymes

The degradation of QS signals by using enzymes is an effective approach to inhibit the communication between bacteria. The AHL-degrading enzymes are categorized into three categories namely: AHL-lactonases, AHL-acylases, and oxidases/oxidoreductases [34].

AHL-lactonases degrade AHL by hydrolyzing ester bonds present in the lactone ring of AHL. The open ring inactivates the AHL signals. These enzymes are metallo-proteins, usually encoded by the *aiiA* (Autoinducer inactivation) gene in many bacterial species [19,35]. It is one of the most extensively reported QQ enzymes. AHL degradation by lactonase activity was reported in many endophytic species of *Enterobacter* isolated from *Ventilago madraspatana* commonly called a woody climber [32]. In a study, a total of 168 isolates associated with tobacco leaves possessed AHL-lactone degradation activity. These bacteria were identified to be members of *Bacillus*, *Lysinibacillus*, *Acinetobacter*, *Pseudomonas*, *Serratia*, and *Myroides* genera [36]. The cell-free lysate of the endophyte *Enterobacter asburiae* PT39 and *Bacillus firmus* PT18 isolated from *Pterocarpus santalinus* showed potent ability to degrade AHL by lactonase activity [37]. AHL lactonase-mediated QQ was reported in four endophytic bacteria isolated from *Cannabis sativa* L. These were identified to be *Bacillus* sp. Strain B3, *Brevibacillus borstelensis* strain B8, *Bacillus megaterium* strain B4, and *Bacillus* sp. Strain B11. They were effective in weakening the AHL-lactone quorum signals like C6-HSL, C8-HSL, C10-HSL, and 3-oxo-C10-HSL of *Chromobacterium violaceum* at different concentrations. Endophytic bacteria were reported to selectively inhibit and modulate all the four QS signals released by *Chromobacterium violaceum* [38]. Endophytic *Microbacterium testaceum* isolated from the leaves of potato was reported to have the *aiiM* gene that has AHL lactonase activity [39]. AiiA lactonase degrading activity was reported in an endophytic *Bacillus thuringiensis* strain KMCL07 isolated from *Madhuca insignis* that inhibited the production of virulence factors and biofilm by *Pseudomonas aeruginosa* PAO1 [40].

AHL-acylases degrade AHL signals to the corresponding fatty acids and homoserine lactone by hydrolyzing the amide bond of AHLs. This enzyme is encoded by the *aiiD* gene in many bacterial species [19]. The presence of HSL acylase was reported in an endophytic *Streptomyces* LPC029 isolated from the plants of Thailand. The enzyme exhibited broad substrate specificity and can break the amide bond present in the acyl chain of HSL that ranges from C_6 to C_{12} carbon atoms [41].

Oxidoreductases/oxidases. These enzymes inactivate AHL by oxidation and reduction process. In the presence of these enzymes, the chemical structures of the signal are modified instead of being degraded. It is usually found in very few bacterial species [19]. It was first discovered in *Rhodococcus erythropolis* bacteria. The *Rhodococcus erythropolis* strain W2 isolated from the tobacco rhizosphere was grown on minimal media supplemented with different types of HSL and was screened for QQ activity using *Escherichia coli* (pSB1075 and pSB401) as biosensor strains. HPLC and LC-MS/MS analyses of 3-oxo-AHL with *Rhodococcus erythropolis* confirmed the presence of oxidase activity [42]. The QQ bacterium *Burkholderia* (GG4 strain) isolated from *Zingiber officinale* reduced 3-oxo-AHLs into 3-hydroxy compounds and was found to produce oxidoreductase activity [43].

The enzymatic inactivation of QS signal molecules other than AHL has also been reported. Many plant pathogens use compounds of the DSF family as QS signals. Two genes were identified as *carA* and *carB* in native microflora associated with plants whose products inhibited the DSF signals. They produced a heterodimeric enzyme made up of two components, CarA and CarB, and the enzyme produces carbamoyl phosphate [19].

5.1.3 Inhibition of QS signal detection

It has been reported that the synthesis of analogs of native signals and already identified inhibitors can also inhibit the QS process by inhibiting signal detection. Binding of such analogs to receptor complexes interrupts with downstream signaling. Synthesis of such nonproductive signal receptor complex competitively blocks binding by the original native signal. Some examples of AHL analogs include lactams, thiolactones, triazolyldihydrofuranones, and urea analogs. Moreover, structures of already identified inhibitor molecules can also be modified, which can result in improved inhibitory activities and broader target specificities. This task has been made easy with high-throughput and computer-aided screening of small molecules that can disrupt QS. Disubstituted imidazolium salts and N-acyl cyclopentylamides are examples of such non-AHL compounds that are reported to be active against LuxR proteins. Besides these, the structures of existing natural AHLs can also be modified with respect to length, saturation of fatty acyl chain, and oxidation states [19]. In one of the studies, many synthetic QS modulators that disrupt QS processes have been identified for the phytopathogen *Pectobacterium carotovora* that causes infection in *Solanum tuberosum* (potato) and for *Pseudomonas syringae* that infects *Phaseolus vulgaris* (green bean) [44].

Analogs have been identified not only for AHL but also for AI-2. In a study, three different types of screening methods were used to identify analogs of AI-2 that can form complexes with LuxP, the receptor of *Vibrio harveyi*. In the first two phases, diol-containing compounds, which are similar to the 2,3-borate diester form of AI-2 (detected by LuxP), were screened. In the last screening, an in silico analysis method was used to test 1.7 million compounds available in commercial compound databases that can act against the LuxP-AI-2 complex. From this study Pyrogallol was found to be the best inhibitor. Similar to AHL and AI-2 analogs, peptide analogs are also known. Inhibitors, which are natural variants of QS signals and act against AgrC receptors of different strains of *Staphylococcus. aureus*, have also been identified [19].

6 Role of quorum quenching to protect plants against phytopathogens

The relationships between plant and bacteria are beneficial to each other. Usually, plants are exposed to infections caused by phytopathogens, many of which are under the control of the QS process. So, inhibiting these processes via QQ can help in protecting plants against the pathogens. Three bacterial strains *Acinetobacter* GG2, *Burkholderia* GG4, and *Klebsiella* Se14 isolated from the rhizosphere of ginger showed AHL degrading activity. Besides AHL inactivation, *Acinetobacter* and *Burkholderia* also showed the production of AHL molecules, thus displaying the QS activity as well. All three isolates were coinoculated with pathogenic *Erwinia carotovora* in potato tubers to observe virulence attenuation in terms of reduction in pectinase activity that depends on the production of 3-oxo-C6-HSL by *Erwinia*. Reduced maceration of potato tissue was observed in coculture experiments. In the same study, virulence was also reported to be reduced in the human pathogen *Pseudomonas aeruginosa* in the presence of these QQ bacteria [43]. QQ activity was reported in bacteria isolated from the rhizosphere and phyllosphere of saffron, fig, and pomegranate plants. It was reported that most of the bacteria associated with these plants showed AHL degrading activity that belonged to the *Pseudomonas* genus. Enzymatic QQ activity was reported in both extracellular and intracellular extracts of *Pseudomonas* and many of them also had nonenzymatic QQ activity. The QQ bacteria that belonged to the *Pseudomonas* genus acted as biocontrol agents and reduced potato rot disease caused by the *Pectobacterium carotovorum* phytopathogen [45].

Pseudomonas segetis strain P6 isolated from the *Salicornia europaea* rhizosphere possessed QQ activity and also reduced soft-rot in potato and carrot caused by *Dickeya solani, Pectobacterium atrosepticum*, and *Pectobacterium carotovorum*. Plant growth-promoting activity was also observed in the P6 strain as it increased height and weight in tomato plants [46]. Similarly, different QQ enzymes degrade specific QS signals of phytopathogens and act as biocontrol agents. HSL acylase reported from endophytic *Streptomyces* LPC029 was observed to reduce the expression of soft-rot disease in potato caused by *Pectobacterium carotovorum* ssp. *carotovorum* [41]. In a study, endophytic bacteria isolated from the xylem of chilli, eggplant, and *olanum torvum* were studied for QQ activity. Five bacterial isolates namely *Stenotrophomonas maltophilia* (XB102), *Pseudomonas aeruginosa* (XB7 and XB122), and *Rhodococcus corynebacterioides* (XB109 and XB115) were reported to degrade the 3-OH-PAME QS signal of *Ralstonia solanacearum* in greenhouse studies. The production of exopolysaccharides and endoglucanase was reduced leading to decreased virulence [28]. Twenty AHL-inactivating endophytic bacteria were isolated from potato tuber (*Solanum tuberosum* L. cv KT3). Out of these, four isolates *Bacillus* sp., *Variovorax* sp., *Variovorax paradoxus*, and *Agrobacterium tumefaciens* possessed AHL lactonase activity, degraded C6-HSL signals produced by *Pectobacterium carotovorum*, and suppressed tuber soft-rot in potatoes [47]. Two QQ endophytes namely, *Bacillus cereus* Si-Ps1 and *Pseudomonas azotoformans* La-Pot3-3 were isolated from the leaves of citrus plants. *Bacillus cereus* Si-Ps1 was reported to have a homolog of the *aiiA* gene possessing lactonase activity that inhibited the QS signals of *Pseudomonas syringae* pv. *syringae* Pss 3289. The study was also conducted in a greenhouse and it was reported that virulence in *Pss* 3289 was reduced on coinoculation of a citrus plant with these endophytes [48].

This strategy can also be used in genetic engineering to insert the QQ gene of interest in plants that might protect it against infection. Reduction in plant disease can also be achieved by artificially inserting the gene of a QQ enzyme into the plant pathogen. In a study, three different genes (*aiiA*, *attM*, and *aiiB*) encoding lactonase enzymes were cloned in to the p6010 plasmid and its derivatives. These plasmids were then introduced into *Erwinia carotovora* subsp. *Atroseptica* 6276 to observe the degradation of acyl HSL molecules produced by *Erwinia*. It was reported that *Erwinia* strains containing these plasmids led to reduced concentration of acyl HSL in the medium. Furthermore, when tubers of *Solanum tuberosum* were experimentally infected with *Erwinia* strains harboring plasmids with lactonase genes, decrease in maceration was observed confirming the reduction in virulence of *Erwinia* on the expression of lactonase genes [49]. This modification helps in crop protection from pathogens and gives higher yield. The *aiiA* gene from *Bacillus thuringiensis* was transformed into endophyte *Burkholderia* sp. KJ006, isolated from rice roots. The engineered strain reduced seedling rot disease in rice caused by *Burkholderia glumae* [50]. In a study, cloning and characterization of the *aiiM* gene encoding AHL lactonase from *Microbacterium testaceum* StLB037 isolated *Solanum tuberosum* leaf surface was reported. The QQ *aiiM* gene, when expressed in *Pectobacterium carotovorum* pathogen, reduced pectinase and soft-rot disease in potato plants [51]. In another study, the *aiiA* gene was inserted and expressed in *Pseudomonas putida* cells. These genetically engineered *Pseudomonas putida* cells offered protection to potato plants against the pathogen *Erwinia carotovora* that causes soft-rot disease [52]. A study revealed the presence of the new gene *aiiE* (autoinducer inactivation gene from *Enterobacter*) from *Enterobacter* sp. CS66 isolated from *Coscinium fenestratum* Gaertntree that showed lactonase activity. Biocontrol potential of this isolate was also studied and it was reported that the *aiiE* gene when expressed in the plant pathogen *Pectobacterium atrosepticum* reduced virulence by decreasing the production of plant cell wall degrading enzymes [53]. For agricultural purposes, these recombinant strategies can prove beneficial to control bacterial infections in plants and consequently will help to increase crop yield. The growth promoting activity of endophytes can also help in lowering the requirement of chemical fertilizers, reducing production cost, and increasing environment sustainability. Hence, the QQ strategy by endophytes has an important biotechnological application to inhibit QS regulated pathogenicity and virulence. The plant-related bacteria also produce volatile organic compounds (VOCs) of low molecular weight. Common examples of these compounds are dimethyl sulfide, dimethyl disulfide (DMDS), and dimethyl trisulfide. These VOCs are involved in inter-kingdom communication, might have antibiotic activity, and also promote plant growth. QQ activities of these have also been reported [17,54]. In a study, VOCs from *Pseudomonas fluorescens* (Q8r1-96 and B-4117) and *Serratia plymuthica* IC1270, repressed tumor growth in tomato caused by *Agrobacterium tumefaciens*. *Serratia plymuthica* IC1270 produced DMDS in high quantities while, *Pseudomonas fluorescens* strains produced 1-undecene in high quantities [55].

7 QQ molecules from plants

There are many plants, which produce chemical compounds similar to QS signals produced in response to bacterial infections. They protect plants against pathogens by inhibiting their QS-regulated gene expression, which results in their virulence behavior. These compounds may interfere with the synthesis of AHL signals, transferring of signals, or reception of signals. These include many cyclic compounds like, vanillin produced by *Vanilla planifolia*, tannic acid produced by *Quercus,* and flavonoids from *Citrus sinensis* and *Citrus sinensis*. Many noncyclic compounds are also reported which include, iberin from *Armoracia rusticana*, ajoene, which is a disulfide produced by *Allium sativum*, acetaldehyde, etc. [56].

Some plants like *Medicago truncatula* and *Pisum sativum* were reported to produce compounds which can mimic AHL signals thereby affecting the AHL-based QS process in bacteria associated with them [57].

Medicago sativa (alfalfa) seeds showed the production of L-canavanine, an arginine analog that inhibited the production of extracellular polysaccharide II in the symbiont *Sinorhizobium meliloti* [58]. This polysaccharide helps in the invasion of plant cells by its symbiont and its production is regulated by QS processes. Many studies have reported that the QS inhibitory activity of plant extracts can be used to reduce the virulence caused by human pathogens and can act as an alternative to antimicrobial agents. This would also help to overcome the ever-growing problem of antimicrobial resistance in pathogens.

8 Conclusions

Quorum sensing is an important process found in the bacterial community that enables them to regulate a number of physiological processes, involved in cross species talk and is important for virulence in many pathogens. Many phytopathogenic species of bacteria produce QS signals like AHL, AI-2, peptides, and DSFs, which regulate the production of virulence factors in them and are important for establishing pathogenicity. One of the important strategies to control plant infections is to inhibit the QS process which leads to reduced virulence. The endophytic microbial community that resides inside plants exhibits this process of quorum quenching and thus protects plant against attack by pathogenic bacteria. Moreover, transgenic plants can also be produced that carry the gene of interest that blocks QS process. These approaches hold promise for integrated disease management in plants and will help to maintain plant health.

Author contributions

CS conceived the idea of the chapter, provided the general concept, provided input for each specific section, and EP and PM wrote the review after collecting the literature and prepared the draft. CS and PC edited, compiled, and finalized the draft. Finally, all the authors read and approved it for publication.

References

[1] N. Anjum, R. Chandra, Endophytic bacteria: optimization of isolation procedure from various medicinal plants and their preliminary characterization, Asian J. Pharm. Clin. Res. 8 (4) (2015) 233–238.

[2] M.D.F. Kahtani, A. Fouda, K.A. Attia, F. Al-Otaibi, A.M. Eid, E.E.-D. Ewais, M. Hijri, M. St-Arnaud, S.E.-D. Hassan, N. Khan, Y.M. Hafez, K.A.A. Abdelaal, Isolation and characterization of plant growth promoting endophytic bacteria from desert plants and their application as bioinoculants for sustainable agriculture, Agronomy 10 (2020) 1325.

[3] R.X. Tan, W.X. Zou, Endophytes: a rich source of functional metabolites, Nat. Prod. Rep. 18 (4) (2001) 448–459.

[4] J. Goutam, R. Singh, R.S. Vijayaraman, M. Meena, Endophytic fungi: carrier of potential antioxidants, in: P. Gehlot, J. Singh (Eds.), Fungi and Their Role in Sustainable Development: Current Perspectives, Springer, Singapore, 2018, pp. 539–551, https://doi.org/10.1007/978-981-13-0393-7_29.

[5] Y. Deng, H. Chen, C. Li, J. Xu, Q. Qi, Y. Xu, Y. Zhu, J. Zheng, D. Peng, L. Ruan, M. Sun, Endophyte *Bacillus subtilis* evade plant defense by producing lantibiotic subtilomycin to mask self-produced flagellin, Commun. Biol. 2 (2019) 368.

[6] D.N. Nair, S. Padmavathy, Impact of endophytic microorganisms on plants, environment and humans, Sci. World J. (2014), https://doi.org/10.1155/2014/250693.

[7] G. Yadav, M. Meena, Bioprospecting of endophytes in medicinal plants of Thar desert: an attractive resource for biopharmaceuticals, Biotechnol. Rep. 30 (2021) e00629, https://doi.org/10.1016/j.btre.2021.e00629.

[8] M. Singh, A. Kumar, R. Singh, K.D. Pandey, Endophytic bacteria: a new source of bioactive compounds, 3 Biotech 7 (5) (2017) 315, https://doi.org/10.1007/s13205-017-0942-z.

[9] P. Mathur, S. Saini, E. Paul, C. Sharma, P. Mehtani, Endophytic fungi mediated synthesis of iron nanoparticles: characterization and application in methylene blue decolorization, Curr. Res. Green Sustain. Chem. 4 (2021), https://doi.org/10.1016/j.crgsc.2020.100053.

[10] M. Meena, A. Zehra, P. Swapnil, Harish, A. Marwal, G. Yadav, P. Sonigra, Endophytic nanotechnology: an approach to study scope and potential applications, Front. Chem. Nanosci. 9 (2021) 613343, https://doi.org/10.3389/fchem.2021.613343 (Impact Factor: 5.221).

[11] X. Fan, T. Ye, Q. Li, P. Bhatt, L. Zhang, S. Chen, Potential of a quorum quenching bacteria isolate *ochrobactrum* intermedium d-2 against soft rot pathogen *Pectobacterium carotovorum* subsp. *carotovorum*, Front. Microbiol. 11 (2020) 898, https://doi.org/10.3389/fmicb.2020.00898.

[12] Q. Jiang, J. Chen, C. Yang, Y. Yin, K. Yao, Quorum sensing: a prospective therapeutic target for bacterial diseases, Biomed. Res. Int. (2019) 1–15, https://doi.org/10.1155/2019/2015978.

[13] A.K. Bhardwaj, K. Vinothkumar, N. Rajpara, Bacterial quorum sensing inhibitors: attractive alternatives for control of infectious pathogens showing multiple drug resistance, Recent Pat. Antiinfect. Drug Discov. 8 (1) (2013) 68–83.

[14] R.T. Pena, L. Blasco, A. Ambroa, B. González-Pedrajo, L. Fernández-García, M. López, M. Tomás, Relationship between quorum sensing and secretion systems, Front. Microbiol. 10 (2019) 1100, https://doi.org/10.3389/fmicb.2019.01100.

[15] A.M.L. Barnard, S.D. Bowden, T. Burr, S.J. Coulthurst, R.E. Monson, G.P.C. Salmond, Virulence and secondary metabolite production in plant soft-rotting bacteria, Philos. Trans. R. Soc. B 362 (2007) 1165–1183.

[16] Y.W. He, J. Wu, J.S. Cha, L.H. Zhang, Rice bacterial blight pathogen *Xanthomonas oryzae* pv. *oryzae* produces multiple DSF-family signals in regulation of virulence factor production, BMC Microbiol. 10 (2010) 187, https://doi.org/10.1186/1471-2180-10-187.

[17] F.A. Ansari, I. Ahmad, Quorum sensing in phytopathogenic bacteria and its relevance in plant health, in: V. Kalia (Ed.), Biotechnological Applications of Quorum Sensing Inhibitors, Springer, Singapore, 2018, pp. 351–370, https://doi.org/10.1007/978-981-10-9026-4_17.

[18] J. Baltenneck, S. Reverchon, F. Hommais, Quorum sensing regulation in phytopathogenic bacteria, Microorganisms 9 (2) (2021) 239.

[19] B. LaSarre, M.J. Federle, Exploiting quorum sensing to confuse bacterial pathogens, Microbiol. Mol. Biol. Rev. 77 (2013) 73–111.

[20] A. Zehra, N.A. Raytekar, M. Meena, P. Swapnil, Efficiency of microbial bio-agents as elicitors in plant defense mechanism under biotic stress: a review, Curr. Res. Microb. Sci. 2 (2021) 100054, https://doi.org/10.1016/j.crmicr.2021.100054.

[21] X. Liu, J. Jia, R. Popat, C.A. Ortori, J. Li, S.P. Diggle, K. Gao, M. Camara, Characterization of two quorum sensing systems in the endophytic *Serratia plymuthica* strain G3: differential control of motility and biofilm formation according to lifestyle, BMC Microbiol. 11 (1) (2011) 26.

[22] W.L. Ng, B.L. Bassler, Bacterial quorum-sensing network architectures, Annu. Rev. Genet. 43 (2009) 197–222.

[23] S.T. Rutherford, B.L. Bassler, Bacterial quorum sensing: its role in virulence and possibilities for its control, Cold Spring Harb. Perspect. Med. 2 (11) (2012) a012427.

[24] M.B. Miller, B.L. Bassler, Quorum sensing in bacteria, Annu. Rev. Microbiol. 55 (2001) 165–199.

[25] F. Rezzonico, T.H. Smits, B. Duffy, Detection of AI-2 receptors in genomes of *Enterobacteriaceae* suggests a role of type-2 quorum sensing in closed ecosystems, Sensors (Basel, Switzerland) 12 (5) (2012) 6645–6665.

[26] M. Meena, P. Swapnil, A. Zehra, M.K. Dubey, M. Aamir, C.B. Patel, R.S. Upadhyay, Virulence factors and their associated genes in microbes, in: H.B. Singh, V.K. Gupta, S. Jogaiah (Eds.), New and Future Developments in Microbial Biotechnology and Bioengineering, Elsevier, 2019, pp. 181–208, https://doi.org/10.1016/B978-0-444-63503-7.00011-5.

[27] T. Hosni, C. Moretti, G. Devescovi, Z.R. Suarez-Moreno, M.B. Fatmi, C. Guarnaccia, S. Pongor, A. Onofri, R. Buonaurio, V. Venturi, Sharing of quorum-sensing signals and role of interspecies communities in a bacterial plant disease, ISME J. 5 (12) (2011) 1857–1870.

[28] G.A. Achari, R. Ramesh, Characterization of bacteria degrading 3-hydroxy palmitic acid methyl ester (3OH-PAME), a quorum sensing molecule of *Ralstonia solanacearum*, Lett. Appl. Microbiol. 60 (5) (2015) 447–455.

[29] J. Lang, D. Faure, Functions and regulation of quorum-sensing in *Agrobacterium tumefaciens*, Front. Plant Sci. 5 (2014) 14.

[30] N.B. Turan, G.Ö. Engin, Quorum quenching, in: D.S. Chormey, S. Bakırdere, N.B. Turan, G.O. Engin (Eds.), Comprehensive Analytical Chemistry, vol. 81, Elsevier, 2018, pp. 117–149.

[31] C.B. Patel, V.K. Singh, A.P. Singh, M. Meena, R.S. Upadhyay, Microbial genes involved in interaction with plants, in: H.B. Singh, V.K. Gupta, S. Jogaiah (Eds.), New and Future Developments in Microbial Biotechnology and Bioengineering, Elsevier, Singapore, 2019, pp. 171–180, https://doi.org/10.1016/B978-0-444-63503-7.00010-3.

[32] R.V. Kumar, R. Singh, P. Mishra, Endophytes as emphatic communication barriers of quorum sensing in gram-positive and gram-negative bacteria—a review, Environ. Sustain. 2 (2019) 455–468.

[33] Y.H. Dong, L.Y. Wang, L.H. Zhang, Quorum-quenching microbial infections: mechanisms and implications, Philos. Trans. R. Soc. Lond. B Biol. Sci. 362 (1483) (2007) 1201–1211.

[34] A. Mookherjee, S. Singh, M.K. Maiti, Quorum sensing inhibitors: can endophytes be prospective sources? Arch. Microbiol. 200 (2) (2017) 355–369.

[35] Y.H. Dong, J.L. Xu, X.Z. Li, L.H. Zhang, AiiA, an enzyme that inactivates the acylhomoserine lactone quorum-sensing signal and attenuates the virulence of *Erwinia carotovora*, Proc. Natl. Acad. Sci. USA 97 (7) (2000) 3526–3531.

[36] A. Ma, D. Lv, X. Zhuang, G. Zhuang, Quorum quenching in culturable phyllosphere bacteria from tobacco, Int. J. Mol. Sci. 14 (7) (2013) 14607–14619.

[37] P.S. Rajesh, V.R. Rai, Quorum quenching activity in cell-free lysate of endophytic bacteria isolated from *Pterocarpus santalinus* Linn., and its effect on quorum sensing regulated biofilm in *Pseudomonas aeruginosa* PAO1, Microbiol. Res. 169 (2014) 561–569.

[38] P. Kusari, S. Kusari, M. Lamshöft, S. Sezgin, M. Spiteller, O. Kayser, Quorum quenching is an antivirulence strategy employed by endophytic bacteria, Appl. Microbiol. Biotechnol. 98 (16) (2014) 7173–7183.

[39] T. Morohoshi, W.Z. Wang, N. Someya, T. Ikeda, Genome sequence of *Microbacterium testaceum* StLB037, an N-acylhomoserine lactone-degrading bacterium isolated from potato leaves, J. Bacteriol. 193 (8) (2011) 2072–2073.

[40] K. Anandan, R.R. Vittal, Quorum quenching activity of AiiA lactonase $_{KMMI17}$ from endophytic *Bacillus thuringiensis* KMCL07 on AHL-mediated pathogenic phenotype in *Pseudomonas aeruginosa*, Microb. Pathog. 132 (2019) 230–242.

[41] S. Chankhamhaengdecha, S. Hongvijit, A. Srichaisupakit, P. Charnchai, W. Panbangred, Endophytic actinomycetes: a novel source of potential acyl homoserine lactone degrading enzymes, Biomed. Res. Int. 4 (2013) 782847, https://doi.org/10.1155/2013/782847.

[42] S. Uroz, S.R. Chhabra, M. Càmara, P. Williams, P.M. Oger, Y. Dessaux, N-acylhomoserine lactone quorum-sensing molecules are modified and degraded by *Rhodococcus erythropolis* W2 by both amidolytic and novel oxidoreductase activities, Microbiology 151 (2005) 3313–3322.

[43] K.G. Chan, S. Atkinson, K. Mathee, C.K. Sam, S.R. Chhabra, M. Cámara, C.L. Koh, P. Williams, Characterization of N-acylhomoserine lactone-degrading bacteria associated with the *Zingiber officinale* (ginger) rhizosphere: co-existence of quorum quenching and quorum sensing in Acinetobacter and Burkholderia, BMC Microbiol. 11 (2011) 51, https://doi.org/10.1186/1471-2180-11-51.

[44] A.G. Palmer, E. Streng, H.E. Blackwell, Attenuation of virulence in pathogenic bacteria using synthetic quorum-sensing modulators undernative conditions on plant hosts, ACS Chem. Biol. 6 (2011) 1348–1356.

[45] M.R. Alymanesh, P. Taheri, S. Tarighi, Pseudomonas as a frequent and important quorum quenching bacterium with biocontrol capability against many phytopathogens, Biocontrol Sci. Tech. 26 (2016) 1719–1735.

[46] M. Rodríguez, M. Torres, L. Blanco, V. Béjar, I. Sampedro, I. Llamas, Plant growth-promoting activity and quorum quenching-mediated biocontrol of bacterial phytopathogens by *Pseudomonas segetis* strain P6, Sci. Rep. 10 (1) (2020) 4121.

[47] N.T. Ha, T.Q. Minh, P.X. Hi, N.T. Thuy, N. Furuya, H. Long, Biological control of potato tuber soft rot using N-acyl-L-homoserine lactone-degrading endophytic bacteria, Curr. Sci. 115 (2018) 1921–1927.

[48] A.S.L. Kiarood, K. Rahnama, M. Golmohammadi, S. Nasrollanejad, Quorum-quenching endophytic bacteria inhibit disease caused by *Pseudomonas syringae pv. syringae* in citrus cultivars, J. Basic Microbiol. 60 (9) (2020) 746–757, https://doi.org/10.1002/jobm.202000038.

[49] A. Carlier, S. Uroz, B. Smadja, R. Fray, X. Latour, Y. Dessaux, D. Faure, The Ti plasmid of *Agrobacterium tumefaciens* harbors an attM-paralogous gene, aiiB, also encoding N-acyl homoserine lactonase activity, Appl. Environ. Microbiol. 69 (8) (2003) 4989–4993.

[50] H.S. Cho, S.Y. Park, C.M. Ryu, J.F. Kim, J.G. Kim, S.H. Park, Interference of quorum sensing and virulence of the rice pathogen *Burkholderia glumae* by an engineered endophytic bacterium, FEMS Microbiol. Ecol. 60 (1) (2007) 14–23.

[51] W.Z. Wang, T. Morohoshi, M. Ikenoya, N. Someya, T. Ikeda, AiiM, a novel class of N-acylhomoserine lactonase from the leaf-associated bacterium *Microbacterium testaceum*, Appl. Environ. Microbiol. 76 (8) (2010) 2524–2530.

[52] Q. Li, H. Ni, S. Meng, Y. He, Z. Yu, L. Li, Suppressing *Erwinia carotovora* pathogenicity by projecting N-acyl homoserine lactonase onto the surface of *Pseudomonas putida* cells, J. Microbiol. Biotechnol. 21 (2011) 1330–1335.

[53] R.P. Shastry, S.K. Dolan, Y. Abdelhamid, R.R. Vittal, M. Welch, Purification and characterisation of a quorum quenching AHL-lactonase from the endophytic bacterium *Enterobacter* sp. *CS66*, FEMS Microbiol. Lett. 365 (9) (2018), https://doi.org/10.1093/femsle/fny054.

[54] Y. Helman, L. Chernin, Silencing the mob: disrupting quorum sensing as a means to fight plant disease, Mol. Plant Pathol. 16 (3) (2015) 316–329.

[55] N. Dandurishvili, N. Toklikishvili, M. Ovadis, P. Eliashvili, R. Giorgobiani, R. Keshelava, M. Tediashvili, A. Vainstein, I. Khmel, E. Szegedi, L. Chernin, Broad-range antagonistic rhizobacteria *Pseudomonas fluorescens* and *Serratia plymuthica* suppress *Agrobacterium* crown gall tumours on tomato plants, J. Appl. Microbiol. 110 (1) (2011) 341–352.

[56] C. Grandclément, M. Tannières, S. Moréra, Y. Dessaux, D. Faure, Quorum quenching: role in nature and applied developments, FEMS Microbiol. Rev. 401 (2015) 86–116.

[57] M. Teplitski, J.B. Robinson, W.D. Bauer, Plants secrete substances that mimic bacterial N-acyl homoserine lactone signal activities and affect population density-dependent behaviors in associated bacteria, Mol. Plant-Microbe Interact. 13 (6) (2000) 637–648.

[58] N.D. Keshavan, P.K. Chowdhary, D.C. Haines, J.E. Gonzalez, L-Canavanine made by *Medicago sativa* interferes with quorum sensing in *Sinorhizobium meliloti*, J. Bacteriol. 187 (2005) 8427–8436.

Chapter 10

Peeking into plant-microbe interactions during plant defense

Shriniketan Puranik[a], Vindhya Bundela[b], Amanda Shylla[c], M. Elakkya[a], Livleen Shukla[a], and Sandeep Kumar Singh[a]

[a]*Division of Microbiology, ICAR-Indian Agricultural Research Institute, New Delhi, India,* [b]*Department of Microbiology, College of Basic Science and Humanities, Govind Ballabh Pant University of Agriculture and Technology, Udham Singh Nagar, Uttarakhand, India,* [c]*DBT-Institute of Bioresources and Sustainable Development, Shillong, Meghalaya, India*

1 Introduction

In nature, plants and microorganisms coexist dynamically. Many studies highlight that plants develop intimate relationship with surrounding microbial communities, among these, pathogens can also colonize within host tissues. Pathogens feed on plants for food, reduce productivity (by 20%–40%), and are a potent threat to global food security. Diseases such as the late blight of potato caused by *Phytophthora infestans* (infamous Irish famine in the 1840s), brown leaf spot of rice caused by *Helminthosporium oryzae* (Bengal famine), and southern corn leaf blight of 1970 by *Helminthosporium maydis* (United States) are historic examples as to how devastating pathogens can be toward agriculture. They induce biotic stress on plants, thereby lethally affecting parts or plants as a whole. Plants-pathogens-environment forms a tripartite complex governing pathogenesis. Broad groups of phytopathogens are bacteria, fungi, nematodes, viruses, phytoplasma, etc. These wide arrays of pathogens produce signals, leading to the development of an induced and actively regulated immune system, thereby helping in the survivability of plants. A localized memory-based immune system withholds further spread of infection, thus reducing pathogenicity of pathogens and protecting plant health [1]. Studies show that plants entirely depend on innate immune systems because they do not usually possess adaptive immune systems as animals do. Thus, the immune system should be fast, dependable, and strong against infections. As a matter of fact, such systems emerge only upon disease incidence [2]. Although plants

Plant-Microbe Interaction—Recent Advances in Molecular and Biochemical Approaches
https://doi.org/10.1016/B978-0-323-91876-3.00012-9

167

demonstrate various kinds of mechanisms, pathogens are smart too, possessing diverse strategies for infecting the hosts. Pathogens such as viruses invade plants via injured host cells, while bacteria by entering through openings such as hair, stomata, and lenticels [3,4]. Fungal pathogens penetrate the host through special structures such as hyphae and plant nematodes form feeding sites by invading plant root cells [5–7]. Also, different pathogens employ various modes of nutrition such as biotrophy, necrotrophy, or hemibiotrophy, either for food or reproduction.

As soon as a pathogen infects a plant, two types of defense are activated. The first barrier involves certain plant cell surface-localized receptor proteins called pattern recognition receptors (PRR) that get activated and identify microbial elicitor molecules such as microbe-associated molecular patterns (MAMPs) or pathogen-associated molecular patterns (PAMPs) such as flagellin [8,9]. This results in pattern/pathogen-triggered immunity (PTI) and expression of defense-related genes. Pathogens produce altered effectors that go unnoticed by PRRs in order to successfully establish infection. The second barrier involves the identification of effectors (virulence factors) by plant intracellular receptor molecules, leading to effector-triggered immunity (ETI). This is mediated by receptor proteins possessing nucleotide-binding domain and leucine-rich repeats (NBLRRs), transcribed/translated by resistance (R) genes [10]. However, ETI is faster and more powerful as compared to PTI, involving a hypersensitive response to prevent pathogen proliferation [11].

Although pathogens attack plants in multiple ways, plants have evolved some strategies for their defense too. Plants have cell walls as the first line of defense that pathogens need to penetrate in order to get entry into the cells. Plants produce some antimicrobial compounds, volatile organic compounds, and regulate some phytohormones upon disease incidence as a counterattack. Some nucleotides and protein molecules such as sRNAs, miRNAs, NBLRRs play a key role in intramolecular pathogen resistance, thus preventing infection and pathogen proliferation. However, these molecules are yet to be studied in depth. A closer look at signal exchange between pathogens and plants provides novel strategies to strengthen plant defense and thereby, tackle disease incidence (Fig. 1).

2 Aspects related to pathogen recognition systems and host defense

Plants are constantly challenged by pathogens that have developed various approaches to protection against them. In response, the microbial invaders have also evolved defense strategies to escape from the plant immune recognition mechanism. This constant evolutionary arms race led to the rapid emergence of various survival strategies in both host and pathogens. H. H. Flor, in his "gene for gene" hypothesis, postulated that for any gene conferring resistance (R) in host, there is a complementary gene for avirulence (avr) in microbial invaders

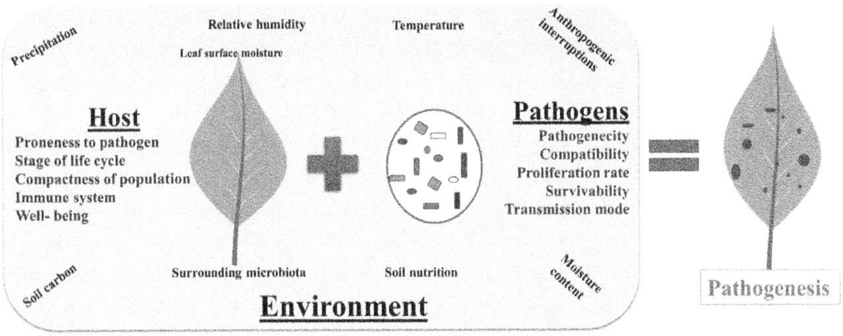

FIG. 1 Factors governing pathogenesis within tripartite interaction.

and vice versa. This experiment attained enormous attention and not only led to the development of resistant varieties but also opened up new approaches to study the pathogen recognition systems; mechanisms such as hypersensitive response to pathogens and complex systems of signal transduction pathway of the plant-microbe interactions [12]. However, today we know that conferring resistance to the pathogen includes both monogenic and multigene resistance. Many concepts and hypotheses have been proposed as an expansion of "gene to gene" hypothesis [13]. The progress in the field of genomics and advancement in functional genomics such as transcriptomics, proteomics, and metabolomics have helped in identifying previously described genes in plant defense systems.

Plants may encounter a wide range of microbes ranging from commensals to pathogens in their vicinity. While biotrophic pathogens such as rust and powdery mildew fungi may only derive food from live hosts and thus, do not kill them. Necrotrophic pathogens causing gray mold, brown rot in stone fruit, white mold, dark leaf spot in cabbage, and soft rot in onion kill the living plant cells and feed on dead tissues. Others such as *Magnaporthe oryzae* and *Pseudomonas syringae* feed on the plant and at later stages kill the plants to be called hemibiotrophs.

The cross talk between plant and pathogen is conceptualized as zig-zag model illustrating overall quantitative output elicited by the host defense system [11]. Conferring immunity or susceptibility in the plant is due to the overall immune response that can be calibrated as [PTI − ETS + ETI], where PTI, ETS, ETI refer to pattern/pathogen-triggered immunity, effector-triggered susceptibility, and effector-triggered immunity, respectively. Generally, there are four phases in such events. In the first phase, plant PRRs sense MAMPs/PAMPs in microbes to induce PTI. In phase 2, the pathogens successfully entering the plant release effector molecules. This results in the restriction of PTI followed by ETS. In the third phase, the plant nucleotide-binding domain containing leucine-rich repeats (NB-LRR/ NLRs) protein recognizes the effector protein, resulting in ETI. In phase 4, pathogens evolve themselves by modification in the effector recognition site or the gene encoding effector molecules, thereby dodging identification by hitherto plant LRR proteins. This again forms ETI. However, natural selection favors plants by evolving the *R* genes. MAMPs, damage-associated molecular patterns (DAMPs), and immunogenic apoplastic effectors are sensed by PRRs, and the intracellular effectors are sensed by NLRs. While PAMPs activate the PTI, the *avr* genes encode effector molecules and NB-LRR proteins stimulate the ETI by either direct or indirect detection mechanism. Ray et al. [14] reported that some molecules such as RPM1-interacting protein 4 (RIN4) mediate both PTI and ETI, thereby playing a vital role.

The activation of immune receptors leads to major downstream signaling events, such as calcium influx, alteration of various ion potential differences across the cellular membranes, reactive oxygen species (ROS) production, dysregulation of immunogenic peptides, and defense hormones. A number of key signaling modules such as mitogen-activated protein kinase cascades (MAPKKK, MAPKK, MAPK), calcium-dependent kinases (CPKs), and heterotrimeric G proteins get activated [15]. Associated with these responses, stimulation of PRRs leads to the strengthening of barriers against pathogens through cytoskeletal remodeling, callose deposition, stomatal closure, and closure of plasmodesmata [16]. Production of jasmonic acid, salicylic acid, and phytoalexins is also induced to terminate further invasions. These proximal signaling events result in transcriptional modulation at various levels including WRKY superfamily in the plant immune system and impart resistance to invaders [17].

Vesicle- and SNARE-mediated production of defense-related proteins consisting of PR-1 are triggered by the exocytosis pathway. The resulting biochemical remodulation leads to the secretion of antimicrobials that are translocated to the external side with the help of some ATP-binding cassette (ABC) transporters. Extracellular generation of ROS by membrane-localized NADPH oxidases (RbohD) is an early MAMP-triggered response. On the contrary, GSL5/PMR4 callose synthase produces and accumulates polyglucan in extracellular space is comparatively a late response [18].

One of the first lines of defense against bacterial invaders is stomatal closure. In response to MAMPs such as flagellin22 (flg22), the osmotic water permeability of the bundle sheath and mesophyll cells were found to be decreased [19]. This is considered an early PTI in response to a pathogen attack (22 denotes the amino acid chain synthesized from a conserved flagellin domain). Several LRR-receptor kinase genes detect the flg22 such as *OsFLS2* in Rice [20], *FLS2* in *Arabidopsis,* etc. [21]. Domain swap approaches have been used to study the different sequences of peptides such as flg22, flg15, flg15-Δ7, flg22-AYA, flg22[Rsol], and their role as ligands for FLS2. These peptides act as either agonists or antagonists, resulting in the induction of MAMP responses in *Arabidopsis* and tomato [22]. FLS3 detect and recognize flgII-28 in tomato, potato, and pepper, activating pattern-triggered immunity (PTI) that results in more sustained production of ROS than the FLS2 [23]. The epitopes of elongation factor Tu (EF-Tu) such as elf26 of *Ralstonia solanacerum* functions as a

PAMP in *Nicotiana* spp. The elf18 are perceived through RLK EFR in Brassicaceae, and EFa50 through an unidentified receptor in rice [24]. Some of the molecules in the peptidoglycan such as the *N*-acetylglucosamine-containing glycan backbone act as MAMP and are detected by the LYM1-LYM3 and LYP4-LYP6. Likewise, the chitins in fungal cell walls are recognized through LysM-type RLKs such as LYSM-containing receptor-like Kinase 5 (LYK5) and by LYK4 partially in *Arabidopsis*. Also, chitin oligomers are sensed by LysM-RLPs such as OsLYP4 and OsLYP6 as well as OsCEBiP (Chitin-elicitor binding protein) in rice [25]. Niehl et al. [26] reported that the dsRNAs of the virus also represent genuine PAMPs in plants eliciting the PTI, inducing a signaling cascade involving SERK1 and a specific dsRNA receptor. This strategy of dsRNA-mediated PTI involves a membrane-mediated process and is found to operate independently of RNA silencing which is a major defense system against the virus.

Some effector molecules such as AvrB, AvrPto, and HopAI1 of *P. syringae* target conserved proteins of both *Arabidopsis* and tomato. The effector molecules such as SEE1 of *Ustilago maydis* in maize and AvrBsT effectors of *Xanthomonas campestris* in pepper interact with SGTI, and the salicylic acid-induced protein kinase (SIPK) phosphorylates SGT1. Subsequently, the signaling cascade of MAPK signaling pathway is activated. In early divergent land plants the NLRs encoded by the host are found to be limited. However, some molecules that contain protein kinases or tetratricopeptide repeat (TPR) domains in α/β-hydrolases in *Marchantia polymorpha* (liverwort), *Physcomitrella patens* (moss) and CONSTANS, CO-like, TOC1 (CCT) or TIF[*F/Y*] XG (TIFY) domains in *Selaginella moellendorffii* (lycophyte) are involved in effector molecule recognition and result in ETI [27].

Plants employ a series of membrane-anchored and intracellular immune receptors to recognize pathogens. MAMPs are present not only in pathogenic invaders but also in commensal and beneficial microbes. A major challenge to plants is to discriminate between a pathogen and a beneficial microbe, thereby exerting an appropriate mechanism that activates the defense system.

3 Types of plant-microbe interaction involved during pathogenesis

Nutrition is of main concern for pathogens to attack plants. Deriving nutrition from plants by pathogens is mainly in three ways. Some can parasitize and complete their life cycle on a living host (termed biotrophy), some by deriving nutrition from dead host tissues after killing them (termed necrotrophy), while others live initially on living host tissues later switching nutrient acquisition from dead tissues, that is from biotrophy to necrotrophy (hemibiotrophs). Particles such as viruses and phytoplasma are obligate biotrophs, needing living host machinery only for replication. Plant parasitic nematodes produce feeding sites and derive nutrition from hosts. Thus, a deep understanding of such a variety of pathogenic nutrient acquisition systems results in the proper management of diseases and improving plant health. Different types of nutrition are discussed below.

3.1 Biotrophism

Plants share their space with the microbes, leading to neutral, mutually beneficial, or detrimental interactions [28,29]. The microbes which are involved in this interaction are called endophytes, symbionts, and pathogens respectively. The interactions that result in the transfer of nutrients, from one partner to the other have been well reported. These "trophic" relationships are commonly employed to classify interactions between microorganisms and plants. Numerous pathogens including fungi, bacteria, and even viruses affect the plant kingdom. These pathogens adopt several strategies to enter, obtain nutrients, and multiply inside the host plant. Pathogens with diverse mechanisms of nutrient uptake infect plants and show differences in immune responses. These have a direct impact on the plant's ability to respond. In most cases, there are three types of host-pathogen associations as per the way of infection in plant host: necrotrophs, biotrophs, and hemibiotrophs. This should be detected early enough, otherwise, the plant may respond incorrectly, exacerbating the damage. Bacteria and fungi use both biotrophic and necrotrophic invasion tactics. This subsection describes bacterial and fungal biotrophy.

Biotrophs grow, proliferate, and obtain energy from living host tissue through an intricate interaction with the living system of plants. They thrive on the nutrients of the living host. In simple terms, microbes are called "biotrophs" when they are involved in an interaction where they get the nutrient from living plants. Biotrophs can be divided into obligate biotrophs (not grown on artificial medium), for example, powdery mildew and rust, and nonobligate biotrophs (can be grown on artificial medium). Plant pathogens produce effector proteins that have the ability to trigger or suppress the plant immune system. Some of them are reported to be Avr proteins (avirulence), *hrp* genes (hypersensitive reaction and pathogenicity), and cell-wall degrading enzymes [30]. Biotrophic pathogens do not kill their host, maintain host viability by causing minimum damage, and also repress host hypersensitive responses (HR) so that they can get the continuous nutrient supply, as in other pathogenic effects these responses lead to programmed cell death [31]. Also, at times, there is a stimulation of salicylic acid-mediated defense responses, other than apoptosis.

Plant fungal infections are responsible for a wide range of illnesses in commercially significant crops. They use a variety of infection techniques and create specialized infection structures such as hyphae, appressoria, and haustoria that pierce plant cytoplasmic membranes. They absorb nutrient sap and produce effector molecules that assist them in invasive proliferation inside host cells. All three forms of pathogenicity have been discovered in fungi (necrotrophy, biotrophy, and hemibiotrophy). Fungi use certain toxins to infect such as cell wall degrading enzymes and secondary metabolites. However, infection methods of biotrophic fungi are thought to be more complicated than those of necrotrophic fungi. There are many biotrophic fungi and effector proteins secreted by them to emphasize the infection process in host plant, for example, *Cladosporium fulvum* which causes tomato leaf mold disease. Effector protein Avr9, a cysteine-rich, 28 amino acids containing protein is involved in HR in tomatoes. During infection *C. fulvum* secretes effector molecules Avr4 and LysM chitin-binding domain effector protein Ecp6 that adhere to chitin, thus preventing their detection by the host. Effector Avr2 is also secreted by pathogens that binds to the cysteine proteases of plants essential for basal defense and suppress the plant immunity. Another example of biotrophic fungus is maize smut caused by *U. maydis* producing different effector proteins such as Pep1 effector, Hum 3 (hydrophobin domain protein), Rsp1 (repetitive secreted protein), and See1 (seedling efficient effector). Pep1 effector, a small secretory protein with 178 amino acids is necessary for the effective invasion of *Zea mays* epidermal cells. It also suppresses plant immunity by inhibiting apoplastic plant peroxidases. Hum 3 and Rsp1 have been demonstrated to be important for cell adherence during infection, while See1 performs triggering of DNA synthesis in foliage cells, critical for tumor proliferation [32]. *Blumeria graminis* causing powdery mildew disease in barley and wheat secrets more than 500 types of candidate-secreted effector proteins (CSEPs), also known as Blumeria effector candidates (BECs). For making haustoria, eight BECs are required including BEC1054 (RNase-like effector). AVRa10 and AVRk1 effectors also aid in infection enhancement. AvrL567 is an effector protein in *Melampsora lini* causing rust in flax (*Linum usitatissiumum*). The L6, L5, and L7 R proteins recognized AvrL567, the first known flax rust effector protein. AvrM, AvrP123, and AvrP4 are three other flax rust effector proteins that play a crucial role in suppressing host defense. *Erysiphe necator*, another obligate biotrophic fungus responsible for causing powdery mildew disease, is completely reliant on photosynthetically active cells to complete the life cycle. Once a conidiospore lands the epidermis of such cells, a lobed appressorium is formed. *E. necator* effector protein targets Mildew resistance Locus O (MLO) proteins to repress host pathogen-triggered immunity. It was found that fungal phytopathogens are also capable of producing diverse small secretory proteins (SSPs), critical for pathogenicity. SSPs tagged as effector proteins, required for pathogen and host attachment, are reported to be secreted more by biotrophs than necrotrophs [33].

   Plant pathogenic bacteria are also responsible for detrimental effects on agricultural crops. Bacterial associations with the plant hosts have an extremely dynamic relationship. Bacteria infiltrate the intercellular spaces of higher plants through natural openings, stomata, and wounds, causing a variety of diseases. The bacteria cause diseases such as Fire blight of apples and pears, Halo blights, cankers, galls, leaf spots, and many other diseases. Classifying pathogenic bacteria on the mode of nutrition sometimes becomes difficult, for example, *P. syringae* can be grouped under all three categories biotrophic, partially necrotrophic, and hemibiotrophic category; *Ralstonia solanacearum* can be placed under both biotrophic and necrotrophic categories. *Xanthomonas* sp. is biotrophic, however, *Erwinia amylovora* comes under the necrotrophic category [34].

The bacterial pathogen in order to infect and aid colonization, proliferation, and development inside the host, produces virulence proteins, cell wall-degrading enzymes, extracellular polysaccharides (EPSs), and certain toxins. Bacterial genera *Pseudomonas, Xanthomonas, Ralstonia,* and *Erwinia* share common features of invading and thriving in intercellular spaces, killing the cells to have expressed *hrp* genes. For pathogenesis, the Hrp protein secretion system plays an important role. Phytopathogenic bacteria synthesize numerous effector proteins such as Avr recognized by the host Resistance gene (*R*), resulting in the construction of HR and avirulence, preventing disease progression. *P. syringae*, has been found to produce roughly 30 effector proteins in the cytosol of plants. *P. syringae* effector protein AvrPto, binds to *Arabidopsis* receptor kinases and inhibits the defense response.

The bacterial EPSs change the defense signal activation, interfere with xylem function and safeguard the bacteria from an unfavorable environment. EPSs produced by *P. syringae* form chlorotic and necrotic signals. *P. syringae* also produce a toxin, coronatine which helps in the penetration of bacteria by stimulating stomatal opening and weakening host defenses. Antimetabolite toxins (mangotoxin, phaseolotoxin, and tabtoxin) are produced by several pathovars of *P. syringae* that impede the production of aromatic amino acids, resulting in interference with the host plant's nitrogen metabolism.

The study of the interaction between pathogen-secreted effector proteins and immunity-triggering signals inside the host cells aids in unraveling disease mechanisms. The development of innovative ways for managing biotrophic plant diseases is vital. Investigations by phytopathologists of such interactions between plants and pathogens might help in the better management of biotic stress.

3.2 Necrotrophism

Necrotrophic pathogens cause a significant effect on yield and economic losses in agricultural production annually. Necrotrophs kill and parasitize the living cells to absorb nutrients from dead tissues as a saprophyte. They produce some toxins and wall-degrading enzymes upon infection of the host tissue through injured sites or dead cells. They are also capable of living as saprophytes that do not need a live host, and thus can be cultured in synthetic media. It is important for a host to be susceptible to toxins and pathogens and should make sure of proper delivery of such toxins at the right time and location, thereby increasing the chances of killing the host. The toxins should be derivatized and compartmentalized, in the right amounts and time in order to effectively bring cellular death. Death of host cells results in the synthesis of secondary metabolite, accumulation of reactive oxygen species, hormones such as ethylene, abscisic acid, salicylic acid, and jasmonic acid in necrotrophic infections. To offset the initial surge of oxidative stress and prevent HR apoptosis, plants must recognize necrotrophs early. In *Arabidopsis*, this role is linked to jasmonic acid signaling, which acts as an antagonistic route to salicylic signaling in this regard [35].

Necrotrophs are further divided into two categories: which include broad-host-range on the basis of toxins they produced such as *Sclerotinia sclerotiorum*, *Botrytis cinerea*, and *Rhizoctonia* spp.; and host-specific pathogens such as *Alternaria*, *Pyrenophora*, *Parastagonospora*, *Cochliobolus*, and *Periconia*. Necrotrophs are less complex and use a variety of virulence tactics to kill and feed nutrients from the infected host cells for their own development and proliferation. They can cause symptoms such as maceration of tissues or cause soft rots, as well as trigger HR-like host cell death.

Fungal necrotrophs invade host cells first, which necessitates multiple steps of conidial attachment, germination, lesion formation, and finally, the tissue becomes soft and sporulation occurs. After the early phase of infection, toxin production, appressoria, haustoria, and hyphae formation, secretion of cell wall-degrading enzymes aid penetration. Some also produce necrotrophic effectors (NEs) that increase ROS, membrane fragmentation, ion/salt leakage, DNA laddering, and dysfunctioning of organelles by targeting the host defense response. This response enables nutrient uptake, an increase in sporulation and biomass of necrotrophs, unlike a sudden and complete shutdown in biotrophism [36]. NEs also determine the range of hosts. Necrotrophic bacteria use a brute-force method to kill the host plant's parenchymatous tissues. They too, like fungi activate HR using *hrp* genes, causing localized cell death in the host plant. This further cell death improves bacterial colonization and nutrient uptake. These genes in a brown spot of the bean caused by *P. syringae* pv. *syringae* play an essential role in the transfer of *Avr* gene-derived signals from the bacteria, inside the host cells. Pathogenic Gram-negative bacteria such as *X. campestris* pv. *vesicatoria*, *E. amylovora*, and *R. solanacearum* have also been found to have *hrp* genes. Necrotrophic fungus *Cochliobolus heterostrophus* in maize causing disease southern corn leaf blight (SCLB) releases a polyketidal T-urf13-T-Toxin which is linear in structure. A related fungus, *C. victoriae* affects oats by causing Victoria blight disease, involving victorin, a cyclic pentapeptide. Victorin is recognized by Locus Orchestrating Victorin effects 1 (LOV1) nucleotide binding, leucine-rich repeat (NLR) protein lead to the stimulation of pathogenicity gene PR-1. It involves the synthesis of camalexin in the host to provide resistance [37]. Bacteria cause soft rot by producing wall degraders such as cellulases, proteases, hemicellulases, and pectic enzymes using a type II secretion system to soften the host tissue and facilitate the uptake of nutrient. Examples of such pectinolytic bacteria in the necrotrophic category such as *Erwinia carotovora*, *E. chrysanthemi*, *Dickeya* sp., *Pectobacterium* sp., and *Pseudomonas viridiflava*.

In a recent study it was investigated that in a wheat receptor-like cytoplasmic kinases (RLCKs) encoding gene *TaRLCK1* triggers host immune response against necrotrophic fungus *Rhizoctonia cerealis* that cause sharp eyespot disease. It is a catastrophic disease of wheat crop globally. *TaRLCK1* gene builds immunity by controlling the expression of ROS-generating/scavenging enzymes, thereby regulating ROS homeostasis in wheat [38]. A study by Sobiczewski et al. [39] reported that the Fire blight of apple and pear disease causal organism *E. amylovora* was necrogenic. Gram-negative Enterobacterium causes extremely destructive disease by inducing HR, which involves the production of ROS such as superoxide radical and H_2O_2, and is mediated by ethylene and jasmonic acid progressing toward programmed cell death. Necrotrophs such as *Parastagonospora nodorum*, *Pyrenophora tritici repentis*, and *Zymoseptoria tritici* cause wheat foliar diseases globally with serious problems. *Sclerotinia sclerotiorum* releases oxalic acid during infection to reduce defensive responses of hosts, while *Alternaria brassicicola* produces enzymes that degrade cell walls and lipases to hamper cellular processes in the host.

At the molecular and cellular levels, interactions between plants and pathogens are evolved and complex processes. Despite this significant gain, the mechanism pertaining to the cause of diseases and the exchange of effectors/elicitor molecules is scantly understood. Interpretation of these complicated relationships can help researchers figure out more about plant resistance pathways. Deciphering the physiological and genetic processes of plant-pathogen interaction, as well as further research into phyto-pathosystems will pave the way for better crop protection and development.

3.3 Hemibiotrophism

Hemibiotrophs are a type of phytopathogens that obtain nutrients in two stages: initial biotrophic phase in which they invade a live cell, later manifesting necrotrophic phase by killing the host cell. Hemibiotrophic plant pathogens include *P. syringae*, *Magnaporthe oryzae*, *Phytophthora infestans*, *Colletotrichum graminicola*, etc. Most taxa of fungi produce intracellular bulged hypha, haustoria, and appressoria that are encased by host cytoplasmic membrane during the initial biotrophic phase and in necrotrophic phase synthesize hydrolytic enzymes and toxins. Reports show that salicylic signaling elicits immune responses in the host and provides resistance against biotrophic and hemibiotrophic phytopathogens [40].

Hemibiotrophic pathogenic fungus such as *Magnaporthe oryzae* causes blast disease in rice. At the initial stage of infection, it penetrates the outer leaf cuticle and epidermis of the host by forming appressorial pegs. This genus produces effector proteins which get accumulated in a lobed structure called biotrophic—an interfacial complex, produce at the penetrating hyphal tip. During the infection process, different effectors proteins are upregulated such as BAS1, PWL2, BAS2, BAS3, BAS4, and AvrPita1. Additionally, AVR-Pii is recorded to decrease the host immunity by hampering the function of Os-NADP-ME (a rice NADP-malic enzyme2 protein) in the host, which aids the production of ROS [41]. Interestingly, the pathogenicity of *M. oryzae* and *F. graminearum* can be decreased by transgenic rice carrying a type of Class III acyl-CoA-binding (ACBPs) protein, OsACBP5. It is homologous to AtACBP3 (of *Arabdiopsis thaliana*) and confers resistance against the pathogen. But, progenic lines obtained due to OsACBP5-OE9 X salicylic acid signaling mutant were found more vulnerable to *M. oryzae* infection [42]. *P. syringae* is another devastating hemibiotroph, that causes diseases in plants, including *Arabidopsis*, tomato (*Solanum lycopersicum*), and tobacco (*Nicotiana tabacum*). It can alter plant morphological and physiological features, also defense-related genes in the plant. In a hemibiotrophic interaction, it was reported jasmonic acid-induced signaling for increased resistance in plants [43].

3.4   Host-viral interactions

Viruses are small intracellular obligate parasites which are not visible under a light microscope, replicate only inside the host cell and cause a number of plant diseases. Plant virus diseases are difficult to control and the most common methods include the selection of resistant varieties, exclusion of virus-infected propagules as seeding materials and use of insecticides for killing transmitting insects. Viruses are composed of nucleoproteins. They have either single or double-stranded DNA or RNA, never both in a single particle. Mostly plant viruses are made of single-stranded ribonucleic acid (ssRNA). They enter passively into the plant cell through the natural opening like wounds, physical injuries or by the vector-mediated transfer which may be insects, nematodes, fungi, and even mites. Upon entry into the host, they control the host machinery to replicate by using their own enzymes such as RNA-dependent RNA polymerase, DNA replicase, or reverse transcriptase, and move to long distances via the vascular system. Plants produce certain defense mechanisms to combat viral attacks, including virus-encoded suppressor proteins, RNA silencing and the development of disease-free tissues. When a virus infects a plant, two types of interactions take place: compatible and noncompatible. In case of compatible interaction, host is not able to recognize the viral particle as foreign, which is favorable for the virus infection process whereas in noncompatible interaction, the host recognizes the virus and obstructs the virus multiplication by a series of defense reactions [4].

Molecular technology can be used in a variety of ways to integrate or build novel resistance elements in plant viral systems. The strategies are particularly valuable in situations where no natural source of resistance has been identified. However, depending on the source of the genes employed, there are primarily three techniques of building genetically engineered resistance: first, pathogen-derived resistance, comprising grouping of viral components, thus disrupting the life cycle of a viral particle; second, pathogen-targeted resistance, which targets genes and gene-products of the virus to make it inactive; third, transferring existing resistance genes into susceptible hosts. Plants acquired both innate (genetically determined) and adaptive immunity when a virus infects the host. Adaptive immunity is based on RNA silencing. In the case of innate immunity, plants locally recognize and interact with viral particles in a nonspecific manner. PAMP-triggered immunity (PTI) inhibits the aggregation of certain viruses (tobamoviruses, turnip crinkle virus), and BAK1 and BKK1 (other components required for external signal perception, predominantly brassinosteroids) are needed for this [44].

Plant viruses use different factors such as protein components and nucleic acids (DNA or RNA) for pathogenesis such as RNA replicase-related proteins/RNA-dependent RNA polymerase (RdRp), Capsid proteins/coat proteins (CPs), and Movement proteins (MPs). The tobacco mosaic virus (TMV) exhibits PAMP property via coat protein since it induces NADPH oxidase activity and causes ROS explosion in Solanaceous crops such as tomato and tobacco. In Cucumber mosaic virus (CMV), PAMP molecules such as CMV 2a, also known as RNA directed-RNA polymerase activate PTI in *Arabidopsis*, leading to an increased release of glucosinolates. Apart from proteins, nucleic acids such as viral dsRNAs can set off PTI by using a procedure discrete from RNA silencing, as described by Niehl et al. [26].

Plant responses to viral infection can be either by cellular stress or developmental anomaly. The defense and stress responses in plants are strikingly similar to the alterations in the expression of gene profiles caused by viral infection. During stress as well as defense like, heat shock proteins (HSPs) and pathogenesis-related (*PR*) genes may be induced. *PR* genes involved in defense are PR-1, β-1 glucanase (PR-2), chitinase (PR-3), thaumatin-like protein (PR-5), glutathione S-transferases (GT), and superoxide dismutase (SOD). It was also observed WRKY family of the different transcription factors gets activated when the virus infects *Arabidopsis thaliana*. Viruses affect plant growth and development by interfering signaling pathways. One such instance is an enhanced expression of SA and PR genes upon inoculation of tobacco plants with TMV. Viruses also affect plants by reducing the amount of chlorophyll protein complex [4].

Plants counteract the virus pathogenesis by adopting strategies such as *R* gene-, RNA silencing-, and Recessive gene-involved defense. There are two sorts of R genes: those that code for components that interfere with the viral infection cycle, and those that code for NLR proteins [45]. An example of an R gene is RTM 1–3 that codes for jacalin repeat lectin and meprin proteins in *A. thaliana* for the inhibition of potyviruses manipulation. RNA silencing (also known as RNA interference, RNAi), another adaptive antiviral mechanism, is an invariant regulatory mechanism of gene expression triggered by dsRNA-triggered gene silencing either by inhibition at transcriptive or post-transcriptive stages by sequence-specific degradation of complementary mRNA transcripts. *A. thaliana* has four RNaseIII-like enzymes called Dicer

(DCLs), 10 Argonaute (AGOs), and six RdRp, which play role in diverse silencing-related pathways. In a recent study, it was determined that some viruses have been implicated in the epigenetic modification via small RNA-mediated transcriptional gene silencing and posttranscriptional gene silencing. In the case of geminiviruses, one of the single-stranded circular DNA viruses causing worldwide damage to crops, undergo methylation changes inside the host cell [46].

Improved knowledge of viral-plant defense manipulation will also shed light on their coevolution, their interaction with host plants and vectors carrying them, thus, providing information about various plant-interacting organisms. Modification of the genome of a plant to express the disruptive component could result in host resistance. Through the use of chemicals, the virus population can be controlled up to some extent, along with managing the vector responsible for their transfer. There is a need to develop disease-resistant cultivars to prevent viral infections.

3.5   Host-nematode interaction

Nematodes, also called roundworms, belong to the Phylum Nematoda or Nemathelminthes and have the highest number of individuals on the planet. They contribute to huge agricultural losses. Root-knot nematodes belonging to the genus *Meloidogyne* (family Heteroderidae) and cyst-forming nematodes of genera *Heterodera* and *Globodera* are some examples of plant-infecting nematodes. These worms have developed highly sophisticated parasitic processes that include the development of specialized "nematode feeding sites" (NFSs) in the roots of the plant. Root-knot nematode feeding structures are called "giant cells," while those in cyst nematodes are called "syncytia." Plant parasitic nematodes (PPNs) cause a number of structural, biochemical, and molecular changes for facilitating and establishing these feeding sites in the cells of plant roots. Nematodes secretion through stylets induces the signaling cascade in the cell of the host.

PPNs interact with the plants to release juveniles after getting a chemotactic response from the root exudates of the host plant. They enter the root cells in the absence of plant responses such as ROS production and deposition of callose. They get the nutrients only via NFSs which are the only nutrient sources for a sedentary life of nematodes. After the development of NFSs, the nematode secretes different cell wall-degrading enzymes, virulence proteins such as Avr, and transcription factors. These secretions contain different effector proteins and lead to successful parasitism. Effectors can reprogram cellular events for induction of NFSs and inhibit host defense mechanism by changing the mechanism of metabolism and development. PPNs activate multiple signaling pathways via nematode-associated molecular patterns (NAMPs), allowing for suitable plant interactions. In response to these effectors, plants activate the basal immune system against nematode, thereby inducing a number of alterations in a plant such as cell cycle change, effective cytoskeleton, and small RNA production. PPNs also lead to the modification of translation and post-translational events, modulating signaling pathways such as silencing pathways, salicylic acid-jasmonic acid pathways, and other phytohormone cycles [47]. Three elicitor peptides are produced by soybean seeds namely, GmPep1, GmPep2, and GmPep3. These molecules stimulate immune responses and curb the proliferation of *M. incognita* and *H. glycines* [48].

NAMPs include ascaroside, PPN-acquired proteins, chitin, and cuticle. In a recent study, it was found that free-living nematode *Caenorhabditis elegans* produce ascarosides pheromone involved in signal activation, developmental process, searching for mating partner and behavioral changes. These signaling cues are the derivatives of 3, 6-dideoxy-L-sugar such as ascarylose, altered by side chains of fatty acid. Ascr#18 is the most abundant ascaroside among the cyst and root-knot nematode [49]. A study by Manohar et al. [50] concluded that Ascr#18 is a signaling molecule triggering the synthesis of a repellant by plants against nematodes to minimize infection. This occurs due to smaller side-chained ascarosides. Nematode can change the message code of ascarosides by peroxisomal β-oxidation.

A variety of microRNAs (miRNAs) in numerous plant species are recently documented to show variations in root cells upon infection since they are required for proper feeding site formation. Furthermore, several conserved miRNA appear to have the same role in the development of feeding sites across diverse plant species. These miRNAs could be prime regulatory molecules that govern expression activities pertaining to the establishment of feeding sites [51]. In some plant-nematode interactions, flavonoids produced by plant roots have been found to alter the development of feeding sites and nematode reproduction. They are vital for the chemotactic behavior of PPNs and may help in plant defense against nematodes [52].

According to another study, Copia-type elements may play a role in regulating soybean cyst nematode resistance (SCN) mediated by the *rhg1-a* resistance gene. SCN-induced epigenetic modifications in this gene may regulate rhg1-a maturation and splicing variants. They reported that the retrotransposon-type tended to be more vulnerable to DNA methylation changes than other types of transposable elements during nematode infection [53]. In consideration of this, DNA methylation and other epigenetic marks may modulate the recruitment of splicing factors to the pre-mRNA of *rhg1-a*, thereby impacting the elongation rate of Pol II and exon inclusion/skipping in mature mRNA, thus impacting defense.

Nematodes are the most abundant and metabolically diversified creatures among soil biodiversity. These interactions are changing as a result of global change, and a greater understanding of them is urgently needed to better predict functional outcomes. Deep sequencing and other emerging molecular technologies could be useful in high-throughput research to provide comprehension of signaling events that control plant-nematode interactions. RNA sequencing of gland cells could be used to learn more about the activity of numerous effectors involved in the signaling cascade. Similarly, genome editing techniques such as CRISPR-Cas9 can be used to investigate different gene expressions pertaining to plant-nematode interactions. Such genome editing technologies are handy in understanding a variety of effectors and R proteins [54].

4   Molecules produced by plants in defense

Pathogens produce a broad range of molecules as chemical cues to establish pathogenicity [55,56]. Plants have evolved to perceive and destroy such chemical cues by employing a number of strategies, of which, some are discussed below.

4.1   Small RNAs (sRNAs)

RNAs are gene regulatory molecules and are classified as coding and noncoding. Coding RNAs include mRNAs (messenger RNAs); noncoding RNAs are divided into rRNAs (ribosomal RNAs), tRNAs (transfer RNAs), and sRNAs (small RNAs) [57]. sRNAs regulate gene expression of different biological processes in higher organisms [58] and include miRNAs (micro-RNAs), siRNAs (short interfering RNAs), piRNAs (piwi interacting RNAs), tasiRNAs (Trans-acting siRNA), rasiRNAs (repeat-associated small interfering RNAs), vsiRNAs (Virus-derived siRNAs), etc. sRNAs are about 18–30 nucleotides long double-stranded RNAs [59]. By targeting chromatin and transcripts, they can be used in regulating gene expression, gene splicing, nucleotide modifications, and protein transport. They were first discovered in *C. elegans* by Fire et al. [60] as a switch for turning off translation [57]. Since then, numerous interesting studies have been made regarding sRNAs and a plethora of these molecular entities have been discovered or artificially designed [61].

sRNAs are mainly involved in gene silencing, also termed as RNAi (RNA interference) mechanism, cosuppression, or quelling [62]. Gene silencing occurs through complementary binding to the target mRNA, leading either to mRNA degradation or translational repression [63]. miRNAs and siRNAs are majorly reported and most studied sRNAs species due to their prominent role in plant development as well as stress responses [64]. In plants, miRNA encoding genes (*MIR*) are found endogenously distributed throughout the genome [65]. siRNAs, on the other hand, can originate endogenously, or from viruses or repetitive elements in the genome [66,67].

Plant and environment interaction involves a collection of signals and molecules, including hormones, volatiles, proteins, and nucleotides [68], of which 30% of gene regulation is controlled by sRNAs [59]. Under normal conditions, plants do not express these sRNAs, but upon encounters with the environmental stressors that threaten the physiology of plants, they either upregulate or downregulate their expression levels depending on the target genes. The sRNAs associated with defense protein silencing seem to be downregulated and those that interact directly with the foreign genome seem to be upregulated. For example, the miR398 levels were found to be decreased after treatment with Cu^+, Fe^+, ozone, and salt in *Arabidopsis*. Also, miR398 levels decreased when *Arabidopsis* leaves were infected with avirulent strains of *P. syringae* pv. *tomato*, Pst DC3000 (effectors avrRpm1 or avrRpt2) [69]. siRNA has evolved as an antiviral defense in plants. An endogenous siRNA, nat-siRNAATGB2, was found to be specifically induced by the bacterial pathogen *P. syringae* carrying effector avrRpt2 [70]. Liu et al. [71] reported a virus-derived siRNA (vsiRNA1) regulated expression of a wheat thioredoxin-like gene (TaAAED1) encoding a negative regulator of reactive oxygen species (ROS) in the chloroplast.

The conservation of various sRNA species across the plant kingdom suggests their dominant role in plant defense mechanisms [72]. With the discovery of diverse sRNA molecules, genetically modified plants are being developed that can adapt themselves to stress situations, especially disease incidence. Advanced research is required to explore and unravel the molecular mechanisms that administer the complex stress responses in plant systems, thereby developing improved varieties of stress-tolerant crops for increased productivity [73].

4.2 Pattern/pathogen recognition receptors (PRRs)

Pattern recognition receptors (PRRs) are immunity-related surface-focalized receptors in hosts that detect Microbe Associated Molecular Patterns or Pathogen Associated Molecular Patterns (MAMPs or PAMPs) released by pathogens [74]. They are radically different domains, extracellular in nature, localized on membranes with high sensitivity and specificity toward MAMP/PAMP moieties [75]. They comprise two types broadly: Receptor-Like Kinase (RLK) type and Receptor-Like Proteins (RLP) type. The former is involved in signal transduction within cells, outside cells (ligand binding) and at transmembrane levels, containing cytosolic kinase property, while the latter does not exhibit intracellular kinase property. They undergo polymerization for signal transduction. Both RLK and RLP types, always functioning in unison, are constituents of structurally complex proteins associated with signal perception and transfer along with regulatory proteins that control the activity of PRRs. Robatzek et al. [76] report that *Arabdiopsis* alone contains 600 and 57 different types of RLK genes and RLPs respectively, of which, many are employed in plant defense. Leucine-rich repeats (LRRs), epidermal growth factor (EFG), lectin-like motifs, lysine motifs (LysMs), etc., are extracellular domains and bind to ligands for pathogen detection. Nucleotide-binding leucine-rich repeats (NBLRRs) are a type of LRRs.

A single plant species can inherit many PRRs that are capable of sensing various groups of MAMPs/PAMPs redundantly, thus having an upper hand in the recognition of pathogens. The stronger the PTI, the more diverse and redundant the PRRs and vice versa. Some plants such as *Arabdiopsis* are reported to sense as many as seven MAMPs of *Psudomanas* by employing different PRRs. Boller and Felix [74] reported that some PRRs are common across plants (e.g., flagellin sensing2 receptor or FLS2), while other plant species contain specific PRRs (e.g., elongation factor18 receptor of EF-Tu or elf18). As pathogens evolve their MAMP epitopes, plants too update their PRRs regularly, so as to defend themselves by rapid identification. Plant species such as Tomato and rice have been reported to detect variations of flagellin sensing receptors and EF-Tu respectively [23,77]. This shows that plants can not only detect different pathogens but also different epitopes of MAMPs produced by the same pathogen.

Many pathogens such as bacteria and fungi are detected by a number of plants via PAMP/ MAMP-PRR interactions, thus activating PTI (pattern/pathogen-triggered immunity). These act as chemical cues during disease incidence and plant defense. A review by Noman et al. [78] provides many examples of these cues. *Xanthomonas compestris* produced axYS22 protein during infection which bound to its plant counterpart Xa21 in rice. *Phytophthora* produces Pep13 molecules as PAMPs in some cases. Flagellin (flg22 epitope) are recognized by FLS2 PRRs in grapes. Similarly, Moroz and Tanaka [79] concluded that FlgII-28 initiated defense system in potato. In *Arabdiopsis*, PRRs such as EFR (elongation factor receptors) detect PAMPs such as EF-Tu (elf18 epitope). Fungal chitin also acts as PAMPs that are identified by a wide range of PRRs, namely AtCERK1 and OsCEBiP in *Arabdiopsis* and *Oryza sativa* respectively. Similarly, these hosts also recognize peptidoglycans produced by bacterial pathogens with the help of PRRs such as AtLYM1 and OsLYP6 respectively [78]. In tomato, LeEIX2 provide resistance to fungal xylanase effectors (Xyn11), thus resisting *Botrytis cinerea* and *Oidium neolycopersici* infections [80].

Of all PRRs, LRRs, particularly NBLRRs are of prime importance in ETI (effector-triggered immunity). ETI is usually linked with localized cell death via hypersensitive response. Pathogens produce avirulent (*avr*) effector molecules to dodge PTI, altering the physiology of plants and protein profile so as to increase infection/virulence. These express R (resistance) genes in the host to code for NBLRRs that trigger ETI [81]. Nucleotide-binding domain is essential for the functioning of these proteins. While ATP is required to activate signaling, LRR moiety helps in the interaction and recognition of effector molecules. NBLRRs can be localized in nucleus, cytoplasm or plasma membrane depending on host plants. So far, *A. thaliana* is reported to contain around 159 NB-LRR genes, of which, *RFO1*, *RPW8*, *WRR4*, etc., are well-documented for plant defense. Similarly, *RCT1* and *QRR1* are reported in *Medicago* sp. Furthermore, *Rpsar-1* is studied in *Phaseolus vulgaris* [78]. Pi-ta and Rpiblb1 are reported in rice and *Solanum bulbocastanum* respectively as resistant proteins against specific pathogens [82,83]. In a recent study, Du et al. [84] concluded that *OsRLR1* plays a role in defense against pathogens such as *Magnaporthe oryzae* and

Xanthomonas oryzae pv. *oryzae*. Wang et al. [85] reported *CsRSF1* and *CsRSF2* to be defense-related genes in *Cucumis sativus* against powdery mildew pathogen *Sphaerotheca fuliginea*.

Thus, PRRs play a crucial role in plant defense and are important to be studied. Genome-wide analysis of these signaling cues along with their mechanisms provides better chances of strengthening plant defense systems and health.

4.3   Cell wall as barrier

Plant tissues are well-equipped with various structural, physical, and chemical barriers. The plant cell wall is the first line of defense against all prospective bacterial and fungal phytopathogens. It acts as an excellent structural barrier against biotic and abiotic stresses [86,87].

Through their interaction chemistry that has coevolved throughout time, the dynamics between plant-pathogen relationships can be understood. For instance, similar pathogen/microbe-associated molecular patterns (PAMPs/MAMPs) of the pathogen can initiate infection recognized by the pattern recognition receptors (PRRs) of the hosts. This PAMPs/MAMPs- PRR interaction leads to the activation of subsequent defense reactions termed pattern-triggered immunity (PTI) [88]. For any establishment of infection, pathogens have to breach the complex network of cellulose microfibrils along with the matrix of pectic polysaccharides of the cell wall. Multiple changes in the cell wall composition may take place in response to the pathogen attack. Hydrophobic coating of cuticle on the cell wall is also reported to serve as a protective layer against the entry of pathogens and is also capable of releasing antimicrobials against the attacking fungi. Upon infection, the cell wall's architecture and integrity often create an imbalance in the cellular ionic concentrations which leads to the generation and accumulation of reactive oxygen species (ROS) at the site of invasion [89]. As a result of ROS accumulation, downstream transcriptional activity is upregulated, resulting in calcium spikes, MAP kinase activity, and calcium-dependent protein kinase expression (CDPK), as reported by Boudko [90]. One of the defensive responses of plants to necrotrophic fungi, bacterial, or insect polygalacturonases (PGs) is the synthesis and overexpression of polygalacturonase-inhibiting proteins (PGIPs). A change in the integrity of cell walls resulting from the impaired expression of specific proteins involved in their biosynthesis or from the invasion of pathogens; activates a response of specific defensive and growth genes. Furthermore, pathogens can induce damage-associated molecular patterns (DAMPs) in plants as well as glucans, fructans, cellulose, pectin, suberin lipids, hydroxyproline-rich glycoproteins (HRGPs), peroxidases, etc. This mechanism is called activating danger/damage-triggered immunity (DTI) [74].

Plant epidermal cells may also grow out in the form of thorns or trichomes, providing resistance against herbivores like pests too. Plant cells also deposit papillae structures at the sites of pathogen detection and act as a physical barrier by restricting the entry of pathogens [91,92]. Harada et al. [93] studied the expression of several pathogenesis-related proteins (PR5 and PR14) and lipid transfer protein (LTP) in leaf trichomes during *Peronospora tabacina* attack on *N. tabacum*. Moreover, these papillae also act as an important reservoir for important antimicrobial metabolites, such as phenolics (lignin and phenolic-polyamines), ROS, peroxidase enzymes, structural proteins (arabino-galactan and HRGPs), and polymers (pectin and xyloglucans). These features represent the plant's innate immune system [94]. Host plant's ability to generate papillae with correct chemical depositions can confer the papillae-mediated penetration resistance in plants against pathogens. Interestingly, plants also seal off the plasmodesmata as roadblocks to prevent the spreading of the pathogen to neighboring cells.

The abovementioned cell wall-associated plant defense mechanisms can halt or stop invading pathogens at an early stage even before it reaches hyper-sensitive response or cell death. The gene-knockout technique appears to be a useful method to understand the transcriptional programming of the genes involved in the plant defense mechanism. The Carbohydrate Microarray Technique is another popular technique that enables rapid and multiplexed analysis of the changes in cell wall composition before or after infection and could be used as a tool to provide new insights into the dynamic nature of host-pathogen interactions.

4.4   Micro-RNAs (miRNAs)

Diseases in plants cause significant losses in global yield every year. Various studies conducted on plant defense strategies have shed light on the role of epi-genetics in contributing to their tolerance toward stress and adaptability. A post-transcriptional mechanism involving small noncoding RNAs is a common approach used by plants to combat pathogens that try to disturb the plant phys-iology [95]. miRNAs (microRNAs)-mediated gene silencing is one of the strat-egies adapted by plants to resist the pathogen-induced diseases. miRNAs are small noncoding RNAs of about 20–24 nucleotides long that function by repres-sing/silencing target genes thereby regulating the expression of genes [96]. miRNA encoding genes (*MIR*) are distributed throughout the genome, most of which are located in the intergenic regions, though they can be rarely found on the intron and exon regions of the plant genome. The biogenesis of miRNAs is similar in both plants and animals with minor differences [65]. In plants, the *MIR* genes are transcribed by RNA polymerase II to form pri-miRNA (primary miRNA) transcripts that are self-complementary and can fold back to form double-stranded hairpin structure. Inside the nucleus, sRNA-like DCL1 (Dicer-like 1) protein cleaves the pri-miRNA into short pre-miRNA (precursor miRNA). Double-stranded RNA binding protein 1 (DRB1), Hyponastic Leaves 1 (HYL1) and Serrate (SE) assistance stabilize the binding and optimize the cleavage of pri-miRNA by DCL1 [97]. Pre-miRNA is further processed by DCL1 to form miRNA (guide)/miRNA* (passenger) duplex. HUA Enhancer 1 (HEN1) methylates the $3'$ overhangs of the duplex to protect it from subsequent degradation [98]. The duplex is transported into the cytoplasm with the help of Exportin-5 ortholog HASTY. The duplex subsequently associates with Argonaute 1 (AGO 1) to form the RNA-induced silencing complex (RISC) chaperoned by Hsc70/Hsp90 and ATP. The passenger strand gets degraded inside the RISC [99]. miRNA-guided gene silencing is dependent on the degree of complementarity between the miRNA and the target gene. Perfect complementarity between miRNA and target gene will result in the degradation of the target mRNA, whereas imperfect complementarity results in translation inhibition [100].

Over the past few years, various studies have shown the crucial role the miRNAs play in plant stress responses. The regulation of miRNAs in the plant while encountering environmental stress is responsible for their flexibility and adaptability. Under normal conditions, they are not detected but when the plant is under stress, the level of expression of miRNAs is altered. The role of miRNAs in plant defense of *Arabdiopsis thaliana* against pathogens such as *P. syringae* has been well documented [101].

During pathogen invasion, miRNAs that target defense proteins seem to be downregulated, while those that directly interact with the exogenous genome seem to be upregulated. miRNAs such as miR319, miR158, miR160, miR167, miR165/166, and miR159 levels were increased while miR390, miR398, miR408, and miR825 levels were decreased in *Arabidopsis* leaves when treated with the virulent *P. syringae* pv. *tomato* PstDC3000. The DCL series (DCL4, DCL2, DCL3) are antiviral in nature, particularly against RNA viruses [102]. In tomato, miR1916 has been reported to provide resistance against *B. cinerea* [103]. Some miRNAs also target hormone-encoding genes suggesting their roles in plant defense signaling by regulating various plant hormone pathways [99,104]. Some miRNAs have also been documented to promote the formation of secondary siRNAs/phasiRNAs by interfering with mRNAs of resistance (R) genes that code for NBLRRs. Artificial miRNAs are also being developed which seem to be a promising tool for controlled disease resistance in closely related species [61].

An important role for miRNAs in plants' biological and metabolic processes includes controlling cell fate and morphology, responding to environmental stresses, signaling, etc. [105]. The miRNA-mediated gene silencing is a beneficial mechanism in agriculture to create genetically modified crop varieties that can withstand adverse environmental conditions. This will not only help to increase crop productivity but also solve the problem of world hunger in future. Thus, analyzing the molecular mechanisms of miRNA regulatory pathways under various stress conditions and identifying the crucial components will enhance our understanding of plant-environment interaction.

4.5 Antimicrobial compounds

To sustain and cope with different environmental conditions plants are known to synthesize both primary and secondary metabolites for their growth and development as well as survival. Primary metabolites are directly produced by plants, i.e., amino acids, lipids, etc., plays an important role in the physiological and morphological growth of the plant. Whereas, secondary metabolites are produced from primary metabolites act as precursors and produce various beneficial effects on the health of living organisms, i.e., alkaloids, flavonoids, etc., which are significantly involved in plant defensive response. These antimicrobials can be constitutive (occur in biologically active forms in healthy plants) or inductive (produced after the recognition of the pathogen's elicitor upon infection) in nature. The preformed metabolites are called "phytoanticipin," while those produced as a response are termed "phytoalexins." A few important functions of secondary metabolites produced by plants often include protection against biotic (pests or pathogens) and combating various abiotic stresses (UV radiation, physical or chemical barriers).

On the bases of biosynthetic origin, secondary metabolites have been classified into three major categories, namely terpenoids, phenolic compounds, and nitrogen-containing and sulfur-containing compounds.

Terpenoids: A large class of secondary metabolites composed of terpenoids has attracted great attention for its physiological functions (e.g., hormones, membrane anchors, stabilizing structures of membranes, etc.), and ecological relevance. (e.g., defense, insect/animal attractant) [106]. The colonization of roots by AM fungi affects secondary metabolism in plants, alteration in terpenoides composition and concentration improvises plant defense against herbivores and pest attacks the concentration and composition of terpenoids, which can enhance both direct and indirect plant defenses against herbivorous insects [107]. Zealexins, a group of sesquiterpenoid phytoalexins is found to have antifungal activity against *Aspergillus* and *Fusarium* spp. Cotton roots are stimulated by *Trichoderma virens* to produce defense response which is the terpenoid synthesis which further act as a biological control against *Rhizoctonia solani* that causes cotton seedling disease [108]. Several terpenoids such as pyrethrins, menthol, camphor, farnesol, and artemisinins are also known to possess antibacterial and antifungal properties.

Phenolic compounds: The diverse functions of phenolic compounds is to maintain plant structural and physiological integrity also protect them against various stress [109]. Phenolic metabolites are well-known for their antimicrobial and nematicidal actions. Some of them such as benzaldehyde, flavones, naringin, genistein, etc., provide resistance against a wide range of fungi. Phenolic compounds such as p-coumaric acid, catechin, caffeic acid, and tannins have a direct effect on fungal growth and sporulation of *Phaeoacremonium aleophilum* and *Phaeomoniella chlamydospore* that caused Petri disease in grapevine and play an important role in the defense mechanism protecting grapevine against the fungi [110]. Lignification is one of the common processes to prevent the growth and development of pathogens upon infection/wounding. Flavonoids such as catechin, galangin, phloretin are found to be effective against various microorganisms owing to their capability to complex with bacterial proteins, thereby, disrupting their cellular structure and physiology and causing the lysis of the cells.

Alkaloids and cyanogenic glycosides: These are sulfur and nitrogen-consisting compounds which are derivatives of tyrosine, lysine, tryptophan, and aspartic acid, which perform a pivotal role in protection against herbivores and pathogens [111]. Alkaloids such as pyrrolizidine, caffeine, dimethyltryptamine, etc., act as potent antimicrobial agents as a part of plant defense mechanisms. Moreover, many plants also contain unusual nonprotein amino acids such as azetidine-2-carboxylic acid and canavanine for defense purposes. *Allium sativum* (garlic) is widely known for its sulfur-containing components, allicin, and its derivatives are endowed with antibacterial, antifungal, antiviral, or antiprotozoal activity. Some thiolsulfinate compounds from onions are also found to possess antimicrobial properties. Thiophenes are other examples of S-containing antimicrobial compounds in plants of Asteraceae. It is also known that thiophenes from *Tagetes* sp. exhibit a broad range of antimicrobial activities [112]. Interestingly, plants also produced different types of defensive proteins, digestive enzyme inhibitors, protease inhibitors, and hydrolytic enzymes that provide effective protection against plant pathogens. In addition to being effective antimicrobial agents themselves or as elicitors of other plant defense responses, secondary metabolites also have tremendous biological potential that can benefit mankind in the form of healthcare, medicines, and industrial applications.

4.6   Phytohormones

Plants have developed a complex and defensive response to various stresses which involves modulation of molecular events, activated by signaling molecules such as phytohormones. Phytohormones are low-molecular-weight molecules that act as growth regulators for plant development and physiological processes, as well as take part in the defensive response system against stresses. Auxins, gibberellins (GAs), cytokinins (CKs), ethylene (ET), abscisic acid (ABA), salicylic acid (SA), jasmonic acid (JA), and brassinosteroids (BRs) are some examples for phytohormones that respond to plant stress via individual signaling pathway and also through signaling cross talk between the pathways either synergistically or antagonistically [113–115]. The cross talk or linking between various phytohormone signaling pathways has been proved using large-scale transcriptome analyses [116]. The production, distribution, and the signal transduction of the phytohormones are affected under stress conditions, leading to morphological, molecular, and physiological changes which enable the plants to withstand unfavorable conditions [117].

Auxins are well-known as a plant growth hormone that helps in cellular elongation and growth, also numerous reports have suggested that auxins promote vulnerability toward disease, combating different biotic stress, etc., for example, Djami-Tchatchou et al. [118] found out that the disruption of auxin signaling resulted in decreased bacterial growth in plants expressing the axr2-1 mutation and this phenotype could be suppressed by introducing the sid2-2 mutation, which impairs SA synthesis. Thus, host auxin signaling is required for normal susceptibility to *P. syringae* (PtoDC3000) and is involved in suppressing SA-mediated defenses. They also investigated that auxin (IAA) promotes PtoDC3000 virulence through a direct effect on the pathogen and found that IAA modulates the expression of virulence genes, both in culture and in planta. There is increasing evidence that cytokinins contribute to biotic stress responses [119]. A cytokinin-activated transcription factor promotes SA defense responses; signaling cascades CK and SA are interconnected [120]. Cytokinin enhances tomato resistance to *X. campestris* pv. *vesicatoria* and *P. syringae* pv. *tomato* through a process that relies on SA and ET signaling [80].

Plant defense responses do not solely depend on any one hormone, rather, all of them work with each other to regulate it. The cross talk between SA, JA, and ET-dependent signaling pathways is thought to be involved in adjusting the defense reaction mechanism, which eventually results in the activation of an optimal combination of defense responses [121]. A number of genes which respond to Jasmonic acids, such as plant defensins, vegetative storage proteins, and lipoxygenase 2 (LOX2), were downregulated by SA. After infection, JA and ET work synergistically to regulate the synthesis of defense genes. Both phytohormones suppress the transcription factor of ethylene EIN3 which helps in resisting necrosis by increasing the number of root hairs [122]. A negative relationship between JA and SA signaling has also been observed against necrotrophic pathogens [123].

Luo et al. [124] studied the integrated transcriptome analysis of plant hormones jasmonic acid (JA) and salicylic acid (SA) in coordination with the growth and defense responses upon fungal infection in Poplar. A total of 943 genes were identified as common responsive genes (CRG) that are generally involved in the processes of stress responses, metabolism, growth, and development. Even they recognized genes which play a vital role during Jasmonic acid/salicylic acid signaling during growth response, fungal attack, and other rust pathogens, *Melampsora larici-populina* (MLP) treatment reflects the pivotal role of jasmonic acid and salicylic acid during the resistance of fungal attack. Also, these genes improve plant metabolic machinery, tolerating different stress, etc.

There are many reports that corroborate phytohormones cross talk hypothesis, helping plants maintain homeostasis and respond to biotic stress. For understanding how plants integrate signals supporting plant defense in a better way, more studies are needed in relation to the molecular aspect of cross talk among plant defense pathways.

4.7   Microbe-triggered plant volatiles (microbe-induced plant volatiles)

The ecological network comprises plants, various insects, pests, and a wide array of microbial pathogens. Plants must be able to recognize these pathogens quickly and precisely respond appropriately. Plants recognize these invaders either by direct contact using molecular entities such as PAMPs /DAMPs or by indirect mechanism using volatile organic compounds (VOCs) as signaling cues. Volatiles represent plants' chemical language, employed in cross talk with various plants and associated microorganisms. These volatiles include green leaf volatiles (GLVs), isoprenoids, and intermediates of shikimic acid cycle, such as methyl salicylate and indole [125,126].

Ethylene, salicylic, and jasmonic acid signaling pathways lead to volatile production. Plant release these volatiles as defense signals against airborne and soilborne pathogens. Depending on the host plant species, microbe-induced plant volatiles (MIPVs) perform diverse functions in Cucumber mosaic virus (CMV) transmission. MIPVs are important in the establishment of pathogenic microorganisms and the vulnerability of the host. MIPVs also increase the settling of vector agents on plants infested with a virus, in case of persistent viruses. A nonpersistent virus such as CMV release effector protein 2b, which modulate the composition of plant volatiles, thus reducing pollinator repellents: 2-carene and β-phellandrene in tomato. Plants utilize MIPV molecules as natural antibiotics at wound sites in addition to their signaling role. Based on their virulence, bacterial elicitors too cause emissions of volatiles in plants. In a study conducted in *Arabdiopsis*, it was found that bacterial pathogens such as *P. syringae* induced the production of plant volatiles such as b-ionone and a-farnesene [127] (Table 1).

In response to pathogenic fungi, plants emit a mixture of volatiles. The ability of a plant to produce volatiles is determined by its sensitivity to virulent microorganisms. Castelyn et al. [139] reported that, upon exposure to stripe rust disease, resistant wheat plants produced VOCs belonging to sesquiterpenes such as bocimene, whereas susceptible wheat plants released oxylipin derivatives, both governed by a single resistant (R) gene of the plant. Oxylipins (C9) have high antifungal activity at natural concentrations and limit the growth of fungus and germination of spores. In *Arabidopsis*, C9-aldehydes limit the growth of *Fusarium oxysporum* and *Botrytis cinerea*. Expression of genes related to defense such as PR1, PR2, and PR4 occurs, upon exposure of susceptible genotype to VOCs of those of resistant ones. The VOC profile of resistant genotypes is dominated by linalool, b-ocimene, limonene, and farnesene, while susceptible plants primarily emit aldehydes (nonanal and decanal). Nerolidol and terpinolene, two of eight microbially produced terpenes from garlic studied by Pontin et al. [138] exhibited deleterious outcomes on mycelium development and sclerotium formation of *Sclerotium cepivorum*. It was noteworthy that Wu et al. [140] reported some volatiles were primarily regulated by jasmonic acid (JA) hormone. Thus, by changing JA signaling, invaders modulate VOCs emission in plants. Microbes use the JASMONATE ZIM DOMAIN (JAZ) to either stimulate or repress JA-dependent VOC production. As soon as JAZ gets degraded, transcription factors dependent on JA are released from JA-dependent volatile synthase genes, thus, getting expressed.

TABLE 1 Plant volatile exuded by microbes and their benefits to plant.

Microbes	Beneficial chemicals	Plant	Benefit in plant mechanism	References
Trichoderma asperellum	6-Pentyl-pyrone	Arabidopsis thaliana	Increase plant defense response also suppress sporulation of A. alternate and B. cinerea	Kottb et al. [128]
Pseudomonas syringae pv.	Esters of (Z)-3-hexenol	Solanum lycopersicum	Enhance resistance of the plant	López-Gresa et al. [129]
P. syringae pv. syringae	Nonanal	Phaesiolus spp.	Enhance resistance in emitting the plant	Yi et al. [130]
Erwinia amylovora	Methyl salicylate	Malus domestica	Vector deterrent	Cellini et al. [131]
F. graminearum	(Z)-3-hexenyl acetate	Triticum aestivum	Resistance enhance	Ameye et al. [132]
Bacillus sp. B55	Dimethyl disulfide	Nicotiana attenuata	Increase reduce sugar availability	Meldau et al. [133]
B. amyloliquefaciens IN937a; B. subtilis GB03	2,3-Butanediol (2,3-BD)	A. thaliana	By using ethylene signaling pathways elicits ISR toward pathogenic microbes	Ryu et al. [134]
Pseudomonas fluorescens	2,4-Diacetylphloroglucinol (DAPG)	S. lycopersicum	Boost development of root	Brazelton et al. [135]
Xanthomonas oryzae pv. oryzae	(−)-Limonene	Oryza sativa	Increased resistance in the emitting plant	Lee et al. [136]
Fusarium spp.	Indole	Zea mays	Enhance resistance	Shen et al. [137]
Sclerotium cepivorum	Mono- and sesquiterpenes	Allium sativum	Increased resistance in the emitting plant	Pontin et al. [138]

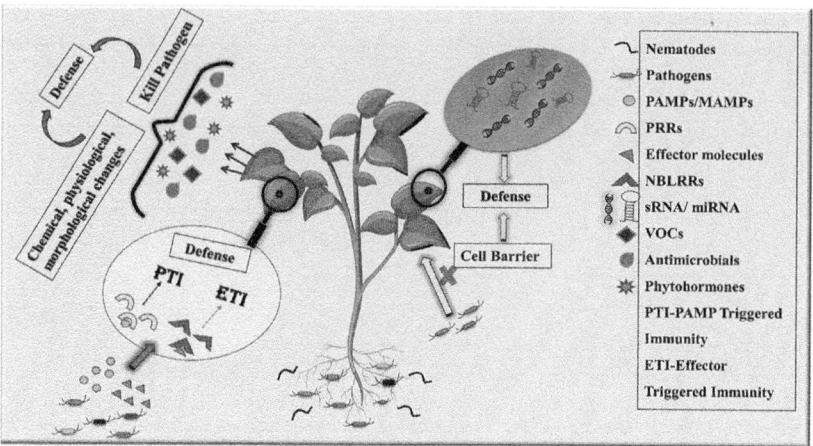

FIG. 2 Various signaling cues involved in cross talk during plant defense.

VOCs play a major role in plant-microbe interaction. Plants signal different cues depending on the pathogen. MIPVs have a huge potential in plant defense. However, their volatility and reactive attribute with various other biomolecules necessitate the development of commercially viable formulations. New technologies for synthesizing volatiles as natural insecticides have recently been studied. To summarize, volatiles are important signaling clues during pathogen attacks (Fig. 2).

5 Conclusions

The population load in present day agriculture is too huge to incur yield losses. Biotic stress, particularly pathogens greatly impacts plant growth, development, immunity, overall health, and ultimately production and productivity. Plants and pathogens exchange important molecules as signaling cues in order to benefit/ harm each other. Their cross talk is very dynamic and complex. However, studies pertaining to this field are scanty. Overhearing such a conversation at biochemical and molecular levels can be very beneficial, especially to combat diseases. Modern molecular genetic tools have helped decipher some of these conversations, but not all. Having a better understanding of the chemical interaction between plants and pathogens helps in the effective management of diseases, thereby avoiding losses.

References

[1] A. Martinez-Medina, V. Flors, M. Heil, B. Mauch-Mani, C.M. Pieterse, M.J. Pozo, J. Ton, N. M. van Dam, U. Conrath, Recognizing plant defense priming, Trends Plant Sci. 21 (2016) 818–822.
[2] J. Imam, P.K. Singh, P. Shukla, Plant microbe interactions in post genomic era: perspectives and applications, Front. Microbiol. 7 (2016) 1488.

[3] A. Kumar, S. Droby, V.K. Singh, S.K. Singh, J.F. White, Entry, colonization, and distribution of endophytic microorganisms in plants, in: Microbial Endophytes, Woodhead Publishing, 2020, pp. 1–33.

[4] S. Yadav, A.K. Chhibbar, Plant–virus interactions, in: Molecular Aspects of Plant-Pathogen Interaction, Springer, Singapore, 2018, pp. 43–77.

[5] C.B. Patel, V.K. Singh, A.P. Singh, M. Meena, R.S. Upadhyay, Microbial genes involved in interaction with plants, in: H.B. Singh, V.K. Gupta, S. Jogaiah (Eds.), New and Future Developments in Microbial Biotechnology and Bioengineering, Elsevier, Singapore, 2019, pp. 171–180, https://doi.org/10.1016/B978-0-444-63503-7.00010-3.

[6] S. Eves-van den Akker, B. Stojilković, G. Gheysen, Recent applications of biotechnological approaches to elucidate the biology of plant–nematode interactions, Curr. Opin. Biotechnol. 70 (2021) 122–130.

[7] I. Sharma, Phytopathogenic fungi and their biocontrol applications, in: Fungi Bio-Prospects in Sustainable Agriculture, Environment and Nano-Technology, Academic Press, 2021, pp. 155–188.

[8] A. Zehra, N.A. Raytekar, M. Meena, P. Swapnil, Efficiency of microbial bio-agents as elicitors in plant defense mechanism under biotic stress: a review, Curr. Res. Microb. Sci. 2 (2021), 100054, https://doi.org/10.1016/j.crmicr.2021.100054.

[9] C. Zipfel, Plant pattern-recognition receptors, Trends Immunol. 35 (2014) 345–351.

[10] J. Dangl, Molecular specificity in the plant immune system, Mol. Biol. Cell 15 (2004) 2a.

[11] J.D. Jones, J.L. Dangl, The plant immune system, Nature 444 (2006) 323–329.

[12] P.P. Singh, A. Kujur, A. Yadav, A. Kumar, S.K. Singh, B. Prakash, Mechanisms of plant-microbe interactions and its significance for sustainable agriculture, in: PGPR Amelioration in Sustainable Agriculture, Woodhead Publishing, 2019, pp. 17–39.

[13] V.J. Higgins, H. Lu, T. Xing, A. Gelli, E. Blumwald, The gene-for-gene concept and beyond: interactions and signals, Can. J. Plant Pathol. 20 (1998) 150–157.

[14] S.K. Ray, D.M. Macoy, W.Y. Kim, S.Y. Lee, M.G. Kim, Role of RIN4 in regulating PAMP-triggered immunity and effector-triggered immunity: current status and future perspectives, Mol. Cells 42 (2019) 503.

[15] D. Tang, G. Wang, J.M. Zhou, Receptor kinases in plant-pathogen interactions: more than pattern recognition, Plant Cell 29 (2017) 618–637.

[16] L. Wang, Tomato Receptors of the Bacterial Cold Shock Protein and the Plant Peptide Signal Systemin (Doctoral dissertation), Universität Tübingen, 2018.

[17] J.M. Zhou, Y. Zhang, Plant immunity: danger perception and signaling, Cell 181 (2020) 978–989.

[18] R. Panstruga, J.E. Parker, P. Schulze-Lefert, SnapShot: plant immune response pathways, Cell 136 (2009) 978–e1.

[19] A. Dalal, Z. Attia, M. Moshelion, Flagellin triggers mesophyll dehydration: An early PTI defense against bacterial establishment in intercellular spaces, bioRxiv (2021). 2020-12.

[20] R. Takai, A. Isogai, S. Takayama, F.S. Che, Analysis of flagellin perception mediated by flg22 receptor OsFLS2 in rice, Mol. Plant-Microbe Int. 21 (2008) 1635–1642.

[21] L. Gómez-Gómez, Z. Bauer, T. Boller, Both the extracellular leucine-rich repeat domain and the kinase activity of FLS2 are required for flagellin binding and signaling in *Arabidopsis*, Plant Cell 13 (2001) 1155–1163.

[22] K. Mueller, P. Bittel, D. Chinchilla, A.K. Jehle, M. Albert, T. Boller, G. Felix, Chimeric FLS2 receptors reveal the basis for differential flagellin perception in Arabidopsis and tomato, Plant Cell 24 (2012) 2213–2224.

[23] S.R. Hind, S.R. Strickler, P.C. Boyle, D.M. Dunham, Z. Bao, I.M. O'Doherty, J.A. Baccile, J. S. Hoki, E.G. Viox, C.R. Clarke, B.A. Vinatzer, Tomato receptor FLAGELLIN-SENSING 3 binds flgII-28 and activates the plant immune system, Nature Plants 2 (2016) 1–8.

[24] X. Fan, Z. Zhao, Y. Li, T. Zhuo, X. Hu, H. Zou, The EF-Tu epitope elf26 of *Ralstonia sola-nacearum* can promote resistance to bacterial wilt disease in nicotiana species, Can. J. Plant Pathol. 40 (2018) 387–398.

[25] S. Ranf, Sensing of molecular patterns through cell surface immune receptors, Curr. Opin. Plant Biol. 38 (2017) 68–77.

[26] A. Niehl, I. Wyrsch, T. Boller, M. Heinlein, Double-stranded RNA s induce a pattern-triggered immune signaling pathway in plants, New Phytol. 211 (2016) 1008–1019.

[27] P. Carella, E. Evangelisti, S. Schornack, Sticking to it: phytopathogen effector molecules may converge on evolutionarily conserved host targets in green plants, Curr. Opin. Plant Biol. 44 (2018) 175.

[28] J. Goutam, R. Singh, R.S. Vijayaraman, M. Meena, Endophytic fungi: carrier of potential antioxidants, in: P. Gehlot, J. Singh (Eds.), Fungi and their Role in Sustainable Development: Current Perspectives, Springer, Singapore, 2018, pp. 539–551, https://doi.org/10.1007/978-981-13-0393-7_29.

[29] G. Yadav, M. Meena, Bioprospecting of endophytes in medicinal plants of Thar Desert: an attractive resource for biopharmaceuticals, Biotechnol Rep. 30 (2021), e00629, https://doi.org/10.1016/j.btre.2021.e00629.

[30] R. Mishra, Fungal and bacterial biotrophy and necrotrophy, in: Molecular Aspects of Plant-Pathogen Interaction, Springer, Singapore, 2018, pp. 21–42.

[31] P. Kumari, M. Meena, R.S. Upadhyay, Characterization of plant growth promoting rhizobac-teria (PGPR) isolated from the rhizosphere of *Vigna radiata* (mung bean), Biocatal. Agric. Biotechnol. 16 (2018) 155–162.

[32] S.A. Gebrie, Biotrophic fungi infection and plant defense mechanism, J. Plant Pathol. Micro-biol. 7 (378) (2016) 2.

[33] K.T. Kim, J. Jeon, J. Choi, K. Cheong, H. Song, G. Choi, S. Kang, Y.H. Lee, Kingdom-wide analysis of fungal small secreted proteins (SSPs) reveals their potential role in host associ-ation, Front. Plant Sci. 7 (2016) 186.

[34] Y. Kraepiel, M.A. Barny, Gram-negative phytopathogenic bacteria, all hemibiotrophs after all? Mol. Plant Pathol. 17 (2016) 313.

[35] S.F.S. Ab Rahman, E. Singh, C.M. Pieterse, P.M. Schenk, Emerging microbial biocontrol strategies for plant pathogens, Plant Sci. 267 (2018) 102–111.

[36] T.L. Friesen, J.D. Faris, Characterization of effector–target interactions in necrotrophic pathosystems reveals trends and variation in host manipulation, Ann. Rev. Phytopathol. 59 (2021) 77–98.

[37] J.D. Faris, T.L. Friesen, Plant genes hijacked by necrotrophic fungal pathogens, Curr. Opin. Plant Biol. 56 (2020) 74–80.

[38] T.C. Wu, X.L. Zhu, L.J. Lu, X.Y. Chen, G.B. Xu, Z.Y. Zhang, The wheat receptor-like cyto-plasmic kinase TaRLCK1B is required for host immune response to the necrotrophic path-ogen *Rhizoctonia cerealis*, J. Integr. Agric. 19 (2020) 2616–2627.

[39] P. Sobiczewski, E.T. Iakimova, A. Mikiciński, E. Węgrzynowicz-Lesiak, B. Dyki, Necro-trophic behaviour of *Erwinia amylovora* in apple and tobacco leaf tissue, Plant Pathol. 66 (5) (2017) 842–855.

[40] X. Di, F.L. Takken, N. Tintor, How phytohormones shape interactions between plants and the soil-borne fungus *Fusarium oxysporum*, Front. Plant Sci. 7 (2016) 170.

[41] R. Singh, S. Dangol, Y. Chen, J. Choi, Y.S. Cho, J.E. Lee, M.O. Choi, N.S. Jwa, *Magnaporthe oryzae* effector AVR-Pii helps to establish compatibility by inhibition of the rice NADP-malic enzyme resulting in disruption of oxidative burst and host innate immunity, Mol. Cells 39 (5) (2016) 426.

[42] S.P. Narayanan, S.C. Lung, P. Liao, C. Lo, M.L. Chye, The overexpression of OsACBP5 protects transgenic rice against necrotrophic, hemibiotrophic and biotrophic pathogens, Sci. Rep. 10 (1) (2020) 1–19.

[43] A. Gupta, M. Bhardwaj, L.S.P. Tran, Jasmonic acid at the crossroads of plant immunity and *Pseudomonas syringae* virulence, Int. J. Mol. Sci. 21 (2020) 7482.

[44] J.P. Carr, A.M. Murphy, T. Tungadi, J.Y. Yoon, Plant defense signals: players and pawns in plant-virus-vector interactions, Plant Sci. 279 (2019) 87–95.

[45] P. Moffett, Transfer and modification of NLR proteins for virus resistance in plants, Curr. Opin. Virol. 26 (2017) 43–48.

[46] C. Wang, C. Wang, J. Zou, Y. Yang, Z. Li, S. Zhu, Epigenetics in the plant–virus interaction, Plant Cell Rep. (2019) 1–8.

[47] M.A. Ali, M.S. Anjam, M.A. Nawaz, H.M. Lam, G. Chung, Signal transduction in plant–nematode interactions, Int. J. Mol. Sci. 19 (6) (2018) 1648.

[48] M.W. Lee, A. Huffaker, D. Crippen, R.T. Robbins, F.L. Goggin, Plant elicitor peptides promote plant defences against nematodes in soybean, Mol. Plant Pathol. 19 (4) (2018) 858–869.

[49] O. Panda, A.E. Akagi, A.B. Artyukhin, J.C. Judkins, H.H. Le, P. Mahanti, S.M. Cohen, P.W. Sternberg, F.C. Schroeder, Biosynthesis of modular ascarosides in *C. elegans*, Angew. Chem. 129 (2017) 4807–4811.

[50] M. Manohar, F. Tenjo-Castano, S. Chen, Y.K. Zhang, A. Kumari, V.M. Williamson, X. Wang, D.F. Klessig, F.C. Schroeder, Plant metabolism of nematode pheromones mediates plant-nematode interactions, Nat. Commun. 11 (2020) 1–11.

[51] S. Jaubert-Possamai, Y. Noureddine, B. Favery, MicroRNAs, new players in the plant–nematode interaction, Front. Plant Sci. 10 (2019) 1180.

[52] S. Chin, C.A. Behm, U. Mathesius, Functions of flavonoids in plant–nematode interactions, Plan. Theory 7 (4) (2018) 85.

[53] T. Hewezi, T. Lane, S. Piya, A. Rambani, J.H. Rice, M. Staton, Cyst nematode parasitism induces dynamic changes in the root epigenome, Plant Physiol. 174 (1) (2017) 405–420.

[54] G. Andolfo, P. Iovieno, L. Frusciante, M.R. Ercolano, Genome-editing technologies for enhancing plant disease resistance, Front. Plant Sci. 7 (2016) 1813.

[55] M. Meena, V. Prasad, R.S. Upadhyay, Evaluation of biochemical changes in leaves of tomato infected with *Alternaria alternata* and its metabolites, Vegetos 30 (2017) 2, https://doi.org/10.5958/2229-4473.2017.00020.9.

[56] M. Meena, V. Prasad, R.S. Upadhyay, Evaluation of *Alternaria alternata* isolates for metabolite production isolated from different sites of Varanasi, India, J. Agric. Res. 2 (1) (2017) 00012.

[57] P. Guleria, M. Mahajan, J. Bhardwaj, S.K. Yadav, Plant small RNAs: biogenesis, mode of action and their roles in abiotic stresses, Genom. Proteom. Bioinform. 9 (2011) 183–199.

[58] P. Peláez, F. Sanchez, Small RNAs in plant defense responses during viral and bacterial interactions: similarities and differences, Front. Plant Sci. 4 (2013) 343.

[59] S. Banerjee, A. Sirohi, A.A. Ansari, S.S. Gill, Role of small RNAs in abiotic stress responses in plants, Plant Gene 11 (2017) 180–189.

[60] A. Fire, S. Xu, M.K. Montgomery, S.A. Kostas, S.E. Driver, C.C. Mello, Potent and specific genetic interference by double-stranded RNA in *Caenorhabditis elegans*, Nature 391 (6669) (1998) 806–811.

[61] R. Kumar, Role of microRNAs in biotic and abiotic stress responses in crop plants, Appl. Biochem. Biotechnol. 174 (1) (2014) 93–115.

[62] Kim., Small RNAs: classification, biogenesis, and function, Mol. Cells 19 (1) (2005) 1–15.

[63] J.K. Lam, M.Y. Chow, Y. Zhang, S.W. Leung, siRNA versus miRNA as therapeutics for gene silencing, Mol. Ther. Nucleic Acids 4 (2015) e252.

[64] D. Yu, X. Ma, Z. Zuo, W. Shao, H. Wang, Y. Meng, Bioinformatics resources for deciphering the biogenesis and action pathways of plant small RNAs, Rice 10 (2017) 1–14.

[65] B. Zhang, Q. Wang, MicroRNA-based biotechnology for plant improvement, J. Cell. Physiol. 230 (2015) 1–15.

[66] R.W. Carthew, E.J. Sontheimer, Origins and mechanisms of miRNAs and siRNAs, Cell 136 (4) (2009) 642–655.

[67] May, Small interfering RNA (siRNA) biogenesis, Reactome, 2009. https://reactome.org/content/detail/R-HSA-426486.

[68] L.E. Rose, E.J. Overdijk, M. van Damme, Small RNA molecules and their role in plant disease, Eur. J. Plant Pathol. 154 (2019) 115–128.

[69] G. Jagadeeswaran, A. Saini, R. Sunkar, Biotic and abiotic stress down-regulate miR398 expression in *Arabidopsis*, Planta 229 (2009) 1009–1014.

[70] S. Katiyar-Agarwal, R. Morgan, D. Dahlbeck, O. Borsani, A. Villegas, J.K. Zhu, H. Jin, A pathogen-inducible endogenous siRNA in plant immunity, Proc. Nat. Acad. Sci. 103 (47) (2006) 18002–18007.

[71] P. Liu, X. Zhang, F. Zhang, M. Xu, Z. Ye, K. Wang, S. Liu, X. Han, Y. Cheng, K. Zhong, T. Zhang, A virus-derived siRNA activates plant immunity by interfering with ROS scavenging, Mol. Plant 14 (7) (2021) 1088–1103.

[72] J. Huang, M. Yang, X. Zhang, The function of small RNAs in plant biotic stress response, J. Integr. Plant Biol. 58 (4) (2016) 312–327.

[73] R.M. Pérez-Clemente, V. Vives, S.I. Zandalinas, M.F. López-Climent, V. Muñoz, A. Gómez-Cadenas, Biotechnological approaches to study plant responses to stress, BioMed Res. Int. 2013 (2013), 654120, https://doi.org/10.1155/2013/654120.

[74] T. Boller, G. Felix, A renaissance of elicitors: perception of microbe-associated molecular patterns and danger signals by pattern-recognition receptors, Annu. Rev. Plant Biol. 60 (2009) 379–406.

[75] S. Ranf, Pattern recognition receptors—versatile genetic tools for engineering broad-spectrum disease resistance in crops, Agronomy 8 (2018) 134.

[76] S. Robatzek, P. Bittel, D. Chinchilla, P. Köchner, G. Felix, S.H. Shiu, T. Boller, Molecular identification and characterization of the tomato flagellin receptor LeFLS2, an orthologue of Arabidopsis FLS2 exhibiting characteristically different perception specificities, Plant Mol. Biol. 64 (2007) 539–547.

[77] T. Furukawa, H. Inagaki, R. Takai, H. Hirai, F.S. Che, Two distinct EF-Tu epitopes induce immune responses in rice and *Arabidopsis*, Mol. Plant-Microbe Int. 27 (2) (2014) 113–124.

[78] A. Noman, M. Aqeel, Y. Lou, PRRs and NB-LRRs: from signal perception to activation of plant innate immunity, Int. Mol. Sci. 20 (8) (2019) 1882.

[79] N. Moroz, K. Tanaka, FlgII-28 is a major flagellin-derived defense elicitor in potato, Mol. Plant-Microbe Int. 33 (2) (2020) 247–255.

[80] R. Gupta, L. Pizarro, M. Leibman-Markus, I. Marash, M. Bar, Cytokinin response induces immunity and fungal pathogen resistance, and modulates trafficking of the PRR LeEIX2 in tomato, Mol. Plant Pathol. 21 (10) (2020) 1287–1306.

[81] J.L. Dangl, D.M. Horvath, B.J. Staskawicz, Pivoting the plant immune system from dissection to deployment, Science 341 (6147) (2013) 746–751.

[82] Y. Chen, Z. Liu, D.A. Halterman, Molecular determinants of resistance activation and suppression by *Phytophthora infestans* effector IPI-O, PLoS Pathog. 8 (3) (2012), e1002595.

[83] Y. Jia, S.A. McAdams, G.T. Bryan, H.P. Hershey, B. Valent, Direct interaction of resistance gene and avirulence gene products confers rice blast resistance, EMBO J. 19 (2000) 4004–4014.

[84] D. Du, C. Zhang, Y. Xing, X. Lu, L. Cai, H. Yun, Q. Zhang, Y. Zhang, X. Chen, M. Liu, X. Sang, The CC-NB-LRR OsRLR1 mediates rice disease resistance through interaction with OsWRKY19, Plant Biotechnol. J. 19 (5) (2021) 1052–1064.

[85] X. Wang, Q. Chen, J. Huang, X. Meng, N. Cui, Y. Yu, H. Fan, Nucleotide-binding leucine-rich repeat genes CsRSF1 and CsRSF2 are positive modulators in the *Cucumis sativus* defense response to *Sphaerotheca fuliginea*, Int. J. Mol. Sci. 22 (2021) 3986.

[86] A. Gupta, S.K. Singh, M.K. Singh, V.K. Singh, A. Modi, P.K. Singh, A. Kumar, Plant growth–promoting rhizobacteria and their functional role in salinity stress management, in: Abatement of Environmental Pollutants, Elsevier, 2020, pp. 151–160.

[87] A. Kumar, H. Verma, V.K. Singh, P.P. Singh, S.K. Singh, W.A. Ansari, A. Yadav, P.K. Singh, K.D. Pandey, Role of *Pseudomonas* sp. in sustainable agriculture and disease management, in: Agriculturally Important Microbes for Sustainable Agriculture, Springer, Singapore, 2017, pp. 195–215.

[88] K. Tsuda, F. Katagiri, Comparing signaling mechanisms engaged in pattern-triggered and effector-triggered immunity, Curr. Opin. Plant Biol. 13 (2010) 459–465.

[89] M. Kobayashi, I. Ohura, K. Kawakita, N. Yokota, M. Fujiwara, K. Shimamoto, N. Doke, H. Yoshioka, Calcium-dependent protein kinases regulate the production of reactive oxygen species by potato NADPH oxidase, Plant Cell 19 (3) (2007) 1065–1080.

[90] D.Y. Boudko, Molecular basis of essential amino acid transport from studies of insect nutrient amino acid transporters of the SLC6 family (NAT-SLC6), J. Insect Physiol. 58 (4) (2012) 433–449.

[91] N.K. Clay, A.M. Adio, C. Denoux, G. Jander, F.M. Ausubel, Glucosinolate metabolites required for an Arabidopsis innate immune response, Science 323 (5910) (2009) 95–101.

[92] M. Meena, S. Samal, *Alternaria* host-specific (HSTs) toxins: an overview of chemical characterization, target sites, regulation and their toxic effects, Toxicol. Rep. 6 (2019) 745–758,-https://doi.org/10.1016/j.toxrep.2019.06.021.

[93] E. Harada, J.A. Kim, A.J. Meyer, R. Hell, S. Clemens, Y.E. Choi, Expression profiling of tobacco leaf trichomes identifies genes for biotic and abiotic stresses, Plant Cell Physiol. 51 (10) (2010) 1627–1637.

[94] R.J. Zeyen, T.L.W. Carver, M.F. Lyngkjaer, Epidermal cell papillae, in: R.R. Belanger, W.R. Bushnell (Eds.), The Powdery Mildew: A Comprehensive Treatise, 2002, pp. 107–125.

[95] C. Rosa, Y.W. Kuo, H. Wuriyanghan, B.W. Falk, RNA interference mechanisms and applications in plant pathology, Annu. Rev. Phytopathol. 56 (2018) 581–610.

[96] S.R. Liu, J.J. Zhou, C.G. Hu, C.L. Wei, J.Z. Zhang, MicroRNA-mediated gene silencing in plant defense and viral counter-defense, Front. Microbiol. 8 (2017) 1801.

[97] D. Bielewicz, J. Dolata, M. Bajczyk, L. Szewc, T. Gulanicz, S.S. Bhat, A. Karlik, A. Jarmolowski, Z. Szweykowska-Kulinska, DRB1 as a mediator between transcription and microRNA processing, bioRxiv (2019).

[98] A.L. Eamens, N.A. Smith, S.J. Curtin, M.B. Wang, P.M. Waterhouse, The *Arabidopsis thaliana* double-stranded RNA binding protein DRB1 directs guide strand selection from microRNA duplexes, RNA 15 (12) (2009) 2219–2235.

[99] S. Bej, J. Basak, MicroRNAs: the potential biomarkers in plant stress response, Am. J. Plant Sci. 5 (2014) 748–759.

[100] A. Valinezhad Orang, R. Safaralizadeh, M. Kazemzadeh-Bavili, Mechanisms of miRNA-mediated gene regulation from common downregulation to mRNA-specific upregulation, Int. J. Genom. 2014 (2014), 970607, https://doi.org/10.1155/2014/970607.

[101] W. Islam, M. Qasim, A. Noman, M. Adnan, M. Tayyab, T.H. Farooq, H. Wei, L. Wang, Plant microRNAs: front line players against invading pathogens, Microb. Pathog. 118 (2018) 9–17.

[102] A. Deleris, J. Gallego-Bartolome, J. Bao, K.D. Kasschau, J.C. Carrington, O. Voinnet, Hierarchical action and inhibition of plant Dicer-like proteins in antiviral defense, Science 313 (5783) (2006) 68–71.

[103] L. Chen, J. Meng, X.L. He, M. Zhang, Y.S. Luan, *Solanum lycopersicum* microRNA1916 targets multiple target genes and negatively regulates the immune response in tomato, Plant Cell Environ. 42 (4) (2019) 1393–1407.

[104] W. Zhang, S. Gao, X. Zhou, P. Chellappan, Z. Chen, X. Zhou, X. Zhang, N. Fromuth, G. Coutino, M. Coffey, H. Jin, Bacteria-responsive microRNAs regulate plant innate immunity by modulating plant hormone networks, Plant Mol. Biol. 75 (2011) 93–105.

[105] G. Sun, MicroRNAs and their diverse functions in plants, Plant Mol. Biol. 80 (2012) 17–36.

[106] C. Kempinski, Z. Jiang, S. Bell, J. Chappell, Metabolic engineering of higher plants and algae for isoprenoid production, in: Biotechnology of Isoprenoids, 2015, pp. 161–199.

[107] E. Sharma, G. Anand, R. Kapoor, Terpenoids in plant and arbuscular mycorrhiza-reinforced defence against herbivorous insects, Ann. Bot. 119 (2017) 791–801.

[108] C.R. Howell, L.E. Hanson, R.D. Stipanovic, L.S. Puckhaber, Induction of terpenoid synthesis in cotton roots and control of *Rhizoctonia solani* by seed treatment with Trichoderma virens, Phytopathology 90 (3) (2000) 248–252.

[109] A. Bhattacharya, P. Sood, V. Citovsky, The roles of plant phenolics in defence and communication during Agrobacterium and Rhizobium infection, Mol. Plant Pathol. 11 (5) (2010) 705–719.

[110] J.A. Del Río, P. Gomez, V. Frías, M.D. Fuster, A. Ortuno, A. Báidez, Phenolic compounds have a role in the defence mechanism protecting grapevine against the fungi involved in petri disease, in: Phenolic Compounds Have a Role in the Defence Mechanism Protecting Grapevine against the Fungi Involved in Petri Disease, 2004, pp. 1000–1008.

[111] R.E. Summons, A.S. Bradley, L.L. Jahnke, J.R. Waldbauer, Steroids, triterpenoids and molecular oxygen, Philos. Trans. R. Soc. B: Biol. Sci. 361 (1470) (2006) 951–968.

[112] M. Wink, Modes of action of herbal medicines and plant secondary metabolites, Medicines 2 (2015) 251–286.

[113] B. Mauch-Mani, F. Mauch, The role of abscisic acid in plant–pathogen interactions, Curr. Opin. Plant Biol. 8 (4) (2005) 409–414.

[114] A. Zehra, M. Meena, M.K. Dubey, M. Aamir, R.S. Upadhyay, Synergistic effects of plant defense elicitors and *Trichoderma harzianum* on enhanced induction of antioxidant defense system in tomato against Fusarium wilt disease, Bot. Stud. 58 (2017) 44, https://doi.org/10.1186/s40529-017-0198-2.

[115] A. Zehra, M. Meena, M.K. Dubey, M. Aamir, R.S. Upadhyay, Activation of defense response in tomato against fusarium wilt disease triggered by *Trichoderma harzianum* supplemented with exogenous chemical inducers (SA and MeJA), Braz. J. Bot. 21 (2017) 1–14, https://doi.org/10.1007/s40415-017-0382-3.

[116] S. Davletova, L. Rizhsky, H. Liang, Z. Shengqiang, D.J. Oliver, J. Coutu, V. Shulaev, K. Schlauch, R. Mittler, Cytosolic ascorbate peroxidase 1 is a central component of the reactive oxygen gene network of Arabidopsis, Plant Cell 17 (1) (2005) 268–281.

[117] F. Eyidogan, M.T. Oz, M. Yucel, H.A. Oktem, Signal transduction of phytohormones under abiotic stresses, in: Phytohormones and Abiotic Stress Tolerance in Plants, Springer, Berlin, Heidelberg, 2012, pp. 1–48.

[118] A.T. Djami-Tchatchou, G.A. Harrison, C.P. Harper, R. Wang, M.J. Prigge, M. Estelle, B.N. Kunkel, Dual role of auxin in regulating plant defense and bacterial virulence gene expression during pseudomonas syringae PtoDC3000 pathogenesis, Mol. Plant-Microbe Int. 33 (8) (2020) 1059–1071.

[119] M. Reusche, J. Klásková, K. Thole, J. Truskina, O. Novák, D. Janz, M. Strnad, L. Spíchal, V. Lipka, T. Teichmann, Stabilization of cytokinin levels enhances Arabidopsis resistance against *Verticillium longisporum*, Mol. Plant-Microbe Int. 26 (8) (2013) 850–860.

[120] D. Wang, K. Pajerowska-Mukhtar, A.H. Culler, X. Dong, Salicylic acid inhibits pathogen growth in plants through repression of the auxin signaling pathway, Curr. Biol. 17 (2007) 1784–1790.

[121] C.M. Pieterse, J. Ton, L.C. Van Loon, Cross-talk between plant signalling pathways: boost or burden, AgBiotechNet 3 (2001) 1–8.

[122] Z. Zhu, F. An, Y. Feng, P. Li, L. Xue, A. Mu, Z. Jiang, J.M. Kim, T.K. To, W. Li, X. Zhang, Derepression of ethylene-stabilized transcription factors (EIN3/EIL1) mediates jasmonate and ethylene signaling synergy in Arabidopsis, Proc. Nat. Acad. Sci. 108 (2011) 12539–12544.

[123] J. Glazebrook, Contrasting mechanisms of defense against biotrophic and necrotrophic pathogens, Annu. Rev. Phytopathol. 43 (2005) 205–227.

[124] J. Luo, W. Xia, P. Cao, Z.A. Xiao, Y. Zhang, M. Liu, C. Zhan, N. Wang, Integrated transcriptome analysis reveals plant hormones jasmonic acid and salicylic acid coordinate growth and defense responses upon fungal infection in poplar, Biomol. Ther. 9 (1) (2019) 12.

[125] M. Meena, P. Swapnil, A. Zehra, M.K. Dubey, R.S. Upadhyay, Antagonistic assessment of *Trichoderma* spp. by producing volatile and non-volatile compounds against different fungal pathogens, Arch. Phytopathol. Plant Protect. 50 (13–14) (2017) 629–648, https://doi.org/10.1080/03235408.2017.1357360.

[126] M. Šimpraga, J. Takabayashi, J.K. Holopainen, Language of plants: where is the word? J. Integr. Plant Biol. 58 (2016) 343–349.

[127] R. Sharifi, S.M. Lee, C.M. Ryu, Microbe-induced plant volatiles, New Phytol. 220 (3) (2018) 684–691.

[128] M. Kottb, T. Gigolashvili, D.K. Großkinsky, B. Piechulla, *Trichoderma* volatiles effecting Arabidopsis: from inhibition to protection against phytopathogenic fungi, Front. Microbiol. 6 (2015) 995.

[129] M.P. López-Gresa, C. Payá, M. Ozáez, I. Rodrigo, V. Conejero, H. Klee, J.M. Bellés, P. Lisón, A new role for green leaf volatile esters in tomato stomatal defence against *Pseudomonas syringe* pv. tomato, Front. Plant Sci. 9 (2018) 1855.

[130] H.S. Yi, M. Heil, R.M. Adame-Alvarez, D.J. Ballhorn, C.M. Ryu, Airborne induction and priming of plant defenses against a bacterial pathogen, Plant Physiol. 151 (2009) 2152–2161.

[131] A. Cellini, V. Giacomuzzi, I. Donati, B. Farneti, M.T. Rodriguez-Estrada, S. Savioli, S. Angeli, F. Spinelli, Pathogen-induced changes in floral scent may increase honeybee-mediated dispersal of *Erwinia amylovora*, ISME J. 13 (2019) 847–859.

[132] M. Ameye, K. Audenaert, N. De Zutter, K. Steppe, L. Van Meulebroek, L. Vanhaecke, D. De Vleesschauwer, G. Haesaert, G. Smagghe, Priming of wheat with the green leaf volatile Z-3-hexenyl acetate enhances defense against *Fusarium graminearum* but boosts deoxynivalenol production, Plant Physiol. 167 (4) (2015) 1671–1684.

[133] D.G. Meldau, S. Meldau, L.H. Hoang, S. Underberg, H. Wünsche, I.T. Baldwin, Dimethyl disulfide produced by the naturally associated bacterium *Bacillus* sp. B55 promotes *Nicotiana attenuata* growth by enhancing sulfur nutrition, Plant Cell 25 (2013) 2731–2747.

[134] C. Ryu, M. Farag, C. Hu, M. Reddy, J. Kloepper, W. Paré, Bacterial volatiles induce systemic resistance in *Arabidopsis*, Plant Physiol. 134 (2004) 1017–1026.

[135] J. Brazelton, E. Pfeufer, T. Sweat, B. Gardener, C. Coenen, 2,4-diacetylphloroglucinol alters plant root development, Mol. Plant-Microbe Interact. 21 (2008) 1349–1358.

[136] G.W. Lee, M.S. Chung, M. Kang, B.Y. Chung, S. Lee, Direct suppression of a rice bacterial blight (*Xanthomonas oryzae* pv. oryzae) by monoterpene (S)-limonene, Protoplasma 253 (3) (2016) 683–690.

[137] Q. Shen, L. Liu, L. Wang, Q. Wang, Indole primes plant defense against necrotrophic fungal pathogen infection, PLoS One 13 (2018), e0207607.

[138] M. Pontin, R. Bottini, J.L. Burba, P. Piccoli, *Allium sativum* produces terpenes with fungistatic properties in response to infection with *Sclerotium cepivorum*, Phytochemistry 115 (2015) 152–160.

[139] H.D. Castelyn, J.J. Appelgryn, M.S. Mafa, Z.A. Pretorius, B. Visser, Volatiles emitted by leaf rust infected wheat induce a defence response in exposed uninfected wheat seedlings, Australas. Plant Pathol. 44 (2) (2015) 245–254.

[140] D. Wu, T. Qi, W.X. Li, H. Tian, H. Gao, J. Wang, J. Ge, R. Yao, C. Ren, X.B. Wang, Y. Liu, Viral effector protein manipulates host hormone signaling to attract insect vectors, Cell Res. 27 (2017) 402–415.

Chapter 11

Understanding plant-plant growth-promoting rhizobacteria (PGPR) interactions for inducing plant defense

Kunal Seth[a], Pallavi Vyas[b], Sandhya Deora[b], Amit Kumar Gupta[b], Mukesh Meena[b], Prashant Swapnil[c], and Harish[b]

[a]*Department of Botany, Govt. Science College, Valsad, Gujarat, India,* [b]*Laboratory of Phytopathology and Microbial Biotechnology, Department of Botany, Mohanlal Sukhadia University, Udaipur, Rajasthan, India,* [c]*School of Basic Sciences, Department of Botany, Central University of Punjab, Bathinda, Punjab, India*

1 Introduction

It has long been recognized that the bacteria are far more abundant within the rhizosphere in comparison to the bulk soil nearby. The rhizosphere is thought to contain 10–1000 folds more bacteria as compared to the bulk soil [1]. The bacterial concentration is higher in rhizosphere as the carbon, which is fixed by the plants in a significant (5%–21%) amount, is excreted in the form of root exudates [2]. The root exudates include nucleotides, amino acids, fatty acids, organic acids, sugars, phenolics, sterols, vitamins, putrescine, and plant growth regulators [3]. Root exudate composition varies based on the plant's physiological state and the microbial compounds present in the rhizosphere. Moreover, the plant not only release root exudate but also takes up the exudate components [4,5]. Several microbes and microbial products in rhizosphere promote plant growth. Microbial community that inhabits rhizosphere normally involves actinobacteria including *Arthrobacter, Actinomyces, Micrococcus,* and *Streptomyces,* proteobacteria namely *Rhizobium, Burkholderia,* and *Pseudomonas,* and firmicutes such as *Bacillus, Neobacillus,* and *Peanibacillus* [6,7]. Rhizosphere microorganisms that serve a diverse role in promoting growth of the plant are referred to as plant growth-promoting rhizobacteria (PGPR). The PGPR can help plants develop in a variety of different ways, both directly and indirectly

Plant-Microbe Interaction—Recent Advances in Molecular and Biochemical Approaches
https://doi.org/10.1016/B978-0-323-91876-3.00010-5

201

[8]. Biofertilization, rhizoremediation, root growth stimulation, and plant stress reduction are certain direct benefits of growth promotion in plants. Plants respond to PGPR by regulating and producing phytohormones, modulating the levels of stress hormones (ethylene levels by ACC deaminase), and growing the availability of key nutrients (iron, nitrogen, phosphorus). Contrarily, reduced disease severity, antibiosis and conflict for niche and resources, and development of systemic resistance within host plant are some of the indirect benefits. To have intended effects, bacteria must be (1) rhizosphere competent, which means they can compete effectively for root exudate nutrients with other microorganisms in rhizosphere and (2) capable of colonizing the root surface proficiently. The most prevalent sites for bacterial colonization are the connections between epidermal cells and the areas where the lateral roots grow. For the biocontrol mechanism such as niche and nutrient competition, and antibiosis, the colonization of root is indeed a basic requirement. The rhizosphere, which is rich in nutrients, is also an ecosystem, where diverse microorganisms and organisms compete for the nutrients that are released. The rhizosphere colonization can be explained by two-step microbiome acquisition model, which proposes that for successful colonization microbial diversity has to pass two filters [9]. The first selection filter is based on cell characteristics of the microbiota and the root exudate composition, which must be utilized by PGPR to colonize within rhizosphere and to compete for spaces. Second selection filter is imposed by the host genetic characteristics, responsible for specific microbial diversity selection. For instance, rhizobia recognize flavonoids produced in the legumes root exudate, which further produces nodulation factors (Nod factors) and in result forms nodule in the host. After nodule formation and colonization, rhizobia help in nitrogen fixation and thus promote plant growth. Barbosa et al. [10] used coinoculation strategy including root exudate of corn plants and *Azospirillum brasilense* strain Ab-V5 with corn seeds. When compared to treatments where just bacteria were inoculated on the seeds, the results showed that the root area of the corns rose by 50% as well as per plant bacterial amount increased by 19%. This suggests that root exudate has a crucial role in the colonization of beneficial bacteria. Motility or chemotaxis is another factor that affects the colonization of rhizospheric bacteria. Chemotaxis is the capability of the cellular components such as pili or flagella to migrate toward a specific chemical stimulus. The hypermobile strains of bacterial mutants are far more competitive phenotypic variants as compared to the reduced motility mutants, which have been found to be poor colonists and have poor attachment to the roots of the host plant [9]. Further plants comprises of innate immune system, that detects the intruding microbes by recognizing microbe-associated molecular patterns (MAMPs) or transmembrane pattern recognition receptors (PRPs), which a wide range of microbes have. The useful microorganism can also be confusingly identified as pathogen, thus prompting an immunological response. However, it has been observed that rhizospheric colonizers have the ability to successfully evade from or suppress plants immune response.

2   Diverse roles of PGPR in promoting plant growth

Plant growth is promoted by certain rhizobacteria in diverse ways (Fig. 1). Bacterial fertilizers provide nutrients to the plant, while also promoting root development to increase water and mineral uptake. For instance, nitrogen-fixing bacteria such as *Rhizobium* and *Bradyrhizobium* form symbiotic association with legumes and convert N_2 into ammonia, which the plant may use as a nitrogen source. *Azospirillum*, on the other hand, is a free-living N_2-fixer that may be used to fertilize wheat, sorghum, and maize. Similarly, low quantities of soluble phosphate can impede plant development or growth. Phosphate-solubilizing bacteria promote plant development by releasing the inorganic and organic phosphates. The organic compounds present in the soil release phosphorus with the help of enzymes including phytases, C-P lyases, phosphatases, and nonspecific phosphatases. Organophosphonate C-P bonds are cleaved by C-P lyases, while the formation of gluconic acid (an organic acid) is responsible for phosphorus removal from the mineral phosphate [11]. Despite the lack of relationship between phosphate solubilization and the capacity of rhizospheric bacteria to colonize, it is noted that phosphate-solubilizing PGPR in the rhizosphere performs better via direct phosphorus intake or by increasing its bioavailability, which encourages better root system growth. Some bacteria produce

FIG. 1   Various plant growth promoting roles of PGPR.

phytostimulators like pyrrolquinoline quinone (PQQ) a cofactor, auxin, and volatiles to boost plant development in pathogen's absence. Under iron limitation, the *Artrobacter agilis* strain UMCV2 (rhizobacterium), that produces dimethyl hexadecyl amine (DMHDA), has been demonstrated to be active for *Medicago truncatula* seedlings growth and development. Many positive effects were induced by DMHDA, including enhanced biomass, ferric reductase activity, and chlorophyll content. Furthermore, DMHDA promoted the absorption of iron under limited conditions by facilitating the extrusion of protons from *M. truncatula* roots, resulting in acidification of the rhizosphere [12]. Later, it was identified that *A. agilis* strain UMCV2 may exist as an endophyte in the plant's internal tissues [13,14]. The PGPR synthesizing an enzyme 1-aminocyclopropane-1-carboxylase (ACC) deaminase operates as stress controllers, reducing the ethylene levels and facilitating the development and growth of plants. These bacteria consume an ethylene precursor, i.e., ACC and transform it to 2-oxobutanoate and ammonia.

3 Role of PGPR in tackling soilborne plant diseases

Every year plant pathogens cause a massive loss of output. To prevent pathogen-caused plant disease, resistant plants and chemicals are frequently utilized. However, not all plants have resistance against disease, and the governments are increasingly discouraging the use of agrochemicals, and fertilizer treated crops are becoming less acceptable among consumers. Moreover, the use of transgenics is still a debatable issue among scientific community and governments. Therefore, an environment friendly method is to use the microbes in form of biocontrol compounds, because it works in vicinity to the plant surface compared to the agrochemicals, majority of which fail to reach the plant. Moreover, unlike agrochemicals, which are meant to resist microbial degradation, biochemicals are degradable in nature. Soils containing the disease-causing pathogen are known as conducive soils and, mixing modest amounts of pathogen-suppressive soils into such conducive soils can make them suppressive. Pathogen-suppressive soils because of their microbiome and activity lead to one of the following conditions: (1) the pathogen fail to establish itself, (2) it establishes itself but unable to cause disease, or (3) it established and have caused the infection initially but later the disease symptoms decline with continued monoculture of the host crop [15]. The *Pseudomonas* spp. is prevalent in strains obtained from naturally disease-suppressive soils [16]. The rhizobacteria's potential to stimulate growth in plants can be because of antibiosis, siderophore-mediated iron competition, or nutrient competition. Nevertheless, root exudates may include compounds that imitate AHL (*N*-acyl homoserine lactones) signals that promote beneficial rhizosphere microflora colonization, while preventing the pathogenic bacteria [17]. The microbial-associated regulation of plant disease on the other hand is a complicated phenomenon involving interactions between pathogens, biocontrol microbe, indigenous microflora, and plant growth substrate (soil or any other substrate). To be effective, the biocontrol microorganisms must be able to function across a wide range of parameters including changes in temperature, pH, and ion concentrations [1]. Although it is challenging to meet these conditions but as our knowledge of biocontrol and selection of suitable strain will increase the efficacy of these biocontrol microbes will definitely improve.

4   Plant defense mechanism by PGPR against biotic stress

Plants adapt to stresses (biotic) by changing the chemistry of their root exudate in order to gather microbiomes, which promote health. The phytopathogens such as protists, insects, nematodes, bacteria, viroids, viruses, and fungi cause biotic stress, which leads to a 15% loss of food globally and 30% decrease in crop output [18,19]. Strategies such as the activation of soil microorganisms that suppress disease and their active selection have been adapted by the diseased and herbivore-affected plants. This is referred as "cry for help" and it explains the soil feedback reaction to plants infection (disease) and also the creation of disease-suppressing soils in a mechanistic way. In addition to protection from pathogenic attack, this adaptive technique benefits the future generations too, which are why these feedback reactions/responses are sometimes referred to as "legacy" or "soil memory" effects [20].

5   Antibiosis

Certain rhizobacteria act via antagonism and produce antifungal and antibacterial metabolites, which kill pathogens. Such bacteria should produce the antimetabolites in appropriate amounts as well as must deliver in right microniche upon surface of roots. To stay competitive, rhizobacteria in rhizosphere use the T6SS secretion system to excrete antibiotic compounds [9]. Some antibiotics produced by *Bacillus* spp. are ribosomal in origin, such as spore-associated antibacterial protein (TasA), sublancin, subtilosin A, subtilin, while polyketide synthases (PKS) and nonribosomal peptide synthases (NRPSs) synthesize rhizoctin, mycobacillin, difficidin, chlorotetain, bacilysin, and bacillaene [21–23]. *Bacillus cereus* can produce zwittermycin A, kanosamine. *Pseudomonas aeruginosa* and *Pseudomonas fluorescens* are able to synthesize antibacterial compounds such as D-gluconic acid, 2-hexyl-5-propyl resorcinol, phenazine-1-carboxylic acid (PCA), 2,4-diacetyl phloroglucinol (DAPG), phenazine-1-carboxamide (PCN), oomycinA, pyrrolnitrin (Prn), pyoluteorin (Plt), viscosinamide, zwittermycin-A, ecomycins, pseudomonic acid, rhamnolipids, cepaciamide A, butyrolactones, kanosamine, aerugine, azomycin, cepafungins, antitumor antibiotic, and karalicins [21,24,25]. Plant defense is supported by volatiles like HCN and 2,3-butanediol, which are produced by *Bacillus* spp. or fungi [26,27]. Moreover, lipopeptides biosurfactants belonging to the iturin, fengycin, and surfactin synthesized by pseudomonads and *Bacillus subtilis* are implicated in biocontrol mechanisms [28,29]. It has been reported that phenazine and rhamno-lipid work coactively to combat *Pythium* spp.-caused soilborne diseases [30]. Many PGPR also synthesize enzymes (protease, β-1,3-glucanase, chitinase, and cellulase), which degrade cell wall of fungus and further restricts fungal growth. For example, β-1,3-glucanase synthesized by *Streptomyces* and *Paeni-bacillus* spp. has been demonstrated to suppress the growth of *Fusarium oxy-sporum*, whereas *Bacillus cepacia* hindered the growth of soilborne fungus *Sclerotium rolfsii* and *Rhizoctonia solani* [31]. Likewise, a rhizospheric bacte-rium, *Bacillus thuringiensis* UM96 synthesize chitinases to combat a gray mold *Botrytis cinerea* [32,33]. Some reports suggest that the antimicrobial substances (iturin A, fengycin, and surfactin) have been found to prevent the fungus *Podo-sphaera fusca* from causing powdery mildew in cucurbits [34]. The release of DIMBOA (2,4-Dihydroxy-7-Methoxy-1,4-Benzoxazine-3-One) in maize root exudate has been found to limit phytopathogen proliferation, while enabling ben-eficial bacteria to thrive [35]. Malic acid in tomato root exudate acts as a chemoat-tractant for *Bacillus subtili*, which then produces bacillomyxin, a phytopathogen inhibitor [36]. The *Pseudomonas* spp. synthesizes 2,4-diacetylphloroglucinol having antifungal activities, in wheat plants infected with *Gaeumannomyces gra-minis* [6,37]. A biocontrol agent for *R. solani* is *Bacillus amyloliquefaciens* SN13 [38]. Some VOCs have antimicrobial properties that antagonize phytopathogens. The dimethyl disulfide (DMDS) is one such chemical generated by *Bacillus* strains that inhibit *B. cinerea* mycelial growth [39].

6   Signal interference

Another biocontrol mechanism involves signal interference, which is based on the breakdown of *N*-acyl homoserine lactones (AHLs). AHLs are quorum-sensing molecule required by bacterial cells to communicate with each other. Bacterial cells can show pathogenicity or express virulence only when there is high cell density. Degradation of AHLs by AHL lactonases or by AHL acy-lases interrupts this process. AHL lactonase break the lactone ring, whereas AHL acylases break the amide links. Degrading AHLs makes biocontrol easier. When sweet basil is infected with *P. aeruginosa* PAO1 and PA14, it produces rosmarinic acid [40]. At higher doses, rosmarinic acid kills bacteria and binds to RhIR (a response regulator), causing premature quorum-sensing signal responses [41].

7   Competition for ferric iron ions

Bacteria have different iron uptake pathways, including iron (II) ferrous-acquisition, iron (III) ferric acquisition and three heme uptake pathways via pyochelin, pyoverdine, siderophores, and different "siderophore privacy" approaches to use ferric ions [42,43]. Crops having rhizosphere with siderophore-producing PGPR can get iron from microbial siderophores [44].

Certain rhizobacteria can inhibit the fungal pathogens by producing high concentration of siderophores under low Fe^{+3} concentrations. Such **PGPR** binds the iron making it less available for pathogen to establish and proliferate. Siderophore-producing bacteria *Burkholderia cenocepacia* strain **XXVI** has shown to exhibit biocontrol properties against *Colletotrichum lindemutianum* ATCC MYA 456, a fungal pathogen [45].

8 Induced systemic resistance

Plants become resistant to harmful bacteria, fungi, and viruses when certain bacteria interact with their roots. Because the rhizobacteria are spatially isolated from the pathogen and stay confined in the rhizosphere, it was inferred that the rhizobacteria decreases disease and/or prevent pathogen development by activating ISR. ISR is a kind of innate immune response that utilizes ethylene and jasmonic acid signaling in the plants. ISR caused by rhizobacteria is analogous to systemic acquired resistance (SAR). Induced resistance makes uninfected plant parts more resistant to a wide range of plant pathogens in both the types. Moreover, the induced resistance (IR) is systemically triggered and spreads to the upper sections of the plants. Interestingly, molecular studies on *Arabidopsis* revealed that both SAR and ISR are interconnected through *NPR1* gene (Non-expressor of Pathogenesis-Related genes 1) (Fig. 2). Activities that contribute to generating plants innate immunity are (i) perceiving stimuli and initiating defensive responses, (ii) defense response via signaling pathways, and (iii) defense priming by rhizobacteria. Following stimulus detection, the defense response begins at the rhizospheric level by eliciting a basal local immunological action, which later progresses to a systemic-defensive response mediated by

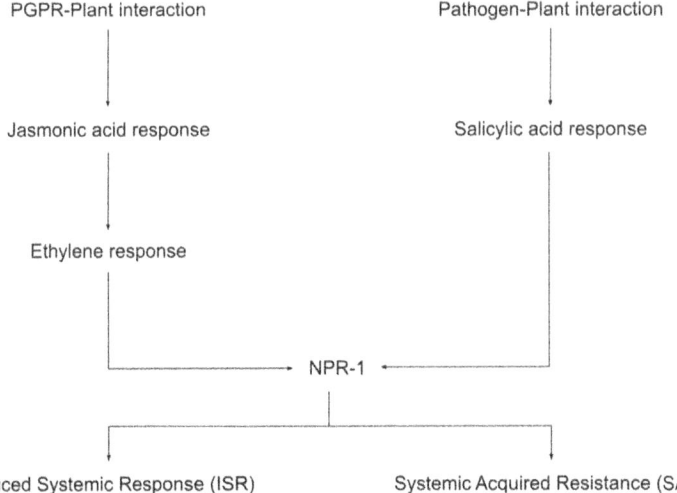

FIG. 2 An overview of plant defense pathway.

hormonal signaling pathways such as JA/ethylene, SA, and other hormone signaling pathways (azelaic acid, methyl jasmonate, and pipecolic acid). However, the phytohormones such as cytokinins, gibberelins, abscisic acid, auxins, and brassinosteroids also play a role in refining and improving the defensive responses through this signaling.

Microbial components that induce ISR include lipopolysaccharides, flagella, salicylic acid, siderophores (e.g., pseudobactins, pyochelin), N-acylhomoserine lactones, cyclic lipopeptides, antifungal factor Phl, certain antifungal metabolites (AFMs), and volatile compounds such as 2,3-butanediol and 3-hydroxy-2-butanone (acetoin) [29,46–49]. ISR mediated by rhizobacteria is a complicated mechanism involving multiple bacterial traits. Purified LPS and WCS417r cell wall preparations containing LPS have shown to cause ISR in radish in the same way as live WCS417r bacteria do [50]. Furthermore, in radish, the ISR was not produced by a mutant of WCS417r (WCS417rOA) O-antigenic side chain of the lipopolysaccharide was absent [50]. In *Arabidopsis*, however, a live bacterial mutant (WCS417rOA) provided normal resistance toward *F. oxysporum* f. sp. *raphani* and *Pst* DC3000 [51]. This suggests that in *Arabidopsis*, in addition to LPS, other bacterial components are also involved in ISR response generation in bacteria. Since the LPS mutant of WCS417r produced ISR under iron-limiting environment, thus the additional factors for ISR were predicted to be iron controlled [52]. The ISR production depends upon plant-rhizobacterium combination, i.e., differential expression of ISR can be seen among different ecotypes as well as among different host-rhizobacterium associations. It has been reported that *P. fluorescens* WCS374r has shown to respond differently in different plant species: radish responds to WCS374r, whereas *Arabidopsis* does not. The *Arabidopsis* on the other hand is WCS358r responsive, but radish and carnation aren't [16,51].

Numerous plant species have been found to have ISR mediated by rhizobacteria, including the model plant *Arabidopsis thaliana* and in others like tomato, cucumber, tobacco, bean, and radish. It has been shown that ISR is induced by *Pseudomonas* sp. strain WCS417R a rhizobacterium in opposition to carnations wilt caused by *Fusarium* [53] and through specific rhizobacteria against cucumbers *Colletotrichum orbiculare* (a fungus) [54]. The increased synthesis of l-malic acid was seen in *A. thaliana* after infected with *Pseudomonas syringae* pv. *tomato* Pst DC3000, which functions as a signal for recruiting the rhizobacterium *B. subtilis* FB17, that further protects the plant via ISR [55]. The bacterium, *P. fluorescens* WCS365 exhibits high chemotaxis toward citric acid, a main component of tomato root exudate, and protects the plant via ISR [56]. The inducing agent which was studied for rhizobacterial-driven ISR in *Arabidopsis* was *P. fluorescens* WCS417r. The colonization of *P. fluorescens* WCS417r bacteria provided resistance against *F. oxysporum* f. sp. *raphani* (fungal pathogen), *Peronospora parasitica* an oomyceteous pathogen and pathogens of bacteria like *Xanthomonas campestris* pv. *Amoracia* and *P. syringae* pv. *tomato* [51,57]. When *Gossypium hirsutum* (cotton) plants were infected with *Bacillus* spp., jasmonic acid and gossypol production increased, resulting in *Spodoptera exigua* larval feeding being decreased. Similarly, *Enterobacter asburiae* BQ9 also increase the production of antioxidant enzymes and genes related to defense, resulting in resistance to yellow leaf curl virus of tomato [58]. Cucumber mosaic virus RNA accumulation was reduced by 91% in the *Nicotiana tabacum* cv leaves (White burley) after PGPR infection with *Peanibacillus lentimorbus* B-30488 [59]. Further, ACC deaminase is synthesized by the bacteria, which protects tomatoes from the fungus *Scelerotium rolfsii*, which causes southern blight disease [60]. Apart from promoting growth, volatile organic compounds also stimulate the ISR. For instance, *B. subtilis* synthesized 2,3-butanediol that increases plants growth and systemic response. After the *Ralstonia solanacearum* (pathogen) infection, when 2,3-butanediol is applied directly to the roots, the PR genes (pathogenesis related) like *CaPAL*, *CaSAR8.2*, *CaPR2* are expressed, and root exudate production increases [61].

9   Deducing the signaling pathway of ISR

In *Arabidopsis*, the pathogen-induced SAR is potent toward many pathogens and also for PR genes activation. The rhizobacteria on the other hand promoted resistance in hosts through pathways regulated by ethylene (ET) and jasmonic acid (JA), and results in systemic response in faraway plant tissues without the participation of PR proteins. The oxidative enzymes (lipoxygenases, polyphenol oxidases, peroxidases), glucanases, antifungal chitinases, low molecular weight phtoalexins, and thaumatins are among the PR proteins [62]. The transgenic *Arabidopsis* carrying transgene NahG fails to accumulate SA resulting in no expression of SAR and PR genes [63], whereas *Arabidopsis* mutants altered in their sensitivity to either ET or JA acquired normal amounts of pathogen-induced SAR [63,64]. This suggests that the pathway for SAR is not reliant on ET or JA for defensive responses. Unlike the SAR induced by pathogen, the WCS417r (rhizobacteria)-driven ISR is neither dependent on the activation of PR genes nor on the salicylic acid buildup. Following WCS417r colonization in the roots, the NahG plants (which doesn't accumulates SA) developed an average level of ISR toward *Pst* DC3000 [51,57]. In another study, the sid1-1 and sid2-1 (SA deficient mutants) exhibited ISR WCS417r-mediated ISR indicating that the ISR mediated by WCS417r is SA independent [65]. Another work on *npr1-1* (SAR regulatory mutant), *etr1-1* (ethylene response mutant), and *jar1-1* (JA response mutant) found that ISR mediated by WCS417r necessitates the responsiveness to ET and JA, but SAR induced by pathogen is NPR1-associated [66]. Similarly, in NahG plants (not accumulating SA), the ACC (1-aminocyclopropane-1-carboxylate) an ethylene precursor and methyl jasmonate (MeJA) were shown to be efficient in developing resistance to *Pst* DC3000. The resistance produced by MeJA was reportedly influenced in *etr1-1*, *jar1-1*, and *npr1-1* plants, while the resistance induced by ACC was reportedly impacted in *npr1-1* and *etr1-1* except *jar1-1* plants [66]. For ISR induced by rhizobacteria, Knoester et al. proposed that at the site of induction an ET-dependent signaling is essential [67]. After treating the roots with WCS417r, *Arabidopsis* mutants deficient in the ethylene-signaling pathway failed to produce ISR against *Pst* DC3000. This clearly suggests that the ISR expression requires a functional ethylene-signaling system. Ethylene-insensitive mutant in roots (*eir1-1*) when infiltrated with WCS417r in leaves was able to mount ISR, however, when roots were administered to bacteria then they were unable to. As a result, it is proposed that ethylene signaling is necessary at the inducer's application site, and that the ethylene-signaling components may be needed for ISR activation in tissues far away to the induction area [68].

These studies evidently state that in order to produce a defensive reaction, such as SAR, the ET and JA reactions are activated, which is NPR1 regulated. NPR1 also influences SAR and the gene expressions associated with it, based on the defensive reaction that is triggered upstream of it. In SAR defense pathway, NPR1 activates *PR* genes downstream of it. The ISR-inducing rhizobacteria strain *Serratia marcescens* was able to protect both NahG (transgenic) and wild-type tobacco plants toward *P. syringae* pv. *Tabaci* and were shown to generate a SA-independent route mediating systemic resistance [69]. However, not every defensive response involving rhizobacteria initiates a SA-independent pathway. By synthesizing SA at the root surface, *P. fluorescens* P3 (a SA overproducing strain) and *P. aeruginosa* 7NSK2 have been found to initiate the SAR pathway dependent on SA [70,71].

The induced disease resistance mechanisms rely heavily on the plant signaling chemicals SA and JA. It has been investigated that how defensive signaling pathways interact. The responses dependent on JA are suppressed by SA, whereas JA and ET can collaborate to induce defensive reactions [72,73]. The probable linkages between the pathways dependent on SA and the ISR (dependent on JA) were investigated by van Wees et al. [74]. When both the pathways were active at the same time, the amount of induced resistance toward *Pst* DC3000 increased. It is noteworthy that the additive effect is owing to the simultaneous stimulation of defensive responses dependent on NPR1, although there is no significant relation between the two. The plants which expressed SAR and ISR did not exhibit greater *PR*-1 and Npr 1 (SAR marker genes) transcript expression [68].

The defense-related genes viz. the genes induced by ET or jasmonic acid (*Pdf1.2, Pall, Hel, Atvsp, Lox1, Lox2*, and *ChiB*) as well as the genes induced by SA (*PR*-1, *PR*-2, and *PR*-5) have been investigated in *Arabidopsis* in search for ISR-related genes. None of the genes listed above were upregulated in ISR-expressing plants [75]. PR-gene expression, on the other hand, accumulates systemically in SAR to levels ranging from 0.3% to 1% of total mRNA and protein concentration [63]. This suggests that in contrary to SAR, ISR does not result in major alterations in expression profile of genes. Furthermore, plants that express ISR are more tolerant to a wide range of pathogens, indicating the possibility of unexplored gene products related to which are responsible for widespread disease resistance.

Increased ethylene and JA synthesis in infected plants is an early symptom of active defense, which coordinate the activation of defense responses. Some defensive genes are activated by ethylene and jasmonic acid in *Arabidopsis* and further makes *Pst* DC3000 resistant [66,75]. However, in ISR mediated by WCS417r in *Arabidopsis* plants, there was no upregulation of genes (such as *Hel*, *Lox1*, *Lox2*, *Pdf1.2*, *Pal1*, *Atvsp*, and *ChiB*) which were responsive to ET or JA [75]. The results show that the ISR mediated by rhizobacteria is dependent on ET and JA due to the increased sensitivity to these hormones rather than increased synthesis of these defense hormones.

10 Adaptive immune response and defense priming

Plants respond to various pathogenic stresses through inducible defense mechanisms. Induced defense involves a wide range of preventive barriers like release of different chemicals, incursions by infectious fungus prevented by cell wall collocations, synthesis of toxic metabolites that assault physiology of pathogens, and formation of preventive physical barriers (Fig. 3). However, the plant's inducible defensive mechanism isn't always enough to keep insects and pathogens at bay. This is solved by the development of more sophisticated defense system to fine tune the inducible immune response in plants, which can be considered a sort of adaptive immunity. The approach that plants use to intensify the immune response upon sensing selected signals in their environment is called priming. Priming provides effective and long-lasting resistance against pathogen and insects. One of the characteristics of immunity is memory formation. Interestingly, the plants immune system recalls pathogen interactions and responds faster and stronger to the future interactions with same pathogen. This ready to fire state of immunity is more specifically referred as defense priming or trained immunity. Noteworthy, the primed state isn't limited to the original pathogen but can protect plant from a wide spectrum of

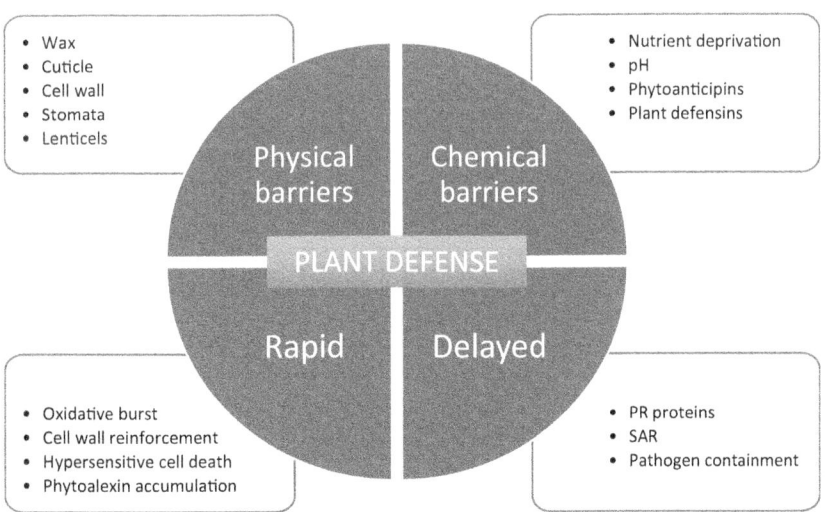

FIG. 3 Different plant defense mechanisms.

pathogens. The *P. syringae* pv. *tomato* inoculation in *Arabidopsis* has shown to pass on to subsequent generation. Furthermore, the offsprings demonstrated enhanced resistance to both, the oomycete *Hyaloperonospore arabidopsis* (obligatory parasite) that causes downy mildew in *Arabidopsis* and to the *P. syringae* pv. tomato [76]. It has been reported that NPR1, MPK3 and MPK6 (MAP kinases), are important for enhanced secondary responses toward pathogens and long-term memory [77–79]. In a study, the *npr1* mutants were unable to acquire transgenerational immunity [76]. Arabidopsis plants generated considerable quantities of MPK3 and MPK6 after being primed with BTH (benzo (1,2,3) thiadiazole-7-carbothionic acid *S*-methyl ester) a synthetic SAR inducer. In BTH primed plants, the higher concentration of above mentioned kinases leads to increased kinase activity and a rapid and more robust activation of the PAL gene following future stress treatment [77]. Recent findings have elucidated the mechanism of defense priming, implying that a primed defensive state may be transmitted epigenetically from plants possessing defense [80]. The *Arabidopsis* plants subjected to localized *P. syringae* infection or priming-inducing treatments with beta-aminobutyric acid (BABA) develop progeny with more resistance to *Pst* DC3000 and *Hyaloperonospora arabidopsidis* [81]. In tomato and *Arabidopsis*, the progeny with high resistance against caterpillar feeding was developed when subjected to insect attack or treated with JA [82]. The explanation for this transgenerational herbivory resistance is linked to a functional COI1 protein involvement and the JA-dependent genes priming [82]. Studies suggest that hypomethylation (a DNA methylation state) is connected to SAR memory transfer through generations. Although chromatin remodeling can have long-term effects on gene transcription, it is a less plausible route for transgenerational immunity as via meiosis the DNA methylation is only reported to be transmitted, making it a more likely approach for epigenetic trait transmission in plants. Reportedly, triple mutant of cytosine methyltransferase 3 (cmt3), domains rearranged methyltransferase 2 (drm2), and domains rearranged methyltransferase 1 (drm1) has decreased DNA methylation, which resembles transgenerational priming of SAR-dependent defense [76]. This evidently suggests that defense priming transmission may be mediated by hypomethylated DNA. The JA-dependent defense production by either JA or fungal infection was found to be linked to the histone H3K36 methylation by SDG8 at promoters of defense genes induced by JA [83]. According to the findings, structural changes allow JA-dependent defense genes to be primed for lengthy periods of time over necrotrophic fungal infections. The priming of SA-dependent defense is coupled with the histone H3 and H4 posttranslational modifications (NPR1-dependent) at the promoters of transcription factor genes which regulates defense [84].

The signals that predict an impeding attack by the pathogens or herbivores trigger defense priming in plants. For instance, localized pathogen attack induces SAR, which results in systemic priming of SA-inducible defense pathways. Likewise, plants infested by herbivores also emit volatile organic compounds (VOCs) and some VOCs activate jasmonic acid-dependent resistance to both systemic plant parts and the adjacent plants also. However, hostile signals are not responsible for all priming reactions. Plant-beneficial rhizobacteria or mycorrhizal fungi can also induce priming, resulting in "induced systemic resistance" effect. The JA-dependent defense priming is associated with priming of ISR. As a result, ISR works effectively against that are resistant to JA-inducible defense. The defense priming, for example, gives fitness advantages in hostile environments, whereas an epigenetic production of defense priming provides the physiological benefits to whole plant populations under long-term biotic assault. From the induced crop plants, choosing the epialleles that induce priming in progenies, such epigenetically regulated priming can be employed for sustainable agriculture [80].

11 Conclusion

The plant-microbiome interactions are ancient and the consequences of prolonged years of coevolution. It is evident that the microbial community has huge potential and is exploited for many beneficial applications to enhance food productivity and sustainable crop management practices. Many unanticipated interactions between plants and microorganisms are also likely to be found in the future. The development of PGPR-based bioformulations can be helpful in promoting plant growth, soil fertility enhancement, and pathogen suppression, in addition to providing green alternatives to traditional agrochemicals. The development of PGPR based bioformulations has been outlined in Fig. 4. By triggering SAR and ISR signals, the PGPR has the ability to produce immunity. Furthermore, many biotechnological discoveries can be used to create versatile synthetic bacterial consortia with enhanced potential and influence on sustainable agriculture and stress management in plants (Tables 1–3).

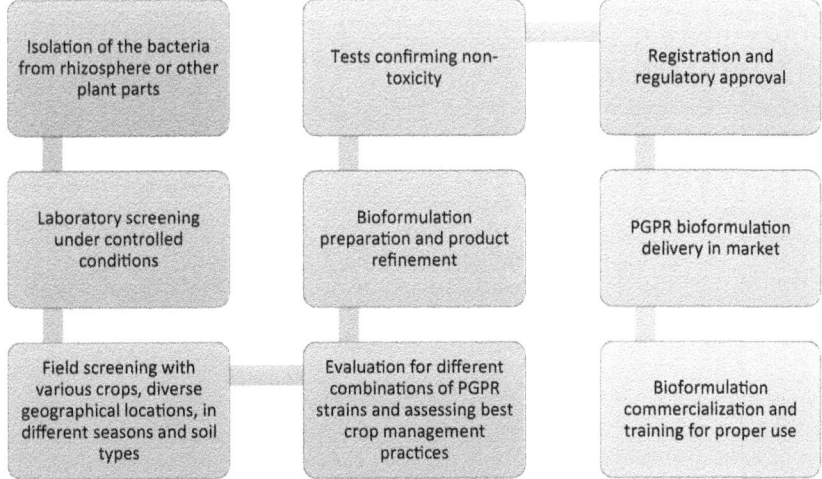

FIG. 4 The development of PGPR-based bioformulation.

TABLE 1 Recent reports on PGPR which induce plant defense system through the production of antibiotics (since last 5 years).

S. no.	Name of host plant	Name of pathogen	Name of PGPR	Compound showing antibiosis	References
1	Tomato	Fusarium oxysporum, Sclerotium rolfsii, Rhizoctonia solani and Pythium ultimum	Delftia lacustris, Bacillus subtilis, and Bacillus cereus	β-1,3-Glucanase	[85]
2	Tomato	Ralstonia solanacearum	Pseudomonas protegens	2,4-Diacetylphloroglucinol (2,4-DAPG)	[86]
3	Watermelon	Meloidogyne incognita and Fusarium oxysporum	Pseudomonas fluorescens	2,4-Diacetylphloroglucinol (2,4-DAPG)	[87]
4	Tomato	Agrobacterium tumefaciens	Bacillus amyloliquefaciens subsp. plantarum	Bacilysin, macrolactin, bacillaene, and difficidin	[88]
5	Sugarcane	Rhizoctonia solani, Fusarium oxysporum, F. moniliforme, Colletotrichum falcatum, Pythium splendens, and Macrophomina phaseolina	Bacillus xiamenensis	Antibiotics and HCN	[89]
6	Rice	Curvularia lunata	Stenotrophomonas malthopilia KJKB5.4, Stenotrophomonas pavanii LMTSA5.4, Alcaligenes faecalis AJ14, and Bacillus cereus AJ34	Bacteriocin	[90]

#	Crop	Pathogens	Biocontrol agent	Compound	Ref.
7	Potato, eggplant, tomato, banana	*Ralstonia solanacearum, Fusarium oxysporum*	*Bacillus velezensis*	Surfactin, iturin, and fengycin	[91]
8	–	*Fusarium oxysporum, Rhizopus microsporus, Alternaria alternata, Penicillium digitatum,* and *Aspergillus niger*	*Pseudomonas aeruginosa*	Pyrrolnitrin	[92]
9	*Pinus* sp.	*Leptographium terebrantis* and *Grosmannia huntii*	*Bacillus velezensis* strains, *Paenibacillus peoriae,* and *B. altitudinis*	Antibiotics	[93]
10	Groundnut	*Aspergillus niger, Aspergillus flavus* and *Fusarium oxysporum*	*Acinetobacter baumannii, Pseudomonas aeruginosa, Bacillus subtilis*	HCN	[94]

TABLE 2 Recent reports on PGPR which induce systemic resistance (ISR) in plants (since last 5 years).

S. no.	Name of host plant	Name of pathogen	Name of PGPR	Compound showing ISR	References
1	Tobacco	Pseudomonas syringae pv. Tabaci, Pectobacterium carotovorum subsp. Carotovorum, Cucumber mosaic virus	Serratia marcescens	Acylhomoserine lactone (AHL)	[26,95]
2	Cabbage	Xanthomonas campestris pv. campestris	Bacillus sp., Fictibacillus solisalsi, Lysinibacillus macrolides	Siderophores	[96]
3	Soybean	Heterodera glycines	Bacillus velezensis, B. mojavensis	Clotjiandin	[97]
4	Tomato	Spodoptera litura	Bacillus endophyticus and Pseudomonas aeruginosa	P-Kaempferol, rutin, indole-3-acetic acid (IAA), salicylic acid (SA), and abscisic acid (ABA)	[98]
5	Tomato	Phytophthora capsici	Pseudomonas fluorescens N04 and Paenibacillus alvei	Phenylpropanoids, benzoic acids, glycoalkaloids, flavonoids,	[99]
6	Tobacco	Potato virus Y	Serratia marcescens	Ubiquitinization of NbHsc70-2 (molecular chaperone proteins)	[100]
7	Arabidopsis	Pseudomonas syringae pv. tomato	Bacillus cereus AR156 ISR	Small RNA (miR472)	[101]

	Plant	Pathogen	Biocontrol agent	Mechanism	Reference
8	Tomato	*Fusarium oxysporum f. sp. lycopersici*	*Streptomyces sp.*, *Pseudomonas fluorescence*	Phenol, Peoxidase, and PAL (phenylalanine-ammonia-lyase)	[102]
9	Tobacco	Cucumber mosaic virus (CMV)	*Paenibacillus lentimorbus B-30488*	Enzymes related to defense such as superoxide dismutase, ascorbate peroxidase, catalase, and guaiacol	[59]
10	Tomato	Tomato spotted wilt virus, PVY (potato virus Y)	*Bacillus amyloliquefaciens* strain MB1600	Salicylic acid dependent	[103]
11	Maize	*Helminthosporium turcicum*	*Bacillus amyloliquefaciens* strain MB1600	Jasmonic acid and Salicylic acid	[104]
12	Rose	*Agrobacterium tumefaciens* C58	*Bacillus velezensis* CLA178	Salicylic acid (SA) or ethylene	[105]
13	Sweet potato	*Fusarium solani* and *Ceratocystis fimbriata*	*Bacillus amyloliquefaciens* YTB1407	Salicylic acid and hydrogen peroxide	[106]
14	Tomato	*Fusarium oxysporum*	*Pseudomonas aeruginosa* PM12	Tyrosine and Eugenol, 3-hydroxy-5-methoxy benzene methanol (HMB)	[107]
15	Rice	*Magnaportha grisea*	*P. pseudoalcaligenes* and *B. pumilus*	Chitin oligomers and lipopolysaccharides	[108]

TABLE 3 Recent reports on PGPR which show systemic acquired resistance (SAR) in plants (since last 5 years).

S. no.	Name of host plant	Name of pathogen	Name of PGPR	Compound showing SAR	References
1	Tobacco	Pectobacterium carotovorum	Serratia rhizosphaerae	Salicylic acid and ethylene	[109]
2	Patchouli (Pogostemon cablin) economically important aromatic plant	Meloidogyne incognita	Pseudomonas putida and Bacillus cereus	Salicylic acid	[110]
3	Tomato	Meloidogyne incognita	Trichoderma spp.	SA (salicylic acid) and JA (jasmonic acid)	[111]
4	Tomato	Cucumber mosaic virus	Bacillus subtills FZB27	Salicylic acid	[112]
5	Tomato	Alternaria solani	Bacillus subtilis	SA (salicylic acid) and jasmonic acid (JA)	[113]
6	Pippermint	Rachiplusia nu	Pseudomonas putida SJ04 and Bacillus amyloliquefaciens GB03	SA (salicylic acid) and jasmonic acid (JA)	[114]

References

[1] B. Lugtenberg, F. Kamilova, Plant-growth-promoting rhizobacteria, Annu. Rev. Microbiol. 63 (2009) 541–556.

[2] H. Marschner, Marschner's Mineral Nutrition of Higher Plants, Academic Press, 2011.

[3] N.C. Uren, Types, amounts, and possible functions of compounds released into the rhizosphere by soil-grown plants, in: The Rhizosphere, CRC Press, 2000, pp. 35–56.

[4] F. Kamilova, L.V. Kravchenko, A.I. Shaposhnikov, N. Makarova, B. Lugtenberg, Effects of the tomato pathogen Fusarium oxysporum f. sp. radicis-lycopersici and of the biocontrol bacterium Pseudomonas fluorescens WCS365 on the composition of organic acids and sugars in tomato root exudate, Mol. Plant-Microbe Interact. 19 (10) (2006) 1121–1126.

[5] B. Lugtenberg, L. van Alphen, Molecular architecture and functioning of the outer membrane of Escherichia coli and other gram-negative bacteria, Biochim. Biophys. Acta 737 (1) (1983) 51–115.

[6] R.L. Berendsen, C.M.J. Pieterse, P.A.H.M. Bakker, The rhizosphere microbiome and plant health, Trends Plant Sci. 17 (8) (2012) 478–486.

[7] C.R. Fitzpatrick, J. Copeland, P.W. Wang, D.S. Guttman, P.M. Kotanen, M.T.J. Johnson, Assembly and ecological function of the root microbiome across angiosperm plant species, Proc. Natl. Acad. Sci. 115 (6) (2018) E1157–E1165.

[8] M. Meena, P. Swapnil, K. Divyanshu, S. Kumar, T. Harish, Y. N., Zehra, A., Marwal, A., Upadhyay, R. S., PGPR-mediated induction of systemic resistance and physiochemical alterations in plants against the pathogens: current perspectives, J. Basic Microbiol. 60 (10) (2020) 828–861.

[9] G. Santoyo, C.A. Urtis-Flores, P.D. Loeza-Lara, M. Orozco-Mosqueda, B.R. Glick, Rhizosphere colonization determinants by plant growth-promoting rhizobacteria (PGPR), Biology 10 (6) (2021) 475.

[10] M.S. Barbosa, E.P. Rodrigues, R. Stolf-Moreira, C.A. Tischer, A.L.M. de Oliveira, Root exudate supplemented inoculant of Azospirillum brasilense Ab-V5 is more effective in enhancing rhizosphere colonization and growth of maize, Environ. Sustain. 3 (2) (2020) 187–197.

[11] H. Rodríguez, R. Fraga, T. Gonzalez, Y. Bashan, Genetics of phosphate solubilization and its potential applications for improving plant growth-promoting bacteria, Plant Soil 287 (1) (2006) 15–21.

[12] M. del Carmen Orozco-Mosqueda, C. Velázquez-Becerra, L.I. Macías-Rodríguez, G. Santoyo, I. Flores-Cortez, R. Alfaro-Cuevas, E. Valencia-Cantero, Arthrobacter agilis UMCV2 induces iron acquisition in Medicago truncatula (strategy I plant) in vitro via dimethylhexadecylamine emission, Plant Soil 362 (1) (2013) 51–66.

[13] M.E. Aviles-Garcia, I. Flores-Cortez, C. Hernández-Soberano, G. Santoyo, E. Valencia-Cantero, La rizobacteria promotora del crecimiento vegetal Arthrobacter agilis UMCV2 coloniza endofíticamente a Medicago truncatula, Rev. Argent. Microbiol. 48 (4) (2016) 342–346.

[14] E. Hernández-Calderón, M.E. Aviles-Garcia, D.Y. Castulo-Rubio, L. Macías-Rodríguez, V. M. Ramírez, G. Santoyo, J. López-Bucio, E. Valencia-Cantero, Volatile compounds from beneficial or pathogenic bacteria differentially regulate root exudation, transcription of iron transporters, and defense signaling pathways in Sorghum bicolor, Plant Mol. Biol. 96 (3) (2018) 291–304.

[15] R.J. Cook, Plant health management: pathogen suppressive soils, in: N.K. Van Alfen (Ed.), Encyclopedia of Agriculture and Food Systems, Elsevier, Amsterdam, Netherlands, 2014, pp. 441–455.

[16] C.M.J. Pieterse, J.A. van Pelt, J. Ton, S. Parchmann, M.J. Mueller, A.J. Buchala, J.-P. Métraux, L.C. van Loon, Rhizobacteria-mediated induced systemic resistance (ISR) in Arabidopsis requires sensitivity to jasmonate and ethylene but is not accompanied by an increase in their production, Physiol. Mol. Plant Pathol. 57 (3) (2000) 123–134.

[17] M. Teplitski, J.B. Robinson, W.D. Bauer, Plants secrete substances that mimic bacterial N-acyl homoserine lactone signal activities and affect population density-dependent behaviors in associated bacteria, Mol. Plant-Microbe Interact. 13 (6) (2000) 637–648.

[18] W.M. Haggag, H.F. Abouziena, F. Abd-El-Kreem, S. el Habbasha, Agriculture biotechnology for management of multiple biotic and abiotic environmental stress in crops, J. Chem. Pharm. Res. 7 (10) (2015) 882–889.

[19] R.N. Strange, P.R. Scott, Plant disease: a threat to global food security, Annu. Rev. Phytopathol. 43 (2005) 83–116.

[20] S.A. Rolfe, J. Griffiths, J. Ton, Crying out for help with root exudates: adaptive mechanisms by which stressed plants assemble health-promoting soil microbiomes, Curr. Opin. Microbiol. 49 (2019) 73–82.

[21] D. Goswami, J.N. Thakker, P.C. Dhandhukia, Portraying mechanics of plant growth promoting rhizobacteria (PGPR): a review, Cogent Food Agric. 2 (1) (2016) 1127500.

[22] V. Leclère, M. Béchet, A. Adam, J.-S. Guez, B. Wathelet, M. Ongena, P. Thonart, F. Gancel, M. Chollet-Imbert, P. Jacques, Mycosubtilin overproduction by Bacillus subtilis BBG100 enhances the organism's antagonistic and biocontrol activities, Appl. Environ. Microbiol. 71 (8) (2005) 4577–4584.

[23] Z. Li, C. Song, Y. Yi, O.P. Kuipers, Characterization of plant growth-promoting rhizobacteria from perennial ryegrass and genome mining of novel antimicrobial gene clusters, BMC Genomics 21 (1) (2020) 1–11.

[24] F.M. Cazorla, S.B. Duckett, E.T. Bergström, S. Noreen, R. Odijk, B.J.J. Lugtenberg, J.E. Thomas-Oates, G.V. Bloemberg, Biocontrol of avocado dematophora root rot by antagonistic Pseudomonas fluorescens PCL1606 correlates with the production of 2-hexyl 5-propyl resorcinol, Mol. Plant-Microbe Interact. 19 (4) (2006) 418–428.

[25] R. Kaur, J. Macleod, W. Foley, M. Nayudu, Gluconic acid: an antifungal agent produced by Pseudomonas species in biological control of take-all, Phytochemistry 67 (6) (2006) 595–604.

[26] C.-M. Ryu, H.K. Choi, C.-H. Lee, J.F. Murphy, J.-K. Lee, J.W. Kloepper, Modulation of quorum sensing in acylhomoserine lactone-producing or-degrading tobacco plants leads to alteration of induced systemic resistance elicited by the rhizobacterium Serratia marcescens 90-166, Plant Pathol. J. 29 (2) (2013) 182.

[27] G. Strobel, Harnessing endophytes for industrial microbiology, Curr. Opin. Microbiol. 9 (3) (2006) 240–244.

[28] I. de Bruijn, M.J.D. de Kock, M. Yang, P. de Waard, T.A. van Beek, J.M. Raaijmakers, Genome-based discovery, structure prediction and functional analysis of cyclic lipopeptide antibiotics in Pseudomonas species, Mol. Microbiol. 63 (2) (2007) 417–428.

[29] M. Ongena, E. Jourdan, A. Adam, M. Paquot, A. Brans, B. Joris, J. Arpigny, P. Thonart, Surfactin and fengycin lipopeptides of Bacillus subtilis as elicitors of induced systemic resistance in plants, Environ. Microbiol. 9 (4) (2007) 1084–1090.

[30] M. Perneel, L. D'hondt, K. de Maeyer, A. Adiobo, K. Rabaey, M. Höfte, Phenazines and biosurfactants interact in the biological control of soil-borne diseases caused by Pythium spp., Environ. Microbiol. 10 (3) (2008) 778–788.

[31] S. Compant, A. Samad, H. Faist, A. Sessitsch, A review on the plant microbiome: ecology, functions, and emerging trends in microbial application, J. Adv. Res. 19 (2019) 29–37.

[32] S.C. Martinez-Absalon, M.D.C. Orozco-Mosqueda, M.M. Martínez-Pacheco, R. Farias-Rodriguez, M. Govindappa, G. Santoyo, Isolation and molecular characterization of a novel strain of Bacillus with antifungal activity from the sorghum rhizosphere, Genet. Mol. Res. 11 (2012) 2665–2673, https://doi.org/10.4238/2012.July.10.15.

[33] S. Martínez-Absalón, D. Rojas-Solís, R. Hernández-León, C. Prieto-Barajas, M.D.C. Orozco-Mosqueda, J.J. Peña-Cabriales, S. Sakuda, E. Valencia-Cantero, G. Santoyo, Potential use and mode of action of the new strain Bacillus thuringiensis UM96 for the biological control of the grey mould phytopathogen Botrytis cinerea, Biocontrol Sci. Tech. 24 (12) (2014) 1349–1362.

[34] D. Romero, A. de Vicente, R.H. Rakotoaly, S.E. Dufour, J.-W. Veening, E. Arrebola, F.M. Cazorla, O.P. Kuipers, M. Paquot, A. Pérez-García, The iturin and fengycin families of lipopeptides are key factors in antagonism of Bacillus subtilis toward Podosphaera fusca, Mol. Plant-Microbe Interact. 20 (4) (2007) 430–440.

[35] B. Guo, Y. Zhang, S. Li, T. Lai, L. Yang, J. Chen, W. Ding, Extract from maize (Zea mays L.): antibacterial activity of DIMBOA and its derivatives against Ralstonia solanacearum, Molecules 21 (10) (2016) 1397.

[36] J. Sasse, E. Martinoia, T. Northen, Feed your friends: do plant exudates shape the root microbiome? Trends Plant Sci. 23 (1) (2018) 25–41.

[37] J.E. Loper, D.Y. Kobayashi, I.T. Paulsen, The genomic sequence of Pseudomonas fluorescens Pf-5: insights into biological control, Phytopathology 97 (2) (2007) 233–238.

[38] S. Srivastava, V. Bist, S. Srivastava, P.C. Singh, P.K. Trivedi, M.H. Asif, P.S. Chauhan, C.S. Nautiyal, Unraveling aspects of Bacillus amyloliquefaciens mediated enhanced production of rice under biotic stress of Rhizoctonia solani, Front. Plant Sci. 7 (2016) 587.

[39] D. Rojas-Solís, E. Zetter-Salmón, M. Contreras-Pérez, M. del Carmen Rocha-Granados, L. Macías-Rodríguez, G. Santoyo, Pseudomonas stutzeri E25 and Stenotrophomonas maltophilia CR71 endophytes produce antifungal volatile organic compounds and exhibit additive plant growth-promoting effects, Biocatal. Agric. Biotechnol. 13 (2018) 46–52.

[40] T.S. Walker, H.P. Bais, E. Déziel, H.P. Schweizer, L.G. Rahme, R. Fall, J.M. Vivanco, Pseudomonas aeruginosa-plant root interactions. Pathogenicity, biofilm formation, and root exudation, Plant Physiol. 134 (1) (2004) 320–331.

[41] A. Corral-Lugo, A. Daddaoua, A. Ortega, M. Espinosa-Urgel, T. Krell, Rosmarinic acid is a homoserine lactone mimic produced by plants that activates a bacterial quorum-sensing regulator, Sci. Signal. 9 (409) (2016) ra1.

[42] P. Cornelis, J. Dingemans, Pseudomonas aeruginosa adapts its iron uptake strategies in function of the type of infections, Front. Cell. Infect. Microbiol. 3 (2013) 75.

[43] Q. Perraud, L. Kuhn, S. Fritsch, G. Graulier, V. Gasser, V. Normant, P. Hammann, I.J. Schalk, Opportunistic use of catecholamine neurotransmitters as siderophores to access iron by Pseudomonas aeruginosa, Environ. Microbiol. 24 (2022) 878–893, https://doi.org/10.1111/1462-2920.15372.

[44] C. Dimkpa, Microbial siderophores: production, detection and application in agriculture and environment, Endocytobiosis Cell Res. 27 (2) (2016) 7–16.

[45] S. de Los Santos-Villalobos, G.C. Barrera-Galicia, M.A. Miranda-Salcedo, J.J. Peña-Cabriales, Burkholderia cepacia XXVI siderophore with biocontrol capacity against Colletotrichum gloeosporioides, World J. Microbiol. Biotechnol. 28 (8) (2012) 2615–2623.

[46] A. Iavicoli, E. Boutet, A. Buchala, J.-P. Métraux, Induced systemic resistance in Arabidopsis thaliana in response to root inoculation with Pseudomonas fluorescens CHA0, Mol. Plant-Microbe Interact. 16 (10) (2003) 851–858.

[47] C.-M. Ryu, M.A. Farag, C.-H. Hu, M.S. Reddy, H.-X. Wei, P.W. Paré, J.W. Kloepper, Bacterial volatiles promote growth in Arabidopsis, Proc. Natl. Acad. Sci. 100 (8) (2003) 4927–4932.

[48] R. Schuhegger, A. Ihring, S. Gantner, G. Bahnweg, C. Knappe, G. Vogg, P. Hutzler, M. Schmid, F. van Breusegem, L.E.O. Eberl, Induction of systemic resistance in tomato by N-acyl-L-homoserine lactone-producing rhizosphere bacteria, Plant Cell Environ. 29 (5) (2006) 909–918.

[49] L.C. van Loon, Plant responses to plant growth-promoting rhizobacteria, in: New Perspectives and Approaches in Plant Growth-Promoting Rhizobacteria Research, Springer, 2007, pp. 243–254.

[50] M. Leeman, J.A. van Pelt, M.J. Hendrickx, R.J. Scheffer, P. Bakker, B. Schippers, Biocontrol of fusarium wilt of radish in commercial greenhouse trials by seed treatment with Pseudomonas fluorescens WCS374, Phytopathology 85 (10) (1995) 1301–1305.

[51] S.C.M. van Wees, C.M.J. Pieterse, A. Trijssenaar, Y.A.M. Van't Westende, F. Hartog, L.C. van Loon, Differential induction of systemic resistance in Arabidopsis by biocontrol bacteria, Mol. Plant-Microbe Interact. 10 (6) (1997) 716–724.

[52] M. Leeman, F.M. den Ouden, J.A. van Pelt, F.P.M. Dirkx, H. Steijl, P. Bakker, B. Schippers, Iron availability affects induction of systemic resistance to Fusarium wilt of radish by Pseudomonas fluorescens, Phytopathology 86 (2) (1996) 149–155.

[53] R. van Peer, G.J. Niemann, B. Schippers, Induced resistance and phytoalexin accumulation in biological control of Fusarium wilt of carnation by Pseudomonas sp. strain WCS 417 r, Phytopathology 81 (7) (1991) 728–734.

[54] G. Wei, J.W. Kloepper, S. Tuzun, Induction of systemic resistance of cucumber to Colletotrichum orbiculare by select strains of plant growth-promoting rhizobacteria, Phytopathology 81 (11) (1991) 1508–1512.

[55] T. Rudrappa, K.J. Czymmek, P.W. Paré, H.P. Bais, Root-secreted malic acid recruits beneficial soil bacteria, Plant Physiol. 148 (3) (2008) 1547–1556.

[56] F. Kamilova, S. Validov, T. Azarova, I. Mulders, B. Lugtenberg, Enrichment for enhanced competitive plant root tip colonizers selects for a new class of biocontrol bacteria, Environ. Microbiol. 7 (11) (2005) 1809–1817.

[57] C.M. Pieterse, S.C. van Wees, E. Hoffland, J.A. van Pelt, L.C. van Loon, Systemic resistance in Arabidopsis induced by biocontrol bacteria is independent of salicylic acid accumulation and pathogenesis-related gene expression, Plant Cell 8 (8) (1996) 1225–1237.

[58] H. Li, X. Ding, C. Wang, H. Ke, Z. Wu, WANG, Y., Liu, H., & Guo, J., Control of tomato yellow leaf curl virus disease by Enterobacter asburiaeBQ9 as a result of priming plant resistance in tomatoes, Turk. J. Biol. 40 (1) (2016) 150–159.

[59] S. Kumar, P.S. Chauhan, L. Agrawal, R. Raj, A. Srivastava, S. Gupta, S.K. Mishra, S. Yadav, P.C. Singh, S.K. Raj, Paenibacillus lentimorbus inoculation enhances tobacco growth and extenuates the virulence of Cucumber mosaic virus, PLoS One 11 (3) (2016) e0149980.

[60] R. Dixit, L. Agrawal, S. Gupta, M. Kumar, S. Yadav, P.S. Chauhan, C.S. Nautiyal, Southern blight disease of tomato control by 1-aminocyclopropane-1-carboxylate (ACC) deaminase producing Paenibacillus lentimorbus B-30488, Plant Signal. Behav. 11 (2) (2016) e1113363.

[61] H.-S. Yi, Y.-R. Ahn, G.C. Song, S.-Y. Ghim, S. Lee, G. Lee, C.-M. Ryu, Impact of a bacterial volatile 2, 3-butanediol on Bacillus subtilis rhizosphere robustness, Front. Microbiol. 7 (2016) 993.

[62] C.M.J. Pieterse, C. Zamioudis, R.L. Berendsen, D.M. Weller, S.C.M. van Wees, P.A.H.M. Bakker, Induced systemic resistance by beneficial microbes, Annu. Rev. Phytopathol. 52 (2014) 347–375.

[63] K. Lawton, K. Weymann, L. Friedrich, B. Vernooij, S. Uknes, J. Ryals, Systemic acquired resistance in Arabidopsis requires salicylic acid but not ethylene, Mol. Plant-Microbe Interact. 8 (6) (1995) 863–870.

[64] K.A. Lawton, S.L. Potter, S. Uknes, J. Ryals, Acquired resistance signal transduction in Arabidopsis is ethylene independent, Plant Cell 6 (5) (1994) 581–588.

[65] C. Nawrath, J.-P. Métraux, Salicylic acid induction–deficient mutants of Arabidopsis express PR-2 and PR-5 and accumulate high levels of camalexin after pathogen inoculation, Plant Cell 11 (8) (1999) 1393–1404.

[66] C.M.J. Pieterse, S.C.M. van Wees, J.A. van Pelt, M. Knoester, R. Laan, H. Gerrits, P.J. Weisbeek, L.C. van Loon, A novel signaling pathway controlling induced systemic resistance in Arabidopsis, Plant Cell 10 (9) (1998) 1571–1580.

[67] M. Knoester, C.M.J. Pieterse, J.F. Bol, L.C. van Loon, Systemic resistance in Arabidopsis induced by rhizobacteria requires ethylene-dependent signaling at the site of application, Mol. Plant-Microbe Interact. 12 (8) (1999) 720–727.

[68] C.M.J. Pieterse, J.A. van Pelt, S. van Wees, J. Ton, K.M. Léon-Kloosterziel, J.J. B. Keurentjes, B.W.M. Verhagen, M. Knoester, I. van der Sluis, P.A.H.M. Bakker, Rhizobacteria-mediated induced systemic resistance: triggering, signalling and expression, Eur. J. Plant Pathol. 107 (1) (2001) 51–61.

[69] C.M. Press, M. Wilson, S. Tuzun, J.W. Kloepper, Salicylic acid produced by Serratia marcescens 90-166 is not the primary determinant of induced systemic resistance in cucumber or tobacco, Mol. Plant-Microbe Interact. 10 (6) (1997) 761–768.

[70] G. de Meyer, K. Capieau, K. Audenaert, A. Buchala, J.-P. Métraux, M. Höfte, Nanogram amounts of salicylic acid produced by the rhizobacterium Pseudomonas aeruginosa 7NSK2 activate the systemic acquired resistance pathway in bean, Mol. Plant-Microbe Interact. 12 (5) (1999) 450–458.

[71] M. Maurhofer, C. Reimmann, P. Schmidli-Sacherer, S. Heeb, D. Haas, G. Défago, Salicylic acid biosynthetic genes expressed in Pseudomonas fluorescens strain P3 improve the induction of systemic resistance in tobacco against tobacco necrosis virus, Phytopathology 88 (7) (1998) 678–684.

[72] C.M.J. Pieterse, L.C. van Loon, Salicylic acid-independent plant defense pathways, Trends Plant Sci. 4 (2) (1999) 52–58.

[73] P. Reymond, E.E. Farmer, Jasmonate and salicylate as global signals for defense gene expression, Curr. Opin. Plant Biol. 1 (5) (1998) 404–411.

[74] S.C.M. van Wees, E.A.M. de Swart, J.A. van Pelt, L.C. van Loon, C.M.J. Pieterse, Enhancement of induced disease resistance by simultaneous activation of salicylate-and jasmonate-dependent defense pathways in Arabidopsis thaliana, Proc. Natl. Acad. Sci. 97 (15) (2000) 8711–8716.

[75] S.C.M. van Wees, M. Luijendijk, I. Smoorenburg, L.C. van Loon, C.M.J. Pieterse, Rhizobacteria-mediated induced systemic resistance (ISR) in Arabidopsis is not associated with a direct effect on expression of known defense-related genes but stimulates the expression of the jasmonate-inducible gene Atvsp upon challenge, Plant Mol. Biol. 41 (4) (1999) 537–549, https://doi.org/10.1023/A:1006319216982.

[76] E. Luna, T.J.A. Bruce, M.R. Roberts, V. Flors, J. Ton, Next-generation systemic acquired resistance, Plant Physiol. 158 (2) (2012) 844–853.

[77] G.J.M. Beckers, M. Jaskiewicz, Y. Liu, W.R. Underwood, S.Y. He, S. Zhang, U. Conrath, Mitogen-activated protein kinases 3 and 6 are required for full priming of stress responses in Arabidopsis thaliana, Plant Cell 21 (3) (2009) 944–953.

[78] H. Cao, S.A. Bowling, A.S. Gordon, X. Dong, Characterization of an Arabidopsis mutant that is nonresponsive to inducers of systemic acquired resistance, Plant Cell 6 (11) (1994) 1583–1592.

[79] X. Dong, NPR1, all things considered, Curr. Opin. Plant Biol. 7 (5) (2004) 547–552.

[80] V. Pastor, E. Luna, B. Mauch-Mani, J. Ton, V. Flors, Primed plants do not forget, Environ. Exp. Bot. 94 (2013) 46–56.

[81] A. Slaughter, X. Daniel, V. Flors, E. Luna, B. Hohn, B. Mauch-Mani, Descendants of primed Arabidopsis plants exhibit resistance to biotic stress, Plant Physiol. 158 (2) (2012) 835–843.

[82] S. Rasmann, M. de Vos, C.L. Casteel, D. Tian, R. Halitschke, J.Y. Sun, A.A. Agrawal, G.W. Felton, G. Jander, Herbivory in the previous generation primes plants for enhanced insect resistance, Plant Physiol. 158 (2) (2012) 854–863.

[83] A. Berr, E.J. McCallum, A. Alioua, D. Heintz, T. Heitz, W.-H. Shen, Arabidopsis histone methyltransferase SET DOMAIN GROUP8 mediates induction of the jasmonate/ethylene pathway genes in plant defense response to necrotrophic fungi, Plant Physiol. 154 (3) (2010) 1403–1414.

[84] M. Jaskiewicz, U. Conrath, C. Peterhänsel, Chromatin modification acts as a memory for systemic acquired resistance in the plant stress response, EMBO Rep. 12 (1) (2011) 50–55.

[85] V. Janahiraman, R. Anandham, S.W. Kwon, S. Sundaram, V. Karthik Pandi, R. Krishnamoorthy, K. Kim, S. Samaddar, T. Sa, Control of wilt and rot pathogens of tomato by antagonistic pink pigmented facultative methylotrophic Delftia lacustris and Bacillus spp, Front. Plant Sci. 7 (2016) 1626, https://doi.org/10.3389/fpls.2016.01626.

[86] R. Rai, R. Srinivasamurthy, P.K. Dash, P. Gupta, Isolation, Characterization and Evaluation of the Biocontrol Potential of Pseudomonas protegens RS-9 Against *Ralstonia solanacearum* in Tomato, 2017.

[87] S.L.F. Meyer, K.L. Everts, B.M. Gardener, E.P. Masler, H.M.E. Abdelnabby, A.M. Skantar, Assessment of DAPG-producing Pseudomonas fluorescens for management of Meloidogyne incognita and Fusarium oxysporum on watermelon, J. Nematol. 48 (1) (2016) 43.

[88] D. Ben Abdallah, O. Frikha-Gargouri, S. Tounsi, Rizhospheric competence, plant growth promotion and biocontrol efficacy of Bacillus amyloliquefaciens subsp. plantarum strain 32a, Biol. Control 124 (2018) 61–67.

[89] Y. Xia, M.A. Farooq, M.T. Javed, M.A. Kamran, T. Mukhtar, J. Ali, T. Tabassum, S. ur Rehman, M.F.H. Munis, T. Sultan, Multi-stress tolerant PGPR Bacillus xiamenensis PM14 activating sugarcane (Saccharum officinarum L.) red rot disease resistance, Plant Physiol. Biochem. 151 (2020) 640–649.

[90] H. Rahma, N. Kristina, Plant growth promoting rhizobacteria (PGPR): as a potential biocontrol for Curvularia lunata Invitro, J. Phys. Conf. Ser. 1940 (1) (2021) 012091.

[91] Y. Cao, Y. Pi, P. Chandrangsu, Y. Li, Y. Wang, H. Zhou, H. Xiong, J.D. Helmann, Y. Cai, Antagonism of two plant-growth promoting Bacillus velezensis isolates against Ralstonia solanacearum and Fusarium oxysporum, Sci. Rep. 8 (1) (2018) 1–14.

[92] B. Uzair, R. Kausar, S.A. Bano, S. Fatima, M. Badshah, U. Habiba, F. Fasim, Isolation and molecular characterization of a model antagonistic Pseudomonas aeruginosa divulging in vitro plant growth promoting characteristics, BioMed Res. Int. 2018 (2018) 6147380, https://doi.org/10.1155/2018/6147380.

[93] P. Devkota, J.W. Kloepper, S.A. Enebak, L.G. Eckhardt, Towards biocontrol of ophiostomatoid fungi by plant growth-promoting rhizobacteria, Biocontrol Sci. Tech. 30 (1) (2020) 19–32.

[94] S. Syed, N.P. Tollamadugu, B. Lian, Aspergillus and Fusarium control in the early stages of Arachis hypogaea (groundnut crop) by plant growth-promoting rhizobacteria (PGPR) consortium, Microbiol. Res. 240 (2020) 126562.

[95] C.R. Silva, R.M. Miller, B.C. Pereira, L. Aveleda, V.A. Marin, Genomic analysis and plant growth-promoting potential of a Serratia marcescens isolated from food, Res. Soc. Dev. 11 (1) (2022). e29611124799.

[96] K. Liu, C. Garrett, H. Fadamiro, J.W. Kloepper, Induction of systemic resistance in Chinese cabbage against black rot by plant growth-promoting rhizobacteria, Biol. Control 99 (2016) 8–13.

[97] N. Xiang, K.S. Lawrence, J.W. Kloepper, P.A. Donald, J.A. McInroy, Biological control of Heterodera glycines by spore-forming plant growth-promoting rhizobacteria (PGPR) on soybean, PLoS One 12 (7) (2017) e0181201.

[98] B. Kousar, A. Bano, N. Khan, PGPR modulation of secondary metabolites in tomato infested with Spodoptera litura, Agronomy 10 (6) (2020) 778.

[99] M.I. Mhlongo, L.A. Piater, P.A. Steenkamp, N. Labuschagne, I.A. Dubery, Metabolomic evaluation of tissue-specific defense responses in tomato plants modulated by PGPR-priming against Phytophthora capsici infection, Plan. Theory 10 (8) (2021) 1530.

[100] M. Ge, M. Gong, Y. Jiao, Y. Li, L. Shen, B. Li, Y. Wang, F. Wang, S. Zhang, J. Yang, *Serratia marcescens*-S3 inhibits Potato virus Y by activating ubiquitination of molecular chaperone proteins NbHsc70–2 in Nicotiana benthamiana, Microb. Biotechnol. 15 (2022) 1178–1188, https://doi.org/10.1111/1751-7915.13964.

[101] C. Jiang, Z. Fan, Z. Li, D. Niu, Y. Li, M. Zheng, Q. Wang, H. Jin, J. Guo, *Bacillus cereus* AR156 triggers induced systemic resistance against *Pseudomonas syringae* pv. tomato DC3000 by suppressing miR472 and activating CNLs-mediated basal immunity in Arabidopsis, Mol. Plant Pathol. 21 (6) (2020) 854–870.

[102] H.Z. Hussein, S.I. Al-Dulaimi, Biological management of fusarium wilt on tomato caused by Fusarium oxysporum f. sp. lycospersici by some plant growth-promoting bacteria, BioRxiv (2020), https://doi.org/10.1101/2020.08.21.262212. Submitted for publication https://www.biorxiv.org/content/biorxiv/early/2020/08/27/2020.08.21.262212.full.pdf.

[103] D. Beris, I. Theologidis, N. Skandalis, N. Vassilakos, Bacillus amyloliquefaciens strain MBI600 induces salicylic acid dependent resistance in tomato plants against tomato spotted wilt virus and potato virus Y, Sci. Rep. 8 (1) (2018) 1–11.

[104] K. Kannan, Induced systemic resistance (ISR) as a mechanism of biocontrol in maize (Zea mays), Crop. Res. 56 (5) (2021) 222–232.

[105] L. Chen, X. Wang, Q. Ma, L. Bian, X. Liu, Y. Xu, H. Zhang, J. Shao, Y. Liu, Bacillus velezensis CLA178-induced systemic resistance of Rosa multiflora against crown gall disease, Front. Microbiol. 11 (2020) 2607.

[106] C.-J. Wang, Y.-Z. Wang, Z.-H. Chu, P.-S. Wang, B.-Y. Liu, B.-Y. Li, X.-L. Yu, B.-H. Luan, Endophytic Bacillus amyloliquefaciens YTB1407 elicits resistance against two fungal pathogens in sweet potato (Ipomoea batatas (L.) Lam.), J. Plant Physiol. 253 (2020) 153260.

[107] S. Fatima, T. Anjum, Identification of a potential ISR determinant from Pseudomonas aeruginosa PM12 against Fusarium wilt in tomato, Front. Plant Sci. 8 (2017) 848.

[108] Y. Jha, B. Dehury, S.P. Kumar, A. Chaurasia, U.B. Singh, M.K. Yadav, U.B. Angadi, R. Ranjan, M. Tripathy, R.B. Subramanian, Delineation of molecular interactions of plant growth promoting bacteria induced β-1,3-glucanases and guanosine triphosphate ligand for antifungal response in rice: a molecular dynamics approach, Mol. Biol. Rep. 49 (2021) 2579–2589, https://doi.org/10.1007/s11033-021-07059-5.

[109] J.-S. Son, Y.-J. Hwang, S.-Y. Lee, S.-Y. Ghim, Serratia rhizosphaerae sp. nov., a novel plant resistance inducer against soft rot disease in tobacco, Int. J. Syst. Evol. Microbiol. 71 (4) (2021) 004788.

[110] B. Borah, M. Hussain, S.B. Wann, B.S. Bhau, Selection and validation of suitable reference genes for quantitative real time PCR analysis of gene expression studies in patchouli under Meloidogyne incognita attack and PGPR treatment, Gene Rep. 19 (2020) 100625.

[111] A. Martínez-Medina, I. Fernandez, G.B. Lok, M.J. Pozo, C.M.J. Pieterse, S.C.M. van Wees, Shifting from priming of salicylic acid-to jasmonic acid-regulated Defenses by Trichoderma protects tomato against the root knot nematode Meloidogyne incognita, New Phytol. 213 (3) (2017) 1363–1377.

[112] A.S. Zene, T. Yongqiang, Z. Jianqiang, The effect of Bacillus bacteria on tomato plants and fruits, Int. J. Agric. Innov. Res. 9 (4) (2021) 271–280.

[113] M. Rasool, A. Akhter, M.S. Haider, Molecular and biochemical insight into biochar and Bacillus subtilis induced defense in tomatoes against Alternaria solani, Sci. Hortic. 285 (2021) 110203.

[114] L. del Rosario Cappellari, J. Chiappero, T.B. Palermo, W. Giordano, E. Banchio, Impact of soil rhizobacteria inoculation and leaf-chewing insect herbivory on Mentha piperita leaf secondary metabolites, J. Chem. Ecol. 46 (7) (2020) 619–630.

Chapter 12

Revitalization of PGPR through integrating nanotechnology for sustainable development in agriculture

Gunja Vasant, Shweta Bhatt, Ragini Raghav, and Preetam Joshi
Department of Biotechnology, Atmiya University, Rajkot, India

1 Introduction

The world population will be approximately 9.6 billion by 2050 as stated by UN report 2013 [1]. The need for enhanced yield of food production with reduced harmful aftereffects on the soil is a challenge for sustainable agriculture. Furthermore, the crops need to be made tolerant to abiotic and biotic factors including salt, drought, disease-causing organisms, and heavy metals. The mentioned desirable properties can be made possible by the use of rhizospheric organisms in soil. The potent organisms present in the soil, which augment plant growth rate without contaminating the environment, are called plant growth-promoting rhizobacteria (PGPR) [2].

The favorable microorganisms colonizing around the rhizoplane, microhabitats, and root endosphere provide plant growth-promoting activities [3,4]. The carbon compounds secreted by the plants into the soil lead to high microbial populations, i.e., approximately a thousand times higher in the rhizospheric soil relative to the bulk soil [4–6]. The plant secretes numerous signal compounds, which attract specific species and regulate their biochemical and genetic activities [7–9]. Thus, the microbial community present in the rhizosphere varies from the bulk soil on account of different root exudates [10]. The PGPR are approximately up to 5% of total rhizospheric bacteria [11,12]. They affect plant growth by direct and indirect mechanisms (Fig. 1). The direct mechanisms include increasing the quantity and absorption of nutrients present in the soil to plants through providing phytohormones (cytokinin, abscisic acid, gibberellins, auxins, and ethylene) [13,14], biological nitrogen fixation, solubilizing nutrients (K, P, Zn) to plant available form, siderophore production [5,15,16].

Plant-Microbe Interaction—Recent Advances in Molecular and Biochemical Approaches
https://doi.org/10.1016/B978-0-323-91876-3.00009-9

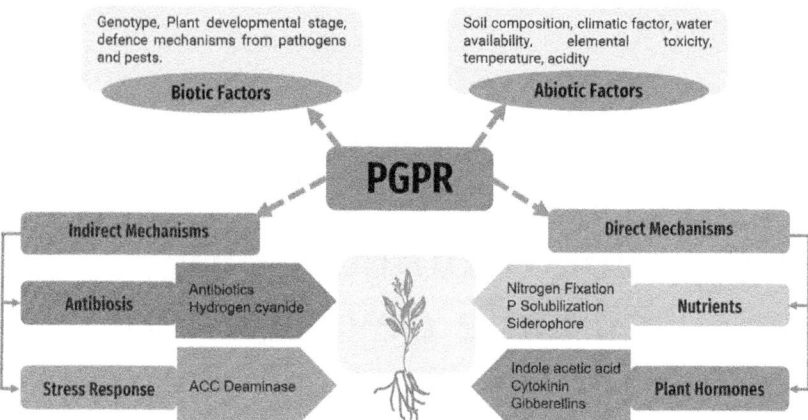

FIG. 1 Schematic representation of plant growth by PGPR.

Further, PGPR exhibit indirect mechanisms including abiotic and biotic stress tolerance [17,18], suppression of plant pathogens [7,16,19], and secretion of various biocontrol specialists such as Volatile Organic Compounds (VOCs). The proclaimed group of PGPR includes bacteria belonging to genera *Acinetobacter, Agrobacterium, Arthrobacter, Azoarcus, Azospirillum, Azotobacter, Bacillus, Bradyrhizobium, Burkholderia, Caulobacter, Chromobacterium, Delftia, Enterobacter, Flavobacterium, Gluconacetobacter, Klebsiella, Mesorhizobium, Micrococcus, Pseudomonas, Rhizobium, Serratia, Streptomyces,* and *Thiobacillus* [5,17,20].

2 Optimal PGPR

A rhizobacterial strain is viewed as an evident PGPR when it exhibits plant development advancing qualities and can upgrade plant development on inoculation. An optimal PGPR follows the indispensable criteria:

(1) It needs to be profoundly rhizosphere-capable and eco-accommodating.
(2) On inoculation, it should colonize the plant in critical number.
(3) It needs to have the option to advance plant development.
(4) It should display a wide range of activity.
(5) It should be viable with different microscopic organisms in the rhizosphere.
(6) It ought to be tolerable toward physicochemical variables like oxidants, temperature, parching, and radiation.

3 Role of PGPR in enhancement of plant growth

The plant growth is enhanced by direct and indirect mechanisms exhibited by PGPR. Plant development is highly affected by an assortment of stresses which can be grouped/categorized into two types—biotic and abiotic. Biotic stress alludes to the plant pathogens and pests, for example, such as fungi, viruses, bacteria, nematodes, insects, while abiotic stress focuses on drought, salinity, concentration of various heavy metals in soils, nutrient deficiency, temperature, and so on [2,16,21,22]. PGPR colonization profoundly improves the stress tolerance in plants and enables enhancement of its growth.

4   PGPR and plant hormones

Phytohormones play an important role in plant growth regulation. They function as molecular signals in response to environmental factors, which may otherwise restrict plant growth or become fatal if uncontrolled [23]. Numerous rhizospheric bacteria are known to secrete hormones and boost the growth of plants, stimulate agricultural production, and alter the stress response. Numerous microorganisms have the competence to produce growth regulators such as indoleacetic acid (IAA), gibberellic acid (GA), cytokinin, and ethylene.

According to Spaepen and Vanderleyden, IAA plays a crucial role in plant growth and its development including primary root elongation, enhancement of root surface area and length [24]. Auxin plays an important role in the beneficial plant-PGPR interaction. PGPR strains producing IAA such as *Azospirillum brasilense* Sp245, *Aeromonas punctata* PNS-1, and *Serratia marcescens* 90–166 stimulate growth and activate morphological changes in *Arabidopsis thaliana* [24].

The process of seed germination, flowering, fruit development, leaf and stem growth involves the hormone gibberellin (GA), a type of phytohormones, which also plays a pivotal role in shoot elongation. Gibberellin-producing PGPR *Enterococcus faecium* LKE12 and *Leifsonia soli* SE134 trigger shoot growth in mutated rice plants deficient in gibberellin synthesis [25]. The gibberellin-producing PGPR strains of *Promicromonospora* sp. SE188 and *Bacillus amyloliquefaciens* RWL-1 result in an augmented amount of gibberellins in the plant.

Cytokinin enhances plant vascular differentiation, cell division, vascular cambium sensitivity and increases root hairs proliferation, but inhibits primary root elongation [26]. Various PGPR strains are synthesizing cytokinin which enhances shoot growth and fruit formation of plants [27,28]. *Bacillus megaterium* UMCV1 was reported to stimulate the growth of lateral roots in *Arabidopsis thaliana*, and the cytokinin receptor genes AHK2 and RPN12 are involved in the mechanism of this stimulation. Cytokinin-producing PGPR strain *Pseudomonas fluorescens* stimulated main roots growth and repressed lateral roots formation in *Brassica napus* [29]. Bacterial cytokinins also have the feature to exhibit plant resistance to biotic and abiotic stresses. For instance, PGPR *Pseudomonas fluorescens* G20-18 synthesizes cytokinin, which improves the resistance of *Arabidopsis thaliana* plants to infection with *Pseudomonas syringae*.

Another hormone is ethylene, which controls many processes including the germination of seed, shoot and root growth, abscission of leaves and fruit ripening. Furthermore, excessive amounts of ethylene result in defoliation, premature senescence, and root and stem growth retardation. This eventually leads to restricted plant growth and development. Several abiotic and biotic stresses such as flood, heavy metals, pathogens lead to synthesis of 1-aminocyclopropane-1-carboxylate (ACC), a precursor of ethylene. The ethylene then causes reduction in root elongation and nitrogen fixation causing premature senescence.

PGPR degrade ACC and assist the growth of the root system. Glick has explained that PGPR producing ACC deaminase and IAA facilitate the growth of plants to a greater extent. Ahmad evidenced that *Pseudomonas* and *Rhizobium* ACC-deaminase-producing strains are able to augment the quality, growth, physiology of mung beans under saline environments.

5 Nutrient availability for plant growth

Various PGPR assist in fixing nitrogen into organic form that can be utilized by the plants. Several collections of soil and root-associated nitrogen-fixing microorganisms have been reported in the literature such as *Azotobacter vinelandii*, *Azospirillum brasilense*, *Acetobacter diazotrophicus*, *Achromobacter insolitus*, *Burkholderia tropica*, *Burkholderia xenovorans*, *Burkholderia silvatlantica*, *Burkholderia caballeronis*, *Bradyrhizobium japonicum*, *Delftia tsuruhatensis*, *Enterobacter sacchari*, *Bacillus megaterium*, *Gluconacetobacter diazotrophicus*, *Stenotrophomonas maltophilia*, *Pseudomonas stutzeri*, *Pseudomonas koreensis*, and *Pseudomonas entomophila*, which colonize different crops and enhances plant growth directly or indirectly. Their activity, however, is influenced by soil type, soil condition, and crop species [2,10,11].

Numerous PGPR are also reported to have the ability to solubilize phosphate and increase the phosphate ions availability and accessibility to the plants. *Kocuria turfanensis* strain 2 M4 PGPR is a phosphate solubilizer, a siderophore producer, and an IAA producer. Kumar et al. [30] have reported that the employment of *Bacillus megaterium*, *Arthrobacter chlorophenolicus*, and *Enterobacter* resulted in a twofold increase in wheat grain yield in greenhouse experiments [30]. PGPR with a phosphate solubilizing capacity including *Bacillus megaterium* [31], *Pseudomonas*, *Delftia* sp., *Azotobacter*, *Xanthomonas* and *Rhodococcus*, *Arthrobacter*, *Serratia*, *Phyllobacterium*, *Chryseobacterium*, and *Gordonia* increased phosphate availability in soil by approximately 30% [32,33]. Furthermore, phosphate deficiency was reported to reduced crop yield by 5%–15% [34]. Phosphate-deficient plants show symptoms such as dark, dull, and reddish colored leaves, necrosis in old leaf tips, and a smaller size of new leaves [35,36]. Employment of phosphate solubilizing bacteria can prove to be highly cost-effective and lead to enhancement of plant growth and development.

Another macronutrient in plant growth is potassium. Inoculation of seeds and seedlings of different plants with potassium solubilizing bacteria (KSB) displays significant enhancement in germination percentage, seedling vigor, plant growth, yield, and K uptake by plants under greenhouse and field conditions.

6   Enzymes by PGPR

The two main hydrolytic enzymes produced by PGPR are chitinase and glucanase. The major components of the fungal cell wall are chitin and beta-glucan; hence, PGPR producing chitinases and beta-glucanases would inhibit fungal growth. *Pseudomonas fluorescens* LPK2 and *Sinorhizobium fredii* KCC5 produce chitinase and beta-glucanases and dictate the fusarium wilt produced by *Fusariumudum*. *Pseudomonas* spp. a PGPR that inhibits *Phytophthora capsici* and *Rhizoctonia solani*, two of the most destructive crop pathogens in the world.

7   Abiotic stress tolerance in plants

Abiotic stress plays a major role in reducing agricultural production. The strength of abiotic stresses changes on the basis of the type of plant factors and type of soils [37]. Sarma and Saikia reported that the *Pseudomonas aeruginosa* strain enhanced the growth of *Vigna radiata* (mung beans) during drought conditions [38]. The stomata of the leaf balance the water content in the leaves and also water uptake by the roots. Ahmad et al. and Naveed et al. reported that the stomatal conductance of leaves in plants inoculated with PGPR was higher than that in plants without PGPR under drought conditions. PGPR increase water use efficiency of plants. Marulanda et al. reported that the *Bacillus megatertum* strain augments the absorption of water by roots under saline conditions. A similar behavior was exhibited by *Pantoea agglomerans* when observed with maize roots. Gabriela et al. used *Azospirillum* for lettuce growth under salinity stress [39]. The results showed that inoculation with *Azospirillum* sp. augments the quality of lettuce and the storage life of lettuce under salt stress, which further increases the yield.

8   Macronutrients and micronutrients

Plants require various minerals throughout their life cycle. Carbon, hydrogen, and oxygen are derived by plants from air; however, thirteen elements are made available to plants from soil. Based on their requirement by plants, they are classified into micro- and macronutrients. Microorganisms play a significant role in enhancing nutrient availability to plant roots by solubilizing minerals. This section enlists various essential macro- and micronutrients, their role and responsibility in plant growth and development [40].

8.1 Potassium

Potassium is involved in numerous biochemical and physiological systems of plants. Potassium is not included in any chemical structure of plants; however, its role in plant development has been widely studied and reported in the literature [41–44]. Potassium is essential for activation of several enzymes (~60 enzymes) involved in the growth and development of plants. Potassium neutralizes various ions in the plant system and hence assists in the maintenance of the pH (7–8), which is crucial for the enzymatic reactions. Potassium is vital for the opening and closing of stomata, which regulate the nutrient transport, photosynthesis, and cooling of plants. Furthermore, potassium aids in the uptake of water by the plant roots by developing a gradient of osmotic pressure with its accumulation. Potassium is also reported to be responsible for the transport of sugars, synthesis of starch and proteins, transportation of water and nutrients in the plant system. It also helps in the enhancement of crop quality and extends the shelf life of fruits and vegetables [44–47].

8.2 Phosphorus

The abundance of phosphorus is essential for plants as it is a key component in several cellular processes such as synthesis of biomolecules (nucleic acids-DNA, RNA), sugar phosphates (intermediates of various metabolic pathways), and energy-rich compounds (adenosine/cytidine/guanosine/uridine-triphosphate and other phosphorylated compounds). Furthermore, phosphorus energizes photosynthesis and respiration making it indispensable for plant survival. Phosphorus is accountable for the maintenance of cell membrane (phospholipids), germination of seeds, formation of roots (morphology, clusters, and architecture), increment in shoot and root length, flowering, and seed formation in plants [35,36,48] (Fig. 2).

8.3 Calcium

Calcium is an important element that regulates growth and development in plants [49]. It has been vividly reported as the second messenger in animal cells; however, its role has been determined to be essential and indispensable in plant cells. It is a crucial component in determining the structural rigidity of the cell wall and maintains selective permeability of the membrane. It has also been reported to promote root hair growth in various plants. The calcium uptake by plants has been reported to protect them against heavy metal toxicity and several pathogenic microorganisms (yeast, bacteria, etc.). Moreover, the role of calcium has been extended to several developmental processes such as pollen tube elongation, cell division, seed germination, apoptosis, stomatal closure, and auxin responses [48–51].

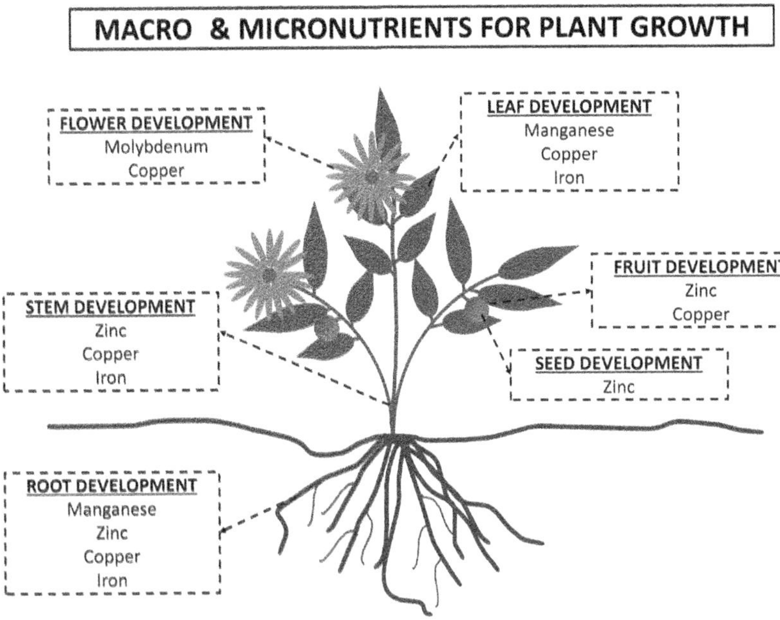

MACRO & MICRONUTRIENTS FOR PLANT GROWTH

FLOWER DEVELOPMENT
Molybdenum
Copper

LEAF DEVELOPMENT
Manganese
Copper
Iron

STEM DEVELOPMENT
Zinc
Copper
Iron

FRUIT DEVELOPMENT
Zinc
Copper

SEED DEVELOPMENT
Zinc

ROOT DEVELOPMENT
Manganese
Zinc
Copper
Iron

FIG. 2 The role and function of various macronutrients and micronutrients in plant growth and development.

8.4 Magnesium

Magnesium performs various biological functions in plant and animal systems by being a dissociable cofactor in enzymes that activate the phosphorylation process. In plant systems, magnesium is the central atom in the tetrapyrrole ring of chlorophyll *a* and *b* present in the leaf chloroplast. Hence, its concentration affects photophosphorylation and the phosphorylation reactions in chloroplast. The magnesium in plant leaves has been directly and indirectly associated with protein synthesis. Furthermore, it has been reported to be critical for maintaining the stability of ribosomal subunits in the plant cells. Magnesium is also required for activation of several metabolic pathways such as lipid metabolism and carbohydrate metabolism. It has been reported that magnesium ions improve the produce and quality of crops.

8.5 Iron

Iron is involved in the synthesis of chlorophyll and maintenance of chloroplast. The concentration of iron and chlorophyll has been reported to be interrelated in green plants. Furthermore, in plant systems, it plays a vital role in several biological processes such as photosynthesis and respiration (energy yielding electron transfer reactions), nitrogen fixation, hormone production, and nutrient uptake mechanisms. Moreover, iron plays a major role in major metabolic processes as it is a constituent of several electron carriers and enzymes. Sufficient amounts of iron result in improved nutritional quality and better yield.

8.6 Zinc

Zinc is a constituent of several enzymes and is required as a cofactor for enzymes such as peroxidases, oxidases, etc. Zinc has also been associated with the regulation of the nitrogen metabolism (utilization of nitrogen in seed formation), multiplication of cells, and photosynthesis in plants [52,53]. In various metabolic pathways, such as starch, carbohydrates, hormones (indoleacetic acid and auxin) and proteins, zinc plays a significant role by aiding the activity of the necessary enzymes. Zinc has also been linked to the maintenance of membrane integrity, formation and turgidity in the leaves in most plants. Furthermore, zinc has also been potent in reducing heavy metal accumulation in plants [54,55].

8.7 Manganese

The manganese in plant cells acts as a cofactor and is beneficial in controlling the conformation of various metalloproteins such as superoxide dismutase, oxalate oxidase, etc. It activates several enzymes, such as phosphokinase and phosphotransferase, by bridging adenosine triphosphate (ATP) with the enzyme complex. There are various metabolic processes which are dependent on divalent manganese such as glycosylation and ROS scavenging. Furthermore, divalent manganese ion itself acts as an antioxidant and supports in the reduction of oxidative damage in plants. Manganese also plays a crucial role in water splitting, chlorophyll production, lignin biosynthesis, and photosynthesis.

8.8 Copper

Copper has been extensively studied for its role in several physiological processes in plants including photosynthesis, electron transport, respiration, metabolism of cell wall, hormones, carbohydrates and nitrogen, and oxidative stress response. At cellular levels, it has been identified to be essential for transcription and protein trafficking, phosphorylation and iron mobilization in plant system. It plays an important role in activation of enzymes such as superoxide dismutase, several oxidases (amino, ascorbate, polyphenol, and mitochondrial cytochrome c oxidase), and laccase. It has been reported to impart disease resistance to several plants, improve the fertility of flowering plants and improve fruit formation.

9   Nanotechnology and PGPR

Nanotechnology is the study and design of materials (with at least one dimension between 1 and 100 nm) and their exploitation in various applications across the environment, agriculture, biomedical, textile, medicine, engineering, etc. [56,57]. The advent of nanotechnology has promised to improve the agricultural sector and has gained immense popularity in the past few decades. Metal and its oxide nanoparticles have gained considerable consideration by researchers due to the high surface to volume ratio and hence enhanced reactivity. Furthermore, nanomaterials can improve the nutrient uptake and utilization by plants over other conventional methods. Moreover, several nanoparticles have been reported in the literature, which can extensively aid plants in their metabolism and improve physicochemical parameters such as root, shoot, dry weight, wet weight, leaf area, etc. Nanoparticles (NPs) augment plant metabolism through their physicochemical properties and hence enhance crop yield and supply nutrients to the soil [58]. Several research groups are exploring the cumulative effect of various nanomaterials with PGPR for crop improvement and higher yield (Table 1).

Nanomaterials are of various types including metal nanoparticles, organic, carbon nanoparticles, and semiconductor nanoparticles [58,89]. The silver [90], titanium, zinc oxide [73], silica [83], calcium, boron [91], gold [67], and zeolite [75] nanoparticles are reported to exhibit plant growth-promoting effects. The plant growth-promoting rhizobacteria (Bacillus sp.) and silver nanoparticles are utilized on *Zea mays* and were reported to show increase in root, shoot growth and inhibit fungal infections too [92]. Timmusk et al. [93] reported that the utilization of Nanotitania (TNs) provides an effectual method for PGPR to stably attach with plant roots and facilitates PGPR for reproducible field applications.

9.1   Silver nanoparticles

Silver nanoparticles with PGPR have been elaborately studied with various plant systems and are being accepted in the agricultural sector. In addition to being highly reactive, these nanoparticles possess antimicrobial and antipest activities. The silver nanoparticles (Ag NPs) in combination with PGPR have been reported to be more effective to increase plant growth; however, their toxicity and underlying risks are still under consideration. Siddiqi and Husen [90] reported the significant impact of silver nanoparticles on fenugreek seedlings. The plant displayed improved physicochemical parameters such as increase in shoot and root length, leaf number, phytochemicals, and diosgenin [90]. Khan and Bano [94] employed three PGPR strains (*Pseudomonas* sp., *Pseudomonas fluorescence*, and *Bacillus cereus*) with silver nanoparticles and evaluated their cumulative effect on maize seeds [94]. The treated plants had enhanced root area and length and growth hormones, such as ABA, IAA, GA, and proline production [94]. Furthermore, Vishwakarma et al. [95] reported that the treatments

TABLE 1 Effect of various nanoparticles on plant with PGPR.

S. no.	Nanoparticles/ nanomaterials	Plant (common and scientific name)	PGPR	Results	References
1.	Molybdenum NPs	Wheat (*Triticum*)	*Bacillus* sp. strain ZH16	Increase in morphological characteristics, nutrients availability and balance of ions in the plants	[59]
2.	Silicon dioxide NPs	Wheat (*Triticum*)	*Azospirillumlipoferum* and *Azospirillum brasilense*, *Bacillus* sp.	Improvement in physicochemical parameters, and yield; improved relative water content, nutrients uptake, antioxidant enzymes—such as catalase, superoxide dismutase and peroxidase increased their upregulation	[60]
3.	Silicon nanoparticles NPs	Lemon balm (*Melissa officinalis* L.)	(*Pseudomonas fluorescens* and *Pseudomonas putida*)	Increment in free radical scavenging activities of plant extracts	[61]
4.	Magnesium oxide NPs	Radish (*Raphanus sativus* L.)	–	Increment secondary metabolite production, total phenolic and dry biomass	[62]
5.	Silver NPs	–	*Azotobacter vinelandii*	Silver nanoparticles display size dependent (10 and 50nm) effect on plant; inhibited the growth of bacteria and induced cell apoptosis, effective against nitrogenase activity and ROS detection	[63]
6.	Silver NPs	–	*Nitrosomonas europaea* ATCC19718	Restricts the biosynthesis of protein, gene expression, and production of energy	[64]
7	Iron NPs	–	*Paracoccus* sp.	Excess amount of iron leads to oxidative damage to cells; iron (II) adhered to cell membranes and changed bionitrification of the microorganism	[65]

8.	Silver nanoparticles and iron oxide nanoparticles	–	Soil microbial activity	Silver NPs reduced soil microbial metabolic activity, nitrification ability and count of the microorganism	[66]
				Iron oxide nanoparticles promotes microbial metabolic activity, nitrification and positively influence on C and N cycle	
9.	Gold nanoparticles	Cow pea (*Vigna unguiculata* L.)	*Pseudomonas monteilii*	Increased growth and IAA production	[67]
10.	Zero valent iron nanoparticles	White willow (*Salix alba* L.)	*P. fluorescens*	Dose-dependent effect of iron nanoparticles; at low concentration root length and leaf area per plant improved; at higher concentration it reduced plant growth and induced stress	[68]
11.	Magnesium oxide NPs	Radish (*R. sativus* L.)	–	Displayed enhanced plant growth, production of secondary metabolites, free radical scavenging activity, and phytoaccumulation of lead	[69]
12.	Zero valent iron nanoparticles	White clover (*Trifolium repens*)	PGPR	Increases photosynthesis, plant growth and phytoremediation performance	[70]
13.	Silver nanoparticles	Wheat (*Triticum*)	*Burkholderia* sp., *Bacillus cereus*, *Bacillus* spp.	Improved sugar production and its translocation to the grains, biocontrol potential against yellow rust	[71]
14.	Graphite and silica nanoparticles	Potato (*Solanum tuberosum*)	*Lysinibacillus* sp., *B. subtilis*, and *P. fluorescens*	Isolated strain reduced the wilt disease caused by *Ralstonia solanacearum*	[72]
15.	Titanium dioxide NPs	Beans (*Phaseolus vulgaris* L.)	*Bacillus subtilis* Vru1	Improved the vegetative growth parameters of plant and metabolites production such as indole-3-acetic acid	[73]

TABLE 1 Effect of various nanoparticles on plant with PGPR—cont'd

S. no.	Nanoparticles/nanomaterials	Plant (common and scientific name)	PGPR	Results	References
16.	Gold nanoparticles		*P. fluorescens, B. subtilis, P. gii,* and *P. putida*	NPs displayed no significant with *P. putida*; significant increase was observed in the case of *P. fluorescens,* and *B. subtilis, Paenibacillus elgii* and displayed a potential to be used as a nanobiofertilizer	[74]
17.	Nanozeolite	Maize (*Zea mays*)	*Bacillus* spp.	Improved growth parameters and crop productivity	[75]
18.	Silver nanoparticles	Onion seedlings (*Allium cepa*)	*Bacillus pumilus* and *Pseudomonas moraviensis*	Increased the sugar and proline contents; enhanced protein content of bulb, decrement in leaf flavonoids and increase in the bulb flavonoid contents	[76]
19.	Molybdenum (Mo) nanoparticles	Chickpea (*Cicer arietinum* L.)	*B. subtilis*	Improved the physiological status of the plant, increasing structural diversity of the microbial community of the rhizosphere through changes in the activity of root exudates	[77]
20.	Iron oxide NPs	(*Brassica napus* L.)		Enhanced growth by reducing ROS damage and improved oxidative defense system	[78]
21.	Iron oxide nanoparticles	Thale cress (*Arabidopsis thaliana*)		Inhibitory effects on development	[79]
22.	Iron nanoparticles	Cow pea (*V. unguiculata* L.)		Increased seedling growth	[80]

No.	Nanoparticle	Plant	Organisms	Effect	Reference
23.	Silicon dioxide nanoparticles	Perennial ryegrass (*Lolium perenne*)		Improved mineral nutritional value and other quality indexes	[81]
24.	Silicon dioxide nanoparticles	Tomato (*Solanum lycopersicum*)		Enhances seed germination	[82]
25.	Silicon dioxide nanoparticles	Maize (*Z. mays*)	*Azotobacter, Bacillus megaterium, B. brevis,* and *P. fluorescens*	Nanoparticles had no toxic effects on microorganisms	[83]
26.	Zinc oxide nanoparticles	Sorghum		Reduced the negative influences on drought stress	[84]
27.	Zinc oxide nanoparticles	*B. napus*		Displayed concentration dependent effect on plant; at lower concentration, enhanced plant growth, while at higher concentration toxicity was observed	[85]
28.	Zinc oxide nanoparticles	*A. cepa* L.		Seed germination was observed to be concentration dependent; at higher concentration of NPs germination rate decreased while at lower concentration seed germination rate increased	[86]
29.	Calcium phosphate nanoparticles	Strawberry		Nano-CaPNPs at 15 ppm improved quality and storability of fruits and gave good appearance with the lowest values of weight loss, and zero decay percentage	[87]
30.	Calcium phosphate nanoparticles	Rice		NPs reduced the amount of fertilizer requirement for the crops thus reducing the fertilizer wastage	[88]

of *Brassica juncea* seedlings with silver nanoparticles and *Bacillus thuringeinsis* KVS25 were observed to significantly reduce the stress in the plant seedlings.

The efficacy of PGPR strains *Bacillus pumilus* and *Pseudomonas moraviensis* with silver nanoparticles on onion bulb weight under salt stress displayed an increase in the sugar content of bulb, root proliferation, and bulb growth [76]. Furthermore, *Pseudomonas moraviensis* with Ag NPs was more effective under saline conditions and had elevated bulb phenolic content (stress related compounds). Bano and Habib in 2020 reported the supplementation of AgNPs with *Bacillus cereus* for enhanced antifungal activity in wheat plants. The cumulative effect of *Bacillus cereus* with AgNPs and salicylic acid effectively reduced the yellow rust in plants [71].

9.2 Zinc oxide nanoparticles

Zinc is a vital micronutrient in the plant cells for the synthesis of tryptophan, which is the precursor of indoleacetic acid, a phytohormone responsible for physiological and biochemical functions [52,53,55]. The effect of zinc oxide nanoparticles (ZnO NPs) on the plants depends on their size, concentration, and the plant species. Recently, the foliar application of ZnO NPs (10 mg/L) led to a higher biomass and photosynthetic rate in the crops. ZnO NPs slightly increased the dry and fresh weight of biomass at a lower concentration. It has been stated that the high concentration of ZnO NPs inhibited root growth. Furthermore, it is reported to have a significant role in the inhibition of chlorophyll biosynthesis, leading to the reduction in photosynthesis efficiency [96].

Dimkpa et al. [84] demonstrated that soil amended with ZnO-NPs mitigated the negative influences of drought stress (40% of field moisture capacity) in sorghum plants [84]. Canola (*Brassica napus*) showed improvement in plant growth with ZnO NPs at 10 mg/L, while a higher concentration (1000 mg/L) resulted in toxic effects [85]. Rahmani et al. [85] reported that on application of ZnO NPs, the seed germination enhanced at lower concentrations, while at higher concentrations of ZnO NPs the germination was limited in onion (*Allium cepa* L.).

9.3 Silicon oxide nanoparticles

Employment of silicon nanoparticles (SiO$_2$ NPs) has been reported to improve the growth performance of plants and attenuate the adverse effects of abiotic stresses and reduces toxicity. Nano-Si at lower concentrations of 1- or 2-mM improved the germination rate of the plants under drought stress. Nanoparticles of silica influenced seed germination, root elongation, and biomass of plants. Silica NPs (10 nm) at 200 mg/kg induced the cucumber plants to alleviate water deficit and soil salinity due to the effect of high silicon and potassium in regulating transpiration and maintaining ion homeostasis.

Under severe drought conditions, SiO_2 NPs at 1 mM improved the mineral nutritional value and other quality indexes in perennial ryegrass [81]. It was reported that the lower concentration of SiO_2 NPs enhances the seed germination of tomato. Nano and bulk SiO_2 particles were nontoxic to PGPRs at very high concentrations (up to 1000 mg/L) in *Bacillus megaterium*, *Bacillus brevis*, *Pseudomonas fluorescens*, and *Azotobacter vinelandii* with various plants [83].

9.4   Iron oxide nanoparticles

Iron nanoparticles have advantageous properties for plant growth. They have inhibitory effects on the development of phytotoxicity. They increase nutrient uptake and transportation. The use of zerovalent iron with PGPR might be suggested as a feasible and environmentally friendly technique to enhance the phytoremediation of heavy metals in contaminated soils. Iron chelates and PGPR had a positive and significant effect on the growth, yield, and physiological characteristics of plants. They also increase the seedling growth. Furthermore, They have the capacity to improve yield, yield components, and oil percentage. They increased photosynthesis and decreased oxidative stress and reduced reactive oxygen species damage in plants.

Palmqvist et al. [78] reported that iron oxide nanoparticles enhanced the growth and agronomic traits by reducing ROS damage and improving the oxidative defense system in *Brassica napus* L. Yang et al. concluded that the impact of iron oxide nanoparticles on *Arabidopsis thaliana* it has inhibitory effects on development. Rahimi et al. [80] showed that iron nanoparticles increased the seedling growth traits in *Vigna unguiculata* (L.) [80].

9.5   Other nanomaterials

The macronutrients and micronutrients have a crucial role in the growth and development of plants. Calcium is a major essential plant element. Synthesized calcium nanoparticles can be exploited for the formulation of new nanogrowth promoters and nanofertilizers in agriculture. Employment of calcium in nanoformulations can potentially reduce the quantity of fertilizers that are applied to the crops. The decreased use of fertilizers can directly and indirectly aid in the reduction of pollution of the environment due to agricultural malpractices [88]. The foliar application of nanofertilizers gives rise to a significant increase in the concentration of various amino acids, increased germination and growth rate of the plant. Furthermore, it has been reported that calcium nanoparticles affect plant height, branch number per plant, pod number per plant, seed number per pod, seed weight (g), and seed yield in several plants.

The foliar spray of calcium phosphate nanoparticles (CaPNPs) in strawberry plants improved the quality and storability of fruits. Furthermore, the appearance of the berries was better and the lowest values of weight loss and zero decay percentage were reported [87]. Calcium Borate nanoparticles

(CaB$_4$O$_7$ NPs) as nanofertilizers were reported to promote shoot and root biomass production by ~twofold compared to untreated plants [91].

Manganese (Mn) is a micronutrient required for growth regulation of the plants. It plays a vital role in photosynthesis, enhances the activity of the electron transport chain in photosynthesis, and reduces oxidative stress [97,98]. Manganese nanoparticles (MnNPs) can be employed as a manganese micronutrient fertilizer or plant growth enhancer. Manganese nanoparticles were biocompatible toward soil microorganisms. MnNP can be employed as a suitable alternative for salts employed in agriculture for the supplementation of manganese in soil and crop management. MnNPs are considered to be an essential constituent of the catalytic center which is responsible for water oxidation at photosystem (PS II) [99]. MnNPs transport electrons to the thylakoid bound electron transport chain (ETC), which produces reducing power and ATP for carbon dioxide assimilation [100].

10   Conclusions

The present chapter indicates the benefits of PGPR, such as biofertilization, biocontrol, and bioremediation, that have a favorable impact on crop productivity. The employment of PGPR (*Bacillus*, *Pseudomonas*, *Azospirillum*, etc.) has significantly improved physicochemical parameters in economically chief crops such as rice, maize, tomatoes, wheat, sugarcane, etc. The formulation prepared of PGPR is a promising alternative to chemical fertilizers which can be employed in sustainable agriculture. The advent of nanotechnology and its inclusion in the agricultural sector has a potential to improve the current biofertilizers and has captivated the interest of various researchers. The interaction of nanomaterials with PGPR can promote and enhance the performance of rhizobacteria and thus has immense potential to be exploited as an environmently friendly fertilizer for the crops. The amalgamation of nanotechnology and PGPR can be employed as a budget- and eco-friendly sustainable alternative to chemical fertilizers for the growth and development of plants.

References

[1] T. Searchinger, C. Hanson, J. Ranganathan, B. Lipinski, R. Waite, R. Winterbottom, A. Dinshaw, R. Heimlich, M. Boval, P. Chemineau, Creating a sustainable food future. A menu of solutions to sustainably feed more than 9 billion people by 2050, in: World Resources Report 2013–14: Interim Findings, World Resources Institute (WRI); World Bank Groupe-Banque Mondiale; United Nations Environment Programme (UNEP); United Nations Development Programme (UNDP); Centre de Coopération Internationale en Recherche Agronomique pour le Développement (CIRAD); Institut National de la Recherche Agronomique (INRA)., 2014, p. 154.

[2] P.N. Bhattacharyya, D.K. Jha, Plant growth-promoting rhizobacteria (PGPR): emergence in agriculture, World J. Microbiol. Biotechnol. 28 (4) (2012) 1327–1350.

[3] K. Hartman, S.G. Tringe, Interactions between plants and soil shaping the root microbiome under abiotic stress, Biochem. J. 476 (19) (2019) 2705–2724.

[4] J. Morgan, G. Bending, P. White, Biological costs and benefits to plant–microbe interactions in the rhizosphere, J. Exp. Bot. 56 (417) (2005) 1729–1739.

[5] D. Goswami, J.N. Thakker, P.C. Dhandhukia, Portraying mechanics of plant growth promoting rhizobacteria (PGPR): a review, Cogent Food Agric. 2 (1) (2016) 1–19. 1127500.

[6] J. Lynch, Microbial interactions in the rhizosphere, Soil Microorganisms 30 (1987) 33–41.

[7] R. Backer, J.S. Rokem, G. Ilangumaran, J. Lamont, D. Praslickova, E. Ricci, S. Subramanian, D.L. Smith, Plant growth-promoting rhizobacteria: context, mechanisms of action, and roadmap to commercialization of biostimulants for sustainable agriculture, Front. Plant Sci. 9 (2018) 1473.

[8] H. Massalha, E. Korenblum, D. Tholl, A. Aharoni, Small molecules below-ground: the role of specialized metabolites in the rhizosphere, Plant J. 90 (4) (2017) 788–807. Wiley Online Library.

[9] M.S. Nelson, M.J. Sadowsky, Secretion systems and signal exchange between nitrogen-fixing rhizobia and legumes, Front. Plant Sci. 6 (2015) 491.

[10] S. Burdman, E. Jurkevitch, Y. Okon, Recent advances in the use of plant growth promoting rhizobacteria (PGPR) in agriculture, in: Microbial Interactions in Agriculture and Forestry, vol. II, Elsevier, 2000, pp. 229–250.

[11] H. Antoun, D. Prévost, Ecology of plant growth promoting rhizobacteria, in: PGPR: Biocontrol and Biofertilization, Springer, 2005, pp. 1–38.

[12] J. Barriuso, B.R. Solano, J.A. Lucas, A.P. Lobo, A. García-Villaraco, F.J.G. Mañero, Ecology, genetic diversity and screening strategies of plant growth promoting rhizobacteria (PGPR), J. Plant Nutr. 4 (2008) 1–17.

[13] E.F. George, M.A. Hall, G.-J. De Klerk, Plant growth regulators III: gibberellins, ethylene, abscisic acid, their analogues and inhibitors; miscellaneous compounds, in: Plant Propagation by Tissue Culture, Springer, 2008, pp. 227–281.

[14] Y.H. Wang, H.R. Irving, Developing a model of plant hormone interactions, Plant Signal. Behav. 6 (4) (2011) 494–500.

[15] B.R. Glick, D.M. Penrose, J. Li, A model for the lowering of plant ethylene concentrations by plant growth-promoting bacteria, J. Theor. Biol. 190 (1) (1998) 63–68.

[16] M. Grover, S. Bodhankar, A. Sharma, P. Sharma, J. Singh, L. Nain, PGPR mediated alterations in root traits: way towards sustainable crop production, Front. Sustain. Food Syst. 4 (2020) 287.

[17] M. Grover, S.Z. Ali, V. Sandhya, A. Rasul, B. Venkateswarlu, Role of microorganisms in adaptation of agriculture crops to abiotic stresses, World J. Microbiol. Biotechnol. 27 (5) (2011) 1231–1240.

[18] M.J. Van Oosten, O. Pepe, S. De Pascale, S. Silletti, A. Maggio, The role of biostimulants and bioeffectors as alleviators of abiotic stress in crop plants, Chem. Biol. Technol. Agric. 4 (1) (2017) 1–12.

[19] A. Beneduzi, A. Ambrosini, L.M. Passaglia, Plant growth-promoting rhizobacteria (PGPR): their potential as antagonists and biocontrol agents, Genet. Mol. Biol. 35 (2012) 1044–1051.

[20] J.K. Vessey, Plant growth promoting rhizobacteria as biofertilizers, Plant Soil 255 (2) (2003) 571–586.

[21] S. Bhatt, R. Raghav, N. Pandhi, Allevation of impacts of salt stress on the growth parameters of groundnut (*Arachis hypogaea* L.) employing halotolerant PGPR, Int. J. Adv. Res. 7 (11) (2019) 331–338.

[22] K. Leontidou, S. Genitsaris, A. Papadopoulou, N. Kamou, I. Bosmali, T. Matsi, P. Madesis, D. Vokou, K. Karamanoli, I. Mellidou, Plant growth promoting rhizobacteria isolated from

halophytes and drought-tolerant plants: genomic characterisation and exploration of phyto-beneficial traits, Sci. Rep. 10 (1) (2020) 1–15.

[23] S. Fahad, S. Hussain, et al., Phytohormones and plant responses to salinity stress: a review, Plant Growth Regul. 75 (2015) 391–404.

[24] S. Spaepen, J. Vanderleyden, Auxin and plant-microbe interactions, Cold Spring Harb. Perspect. Biol. 3 (4) (2011) a001438.

[25] S.-M. Kang, A.L. Khan, Y.-H. You, J.-G. Kim, M. Kamran, I.-J. Lee, Gibberellin production by newly isolated strain *Leifsonia soli* SE134 and its potential to promote plant growth, J. Microbiol. Biotechnol. 24 (1) (2014) 106–112.

[26] C.K. Jha, M. Saraf, Plant growth promoting rhizobacteria (PGPR): a review, J. Agric. Res. Dev. 5 (2) (2015) 108–119.

[27] J. Barea, M.E. Brown, Effects on plant growth produced by *Azotobacter paspali* related to synthesis of plant growth regulating substances, J. Appl. Bacteriol. 37 (4) (1974) 583–593.

[28] F. Liu, S. Xing, H. Ma, Z. Du, B. Ma, Cytokinin-producing, plant growth-promoting rhizobacteria that confer resistance to drought stress in *Platycladus orientalis* container seedlings, Appl. Microbiol. Biotechnol. 97 (20) (2013) 9155–9164.

[29] R. Pallai, R.K. Hynes, B. Verma, L.M. Nelson, Phytohormone production and colonization of canola (*Brassica napus* L.) roots by *Pseudomonas fluorescens* 6-8 under gnotobiotic conditions, Can. J. Microbiol. 58 (2) (2012) 170–178.

[30] A. Kumar, B. Maurya, R. Raghuwanshi, Isolation and characterization of PGPR and their effect on growth, yield and nutrient content in wheat (*Triticum aestivum* L.), Biocatal. Agric. Biotechnol. 3 (4) (2014) 121–128.

[31] M.M. Alzoubi, M. Gaibore, The effect of phosphate solubilizing bacteria and organic fertilization on availability of Syrian rock phosphate and increase of triple superphosphate efficiency, World J. Agric. Sci. 8 (5) (2012) 473–478.

[32] S.B. Sharma, R.Z. Sayyed, M.H. Trivedi, T.A. Gobi, Phosphate solubilizing microbes: sustainable approach for managing phosphorus deficiency in agricultural soils, Springerplus 2 (1) (2013) 1–14.

[33] P.A. Wani, A. Zaidi, A.A. Khan, M.S. Khan, Effect of phorate on phosphate solubilization and indole acetic acid releasing potentials of rhizospheric microorganisms, Ann. Plant Prot. Sci. 13 (1) (2005) 139–144.

[34] V. Shenoy, G. Kalagudi, Enhancing plant phosphorus use efficiency for sustainable cropping, Biotechnol. Adv. 23 (7–8) (2005) 501–513.

[35] H. Malhotra, S. Sharma, R. Pandey, Phosphorus nutrition: plant growth in response to deficiency and excess, in: Plant Nutrients and Abiotic Stress Tolerance, Springer, 2018, pp. 171–190.

[36] K.G. Raghothama, Phosphorus and plant nutrition: an overview, in: Phosphorus: Agriculture and the Environment, vol. 46, John Wiley & Sons, 2005, pp. 353–378.

[37] S. Bhatt, N. Pandhi, R. Raghav, Improved salt tolerance and growth parameters of groundnut (*Arachis hypogaea* L.) employing halotolerant *Bacillus cereus* SVSCD1 isolated from Saurashtra Region, Gujarat, Ecol. Environ. Conserv. 26 (2020) S199–S212.

[38] R.K. Sarma, R. Saikia, Alleviation of drought stress in mung bean by strain *Pseudomonas aeruginosa* GGRJ21, Plant Soil 377 (1) (2014) 111–126.

[39] G. Fasciglione, E.M. Casanovas, V. Quillehauquy, A.K. Yommi, M.G. Goni, S.I. Roura, C.A. Barassi, *Azospirillum* inoculation effects on growth, product quality and storage life of lettuce plants grown under salt stress, Sci. Hortic. 195 (2015) 154–162.

[40] M. İpek, A. Eşitken, The actions of PGPR on micronutrient availability in soil and plant under calcareous soil conditions: an evaluation over Fe nutrition, in: Plant-Microbe Interactions in Agro-Ecological Perspectives, Springer, 2017, pp. 81–100.

[41] M. Ihtisham, A. Noori, S. Yadav, M. Sarraf, P. Kumari, M. Brestic, M. Imran, F. Jiang, X. Yan, A. Rastogi, Silver nanoparticle's toxicological effects and phytoremediation, Nanomaterials 11 (9) (2021) 2164.

[42] S. Perrenoud, in: S. Perrenoud (Ed.), Potassium and Plant Health, International Potash Institute, 1977.

[43] K. Prajapati, H. Modi, The importance of potassium in plant growth—a review, Indian J. Plant Sci. 1 (2-3) (2012) 177–186.

[44] J. Sardans, J. Peñuelas, Potassium control of plant functions: ecological and agricultural implications, Plan. Theory 10 (2) (2021) 419.

[45] K. Mengel, E.A. Kirkby, Potassium in crop production, Adv. Agron. 33 (1980) 59–110.

[46] R. Mikkelsen, The importance of potassium management for horticultural crops, Indian J. Fertil. 13 (11) (2017) 82–86.

[47] H.L.S. Tandon, G.S. Sekhon, Potassium Research and Agricultural Production in India, Fertiliser Development and Consultation Organisation, 1988.

[48] Z. He, X. Yang, B.A. Kahn, P.J. Stoffella, D.V. Calvert, Plant nutrition benefits of phosphorus, potassium, calcium, magnesium, and micronutrients from compost utilization, in: P.J. Stoffella, B.A. Kahn (Eds.), Compost Utilization in Horticultural Cropping Systems, CRC Press, LLC, Boca Raton, FL, 2001, pp. 307–317.

[49] R.W. Jones, O. Lunt, The function of calcium in plants, Bot. Rev. 33 (4) (1967) 407–426.

[50] W.A. Albrecht, Nutritional role of calcium in plants, Plant Soil 33 (1) (1970) 361–382.

[51] P.K. Hepler, R.O. Wayne, Calcium and plant development, Annu. Rev. Plant Physiol. 36 (1) (1985) 397–439.

[52] A. Camp, Zinc as a nutrient in plant growth, Soil Sci. 60 (2) (1945) 157–164.

[53] B. Hafeez, Y. Khanif, M. Saleem, Role of zinc in plant nutrition-a review, J. Exp. Agric. Int. 3 (2) (2013) 374–391.

[54] L. Rudani, P. Vishal, P. Kalavati, The importance of zinc in plant growth—a review, Int. Res. J. Nat. Appl. Sci. 5 (2) (2018) 38–48.

[55] T. Tsonev, F.J. Cebola Lidon, Zinc in plants-an overview, Emir. J. Food Agric. 24 (4) (2012) 322–333.

[56] P. Mathur, R. Trivedi, P. Joshi, Abrus precatorius L.: a review from ethno to nano applications, Asian Agri-Hist. 23 (4) (2019) 245–259.

[57] S.E. McNeil, Nanotechnology for the biologist, J. Leukoc. Biol. 78 (3) (2005) 585–594.

[58] M.H. Siddiqui, M.H. Al-Whaibi, M. Firoz, M.Y. Al-Khaishany, Role of nanoparticles in plants, in: Nanotechnology and Plant Sciences: Nanoparticles and Their Impact on Plants, Springer, 2015, pp. 19–35.

[59] T. Ahmed, M. Noman, M. Rizwan, S. Ali, U. Ijaz, M.M. Nazir, H.A.S. Al Haithloul, S.M. Alghanem, A.M. Abdul Majeed, B. Li, Green molybdenum nanoparticles-mediated biostimulation of Bacillus sp. strain ZH16 improved the wheat growth by managing in planta nutrients supply, ionic homeostasis and arsenic accumulation, J. Hazard. Mater. 423 (2022) 127024.

[60] N. Akhtar, N. Ilyas, R. Hayat, H. Yasmin, A. Noureldeen, P. Ahmad, Synergistic effects of plant growth promoting rhizobacteria and silicon dioxide nano-particles for amelioration of drought stress in wheat, Plant Physiol. Biochem. 166 (2021) 160–176.

[61] M. Hatami, P. Khanizadeh, F. Bovand, A. Aghaee, Silicon nanoparticle-mediated seed priming and Pseudomonas spp. inoculation augment growth, physiology and antioxidant metabolic status in Melissa officinalis L. plants, Ind. Crop. Prod. 162 (2021) 113238.

[62] F. Hussain, F. Hadi, F. Akbar, Magnesium oxide nanoparticles and thidiazuron enhance lead phytoaccumulation and antioxidative response in Raphanus sativus L, Environ. Sci. Pollut. Res. 26 (29) (2019) 30333–30347.

[63] L. Zhang, L. Wu, Y. Si, K. Shu, Size-dependent cytotoxicity of silver nanoparticles to *Azotobacter vinelandii*: growth inhibition, cell injury, oxidative stress and internalization, PLoS One 13 (12) (2018) e0209020.

[64] Z. Yuan, J. Li, L. Cui, B. Xu, H. Zhang, C.-P. Yu, Interaction of silver nanoparticles with pure nitrifying bacteria, Chemosphere 90 (4) (2013) 1404–1411.

[65] C. Jiang, X. Xu, M. Megharaj, R. Naidu, Z. Chen, Inhibition or promotion of biodegradation of nitrate by *Paracoccus* sp. in the presence of nanoscale zero-valent iron, Sci. Total Environ. 530 (2015) 241–246.

[66] S. He, Y. Feng, J. Ni, Y. Sun, L. Xue, Y. Feng, Y. Yu, X. Lin, L. Yang, Different responses of soil microbial metabolic activity to silver and iron oxide nanoparticles, Chemosphere 147 (2016) 195–202.

[67] J. Panichikkal, R. Thomas, J.C. John, E. Radhakrishnan, Biogenic gold nanoparticle supplementation to plant beneficial *Pseudomonas monteilii* was found to enhance its plant probiotic effect, Curr. Microbiol. 76 (4) (2019) 503–509.

[68] S. Mokarram-Kashtiban, S.M. Hosseini, M.T. Kouchaksaraei, H. Younesi, The impact of nanoparticles zero-valent iron (nZVI) and rhizosphere microorganisms on the phytoremediation ability of white willow and its response, Environ. Sci. Pollut. Res. 26 (11) (2019) 10776–10789.

[69] J.S. Salas-Leiva, A. Luna-Velasco, D.E. Salas-Leiva, Use of magnesium nanomaterials in plants and crop pathogens, J. Nanopart. Res. 23 (12) (2021) 1–34.

[70] A.D. Zand, A.M. Tabrizi, A.V. Heir, The influence of association of plant growth-promoting rhizobacteria and zero-valent iron nanoparticles on removal of antimony from soil by *Trifolium repens*, Environ. Sci. Pollut. Res. 27 (34) (2020) 42815–42829.

[71] A. Bano, Interactive effects of Ag-nanoparticles, salicylic acid, and plant growth promoting rhizobacteria on the physiology of wheat infected with yellow rust, J. Plant Pathol. 102 (4) (2020) 1215–1225.

[72] L. Djaya, N. Istifadah, S. Hartati, I.M. Joni, In vitro study of plant growth promoting rhizobacteria (PGPR) and endophytic bacteria antagonistic to *Ralstonia solanacearum* formulated with graphite and silica nano particles as a biocontrol delivery system (BDS), Biocatal. Agric. Biotechnol. 19 (2019) 101153.

[73] R. Saberi-Rise, M. Moradi-Pour, The effect of *Bacillus subtilis* Vru1 encapsulated in alginate–bentonite coating enriched with titanium nanoparticles against *Rhizoctonia solani* on bean, Int. J. Biol. Macromol. 152 (2020) 1089–1097.

[74] S.K. Shukla, R. Kumar, R.K. Mishra, A. Pandey, A. Pathak, M. Zaidi, S.K. Srivastava, A. Dikshit, Prediction and validation of gold nanoparticles (GNPs) on plant growth promoting rhizobacteria (PGPR): a step toward development of nano-biofertilizers, Nanotechnol. Rev. 4 (5) (2015) 439–448.

[75] P. Khati, P. Bhatt, R. Kumar, A. Sharma, Effect of nanozeolite and plant growth promoting rhizobacteria on maize, 3 Biotech 8 (3) (2018) 1–12.

[76] S. Jahangir, K. Javed, A. Bano, Nanoparticles and plant growth promoting rhizobacteria (PGPR) modulate the physiology of onion plant under salt stress, Pak. J. Bot. 52 (4) (2020) 1473–1480.

[77] M.M. Raffi, A. Husen, Impact of fabricated nanoparticles on the rhizospheric microorganisms and soil environment, in: Nanomaterials and Plant Potential, Springer, 2019, pp. 529–552.

[78] N.M. Palmqvist, G.A. Seisenbaeva, P. Svedlindh, V.G. Kessler, Maghemite nanoparticles acts as nanozymes, improving growth and abiotic stress tolerance in Brassica napus, Nanoscale Res. Lett. 12 (1) (2017) 1–9.

[79] S. Bombin, M. LeFebvre, J. Sherwood, Y. Xu, Y. Bao, K.M. Ramonell, Developmental and reproductive effects of iron oxide nanoparticles in *Arabidopsis thaliana*, Int. J. Mol. Sci. 16 (10) (2015) 24174–24193.

[80] H. Rahimi, A. Ghasemi, R. Mozaffarinia, M. Tavoosi, On the magnetic and structural properties of neodymium iron boron nanoparticles, J. Supercond. Nov. Magn. 29 (8) (2016) 2041–2051.

[81] S. Mahdavi, M. Kafi, E. Fallahi, M. Shokrpour, L. Tabrizi, Water stress, nano silica, and digoxin effects on minerals, chlorophyll index, and growth in ryegrass, Int. J. Plant Prod. 10 (2) (2016) 251–264.

[82] Z. Iqbal, M.I. Ansari, A. Ahmad, Z. Haque, M.S. Iqbal, Impact of nanomaterials stress on plants, in: Nanobiotechnology, Springer, 2021, pp. 499–526.

[83] G. Karunakaran, R. Suriyaprabha, P. Manivasakan, R. Yuvakkumar, V. Rajendran, P. Prabu, N. Kannan, Effect of nanosilica and silicon sources on plant growth promoting rhizobacteria, soil nutrients and maize seed germination, IET Nanobiotechnol. 7 (3) (2013) 70–77.

[84] C.O. Dimkpa, U. Singh, P.S. Bindraban, W.H. Elmer, J.L. Gardea-Torresdey, J.C. White, Zinc oxide nanoparticles alleviate drought-induced alterations in sorghum performance, nutrient acquisition, and grain fortification, Sci. Total Environ. 688 (2019) 926–934.

[85] F. Rahmani, A. Peymani, E. Daneshvand, P. Biparva, Impact of zinc oxide and copper oxide nano-particles on physiological and molecular processes in *Brassica napus* L, Indian J. Plant Physiol. 21 (2) (2016) 122–128.

[86] S. Laware, S. Raskar, Influence of zinc oxide nanoparticles on growth, flowering and seed productivity in onion, Int. J. Curr. Microbiol. Sci. 3 (7) (2014) 874–881.

[87] S. Zakaria, M.E. Ragab, A. Abou EL-Yazied, M.A. Rageh, K.Y. Farroh, T.A. Salaheldin, Improving quality and storability of strawberries using preharvest calcium nanoparticles application, Middle East J. Agric. 7 (3) (2018) 1023–1040.

[88] H. Upadhyaya, L. Begum, B. Dey, P. Nath, S. Panda, Impact of calcium phosphate nanoparticles on rice plant, J. Plant Sci. Phytopathol. 1 (2017) 1–10.

[89] A. Farooqui, H. Tabassum, A. Ahmad, A. Mabood, A. Ahmad, I.Z. Ahmad, Role of nanoparticles in growth and development of plants: a review, Int J Pharm. Bio. Sci 7 (4) (2016) 22–37.

[90] K.S. Siddiqi, A. Husen, Plant response to silver nanoparticles: a critical review, Crit. Rev. Biotechnol. 42 (7) (2021) 973–990.

[91] S. Meier, F. Moore, A. Morales, M.-E. González, A. Seguel, C. Meriño-Gergichevich, O. Rubilar, J. Cumming, H. Aponte, D. Alarcón, Synthesis of calcium borate nanoparticles and its use as a potential foliar fertilizer in lettuce (Lactuca sativa) and zucchini (Cucurbita pepo), Plant Physiol. Biochem. 151 (2020) 673–680.

[92] P. Kumar, V. Pahal, A. Gupta, R. Vadhan, H. Chandra, R.C. Dubey, Effect of silver nanoparticles and *Bacillus cereus* LPR2 on the growth of Zea mays, Sci. Rep. 10 (1) (2020) 1–10.

[93] S. Timmusk, G. Seisenbaeva, L. Bchers, Titania (TiO2) nanoparticles enhance the performance of growth-promoting rhizobacteria, Sci. Rep. 8 (1) (2018) 617.

[94] N. Khan, A. Bano, Role of plant growth promoting rhizobacteria and Ag-nano particle in the bioremediation of heavy metals and maize growth under municipal wastewater irrigation, Int. J. Phytoremediation 18 (3) (2016) 211–221.

[95] K. Vishwakarma, V.P. Singh, S.M. Prasad, D.K. Chauhan, D.K. Tripathi, S. Sharma, Silicon and plant growth promoting rhizobacteria differentially regulate AgNP-induced toxicity in *Brassica juncea*: implication of nitric oxide, J. Hazard. Mater. 390 (2020) 121806.

[96] G.R. Rout, P. Das, Effect of metal toxicity on plant growth and metabolism: I. Zinc, in: Sustainable Agriculture, Springer, 2009, pp. 873–884.

[97] S. Alejandro, S. Höller, B. Meier, E. Peiter, Manganese in plants: from acquisition to subcellular allocation, Front. Plant Sci. 11 (2020) 300.

[98] D.T. Clarkson, The uptake and translocation of manganese by plant roots, in: Manganese in Soils and Plants, Springer, 1988, pp. 101–111.

[99] S. Pradhan, P. Patra, S. Das, S. Chandra, S. Mitra, K.K. Dey, S. Akbar, P. Palit, A. Goswami, Photochemical modulation of biosafe manganese nanoparticles on *Vigna radiata*: a detailed molecular, biochemical, and biophysical study, Environ. Sci. Technol. 47 (22) (2013) 13122–13131.

[100] R.L. Houtz, R.O. Nable, G.M. Cheniae, Evidence for effects on the in vivo activity of ribulose-bisphosphate carboxylase/oxygenase during development of Mn toxicity in tobacco, Plant Physiol. 86 (4) (1988) 1143–1149.

Chapter 13

Potential scope and prospects of plant growth-promoting microbes (PGPMs) in micropropagation technology

Sagar Teraiya, Dhaval Nirmal, and Preetam Joshi
Department of Biotechnology, Atmiya University, Rajkot, India

1 Introduction

Plant tissue culture technology utilizes the ability of a single cell or a group of plant cells to transform into a whole plant when grown under controlled environmental conditions. This interesting idea of in vitro culturing of the plant cell was put forward by Gottlieb Haberlandt in 1902 in the form of a postulate "totipotentiality," which later on led to significant discoveries in biology. One important aspect of plant tissue culture is micropropagation, which is being exploited by a large number of researchers and business firms. The primary use of micropropagation is large-scale production of plants, ranging from nursery stock species (like rhododendron or rose) to ornamentals (like gerbera or carnation), fruits (like banana or raspberries), and vegetables and crops (like cauliflower, potato, or pointed gourd). In the last two decades, there has been a significant growth in micropropagation-based industries, and these industries have been internationally acknowledged as one of the significant tools for the direct application of this technology in the agriculture field. Other important applications of tissue culture technology are conservation of endangered plants, in vitro production of secondary metabolite, crop improvement, and development of new varieties through transgenic approach. Besides its several advantages, this technique has many challenges. For example, any micropropagation system must produce large numbers of genetically uniform plants that maintain the genetic truthfulness (i.e., genetic fidelity). Moreover, the technique involves the use of certain chemical sterilizers, plant growth regulators (PGRs), and sometimes antifungal agents and antibiotics to control the contamination. Most of these chemicals are very costly and therefore limit the profitability to end

Plant-Microbe Interaction—Recent Advances in Molecular and Biochemical Approaches
https://doi.org/10.1016/B978-0-323-91876-3.00017-8

249

users. Similarly, it requires certain costly instruments and a sophisticated setup and skilled manpower, which further increase the production cost of plants. Other important challenges in micropropagation are the low in vitro multiplication rate, loss of plantlets due to contamination, increased susceptibility toward pathogens pre- and postfield transfer, mixotrophic behavior of plantlets during culture conditions, and low survival of plantlets during the hardening and acclimatization stage.

To address these challenges and cut the production cost, plant growth-promoting microbes (PGPMs) can be used as an effective tool. PGPMs could be an effective agent for the promotion of growth, uptake of nutrients from soil, and sometimes can be an alternative source of nitrogen fertilizer for plants. After confirmation of the role of microbes in soil fertility and plant growth, PGPMs have gained a lot of attention from many soil scientists and agriculture biotechnologists. PGPMs promote plant growth in many ways, for example, they may produce plant hormones [1–3] or growth-stimulating biomolecules, viz., vitamins and related products [4], by suppressing the growth of pathogens by different mechanisms [5]. Nowadays, PGPMs have received a lot of attention, particularly in the field of crop improvement, and many related articles got published in the last two decades. However, research into the application of PGPMs in plant tissue culture has not gained much popularity just because it is a general notion that the presence of microbes in the tissue culture growth medium is deleterious and is considered as a can of worms, which not only limit the establishment of culture but also leads to further obstacles in subsequent stages [6]. Hence, most of the focus in tissue culture is on how to get rid of microbes despite the fact that many PGPMs can be beneficial at different stages of tissue culture. However, many PGPMs can be beneficial at different stages of tissue culture, viz., at the stage of in vitro rooting, in vitro shoot multiplication and elongation, and the acclimatization stage. Moreover, they can provide a defense against biotic (pathogens) and abiotic (temperature, salinity, heavy metals, etc.) stress that arises during the hardening and acclimatization stage. PGPMs can act as a nostrum for sustainable agriculture, if used judiciously, and therefore promoting the use of PGPMs in tissue culture is advantageous as well as challenging. This chapter is mainly focused on the potential possibilities of PGPMs in the advancement of micropropagation technology.

2 Plant tissue culture

Plant tissue culture system is a method in which a whole plant, a plant part (generally a 1–2 cm portion of a leaf/node/internode/cotyledon or any other suitable plant part), or even sometime a single cell is taken and allowed to grow under controlled aseptic environmental conditions. The tissue culture setup is optimized to provide all macro- and micronutrients, carbon as a source of energy, phytohormones for division and differentiation and, of course, water, which is necessary for the growth of plants. All these requirements are provided in the form of a basal growth medium. In addition, environmental factors, viz., light, temperature, and humidity are also maintained optimally in a way that supports better in vitro growth and multiplication. Further, the plant development and differentiation can then be controlled by providing plant growth regulators, viz., auxin, cytokinin, gibberellins, etc. Regeneration of a plant or plant part (often regarded as an explant) under in vitro conditions relies on the concept of totipotency, originally proposed by Haberlandt in 1902. The explant is any plant part (generally 1–2 cm in size, viz., a nodal segment, an internode, a leaf segment, an immature embryo, a pollen grain, a seed, an ovule, an anther, etc.), which is used as an initial material for establishment purpose. Micropropagation may also be regarded as the method of taking explants, putting aseptically this explant on a suitable growth medium and allowing it to undergo differentiation and develop into a whole new plant [7]. The plant part (i.e., explant) is allowed to grow in a culture vessel filled with synthetic growth medium under aseptic conditions in a chamber where all the environmental conditions are kept at the optimum level. In addition to large-scale production of plants, micropropagation technology is also a key step in transgenic plant development in which the regeneration of novel plants from genetically engineered cells is carried out. Micropropagation of plants can be achieved by four different pathways, namely: (a) enhanced axillary branching; (b) adventitious shoot bud differentiation; (c) callus organogenesis; and (d) somatic embryogenesis. In the case of enhanced axillary branching, the explant contains preexisting axillary shoot buds, while in callus organogenesis and adventitious shoot bud formation, the shoots are formed de novo by the process of organogenesis. During the process of somatic embryogenesis, bipolar somatic embryos are formed that have the competency to develop into a complete plant. In any chosen pathway of micropropagation, a sequence of events is involved to achieve success (Fig. 1). Micropropagation, in contrast to conventional propagation methods, is a multistage process in which every stage is important to realize the goal of producing plants in culture.

Notwithstanding the advantage or disadvantage of various methods of micropropagation, each method involves five different stages to achieve the goal. These stages are as follows:

Stage 0:. Management of donor plant/s (source of explant)
Stage 1:. Aseptic establishment and initiation of cultures
Stage 2:. Shoot multiplication and/or elongation
Stage 3:. In vitro rooting of shoots
Stage 4:. Hardening, acclimatization, and transplantation in soil.

The first four stages described above are carried out in a highly controlled manner where the main concern is to avoid any kind of microbial contamination; hence, a high levels of aseptic conditions is maintained which results in zero contact of regenerated micropropagules with the common microbiota of the environment. The outcome of this is that the regenerated plants become more

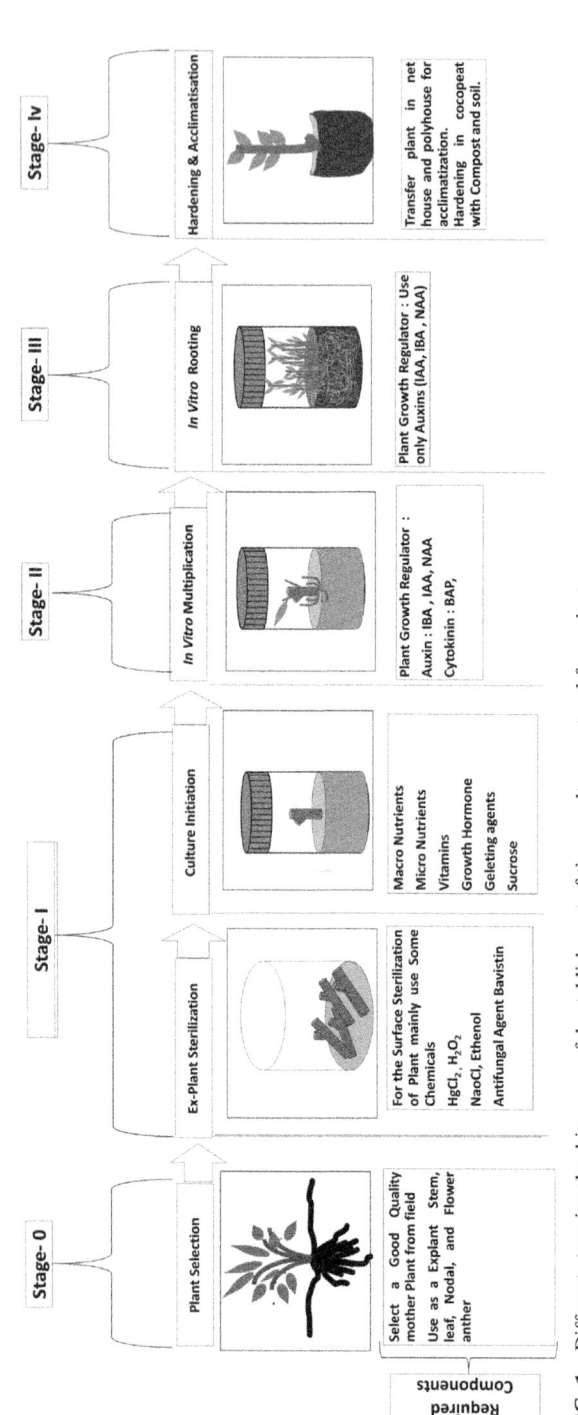

FIG. 1 Different steps involved in successful establishment of tissue culture protocol for a plant.

vulnerable and, when transferred to the soil, become more sensitive to infections as well as the harsh environmental conditions. This makes it imperative that the controlled exposure of some beneficial microorganism may positively affect the in vitro and ex vitro growth of plants during tissue culture. Although there have been few reports where the beneficial effects of these PGPMs during in vitro culture conditions have been reported, a thorough study needs to be done to explore the potential possibilities of PGPMs in micropropagation technology.

3 Challenges in plant tissue culture

Plant tissue culture technology sometimes fails to translate at the commercial level for large-scale production due to certain limitations. Some of the problems that are encountered during large-scale micropropagation are discussed here:

Higher production cost: Due to the requirement of sophisticated instruments, production setup and skilled labor, the production cost of plants increases in tissue culture. Moreover, raw materials like glassware and chemicals (viz., agar, sucrose, plant hormones, and other media components) make this technology a costly affair. In some cases, the unit cost per plant becomes exorbitant. This has restricted the growth of these industries in developing countries like India [8].

Low multiplication rate: In some plant species, the multiplication rate in tissue culture is less than threefold, which makes it nonviable technology. The high multiplication rate is an important primary concern, particularly during the commercialization phase. The high multiplication rate lessens the number of cycles required for subculturing in mass cloning and thus cuts the labor cost. The high multiplication rate also partially compensates for the loss that occurs due to the contamination at different culture stages.

Loss of culture due to contamination: In tissue culture, contamination is a major problem, which sometimes wipes out the hard work of months. The major contaminants in tissue culture are bacteria and fungi, which are either present in explants or may arise due to handling error. Whatever the reason, the contaminants are responsible for the huge loss of plantlets which ultimately result in further economic loss.

Hyperhydricity: Shoots grown in vitro are exposed to a unique microenvironment which is nutrient rich and has high humid conditions. Sometimes these cultural conditions induce morphological, anatomical, and physiological abnormalities in micropropagules. Hyperhydricity or vitrification is a physiological deformity that results in excessive hydration, low lignification, nonfunctional stomata, and poor mechanical strength in shoots. The result of this is poor regeneration abilities in such plants which require intensive care and hardening and acclimatization before soil transfer.

Susceptibility to diseases: The tissue culture grown plants are more susceptible to the soil microflora and do not show sufficient resistance against bacterial and fungal pathogen. One of the reasons behind this response is their sudden exposure (mainly the roots) to the microbes present in the soil and outer environment. Under natural conditions, plants are continuously exposed to various microflora which directly or indirectly induce various defense mechanisms in plants, which finally leads to the development of resistance against pathogens. If the natural defense mechanism of plantlets is induced during the culture conditions against different pathogens, at least for the time when they are most susceptible, the problem of quick susceptibility to the infection can be reduced to a great extent [9,10].

Acclimatization of micropropagated plants: During tissue culture, there is a loss of a significant number of plants when transferred to field conditions. The shoots grown in tissue culture are continuously exposed to a unique and lavish microenvironment where there are minimal stress conditions. Moreover, there is a continuous supply of sucrose in the medium which makes the plants partially heterotrophic in nature. All these conditions contribute to a physiological and anatomical transformation in the plants like poor development of cuticles, raised and nonfunctional stomata, poorly developed internal anatomy, less efficient photosystem, etc. Finally, when the plant is transferred to the field, it fails to tolerate the sudden shock of outer stressful conditions and strives to survive [11].

4   Plant growth-promoting microbes

Plant growth-promoting microbes (PGPMs) are a special heterogenous group of microbes which are considered advantageous for the plants in terms of being not only a growth promoter but also a savior against biotic and abiotic stress. PGPMs are generally found near the rhizospheric zone of the roots of plants or inside the plant tissue (in the case of certain endophytes) and exert their beneficial effects through several mechanisms. Some of these mechanisms include biological fixation of nitrogen, solubilization of phosphate, alleviation of stress through modulation of ACC deaminase expression, production of siderophore, synthesis of plant growth regulators, etc. Moreover, they also act as biocontrol agents against several pathogens. PGPMs are further classified into three categories on the basis of their mode of action:

(a) **Biofertilizers**: This group of PGPMs acts through the direct mechanism of PGP and contributes to plant growth through solubilization of minerals (like phosphate, potassium, and zinc) and also through the biological fixation of nitrogen.

(b) **Biostimulants**: This group includes PGPMs which enhance plant growth through the biosynthesis of phytohormones, organic compounds, and certain enzymes. This class of microbes may act either through a direct mechanism or through an indirect mechanism.

(c) **Biocontrol agents**: The PGPMs of this group provide protection to the plants against pathogens by synthesizing certain antimicrobial compounds or by challenging the pathogens for available space and nutrients.

Few of the PGPMs exhibit more than two mechanisms of growth promotion and hence may be categorized in two groups in the above classification [12]. Furthermore, on the basis of the type of microorganisms, PGPMs can be of two types:

(a) **Plant growth-promoting fungi (PGPF)**: The growth-promoting effect of several rhizospheric fungi has been reported. These PGPF include a number of species belonging to different genera of fungi. These fungi mostly belong to the arbuscular mycorrhizal fungi (AMF) family. The life cycle of these AMF cannot complete without the plant host; hence, they are termed as obligate biotrophs and are grouped in the phylum Glomeromycota. The phylum Glomeromycota includes 10 important families and the most prominent genera of this phylum include Glomus, Acaulospora, and Gigaspora. Besides this, the other PGPF include species of the genera *Trichoderma, Aspergillus, Penicillium, Fusarium, Piriformospora, Phoma*, and *Rhizoctonia*, which have the innate ability to enhance the growth of plants [13]. These PGPF are found in soil as well as other natural habitats and exert their beneficial effects on plants by improving the plant nutrition, soil fertility and providing resistance against pathogens.

(b) **Plant growth-promoting (rhizo) bacteria (PGPB or PGPR)**: PGPB represent 2%–5% of the rhizospheric bacteria, classified mainly into four groups: (a) free-living bacteria, (b) associative bacteria, (c) endophytic bacteria, and (d) nodule-forming bacteria (symbiotic). Similar to PGPF, these bacteria also have proved their potentiality as biofertilizers, biostimulants, and/or biocontrol agents. On the basis of their location in the host plant, PGPB can be classified into two groups: (i) extracellular plant growth-promoting rhizobacteria (ePGPR) and (ii) intracellular plant growth-promoting rhizobacteria (iPGPR). The ePGPRs may acquire the space on the surface of the root/or on the rhizoplane/or in the intercellular space of the root cortex. In contrast, iPGPRs are commonly found inside the cells of the nodule (a specific compacted tissue found in roots). Examples of ePGPR include *Azotobacter, Erwinia, Arthrobacter, Azospirillum, Burkholderia, Bacillus, Chromobacterium, Flavobacterium, Caulobacter, Micrococcus, Serratia, Pseudomonas*, etc. iPGPR mainly include certain endophytes and species of *Frankia*, both of which can fix environmental nitrogen symbiotically with the higher plants. The examples of some potent endophytes are *Azorhizobium, Mesorhizobium, Allorhizobium, Bradyrhizobium*, and *Rhizobium* of the family *Rhizobiaceae*. The members of this family invade the roots of plants, particularly the members of the Leguminosae family, form nodules, and fix the atmospheric nitrogen. PGPB are generally used for the promotion of growth, uptake of nutrient from soil and sometimes as a substitute of N-fertilizers of nonleguminous crops. PGPR have also proved to be an effective tool against several plant pathogen, they act as a biocontrol agent by secreting some important antibiotics [4].

Recently, research on PGPF and PGPB for crop improvement is gaining importance and many researchers are getting attracted toward this fascinating area. However, the application of these microbes in micropropagation technology is limited. Nevertheless, encouraging results from various research findings suggest that these PGPM strains can successfully be used in micropropagation technology to produce more vigor and resistant plants.

5 Application of PGPM in micropropagation technology

In tissue culture, sterilization of explants is carried out to remove all microbes during the establishment stage. Moreover, strict aseptic conditions are maintained throughout all growth room conditions considering the microbes as a potential enemy. After establishing the advantageous role of PGPM in plant growth and protection, the perspective of complete removal of microbes from tissue culture has shifted and is restricted to only harmful microbes. In fact, more attention is paid to the right utilization of PGPM at different culture stages during in vitro growth conditions. This idea of using beneficial microbes during in vitro conditions was conceptualized by Nowak in 1998 and was termed as "Biotization." As stated by Gosal et al. [14], "Biotization is the metabolic response of in vitro grown plant material to a microbial inoculum(s), leading to development and physiological changes enhancing biotic and abiotic resistance of the derived propagules." The process of biotization can be carried out at any stage of tissue culture which can generally be contingent on the objective of the researcher or the nature of the problem. For example, at the establishment stage (stage I), multiplication stage (stage II), and rooting stage (stage III), those PGPMs are added which act as biostimulants and stimulate overall growth, multiplication, rooting or competency in propagules, while at stage IV (hardening and acclimatization stage), the main choice is those PGPMs that stimulate the resistance and photosynthetic efficiency of plantlets. Besides being potential biostimulator and biocontrol agents, certain PGPMs (e.g., *Frankia*, *Bradyrhizobium*, *Rhizobium*, *Azofobacler*, *Bacillus*, and *Xanthomonas*) play an important role in improving the physical properties of soil [15]. The schematic representation of the Biotization approach and its advantages during tissue culture is presented in Fig. 2.

5.1 Biotization with plant growth-promoting fungi

It has been observed that prolonged exposure of plantlets to in vitro conditions makes their roots unresponsive to water absorption, which leads to water stress

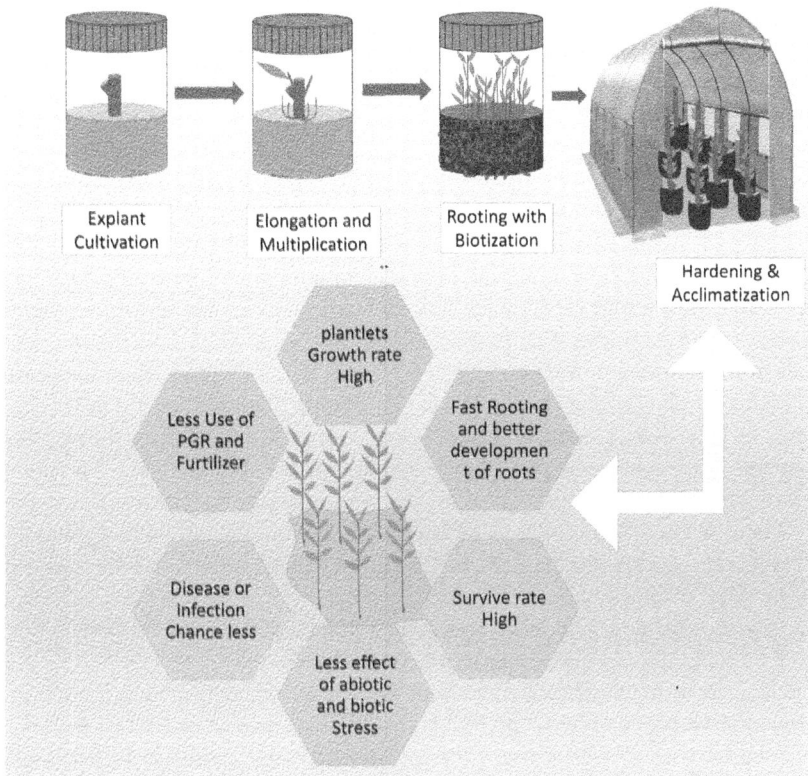

FIG. 2 Schematic representation of the Biotization approach and its advantages during tissue culture.

at a later stage. Inoculation of such plants with AM fungi during tissue culture may be beneficial to overcome this problem [16]. It has also been testified that inoculation of AM fungi during culture conditions helps the plants in nutrient availability, increased growth and resistance to pathogens after transplantation to the soil [17]. The other experimentally proved advantages of AM fungi under in vivo conditions include the ability to utilize the available Phosphate present in the soil through their hyphae [18]. Moreover, the PGPF help in better growth of plants [19] as well as higher production of secondary metabolites and related compounds such as alkaloids, phenolics, plant-based sterols, vitamins, lignans, terpenes, etc. These compounds are valued from the human health perspective as well as provide tolerance to the plants against various biotic and abiotic stresses [20]. Moreover, the PGPF also play a significant role in enhancing the production of several enzymes [21], stimulating the photosynthesis process [22], and improving the fertility of the soil [23]. According to Streletskii et al.

[24], fungi produce plant hormones, and these hormones regulate the development of plants by activating signaling pathways throughout the biotic and/or abiotic stresses. Considering the above properties, PGPFs have been tried during tissue culture for overcoming the existing problem of post vitro survival. Tissue culture raised plantlets of the wood-apple were allowed to get colonized with the root fungus *Piriformospora indica* during stage III (in vitro rooting stage) and stage IV (hardening and acclimatization stage), and significant growth was observed in terms of shoot number, shoot length, root length, leaf number, leaf area, and fresh weight. Moreover, the survival percentage and performance of plants after field transfer were also significantly increased [25]. Similar results were also observed in tissue culture raised *Terminalia bellerica* by Suthar and Purohit [26]. Since the AM fungi significantly upsurge mineral uptake, their role is specifically important during stage IV (i.e., hardening and acclimatization stage). AMF can be an effective tool to address a common problem in tissue culture derived plants, i.e., mineral absorption, since the AM fungi have very well-developed arbuscules and hyphae which can easily transfer nutrients (particularly the phosphate) from the soil to the plant [27]. The main reason behind the poor survival of micropropagated plants, post transfer to the soil, is absence of their microsymbiont partner, and this can be mitigated by inoculating the plantlets with PGPG at the hardening stage. This was proved in the case of in vitro grown hydrangea plants where the post survival rate was attained up to 100% when inoculated with AMF *Glomus intraradices* at the hardening stage [28]. Similar results were observed in *Quercus suber*, where inoculation at the hardening stage with *Pisolithus tinctorius* and *Scleroderma polyrhizum* resulted in better growth and performance of plants [29]. Likewise, inoculation with *Piriformospora indica* showed better results in tobacco and brinjal [30]. Besides the biostimulatory effect, the biocontrol action of certain fungi has also been reported in tissue cultured plants [31]. A comprehensive list of some successful biotization with PGPFs is presented in Table 1.

Apart from having many advantages, there are also some challenges in using these PGPFs during micropropagation. The major challenge is to prepare a pure fungal inoculum without any contamination. When such contaminated mix culture is exposed to plants, it may cause significant damage to the plants. Secondly, sometimes germination and growth of fungal spores on the Murashige and Skoog medium is quite difficult as it is not a favorable choice to grow [45]. To overcome this problem, the modification and optimization of the MS medium, suitable for coculture of the plant cell and PGPF, can be done.

5.2 Biotization with plant growth-promoting (rhizo)bacteria (PGPB or PGPR)

The beneficial effect and role of certain bacteria that can enhance plant growth and add to productivity were known for over a century. Over time, their application in plant tissue culture was first proved by Digat et al. [46] in Primrose.

TABLE 1 Effect of different PGPF (plant growth-promoting fungi) on plant propagules when used as Biotization agents during tissue culture process.

Name of fungi	Investigational plant	Micropropagation stage at which fungi was inoculated	Observed effect	References
Piriformospora indica	Chlorophytum sp.	Rooting	Increased root length, shoot length, fresh and dry weight, and leaf length and area	[14]
AM fungus Gigaspora rosea	Pyrus sp. clone HW 609	Rooting	Increased shoot length	[32]
Piriformospora indica	Boswellia serrata Roxb	Acclimatization	Increased shoot length, increase root length, increased leaf no., increased fresh weight, and dry weight	[33]
Trichoderma viride	Albizia amara	Acclimatization	Increased shoot length, root length, and leaf number	[34]
VAM fungi Glomus fasciculatim	Banana	Acclimatization	Increased shoot length, root biomass (fresh and dry weight)	[35]
Endomycorrhizal Fungi	Helleborus niger L.	Acclimatization	Increased dry weight of plant, dry weight of root	[36]
Piriformospora indica	Vernonia divergens	Rooting	Increased shoot-root length, shoot-root dry weight, leaf numbers. Increased anticancerous properties of plant extract	[37]
Piriformospora indica	Tinospora cordifolia, Vernonia divergens and Mucuna pruriens	Rooting	Increased shoot length, shoot numbers	[38]

Continued

TABLE 1 Effect of different PGPF (plant growth-promoting fungi) on plant propagules when used as Biotization agents during tissue culture process—cont'd

Name of fungi	Investigational plant	Micropropagation stage at which fungi was inoculated	Observed effect	References
Piriformospora indica	Terminalia bellerica Roxb.	Acclimatization	Increased biomass and root system	[26]
Serendipita indica	Hordeum vulgare L.	Acclimatization	Increased photosynthesis	[39]
Gigaspora margarita, Glomus etunicatumus	Scutellaria integrifolia	Acclimatization	Increase plant growth and survival	[40]
Glomus intraradice	Prunus sp.	Rooting	Increased shoot length and root number	[41]
Glomus mosseae	Juglans sp.	Rooting	Increase plant survive rate	[42]
Glomus intraradices	Fragaria x ananassa	Rooting	Extensive root systems and better shoot growth	[43]
Glomus coronatum	Prunus cerasifera	Acclimatization	Better plant growth and survival rate	[44]

They observed that when microshoots of Primrose (at the rooting stage and hardening stage) were exposed to *Pseudomonas putida* and *Pseudomonas fluorescens*, they exerted positive effects on growth and survival. A few years later, Elmeskaoui et al. [43] proved that biotization also improves the photosynthetic efficiency in in vitro grown plants, which leads to increased biomass accumulation. The biostimulatory action of PGPR is generally through the production of phytohormones. The production of auxins and cytokinins is a common phenomenon in PGPRs which is reported in more than 80% of rhizobacteria. The phytohormones produced by PGPBs mitigate the insufficient endogenous quantity of these hormones in microshoots during culture conditions [47]. The effects of phytohormones produced by a variety of PGPBs (viz., *Bradyrhizobium, Rhizobium, Bacillus, Microbacterium, Rhodococcus*, etc.) during tissue culture have been studied by Spaepen and Vanderleyden [48]. Rodríguez-Romero [49] studied the combined effect of PGPF and PGPR during the hardening phase of micropropagation of banana. They took *Glomus manihotis* (AMF) and the rhizobacteria consortium of *Bacillus* spp. and inoculated them with stage III plantlets, alone and in combination, and observed that the combined application of fungi and bacteria proved to be an effective inoculant and the resultant plants showed better growth in terms of more fresh weight, dry weight, shoot length, leaf area and required less time to become ready to get transplanted in the soil. Both the fungal and bacterial partners did not show any antagonistic effect toward each other. The positive effect of biotization was also reported in the banana plantlets at the rooting stage. In a study carried out by Mia [50], rhizobacteria were inoculated at the in vitro rooting stage of banana and the results were encouraging as a significant increase in the root length, root number, and root biomass was observed. Moreover, if exposure to PGPR continues during the subsequent steps of hardening and soil transfer, it results in the early attainment of the reproductive stage. Flowering takes place 3 weeks before compared to normal noninoculated plants and the yield also increases up to 51%. Similar results were also reported in potato micropropagation. When potato microcuttings were inoculated with *Azospirillum brasilense* during the in vitro rooting stage on a hormone-free medium, it significantly increased the IAA production, which ultimately led to the development of a solid root system and better survival in post vitro transfer [51]. Moreover, in another study, the *Azospirillum brasilense* strain Sp245 was inoculated at the hardening stage of potato, which not only results in an increase in the post vitro survival rate of plantlets (1.5 times) but also increases the weight of tubers by 30% [52]. The inoculation of rhizobacteria on a modified MS medium was studied on in vitro grown banana plantlets by Mahmood et al. [53]. The modification was done in terms of addition of salt (0.2% Sodium chloride), and it resulted not only in better growth and performance but also increased the synthesis of protein and chlorophyll. Likewise, biotization with *Azorhizobium caulinodans* in the rice plant gave better performance of plants in terms of biomass accumulation and grain yield as compared to uninoculated plants [54]. Other successful biotization trials were reported in many food crops such as wheat, where inoculation of *Azotobacter chroococcum* resulted in better root length [55]; potato and strawberry, where *Pseudomonas aureofaciens* was used which resulted in better growth [56]; watermelon, where the pseudomonas strain resulted in increased root length [57]; and maize, where the use of *Streptomyces griseorubens* and *Norcardiopsis alba* resulted in better growth of plantlets under phosphorus-deficient soil [58]. The addition of PGPR also resulted in the uptake of phosphorus in banana and rapeseed [59]. In banana, besides nutrient uptake, several physiological processes such as photosynthesis, stomatal conductance, and proline accumulation are also positively got affected by PGPR [60]. More recently, Lim et al. [61] testified that in the palm, biotized with *Herbaspirillum seropedicae* induces embryogenic callus formation and proliferation.

Biotization with PGPR have also proved to be an effective method to induce resistance against pathogens. The defense mechanism in plants are greatly affected by ethylene. Moreover, during culture conditions, attributes of growth and senescence are controlled by ethylene production, which is indirectly modulated through other growth hormones like auxins and cytokinins [62,63]. Ethylene induces defense in plants by activating several complex pathways, which ends in production of important defense molecules like jasmonic acid (JA), salicylic acid (SA), and abscisic acid [9,10]. Similarly, phenolic compounds and other secondary metabolites also play a significant role in pigmentation and provide protection against pathogens [64]. Biotization with *Pseudomonas* spp in oregano cultures resulted in an elevated level of phenolic compounds and chlorophyll [57]. Certain toxic compounds, antibiotics, and hydrolytic enzymes have also shown a negative effect on the growth of pathogens. These compounds act by either degrading the cell wall of pathogens or by suppressing pathogenic molecules [65]. Many PGPB synthesize the ACC (1-aminocyclopropane-1-carboxylate) deaminase enzyme. This bacterial enzyme promotes plant growth by decreasing the plant ethylene concentration. This enzyme converts amino cyclopropane carboxylate (ACC) into ammonia and α-ketobutyrate (α-KB), which leads to the scarcity of ACC, the prime precursor of ethylene in the cell. The role of ACC deaminase producing PGPR in alleviating stress and inducing resistance through reduced ethylene production has already been reported [66]. Beside this, many PGPB also synthesize certain volatile compounds which can promote callus organogenesis [67], enhance the photosynthesis efficiency [68], and offer better defense against abiotic stresses [69]. PGPMs also play a critical role in Induced Systemic Resistance (ISR) and hence provide protection against pathogens as well as insects. Many bacteria and fungi (like *Bacillus*, *Pseudomonas*, and *Trichoderma*) prepare the plant for any future attack of pathogen and save the energy of the plant which may get wasted unnecessarily otherwise during infection [70]. A comprehensive list of some successful biotization with PGPBs is presented in Table 2.

TABLE 2 Effect of different PGPBs (plant growth-promoting bacteria) on plant propagules when used as Biotization agents during tissue culture process.

Name of bacteria	Investigational plant	Micropropagation stage at which bacteria was inoculated	Observed effect	References
Methylobacterium salsuginis	Banana	Acclimatization	Increased plant height, girth and number of leaves, root length, lateral root, and biomass	[71]
Bacillus megaterium MiR-4	Vigna radiata (L.) R. Wilczek	Rooting	Increased shoot length and root length	[72]
Pseudomonas putida	Mentha piperita	Multiplication	Increased biomass and increased essential oil production	[73]
Pseudomonas putida	Pennisetum glaucum and Zea mays	Multiplication	Increased plant resistance against stress condition	[74]
Azospirillum brasilense Sp245	Prunus cerasifera	Acclimatization	Increased stem length and node number	[75]
Acinetobacter lwofii Acinetobacter haemoliticus Pseudomonas sp.	Crocus sativus L.	Seed germination	Increased seed germination rate and decrease germination time	[76]
Azospirillum brasilense (Sp245, S27, and SR8)	Solanum tuberosum L.	Multiplication and acclimatization	Increased shoot length and shoot number in in vitro and ex vitro condition	[51]

Continued

TABLE 2 Effect of different PGPBs (plant growth-promoting bacteria) on plant propagules when used as Biotization agents during tissue culture process—cont'd

Name of bacteria	Investigational plant	Micropropagation stage at which bacteria was inoculated	Observed effect	References
Bacillus subtilis M3 Trichoderma harzianum Gliocladium catenulatum	Fragaria × ananssa	Acclimatization	Overall growth promotion and disease resistance	[77]
Bacillus consortium (INR7, T4 and INR 937b)	Banana "Grande Naine"	Hardening	Increase all over growth of plant	[78]
Pseudomonas putida strains G2-8 and GI I-32	Glycine mux L.	Callus	Increased biomass of callus	[79]
Bacillus amyloliquifaciens UCMB5113	Arabidopsis thaliana (L.)	Rooting	Increased lateral root outgrowth and elongation and root hair formation	[80]
Bacillus	Platycladus orientalis	Multiplication	Increased biomass in water stress Condition	[81]
Pseudomonas putida	Arabidopsis thaliana	Rooting	Increased shoot number and root number	[82]

Herbaspirillum seropedicae strain Z78	Calluse Multiplication	Increased biomass of callus	[61]
Pseudomonas fluorescens	Rooting	Increase root length shoot dry weight, and leaf length	[14]
Agrobacterium rhizogene	Rootsing	Increase root and enhance alkannin and shikonin derivatives in hairy roots	[83]
Trichoderma viride, *P. fluorescens*	Acclimatization	Enhanced stress tolerance, increased root length and number, increased shoot length and higher leaf number and biomass	[34]
Pseudomonas fluorescens isolates	Acclimatization	Increased resistance and vigourness	[84]
Bacillus pumilus	Seed germination and acclimatization	Stem elongation	[85]
Methylobacterium salsuginis TNMB03	Acclimatization	Increased plant growth	[86]
Bacillus sp.	Acclimatization	Increased plant growth in stress condition and better survival	[87]
Serratia marcescens *Brevibacillus parabrevis*	Multiplication	Overall growth and better survival	[88]
Rhizophagus irregularis MUCL 41833	Rooting	Increased root number and root length	[89]

Plant species column (leftmost reading): *Elaeis guineensis* Jacq; *Chlorophytum* sp.; *Arnebia hispidissima*; *Albizia amara*; *Arachis hypogaea* L.; *Alnus glutinosa*; Banana "Grande Naine"; *Sorghum bicolor* (L.); *Guadua chacoensis*; *Medicago truncatula*

5.3 Biotization to elevate in vitro secondary metabolite production

Higher plants synthesize a variety of secondary metabolites (SM) such as alkaloids, flavonoids, steroids, terpenoids, quinones, lignans, and anthocyanins. These SM are immensely valued products, generally used as pharmaceuticals, agrochemicals, flavors, fragrances, colors, biopesticides, and food additives. For plants, these SM have no significant role in vital metabolic pathways for survival, but play an important role in the interaction of the plant with its environment and also act as defense chemicals [90]. Generally, these SM are accumulated in plants in a very low amount (less than 1%). Moreover, their synthesis is dependent on the physiological conditions (particularly stress conditions) and developmental stage of the plant [91]. Considering their immense economical values, their production through tissue culture methods was promoted. However, since in tissue culture, plant cells are grown under lavish environmental conditions, which do not favor SM synthesis, their accumulation further decreases. Several biotechnological approaches have been applied to increase SM production under in vitro conditions, but elicitation is recognized as the most viable technique for increasing the production of desirable SM from cell, organ, and plant culture [92,93]. The strategy through which SM production is stimulated through the involvement of any biotic or abiotic factor is called "elicitation" and the factor is called the "elicitor." Elicitors may be formed inside or outside plant cells and can be endogenous or exogenous in nature. Depending on their origin, they are classified as biotic or abiotic elicitors. Abiotic elicitors include UV irradiation, salts of heavy metals, and some other chemicals (like jasmonic acid, salicylic acid, etc.), while biotic elicitors may include chitin, chitosan, or glucans present in fungal cell wall materials, glycoprotein present in bacteria, and low molecular weight organic acids. Sometimes the entire microorganism (which may be a pathogen also) can act as elicitors. Several PGPB and PGPF have also been proved to be potential elicitors and their role in increasing SM production has been established. Inoculation of plant with PGPM (biotic elicitors) may significantly induce higher production of SM during tissue culture conditions. Several studies on the effect of PGPRs on higher SM production were carried out and it was found that PGPRs induce SM production through the ISR (induced systemic resistance) mechanism [94]. PGPR act as a potent activator of the key enzymes that are involved in the biosynthetic pathways of secondary metabolites [95].

PGPR also induce biosynthesis of certain other chemicals (like jasmonic acid and salicylic acid) in plants which acts as a transducer for elicitor signaling pathways and ultimately leads to the accumulation of secondary metabolites in plants [96]. Similarly, several PGPFs (mostly AM fungi) also induce increased production of SM when inoculated during culture conditions. The effect of different PGPMs on secondary metabolite production is depicted in Table 3.

TABLE 3 Effect of different PGPMs (plant growth-promoting microbes) on secondary metabolite production.

Name of PGPMs (bacteria/fungi)	Investigational plant	Compounds	References
Phaseolus vulgaris (F)	Colletotrichum lindemuthianum	Krevitone	[97]
Coriolus versicolor (F)	Rhodiola sachalinensis	Salidroside	[98]
Trichoderma viride (F)	Catharanthus roseus	Ajmalicine	[93]
Fusarium oxysporum (F)	Hypericum perforatum	Gymnemic acid	[99]
Trichoderma atroviride (B)	Salvia miltiorrhiza	Tanshinone	[100]
Bacillus polymyxa (B)	Stevia rebaudiana	Stevioside	[23]
Bacillus subtilis (B)	Ocimum basilicum	Eugeno	[101]
Pseudomonas fluorescens (B)	Catharanthus roseus	Ajmalicine	[93]
Azospirillum brasilense (B)	Origanum × majoricum	Thymol	[102]
Bradyrhizobium sp. (B)	Origanum majorana L.	Trans-sabinene hydrate	[103]
Datura stramonium (F)	Penicillium chrysogenum	Lubimin	[104]
Pseudomonas aeruginosa (B)	Scopolia parviflora	Scopolamine.	[105]
Pseudomonas fluorescens (B)	Rubus fruticosus	Phenolic compounds, flavonoids and anthocyanins	[106]
Pseudomonas fluorescens (B)	Glycine max	Isoflavone	[107]
Agrobacterium rhizogenes (B)	Althaea officinalis	Phenolics and flavonoids	[108]

Continued

TABLE 3 Effect of different PGPMs (plant growth-promoting microbes) on secondary metabolite production—cont'd

Name of PGPMs (bacteria/fungi)	Investigational plant	Compounds	References
Rhizobium radiobacter (B)	Hypericum perforatum	Xanthon	[109]
Glomus mosseae (F)	Andrographispaniculata	Andrographolide	[110]
Pythium aphanidermatum (F)	Coleus blume	Rosmarinic acid	[111]
Pythium aphanidermatum (F)	Daucus carota	p-Hydroxybenzoic acid	[112]
Yeast (F)	Hypericum perforatum	Hypericin	[113]
Fusarium sp. (F)	Euphorbia pekinensis	Euphol	[114]

6 Future prospects of biotization

Despite their immense potential, the application of PGPMs in tissue culture has not been exploited thoroughly. One of the reasons behind this is the response of PGPM, which varies not only from plant to plant but also at the explant level (e.g., root, stem, leaf, etc.) [12]. Further research is required to select efficient PGPMs as well as the development of an efficient protocol so that these organisms can be effectively used in tissue culture. Understand the signal recognition and transduction during natural conditions and culture conditions, which leads to association between the plant partner and microbial partner, is also a challenge. The combined use of more than two organisms can also be a viable option for better results [115]. Moreover, an efficient technique for the inoculation of PGPMs at different stages of tissue culture and a mechanism to control the population of microbes without affecting plant growth as well as potency of microbes should also be developed. Bio-nanotechnology can be used to address this problem and ready-to-use effective formulation of PGPMs can be developed [116]. Currently, very few reports are there on the use of bio-nanotechnology in tissue culture; hence, it will be a bright field to investigate and surely will add new development in the biotization process. Another important challenge in biotization is the low potency, specificity, and neutral response toward certain plant species. In this respect, the transgenic approach to develop a highly vigorous strain can be adopted in order to achieve a specific objective. The recent advancement taking place in biotechnology (such as functional genomics, bioinformatics, signaling in the rhizosphere, etc.) can be used as an effective tool in the development of transgenic microorganisms to confer better utilization of these PGPMs as biotization agents in tissue culture.

7 Conclusions

Micropropagation is an important tool for the large-scale production of elite germplasm of many economically important plants. However, due to certain limitations (like the high production cost and loss of plants during the acclimatization phase), the technique has not reached the planned success point. PGPMs are considered as potentially advantageous microbes for the plants in terms of better growth and providing protection against biotic and abiotic stress. The biofertilizer, biostimulation, and biocontrol properties of these PGPMs can be exploited to overcome the existing problems in tissue culture. An efficient consortium of PGPMs and the inoculation method can be developed which will not only decrease the production cost (by replacing the costly synthetic phytohormones) but also increase the survival percentage of plants after field transfer. However, intensive care should be taken to ensure that any vigorous plant/human pathogen should not contaminate the culture, particularly when the plant is used as raw food, as certain pathogens can stably survive in the tissue for a prolonged period both under in vitro as well as ex vitro conditions. In addition, a lot of research needs to be carried out in order to identify suitable PGPMs and to develop formulations for appropriate application in plant tissue culture. Moreover, the molecular understanding of relationship between plants and PGPMs will open new avenues in this prospective field of biotization.

References

[1] Y. Okon, C.A. Labandera-Gonzalez, Agronomic applications of *Azospirillum*: an evaluation of 20 years worldwide field inoculation, Soil Biol. Biochem. 26 (12) (1994) 1591–1601,- https://doi.org/10.1016/0038-0717(94)90311-5.

[2] T.M. Tien, M.H. Gaskins, D.H. Hubbell, Plant growth substances produced by *Azospirillum brasilense* and their effect on the growth of pearl millet (*Pennisetum americanum L.*), Appl. Environ. Microbiol. 37 (5) (1979) 1016–1024, https://doi.org/10.1128/AEM.37.5.1016-1024.1979.

[3] Y.G. Yanni, R.Y. Rizk, F.K. Abd El-Fattah, A. Squartini, V. Corich, A. Giacomini, F. de Bruijn, J. Rademaker, J. Maya-Flores, P. Ostrom, M. Vega-Hernandez, F.B. Dazzo, The beneficial plant growth-promoting association of *Rhizobium leguminosarum bv. trifolii* with rice roots, Funct. Plant Biol. 28 (9) (2001) 845–870, https://doi.org/10.1071/PP01069.

[4] U. Riaz, G. Murtaza, W. Anum, T. Samreen, M. Sarfraz, M.Z. Nazir, Plant growth-promoting rhizobacteria (PGPR) as biofertilizers and biopesticides, in: Microbiota and Biofertilizers, Springer, Cham, 2021, pp. 181–196.

[5] S. Ehteshamul-Haque, A. Ghaffar, Use of rhizobia in the control of root rot diseases of sunflower, okra, soybean and mungbean, J. Phytopathol. 138 (2) (1993) 157–163, https://doi.org/10.1111/J.1439-0434.1993.TB01372.X.

[6] U. Chauhan, A.K. Singh, D. Godani, S. Handa, P.S. Gupta, S. Patel, P. Joshi, Some natural extracts from plants as low-cost alternatives for synthetic PGRs in rose micropropagation, J. Appl. Hortic. 20 (2) (2018) 103–111, https://doi.org/10.37855/jah.2018.v20i02.19.

[7] S.S. Bidabadi, S. Mohan Jain, Cellular, molecular, and physiological aspects of *in vitro* plant regeneration, Plan. Theory 9 (6) (2020) 702.

[8] P. Joshi, S. Vyas, S.D. Purohit, Photosynthetic performance of shoots of *Feronia limonia* grown *in vitro* under carbon dioxide enriched environment, in: IV International Symposium on Acclimatization and Establishment of Micropropagated Plants, vol. 865, 2008, pp. 225–230.

[9] A.K. Singh, A. Singh, P. Joshi, Combined application of chitinolytic bacterium *Paenibacillus* sp. D1 with low doses of chemical pesticides for better control of *Helicoverpa armigera*, Int. J. Pest Manag. 62 (3) (2016) 222–227.

[10] A.K. Singh, T. Tala, M. Tanna, H. Patel, P. Sudra, D. Mungra, P. Joshi, Exogenous supply of salicylic acid results into better growth of banana propagules under *in vitro* conditions, Int. J. Rec. Sci. Res. 7 (3) (2016) 9488–9493.

[11] P. Joshi, S.D. Purohit, Genetic stability in micro-clones of 'Wood-Apple' derived from different pathways of micropropagation as revealed by RAPD and ISSR markers, in: VII International Symposium on *In Vitro* Culture and Horticultural Breeding, vol. 961, 2011, pp. 217–224.

[12] K. Sunita, I. Mishra, J. Mishra, J. Prakash, N.K. Arora, Secondary metabolites from halotolerant plant growth promoting rhizobacteria for ameliorating salinity stress in plants, Front. Microbiol. 11 (2020) 2619.

[13] M. Hossain, F. Sultana, S. Islam, Plant growth-promoting fungi (PGPF): phytostimulation and induced systemic resistance, in: Plant-Microbe Interactions in Agro-Ecological Perspectives, Springer, New York, 2017, pp. 135–191.

[14] S.K. Gosal, A. Karlupia, S.S. Gosal, I.M. Chhibba, A. Varma, Biotization with *Piriformospora indica* and *Pseudomonas fluorescens* improves survival rate, nutrient acquisition, field performance and saponin content of micropropagated *Chlorophytum* sp, Indian J. Biotechnol. 9 (2010) 289–297.

[15] D. Egamberdieva, S. Wirth, S.D. Bellingrath-Kimura, J. Mishra, N.K. Arora, Salt-tolerant plant growth promoting rhizobacteria for enhancing crop productivity of saline soils, Front. Microbiol. 10 (2019) 2791.

[16] M.K. Rai, Current advances in mycorrhization in micropropagation, In Vitro Cell. Dev. Biol. Plant 37 (2) (2001) 158–167, https://doi.org/10.1007/S11627-001-0028-8.

[17] P.E. Akin-Idowu, D.O. Ibitoye, O.T. Ademoyegun, Tissue culture as a plant production technique for horticultural crops, Afr. J. Biotechnol. 8 (16) (2009) 3782–3788, https://doi.org/10.4314/AJB.V8I16.62060.

[18] V. Bianciotto, P. Bonfante, Arbuscular mycorrhizal fungi: a specialised niche for rhizospheric and endocellular bacteria, Antonie Leeuwenhoek 81 (1–4) (2002) 365–371, https://doi.org/10.1023/A:1020544919072.

[19] S.C. Wu, Z.H. Cao, Z.G. Li, K.C. Cheung, M.H. Wong, Effects of biofertilizer containing N-fixer, P and K solubilizers and AM fungi on maize growth: a greenhouse trial, Geoderma 125 (1–2) (2005) 155–166.

[20] S. Gianinazzi, A. Gollotte, M.N. Binet, D. van Tuinen, D. Redecker, D. Wipf, et al., Agroecology: the key role of arbuscular mycorrhizas in ecosystem services, Mycorrhiza 20 (8) (2010) 519–530, https://doi.org/10.1007/s00572-010-0333-3.

[21] M.L. Adriano-Anaya, M. Salvador-Figueroa, J.A. Ocampo, I. García-Romera, Hydrolytic enzyme activities in maize (*Zea mays*) and sorghum (*Sorghum bicolor*) roots inoculated with *Gluconacetobacter diazotrophicus* and *Glomus intraradices*, Soil Biol. Biochem. 38 (5) (2006) 879–886.

[22] Q.S. Wu, R.X. Xia, Arbuscular mycorrhizal fungi influence growth, osmotic adjustment and photosynthesis of citrus under well-watered and water stress conditions, J. Plant Physiol. 163 (4) (2006) 417–425.

[23] F. Vafadar, R. Amooaghaie, M. Otroshy, Effects of plant-growth-promoting rhizobacteria and arbuscular mycorrhizal fungus on plant growth, stevioside, NPK, and chlorophyll content of *Stevia rebaudiana*, J. Plant Interact. 9 (1) (2014) 128–136.

[24] R.A. Streletskii, A.V. Kachalkin, A.M. Glushakova, A.M. Yurkov, V.V. Demin, Yeasts producing zeatin, PeerJ 7 (2) (2019) e6474, https://doi.org/10.7717/PEERJ.6474.

[25] S. Vyas, R. Nagori, S.D. Purohit, Root colonization and growth enhancement of micropropagated *Feronia limonia* (L.) Swingle by *Piriformospora indica*—a cultivable root endophyte, Int. J. Plant Dev. Biol. 2 (2) (2008) 128–132.

[26] R.K. Suthar, S.D. Purohit, Root colonization and improved growth performance of micropropagated *Terminalia bellerica* Roxb. plantlets inoculated with *Piriformospora indica* during ex vitro acclimatization, Int. J. Plant Dev. Biol. 2 (2) (2008) 133–136.

[27] M. Chen, M. Arato, L. Borghi, E. Nouri, D. Reinhardt, Beneficial services of arbuscular mycorrhizal fungi—from ecology to application, Front. Plant Sci. 9 (2018) 1270, https://doi.org/10.3389/FPLS.2018.01270/BIBTEX.

[28] A. Varma, H. Schüepp, Infectivity and effectiveness of *Glomus intraradices* on micropropagated plants, Mycorrhiza 5 (1) (1994) 29–37, https://doi.org/10.1007/BF00204017.

[29] J. Díez, J.L. Manjón, G.M. Kovács, C. Celestino, M. Toribio, Mycorrhization of vitro plants raised from somatic embryos of cork oak (*Quercus suber* L.), Appl. Soil Ecol. 15 (2) (2000) 119–123, https://doi.org/10.1016/S0929-1393(00)00087-1.

[30] N.S. Sahay, A. Varma, A biological approach towards increasing the rates of survival of micropropagated plants, Curr. Sci. 78 (2) (2000) 126–129.

[31] S. Harish, M. Kavino, N. Kumar, D. Saravanakumar, K. Soorianathasundaram, R. Samiyappan, Biohardening with plant growth promoting rhizosphere and endophytic bacteria induces systemic resistance against Banana bunchy top virus, Appl. Soil Ecol. 39 (2) (2008) 187–200, https://doi.org/10.1016/J.APSOIL.2007.12.006.

[32] C. Cordier, M.C. Lemoine, P. Lemanceau, V. Gianinazzi-Pearson, S. Gianinazzi, The beneficial rhizosphere: a necessary strategy for microplant production, Acta Hortic. 530 (2000) 259–268.

[33] R.K. Suthar, S.D. Purohit, Biopriming of micropropagated *Boswellia serrata* Roxb. Plantlets—role of endophytic root fungus *Piriformospora indica*, Indian J. Biotechnol. 11 (3) (2012) 304–308. http://nopr.niscair.res.in/handle/123456789/14571.

[34] G. Indravathi, P.S. Babu, Enhancing acclimatization of tissue cultured plants of *Albizia amara* by biotization, Int. J. Sci. Res. Biol. Sci. 6 (4) (2019) 43–50, https://doi.org/10.26438/IJSRBS/V6I4.4350.

[35] V. Mandhare, A. Suryawanshi, Biotization of banana tissue cultured plantlets with VAM fungi, Agric. Sci. Dig. 25 (2005) 65–67.

[36] A. Susek, J.P. Guillemin, M.C. Lemoine, A. Gollotte, A. Ivancic, J. Caneill, S. Gianinazzi, Effect of rhizosphere bacteria and endomycorrhizal fungi on the growth of Christmas Rose (*Helleborus niger L.*), Eur. J. Hortic. Sci. 75 (2) (2010) 85.

[37] R. Kumar, A. Prakash, S. Kumari, N. Kumari, S. Kumar, Cytotoxicity of ethanol extracts of *in vivo, in vitro* and biotized grown plants of *Vernonia divergens* on EAC cell lines, Int. J. Pharma. Phyto. Res. 6 (4) (2014) 678–684.

[38] R. Kumar, S. Kumar, Antibacterial, antidiabetic and anticancer activities of natural products of some medicinal plants of Muzaffarpur district, J. Emerg. Technol. Innov. Res 5 (11) (2018) 697–702.

[39] M. Sepehri, M.R. Ghaffari, M. Khayam Nekoui, E. Sarhadi, A. Moghadam, B. Khatabi, G. Hosseini Salekdeh, Root endophytic fungus *Serendipita indica* modulates barley leaf blade proteome by increasing the abundance of photosynthetic proteins in response to salinity, J. Appl. Microbiol. 131 (4) (2021) 1870–1889, https://doi.org/10.1111/JAM.15063.

[40] N. Joshee, S.R. Mentreddy, A.K. Yadav, Mycorrhizal fungi and growth and development of micropropagated *Scutellaria integrifolia* plants, Ind. Crop. Prod. 25 (2) (2007) 169–177, https://doi.org/10.1016/J.INDCROP.2006.08.009.

[41] C. Cordier, A. Trouvelot, S. Gianinazzi, V. Gianinazzi-Pearson, Arbuscular mycorrhiza technology applied to micropropagated *Prunus avium* and to protection against *Phytophthora cinnamomi*, Agronomie 16 (10) (1996) 679–688, https://doi.org/10.1051/AGRO:19961013.

[42] R. Dolcet-Sanjuan, E. Claveria, A. Camprubí, V. Estaún, C. Calvet, Micropropagation of walnut trees (*Juglans regia L*) and response to arbuscular mycorrhizal inoculation, Agronomie 16 (10) (1996) 639–645.

[43] A. Elmeskaoui, J.P. Damont, M.J. Poulin, Y. Piché, Y. Desjardins, A tripartite culture system for endomycorrhizal inoculation of micropropagated strawberry plantlets *in vitro*, Mycorrhiza 5 (5) (1995) 313–319, https://doi.org/10.1007/BF00207403.

[44] P. Fortuna, S. Citernesi, S. Morini, M. Giovannetti, F. Loreti, Infectivity and effectiveness of different species of arbuscular mycorrhizal fungi in micropropagated plants of Mr S 2/5 plum rootstock, Agronomie 12 (10) (1992) 825–829, https://doi.org/10.1051/AGRO:19921015.

[45] K.L. Rana, D. Kour, I. Sheikh, A. Dhiman, N. Yadav, A.N. Yadav, A.K. Saxena, Endophytic fungi: biodiversity, ecological significance, and potential industrial applications, in: Recent Advancement in White Biotechnology Through Fungi, Springer, Cham, 2019, pp. 1–62, https://doi.org/10.1007/978-3-030-10480-1_1.

[46] B. Digat, V. Hermelin, P. Brochard, M. Touzet, Interest of bacterized synthetics substrates milcap® for *in vitro* culture, Acta Hortic. 212 (1987) 375–378, https://doi.org/10.17660/ACTAHORTIC.1987.212.57.

[47] R. de Souza, A. Ambrosini, L.M.P. Passaglia, Plant growth-promoting bacteria as inoculants in agricultural soils, Genet. Mol. Biol. 38 (4) (2015) 401–419, https://doi.org/10.1590/S1415-475738420150053.

[48] S. Spaepen, J. Vanderleyden, Auxin and plant-microbe interactions, Cold Spring Harb. Perspect. Biol. 3 (4) (2011) 1–13, https://doi.org/10.1101/CSHPERSPECT.A001438.

[49] A.S. Rodríguez-Romero, E. Badosa, E. Montesinos, M.C. Jaizme-Vega, Growth promotion and biological control of root-knot nematodes in micropropagated banana during the nursery stage by treatment with specific bacterial strains, Ann. Appl. Biol. 152 (1) (2008) 41–48,- https://doi.org/10.1111/J.1744-7348.2007.00189.X.

[50] M.A.B. Mia, Beneficial Effects of Rhizobacterial Inoculation on Nutrient Uptake, Growth and Yield of Banana *(Musa spp.)*, (Doctoral dissertation, Ph. D. thesis,, Faculty of Agriculture, UPM, Malaysia, 2002.

[51] K.Y. Kargapolova, G.L. Burygin, O.V. Tkachenko, N.V. Evseeva, Y.V. Pukhalskiy, A.A. Belimov, Effectiveness of inoculation of *in vitro*-grown potato microplants with rhizosphere bacteria of the genus *Azospirillum*, Plant Cell Tissue Organ Cult. 141 (2) (2020) 351–359,- https://doi.org/10.1007/S11240-020-01791-9.

[52] O.V. Tkachenko, N.V. Evseeva, N.V. Boikova, L.Y. Matora, G.L. Burygin, Y.V. Lobachev, S.Y. Shchyogolev, Improved potato microclonal reproduction with the plant growth-promoting rhizobacteria *Azospirillum*, Agron. Sustain. Dev. 35 (3) (2015) 1167–1174.

[53] M. Mahmood, Z.A. Rahman, H.M. Saud, Z. Shamsuddin, S. Subramaniam, Responses of banana plantlets to rhizobacteria inoculation under salt stress condition, Am. Eurasian J. Sustain. Agric. 3 (3) (2009) 290–305.

[54] M. Senthilkumar, M. Madhaiyan, S.P. Sundaram, H. Sangeetha, S. Kannaiyan, Induction of endophytic colonization in rice (*Oryza sativa* L.) tissue culture plants by *Azorhizobium caulinodans*, Biotechnol. Lett. 30 (2008) 1477–1487, https://doi.org/10.1007/s10529-008-9693-6.

[55] D. Andressen, I. Manoochehri, S. Carletti, B. Llorente, M. Tacoronte, M. Vielma, Optimization of the *in vitro* proliferation of jojoba (*Simmondsia chinensis* Schn.) by using rotable central composite design and inoculation with rhizobacteria, Bioagro 21 (2009) 41–48.

[56] N.S. Zakharchenko, V.V. Kochetkov, Y.I. Buryanov, A.M. Boronin, Effect of rhizosphere bacteria *Pseudomonas aureofaciens* on the resistance of micropropagated plants to phytopathogens, Appl. Biochem. Microbiol. 47 (2011) 661–666.

[57] J. Nowak, Benefits of *in vitro* "biotization" of plant tissue cultures with microbial inoculants, In Vitro Cell. Dev. Biol. Plant 34 (2) (1998) 122–130, https://doi.org/10.1007/BF02822776.

[58] A. Soumare, A.G. Diedhiou, M. Thuita, M. Hafidi, Y. Ouhdouch, S. Gopalakrishnan, L. Kouisni, Exploiting biological nitrogen fixation: a route towards a sustainable agriculture, Plan. Theory 9 (8) (2020) 1011, https://doi.org/10.3390/PLANTS9081011.

[59] Y. Bashan, M. Moreno, E. Troyo, Growth promotion of the seawater-irrigated oilseed halophyte *Salicornia bigelovii* inoculated with mangrove rhizosphere bacteria and halotolerant *Azospirillum* spp, Biol. Fertil. Soils 32 (4) (2000) 265–272, https://doi.org/10.1007/S003740000246.

[60] M.A.B. Mia, Z.H. Shamsuddin, W. Zakaria, M. Marziah, Growth and physiological attributes of hydroponically-grown bananas inoculated with plant growth promoting rhizobacteria, Trans. Malaysian Soc. Plant Physiol. 9 (2000) 324–327.

[61] S.L. Lim, S. Subramaniam, I. Zamzuri, H.G. Amir, Biotization of *in vitro* calli and embryogenic calli of oil palm (*Elaeis guineensis Jacq.*) with diazotrophic bacteria *Herbaspirillum*

seropedicae (Z78), Plant Cell Tissue Organ Cult. 127 (1) (2016) 251–262, https://doi.org/10.1007/S11240-016-1048-8.

[62] N. Iqbal, N.A. Khan, A. Ferrante, A. Trivellini, A. Francini, M.I.R. Khan, Ethylene role in plant growth, development and senescence: interaction with other phytohormones, Front. Plant Sci. 8 (2017) 475, https://doi.org/10.3389/FPLS.2017.00475.

[63] N.A. Khan, M.R. Mir, R. Nazar, S. Singh, The application of ethephon (an ethylene releaser) increases growth, photosynthesis and nitrogen accumulation in mustard (*Brassica juncea* L.) under high nitrogen levels, Plant Biol. 10 (5) (2008) 534–538, https://doi.org/10.1111/J.1438-8677.2008.00054.X.

[64] V. Lattanzio, V.M.T. Lattanzio, A. Cardinali, Role of phenolics in the resistance mechanisms of plants against fungal pathogens and insects, Phytochem. Adv. Res. 661 (2) (2006) 23–67.

[65] S. Compant, B. Duffy, J. Nowak, C. Clément, E.A. Barka, Use of plant growth-promoting bacteria for biocontrol of plant diseases: principles, mechanisms of action, and future prospects, Appl. Environ. Microbiol. 71 (9) (2005) 4951–4959, https://doi.org/10.1128/AEM.71.9.4951-4959.2005.

[66] S. Gupta, S. Pandey, ACC deaminase producing bacteria with multifarious plant growth promoting traits alleviates salinity stress in French bean (*Phaseolus vulgaris*) plants, Front. Microbiol. 10 (2019) 1506, https://doi.org/10.3389/FMICB.2019.01506.

[67] S. Gopinath, K.S. Kumaran, M. Sundararaman, A new initiative in micropropagation: airborne bacterial volatiles modulate organogenesis and antioxidant activity in tobacco (*Nicotiana tabacum* L.) callus, In Vitro Cell. Dev. Biol. Plant 51 (5) (2015) 514–523,-https://doi.org/10.1007/S11627-015-9717-6.

[68] X. Xie, H. Zhang, P. Pare, Sustained growth promotion in *Arabidopsis* with long-term exposure to the beneficial soil bacterium *Bacillus subtilis* (GB03), Plant Signal. Behav. 4 (10) (2009) 948–953.

[69] T. Orlikowska, K. Nowak, B. Reed, Bacteria in the plant tissue culture environment, Plant Cell Tissue Organ Cult. 128 (3) (2017) 487–508, https://doi.org/10.1007/S11240-016-1144-9.

[70] L.K.T. Al-Ani, A.M. Mohammed, Versatility of Trichoderma in plant disease management, in: Molecular Aspects of Plant Beneficial Microbes in Agriculture, Academic Press, Cambridge, 2020, pp. 159–168, https://doi.org/10.1016/B978-0-12-818469-1.00013-4.

[71] P. Pushpakanth, R. Krishnamoorthy, R. Anandham, M. Senthilkumar, Biotization of tissue culture banana plantlets with *Methylobacterium salsuginis* to enhance the survival and growth under greenhouse and open environment condition, J. Environ. Biol. 42 (6) (2021) 1452–1460.

[72] B. Ali, A.N. Sabri, K. Ljung, S. Hasnain, Quantification of indole-3-acetic acid from plant associated *Bacillus* spp. and their phytostimulatory effect on Vigna radiata (L.), World J. Microbiol. Biotechnol. 25 (3) (2008) 519–526, https://doi.org/10.1007/S11274-008-9918-9.

[73] M.V. Santoro, L.R. Cappellari, W. Giordano, E. Banchio, Plant growth-promoting effects of native *Pseudomonas* strains on *Mentha piperita* (peppermint): an *in vitro* study, Plant Biol. 17 (6) (2015) 1218–1226, https://doi.org/10.1111/PLB.12351.

[74] T. Patel, M. Saraf, Biosynthesis of phytohormones from novel rhizobacterial isolates and their in vitro plant growth-promoting efficacy, J. Plant Interact. 12 (1) (2017) 480–487.

[75] L. Vettori, A. Russo, C. Felici, G. Fiaschi, S. Morini, A. Toffanin, Improving micropropagation: effect of *Azospirillum brasilense* Sp245 on acclimatization of rootstocks of fruit tree, J. Plant Interact. 5 (4) (2010) 249–259.

[76] J.A. Parray, A.N. Kamili, Z.A. Reshi, R.A. Qadri, S. Jan, Interaction of rhizobacterial strains for growth improvement of *Crocus sativus* L. under tissue culture conditions, Plant Cell Tissue Organ Cult. 121 (2) (2015) 325–334.

[77] M. Vestberg, S. Kukkonen, K. Saari, P. Parikka, J. Huttunen, L. Tainio, S. Gianinazzi, Micro-bial inoculation for improving the growth and health of micropropagated strawberry, Appl. Soil Ecol. 27 (3) (2004) 243–258.

[78] M. del Carmen Jaizme-Vega, A.S. Rodríguez-Romero, M.S.P. Guerra, Potential use of rhi-zobacteria from the *Bacillus* genus to stimulate the plant growth of micropropagated bananas, Fruits 59 (2) (2004) 83–90.

[79] C.P. Chanway, L.M. Nelson, Tissue culture bioassay for plant growth promoting rhizobac-teria, Soil Biol. Biochem. 23 (4) (1991) 331–333.

[80] S. Asari, D. Tarkowská, J. Rolčík, O. Novák, D.V. Palmero, S. Bejai, J. Meijer, Analysis of plant growth-promoting properties of *Bacillus amyloliquefaciens* UCMB5113 using *Arabi-dopsis thaliana* as host plant, Planta 245 (1) (2017) 15–30, https://doi.org/10.1007/S00425-016-2580-9.

[81] F. Liu, S. Xing, H. Ma, Z. Du, B. Ma, Cytokinin-producing, plant growth-promoting rhizobacteria that confer resistance to drought stress in *Platycladus orientalis* container seedlings, Appl. Micro-biol. Biotechnol. 97 (20) (2013) 9155–9164, https://doi.org/10.1007/S00253-013-5193-2.

[82] E. Arslan, Ö. Akkaya, Biotization of *Arabidopsis thaliana* with *Pseudomonas putida* and assessment of its positive effect on *in vitro* growth, In Vitro Cell. Dev. Biol. Plant 56 (2) (2020) 184–192.

[83] B. Singh, R.A. Sharma, Yield enhancement of phytochemicals by *Azotobacter chroococcum* biotization in hairy roots of *Arnebia hispidissima*, Ind. Crop. Prod. 81 (2016) 169–175, https://doi.org/10.1016/J.INDCROP.2015.11.068.

[84] P. Kalaiarasan, P.L. Lakshmanna, R. Samiyappan, Biotization of *Pseudomonas fluorescens* isolates in groundnut (*Arachis hypogaea* L.) against root-knot nematode, *Meloidogyne are-naria*, Indian J. Nematol. 36 (1) (2006) 1–5.

[85] F.J. Gutiérrez-Mañero, B. Ramos-Solano, A. Probanza, J. Mehouachi, F.R. Tadeo, M. Talon, The plant-growth-promoting rhizobacteria *Bacillus pumilus* and *Bacillus licheniformis* pro-duce high amounts of physiologically active gibberellins, Physiol. Plant. 111 (2) (2001) 206–211, https://doi.org/10.1034/J.1399-3054.2001.1110211.X.

[86] M. del Rosario Espinoza-Mellado, E.O. López-Villegas, M.F. López-Gómez, A.V. Rodríguez-Tovar, M. García-Pineda, A. Rodríguez-Dorantes, Biotization and *in vitro* plant cell cultures: plant endophyte strategy in response to heavy metals knowledge in assisted phytoremediation, in: Microbe Mediated Remediation of Environmental Contaminants, Woodhead Publishing, UK, 2021, pp. 27–36, https://doi.org/10.1016/B978-0-12-821199-1.00003-1.

[87] M. Umapathi, C.N. Chandrasekhar, A. Senthil, T. Kalaiselvi, R. Santhi, R. Ravikesavan, Effect of bacterial endophytes inoculation on morphological and physiological traits of sor-ghum (*sorghum bicolor* (L.) Moench) under drought, J. Pharmacogn. Phytochem. 10 (2) (2021) 1021–1028.

[88] C. Belincanta, G. Botelho, T.S. Ornellas, J. Zappelini, M.P. Guerra, Characterization of the endophytic bacteria from *in vitro* cultures of *Dendrocalamus asper* and *Bambusa oldhamii* and assessment of their potential effects in *in vitro* co-cultivated plants of *Guadua chacoensis* (Bambusoideae, Poaceae), In Vitro Cell. Dev. Biol. Plant (2021) 1–11.

[89] R. El Hilali, R. Bouamri, P. Crozilhac, M. Calonne, S. Symanczik, L. Ouahmane, S. Declerck, *In vitro* colonization of date palm plants by *Rhizophagus irregularis* during the rooting stage, Symbiosis 84 (1) (2021) 83–89, https://doi.org/10.1007/S13199-021-00768-2.

[90] A. Ramakrishna, G.A. Ravishankar, Influence of abiotic stress signals on secondary metab-olites in plants, Plant Signal. Behav. 6 (11) (2011) 1720–1731, https://doi.org/10.4161/PSB.6.11.17613.

[91] K.M. Oksman-Caldentey, D. Inzé, Plant cell factories in the post-genomic era: new ways to produce designer secondary metabolites, Trends Plant Sci. 9 (9) (2004) 433–440, https://doi.org/10.1016/J.TPLANTS.2004.07.006.

[92] Z. Angelova, S. Georgiev, W. Roos, Elicitation of plants, Biotechnol. Biotechnol. Equip. 20 (2006) 72–83.

[93] A. Namdeo, S. Patil, D.P. Fulzele, Influence of fungal elicitors on production of ajmalicine by cell cultures of *Catharanthus roseus*, Biotechnol. Prog. 18 (1) (2002) 159–162, https://doi.org/10.1021/BP0101280.

[94] S. Sekar, D. Kandavel, Interaction of plant growth promoting rhizobacteria (PGPR) and endophytes with medicinal plants—new avenues for phytochemicals, J. Phytol. 2 (7) (2010) 91–100.

[95] C. Chen, R.R. Bélanger, N. Benhamou, T.C. Paulitz, Defense enzymes induced in cucumber roots by treatment with plant growth-promoting rhizobacteria (PGPR) and *Pythium aphanidermatum*, Physiol. Mol. Plant Pathol. 56 (1) (2000) 13–23, https://doi.org/10.1006/PMPP.1999.0243.

[96] M.J. Mueller, W. Brodschelm, E. Spannagl, M.H. Zenk, Signaling in the elicitation process is mediated through the octadecanoid pathway leading to jasmonic acid, Proc. Natl. Acad. Sci. 90 (16) (1993) 7490–7494, https://doi.org/10.1073/PNAS.90.16.7490.

[97] R.A. Dixon, P.M. Dey, D.L. Murphy, I.M. Whitehead, Dose responses for *Colletotrichum lindemuthianum* elicitor-mediated enzyme induction in French bean cell suspension cultures, Planta 151 (3) (1981) 272–280, https://doi.org/10.1007/BF00395180.

[98] X. Zhou, Y. Wu, X. Wang, B. Liu, H. Xu, Salidroside production by hairy roots of Rhodiola sachalinensis obtained after transformation with Agrobacterium rhizogenes, Biol. Pharm. Bull. 30 (3) (2007) 439–442, https://doi.org/10.1248/BPB.30.439.

[99] B. Chodisetti, K. Rao, S. Gandi, A. Giri, Improved gymnemic acid production in the suspension cultures of *Gymnema sylvestre* through biotic elicitation, Plant Biotechnol. Rep. 7 (4) (2013) 519–525, https://doi.org/10.1007/S11816-013-0290-3.

[100] Q. Ming, C. Su, C. Zheng, M. Jia, Q. Zhang, H. Zhang, K. Rahman, T. Han, L. Qin, Elicitors from the endophytic fungus *Trichoderma atroviride* promote *Salvia miltiorrhiza* hairy root growth and tanshinone biosynthesis, J. Exp. Bot. 64 (18) (2013) 5687–5694, https://doi.org/10.1093/JXB/ERT342.

[101] E. Banchio, X. Xie, H. Zhang, P.W. Paré, Soil bacteria elevate essential oil accumulation and emissions in sweet basil, J. Agric. Food Chem. 57 (2) (2009) 653–657, https://doi.org/10.1021/JF8020305.

[102] B. Erika, P.C. Bogino, M. Santoro, L. Torres, J. Zygadlo, W. Giordano, Systemic induction of monoterpene biosynthesis in *Origanum* × *majoricum* by soil bacteria, J. Agric. Food Chem. 58 (1) (2009) 650–654, https://doi.org/10.1021/JF9030629.

[103] E. Banchio, P.C. Bogino, J. Zygadlo, W. Giordano, Plant growth promoting rhizobacteria improve growth and essential oil yield in *Origanum majorana* L, Biochem. Syst. Ecol. 36 (10) (2008) 766–771.

[104] I.M. Whitehead, A.L. Atkinson, D.R. Threlfall, Studies on the biosynthesis and metabolism of the phytoalexin, lubimin and related compounds in *Datura stramonium* L, Planta 182 (1) (1990) 81–88, https://doi.org/10.1007/BF00239988.

[105] H.Y. Jung, S.M. Kang, Y.M. Kang, M.J. Kang, D.J. Yun, J.D. Bahk, J.K. Yang, M.S. Choi, Enhanced production of scopolamine by bacterial elicitors in adventitious hairy root cultures of *Scopolia parviflora*, Enzym. Microb. Technol. 33 (7) (2003) 987–990, https://doi.org/10.1016/S0141-0229(03)00253-9.

[106] D. García-Seco, A. Bonilla, E. Algar, A. García-Villaraco, J.G. Mañero, B. Ramos-Solano, Enhanced blackberry production using *Pseudomonas fluorescens* as elicitor, Agron. Sustain. Dev. 33 (2) (2012) 385–392, https://doi.org/10.1007/S13593-012-0103-Z.

[107] E. Algar, F.J. Gutierrez-Mañero, A. Bonilla, J.A. Lucas, W. Radzki, B. Ramos-Solano, *Pseudomonas fluorescens* N21.4 metabolites enhance secondary metabolism isoflavones in soybean (*Glycine max*) Calli cultures, J. Agric. Food Chem. 60 (44) (2012) 11080–11087, https://doi.org/10.1021/JF303334Q.

[108] P. Tavassoli, A. Safipour Afshar, Influence of different *Agrobacterium rhizogenes* strains on hairy root induction and analysis of phenolic and flavonoid compounds in marshmallow (*Althaea officinalis* L.), 3 Biotech 8 (8) (2018) 1–8.

[109] G. Franklin, L.F.R. Conceição, E. Kombrink, A.C.P. Dias, Xanthone biosynthesis in *Hypericum perforatum* cells provides antioxidant and antimicrobial protection upon biotic stress, Phytochemistry 70 (1) (2009) 60–68, https://doi.org/10.1016/J.PHYTOCHEM.2008.10.016.

[110] J. Arpana, D.J. Bagyaraj, Response of kalmegh to an arbuscular mycorrhizal fungus and a plant growth promoting rhizomicroorganism at two levels of phosphorus fertilizer, Am.-Eurasian J. Agric. Environ. Sci. 2 (2007) 33–38.

[111] E. Szabo, A. Thelen, M. Petersen, Fungal elicitor preparations and methyl jasmonate enhance rosmarinic acid accumulation in suspension cultures of *Coleus blumei*, Plant Cell Rep. 8 (6) (1999) 485–489, https://doi.org/10.1007/S002990050608.

[112] J.P. Schnitzler, J. Madlung, A. Rose, H. Ulrich Seitz, Biosynthesis of p-hydroxybenzoic acid in elicitor-treated carrot cell cultures, Planta 188 (4) (1992) 594–600, https://doi.org/10.1007/BF00197054.

[113] A. Kirakosyan, H. Hayashi, K. Inoue, A. Charchoglyan, H. Vardapetyan, Stimulation of the production of hypericins by mannan in *Hypericum perforatum* shoot cultures, Phytochemistry 53 (3) (2000) 345–348, https://doi.org/10.1016/S0031-9422(99)00496-3.

[114] F.K. Gao, Y.H. Yong, C.C. Dai, Effects of endophytic fungal elicitor on two kinds of terpenoids production and physiological indexes in *Euphorbia pekinensis* suspension cells, J. Med. Plant Res. 5 (18) (2011) 4418–4425, https://doi.org/10.5897/JMPR.9000519.

[115] A. Soumare, K. Boubekri, K. Lyamlouli, M. Hafidi, Y. Ouhdouch, L. Kouisni, Efficacy of phosphate solubilizing *Actinobacteria* to improve rock phosphate agronomic effectiveness and plant growth promotion, Rhizosphere 17 (2020) 100284, https://doi.org/10.1016/j.rhisph.2020.100284.

[116] S. Kumari, I. Mishara, J. Mishara, J. Prakash, N.K. Arona, Secondary metabolites from halotolerant plant growth promoting rhizobacteria for ameliorating salinity stress in plants, Front. Microbiol. 11 (2020), 567768, https://doi.org/10.3389/FMICB.2020.567768.

Chapter 14

Advantageous features of plant growth-promoting microorganisms to improve plant growth in difficult conditions

Mukesh Meena[a], Garima Yadav[a], Priyankaraj Sonigra[a], Adhishree Nagda[a], Tushar Mehta[a], Prashant Swapnil[b], Avinash Marwal[c], and Andleeb Zehra[d]

[a]*Laboratory of Phytopathology and Microbial Biotechnology, Department of Botany, Mohanlal Sukhadia University, Udaipur, Rajasthan, India*, [b]*School of Basic Sciences, Department of Botany, Central University of Punjab, Bathinda, Punjab, India*, [c]*Department of Biotechnology, Mohanlal Sukhadia University, Udaipur, Rajasthan, India*, [d]*Laboratory of Mycopathology and Microbial Technology, Department of Botany, Centre of Advanced Study in Botany, Institute of Science, Banaras Hindu University, Varanasi, Uttar Pradesh, India*

1 Introduction

The microbial community is a major uncovered pool of the Earth's biodiversity, of which only a tiny fraction is known [1]. Microbes comprise the major part of life on the planet and also play a vital role in the life of other living organisms [2]. Nowadays, climate change is appearing as a major issue because it affects life forms on the Earth. These changes are usually seen in the morphology, metabolic rate, photosynthetic rate, activity, and interactions of plant roots with their surrounding plants and microorganisms [3,4]. Growth and development of plants diminishes due to disturbed conditions of soils such as water deficiency, acidity, salinity, shortage of organic and inorganic nutrients mainly essential micronutrients (P, Ca, Mg, and molybdenum). Microorganisms that live in this type of soil enable plants to cope with these conditions [5–9]. The plant roots affect the rhizosphere through the exudation of substances in soil and influence microbial activity. Plant roots offer niches for a multitude of microorganisms, including viruses, fungi, bacteria, protists, and nematodes. Among different phyla, some of the phyla that are included in the plant growth-promoting microbe family are Cyanobacteria, Actinobacteria, Firmicutes, Bacteroidetes, *Bacillus* spp.,

Plant-Microbe Interaction—Recent Advances in Molecular and Biochemical Approaches
https://doi.org/10.1016/B978-0-323-91876-3.00019-1
279

Pseudomonas spp., and Proteobacteria [10–13]. Microbes are attached to the root surface and avail possible benefits from the plant. Generally, plants secrete a carbon-rich source for their surrounding microbes present in the rhizosphere [14]. Plant exudes help in co-associations and interaction between plants and microbes. These types of interaction have important roles in plant health and productivity promotion in natural environments [15]. Under pressure of biotic and abiotic stressors, plants exude different types of chemical molecules which are sensed by the surrounding microbes. In response, microbes exude some other types of chemical compounds which initiate defense responses in plants [16]. PGP microbes can be grouped based on their location in plant parts such as near the roots (rhizospheric microbes), leaves (epiphytic microbes), and inside the tissue (endophytic microbes) [9,17–20]. Microbes enter plant via root hairs and attach to the apical root zone and colonize the plant root. They might be transmitted vertically and horizontally means from parent to offspring and among individuals, respectively. Some well-known examples of rhizospheric and phyllosperic microbial species are as follows: *Arthrobacter, Bacillus, Enterobacter, Paenibacillus, Aspergillus, Penicillium, Flavobacterium, Haloferax, Methylobacterium, Piriformospora, Rhizobium, Pseudomonas, Achromobacter, Micromomospora, Streptomyces, Micrococcus, Serratia*, etc. [21]. The microbiomes linked with different plants perform multifunctions such as synthesis of phytohormones (cytokinins, indole-3-acetic acid (IAA), abscisic acid, and gibberellins), siderophores, antibiotics (pyrrolnitrin, pyoluteorin, pyocyanin, phenazine-1-carboxylic acid, 2,4-diacetylphloroglucinol, kanosamine, and neomycin A), hydrocyanic acid (HCN), ACC deaminase, ammonia, lytic enzymes (chitinase, β-1,3-glucanase, lipase protease), micronutrient solubilization (P, K, Fe, and Zn), and biological N_2 fixation [22]. Moreover, PGP microbes indirectly hinder disease development through the production of secondary metabolites and also stimulate induced systemic resistance (ISR) in plants [23]. The use of PGP microbes as biofertilizer and biopecticide decrease the use of chemical fertilizers which support a healthy environment. The Earth involves a completely different and extreme environmental condition with numerous microorganisms which have the ability to grow in that particular environment (ecosystem). These microorganisms can survive in extreme low and high temperatures, pH, low water availability hypersalinity, and hence are referred to as extremophiles. As extremophiles can face adverse environment, they can also be used for plant growth promotion under stressful condition. There are some potent enzymes that have been isolated from extremophiles, such as cellulase, pectinase, amylase, lipase, laccase, chitinase, protease, xylanase, β-glucosidase, and β-galactosidase, and used as biocontrol agents in agriculture [24,25].

2 Plant growth-promoting microorganisms (PGPMs) and plant growth

Plant-microbes interaction is a crucial step for growth of plants and their development as well as soil health. Basically, there are three modes of plant-microbe interactions, i.e., epiphytic, endophytic, and rhizospheric. The rhizospheric microbes are capable of attaching to the root surfaces to gain a lot of benefits from root exudates. The type of microorganism in the rhizospheric region depends upon the type of soil, its pH, moisture, temperature, age and plant conditions [26]. Genera like *Aspergillus, Acinetobacter, Arthrobacter, Azospirillum, Burkholderia, Bacillus, Erwinia, Enterobacter, Haloarcula, Halobacterium, Flavobacterium, Haloferax, Methylobacterium, Piriformospora, Pseudomonas, Rhizobium, Penicillium,* and *Serratia* have been identified from various crop plants [27]. Diffused parts of leaves contain the essential nutrient factors such as amino acids, fructose, sucrose, and glucose; these types of specialized habitats may provide a niche for processes like nitrogen fixation and substance secretions which have the capacity of plant growth promotion [9]. Many phyllospheric microbes have been identified in which *Agrobacterium, Bacillus, Burkholderia, Microbiospora, Rhizobium,* and *Xanthomonas* are the dominant genera. These phyllospheric microbes are also involved in controlling airborne pathogens [28,29]. When we talk about endophytic microflora, the representative groups are Fungi like *Trichoderma, Curvularia,* etc., Actinomycetes, bacterial genera like *Azoarcus, Collimonas, Enterobacter, Planomonospora,* etc. These endophytic microfloras show various activities like antimicrobial, insecticidal, antioxidant, etc. that are beneficial for plant health [30]. In general, studies show that representative microorganisms from archaea (Euryarchaeota) bacteria (*Acidobacteria, Bacteroidetes, Actinobacteria, Deinococcus-Thermus, Proteobacteria,* and *Firmicutes*), and fungal genera belonging to Ascomycota and Basidiomycota are involved in the interaction between plants and microbes and support plant growth and development. These plant growth-promoting microorganisms increase yields of agricultural products, maintain the nutrients via regulating biogeochemical cycles, and perform the function of homeostasis inside the root ecosystem. Factors such as drought, salinity, extreme temperature events, flooding, heavy metal pollution, and ultraviolet irradiation create abiotic stress in plants [31]. Heavy metals impart toxic impacts on all living beings blocking the functional groups of organic compounds and disrupting the essential metabolic pathways [32,33]. PGPMs have the capacity to alleviate this stress by providing minerals and ions such as nitrogen, phosphorus, iron, etc.; similarly free living or symbiotic microbes also synthesize some compounds which enhance plant growth in harsh conditions; they induce synthesis of plant growth regulators or plant hormones such as IAA, gibberellins, cytokinins, abscisic acid, and ethylene [34]. Under biotic stress, plant associated microbes assist some more functions in plants: they compete with pathogens for essential nutrients predominantly for iron; they biocontrol the activity of pathogen by producing antibiotics, synthesis of lytic enzymes of the fungal cell wall and of host systemic response (ISR) induction in plants [35,36]. PGPMs can reduce the accessibility of iron ions for their competitors by producing siderophores [37]. There are several other beneficial functions performed by PGPMs that are summarized in Table 1.

TABLE 1 General description of plant growth-promoting microorganisms (PGPMs) and their plant growth-promoting functions or attributes.

PGPMs	Type of stress	Plant growth-promoting attributes	References
Pseudomonas, Bacillus, Methylobacterium	Cold and biotic stress	Minerals (P, K, Zn) solubilization, ACC deaminase and antagonistic activity and production of IAA and siderophores	Verma et al. [38]
Arthrobacter methylotrophus IARI-HHS1-25	Cold stress	ACC deaminase activity, phosphate solubilization, production of siderophores, and IAA	Yadav et al. [9]
Bordetella bronchiseptica IARI-HHS2-29	Cold stress	Production of IAA, siderophores, and phosphate solubilization	Verma et al. [39]
Piriformospora indica	Salinity, drought, and heavy metal toxicity stress	Production of IAA and antioxidant enzymes	Gill et al. [40]
Trichoderma sp.	Biotic stress	Production of IAA and siderophore, P solubilization; biocontrol	Rana et al. [22]
Acremonium sp., Talaromyces flavus, Penicillium simplicissimum, Leptosphaeria sp.	Biotic stress	Production of antagonistic substances, expression of antioxidative genes	Yuan et al. [41]
Enterobacter, Herbaspirillum, Pantoea	Abiotic and biotic stress	Nitrogen fixation, mineral solubilization, production of phytohormone, siderophores, and antagonistic activity	Suman et al. [19,20]
Bacillus megaterium, Trichoderma longibrachiatum, T. simmonsii	Draught and salinity stress	Improves seed germination by potassium uptake	Bakhshandeh [42]
Penicillium sp. LWL3, Phoma glomerata LWL2	Draught and salinity stress	By increase production of IAA and gibberellins	Sattiraju et al. [43]
Alkalibacillus, Bacillus, Haloalkalibacillus	Alkalinity stress	Production of ACC deaminase, IAA, and increasing K levels	Msimbira and Smith [44]

3 Some advantageous features of plant growth-promoting microorganisms

The microbial together with the plants not only have the capability to synthesize phytohormones such as IAA, gibberellic acids, and cytokines and solubilize micronutrients but also elicit plant defense mechanisms against pathogens to sustain the growth of plant throughout the harsh environmental circumstances [45]. These beneficial microbes also play a crucial role in N_2 fixation, production of siderophores, antagonistic substances, antifungal, antibiotic, or antibacterial agents and Fe-chelating compounds which results in increase in plant growth the bioavailability of minerals and nutrients (Fig. 1) [9].

4 Biological nitrogen fixation

The implementation of nitrogen-fixing microorganisms as biofertilizers has appeared as an immensely resourceful and eco-friendly approach for accelerating the growth and production of crop plants since nitrogen is the chief limiting factor for plant growth [45]. Various nitrogen-fixing bacteria like *Azoarcus*, *Arthrobacter*, *Bacillus*, *Azospirillum*, *Azotobacter*, *Enterobacter*, *Klebsiella*, *Gluconacetobacter*, *Serratia*, *Herbaspirillum*, and *Pseudomonas* have been identified to fix atmospheric nitrogen during the low-temperature state [46,47]. Choudhury and Kennedy [48] recorded that the application of *Azolla* and cyanobacteria along with different N_2-fixing microorganisms such as *Burkholderia*, *Azotobacter*, *Azospirillum*, *Clostridium*, and *Herbaspirillum* in the rice field can reduce the 30%–50% requirement of the urea. Pham et al. [49] reported a significant enhancement in the growth of rice seedlings inoculated with rhizospheric *Pseudomonas stutzeri* A15, which helps in biological fixation of atmospheric nitrogen in rice [45]. Verma et al. [50] isolated a total of 395 bacilli from a wheat producing area in the northern hills of India. This study for the first time confirmed the existence of *Bacillus endophyticus*, *Planococcus citreus*, *Staphylococcus succinus*, *Paenibacillus xylanexedens*, *Sporosarcina* sp., and *Planomicrobium okeanokoites* in the wheat rhizospheric region with plant growth-promoting traits. In this study, 55 distinct bacilli were recognized by using phylogenetic analysis based upon 16S rRNA gene sequencing, which was further categorized in 5 families, Bacillaceae (68%), Paenibacillaceae (15%), Planococcaceae (8%), Staphylococcaceae (7%), and Bacillales incertae sedis (2%), which comprises 8 genera, viz., *Bacillus*, *Lysinibacillus*, *Exiguobacterium*, *Paenibacillus*, *Staphylococcus*, *Planococcus*, *Planomicrobium*, and *Sporosarcina*. Among all the studied bacteria, *Bacillus circulans* (45.3 ± 1.5 nmol ethylene h^{-1} mg^{-1} protein), *Lysinibacillus fusiformis* (43.5 ± 1.6 nmol ethylene h^{-1} mg^{-1} protein), *Lysinibacillus sphaericus* (19.5 ± 1.0 nmol ethylene h^{-1} mg^{-1} protein), and *Bacillus barbaricus* (18.2 ± 1.6 nmol ethylene h^{-1} mg^{-1} protein) showed the ability to fix the atmospheric nitrogen in the low-temperature state of the hilly area [47].

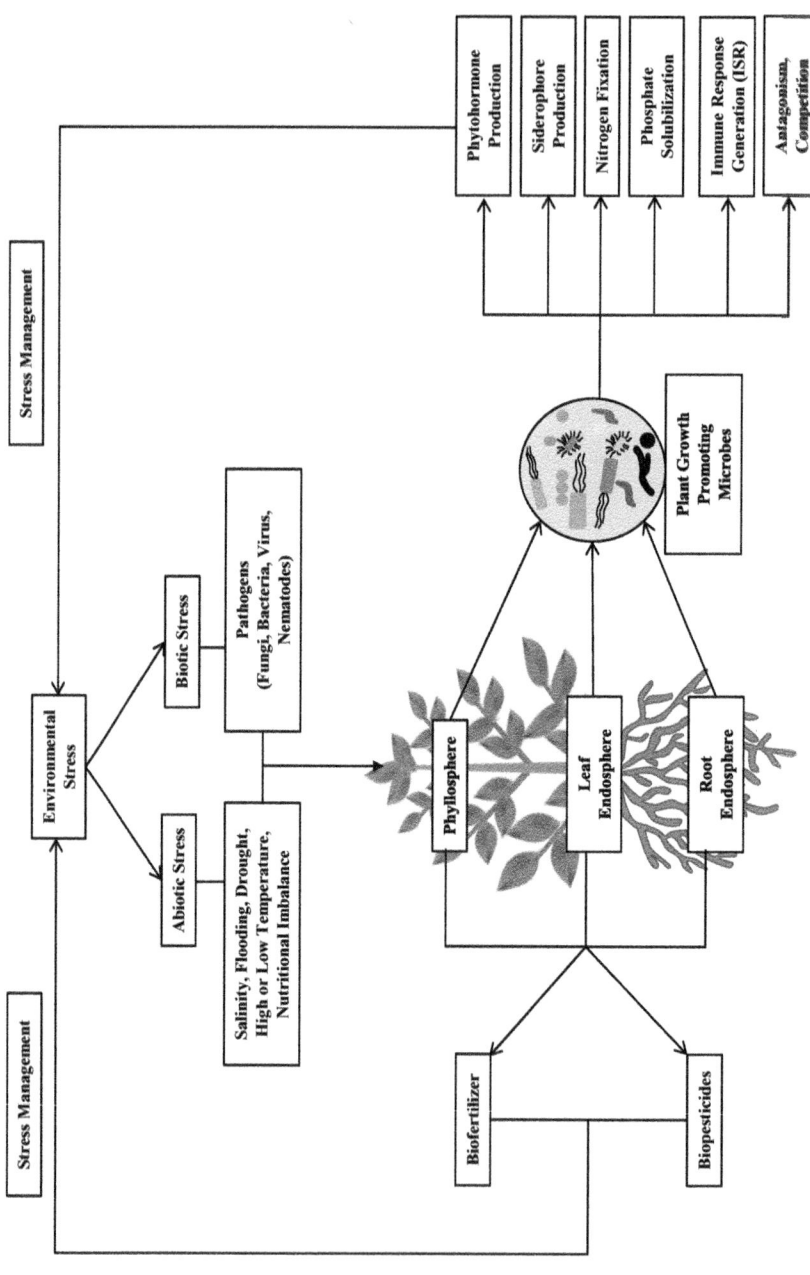

FIG. 1 Significant contribution of plant growth-promoting microbes in plant growth in different stress condition.

5 Production of phytohormone

Several soil microorganisms comprising fungi, bacteria, and algae typically synthesize plant growth hormones such as auxins, cytokinins, and gibberellins, which may affect the physiological growth and development of plants [51]. Auxin is generally produced by all plant-associated microorganisms, which include endophytic, epiphytic, as well as rhizospheric, while gibberellin is produced mostly by the root-associated microorganisms. Auxins are a class of indole derivatives that play several growth-promoting roles in plants, like regulation of fruit ripening, induction of root formation, and promotion of cell division. Indole acetic acid (IAA) is the well-studied form of auxin [45]. There are numerous bacterial auxins that have been well studied. The capability of bacteria to synthesize IAA may have developed as the plant-microbe association evolved since IAA does not serve as a hormone in bacteria. Several plant-associated microorganisms show the ability to produce these phytohormones; that is why IAA may possibly be used to increase plant growth or decrease weed development. Complexes having adenine in their backbone and changes at the N-6 atom of the purine ring represent the Cytokinin's. This phytohormone plays a crucial role in the growth and development of the plant by inducing cell division, promoting seed germination and activation of dormant buds. It also regulates the production of nucleic acids, chloroplast proteins, and chlorophyll of the developing leaf at the early stages. Cytokinins are produced by both beneficial and pathogenic bacterial species [47]. Verma et al. [29] reported endophytic, epiphytic, and rhizospheric bacteria from wheat microbiomes from five distinct locations of central India. Among all 89 bacteria which were isolated from the phyllosphere, identified as *Arthrobacter, Corynebacterium, Bacillus, Methylobacterium, Psychrobacter, Paenibacillus*, and *Pseudomonas*, and 37 were endophytic bacteria which belongs to genera such as *Delftia, Stenotrophomonas, Micrococcus*, and *Pseudomonas*. Rest of rhizospheric bacteria were recognized as *Stenotrophomonas, Acinetobacter, Duganella, Bacillus, Exiguobacterium, Lysinibacillus, Kocuria, Micrococcus, Paenibacillus, Serratia, Pantoea* and *Pseudomonas*, in which 12% isolates of the total of isolates were capable of IAA production. Tabatabaei et al. [52] observed the in vitro synthesis of indole-3-acetic acid in the wheat upon inoculation with four indole-3-acetic acid-synthesizing *Pseudomonas* isolates [45].

6 Solubilization of minerals

Plants obtain phosphorus from soil in the form of phosphate anion. As compared to other macronutrients, it exhibits minimum mobility in plants. Phosphorus solubilizing microbes (PSMs) act as an important part of phosphorus nutrition by escalating its accessibility to plants through leaching out it from the organic and inorganic soil phosphorus reservoir through the procedure of solubilization and mineralization [53]. Phosphorus is the chief macronutrient for the growth and development of plants, and phosphorus solubilizing microbes provide a good biological system adequate for the solubilization of inorganic phosphorus of soil into an orthophosphate-like accessible form and make it accessible to the plants. Most of the microbiome from maize, wheat, legumes, and rice have the capability to solubilize inorganic phosphorus in plate assays, belonging to different genera of *Azotobacter, Burkholderia, Halolamina, Enterobacter, Pantoea, Pseudomonas*, and *Citrobacter* [54]. Potassium (K) is also one of the important determinants of plant nutrition, since it plays a crucial role in the accomplishment of important biological functions for the development and growth of the plant. Potassium is generally plentiful in soil. The total potassium amount in top soil ranges from 3000 to 100,000 kg/ha. Some microorganisms, for instance, *Rhizobium* sp., *Acidithiobacillus ferrooxidans, Azotobacter* sp., *Arthrobacter* sp., *Bacillus mucilaginosus, Frateuria* sp., *Bacillus edaphicus, Klebsiella* sp., *Paenibacillus* sp., and *Pseudomonas* sp., etc., have been reported to be used for the mobilization of insoluble potassium in soil into a plant-accessible nutrient pool [51]. Verma et al. [50] detected the mineral solubilization capacity in 55 out of 395 bacilli isolated from a wheat microbiome, where of the 55 bacilli, 40, 39, and 18 strains showed solubilization of zinc, phosphorus, and potassium, respectively. Among phosphorus, potassium, and zinc solubilizers, *Paenibacillus polymyxa*, *Planococcus salinarum* and *Bacillus pumilus* exhibited the maximum solubilization of phosphorus, potassium, and zinc, respectively [47].

7 As biocontrol agents

Biocontrol agents are the microorganisms that exhibit disease suppressive activity by secreting numerous compounds, viz., DAPG, phenazine, viscosinamide, pyoluteorin, and tensin, thus indirectly stimulating the plant growth [12]. The microscopic biocontrol agents are mainly algae, bacteria, and fungi which are used to apply on agricultural products to protect from pathogens, weeds, and insects [55,56]. When applied to plants roots, biocontrol agents induce systemic resistance, thus suppressing the disease throughout the plant [57,58]. For control of microbial diseases, bacteria are generally applied on the roots and seeds [51]. The most widely used bacteria are *Paenibacillus, Azospirillum, Pseudomonas, Streptomyces, Bacillus, Enterobacter*, and *Azotobacter* [59–61]. Many advantageous fungi are beneficial in biocontrol by producing a huge number of degradative enzymes like proteases, glucanases, and chitinases, thus parasitizing sclerotia, hyphae, or spores of pathogenic fungi. Biocontrol agents are environmentally friendly, less toxic, and more cost-efficient than chemical pesticides [62]. A recent study was done on hybrid maize to reduce the late wilt disease symptoms by *Trichoderma*; the isolates enhanced the plant's growth parameters and decreased the host pathogen's DNA [63]. Shi et al. [64] applied *Aureobasidium pullulans* on tomato to reduce the influence of postharvest fruit pathogens and reported a decrease in *Penicillium, Mycosphaerella, Cladosporium*, and *Alternaria*.

8 In phytoremediation

Plant toxicity by heavy metals is generated by displacement of essential components in biomolecules, modification of proteins/enzymes, structure and function of plasma membrane and membrane transporters, and by blocking essential biological functions of molecules [65]. Plant growth-promoting microorganisms can enhance mineral and metal mobilization by producing metal-specific ligands and organic acids which results in heavy metal uptake and increased nutrient levels and subsequently an improvement in the phytoextraction process by the plants [66,67]. Some PGPRs, when coinoculated with other PGPRs, can reduce the oxidative stress caused by some metals to the plants by increasing the metal intake during phytoextraction; thus, PGPRs have the ability to provide balanced nutrition to the host plant [68,69]. Many studies have reported microbial strains that are effective in the phytosolubilization or phytoextraction of heavy metals, for example, *Rhodococcus erythropolis*, *Chryseobacterium humi*, *Ralstonia eutropha*, and *Bradyrhizobium* sp. [70].

9 ACC deaminase activity

Under stressful conditions, plants produce stress hormone ethylene by its precursor 1-aminocyclopropane-1-carboxylic acid [71]. It has an important role in various physiological processes like nitrogen fixation, photosynthesis, and respiration [72]. The high amount of ethylene has detrimental effects on the growth and development of plants and can cause a reduction in chlorophyll content, leaf abscission, epinasty, aging, and the ultimate death of the plant [73,74]. Many PGPMs with ACC deaminase activity hydrolyze the precursor 1-aminocyclopropane-1-carboxylic acid, thus exhibiting a crucial role in abiotic and biotic stress management [75]. Prominent ACC deaminase producers under abiotic stress are *Pseudomonas*, *Methylobacterium*, *Rhizobium*, *Bacillus*, *Burkholderia*, and *Enterobacter* [76].

10 PGPMs as biofertilizers and biopesticides

Plant nutrients are essential components for better plant growth and for sustainable agriculture. Chemical fertilizers are hazardous for the abiotic as well as biotic components of the ecosystem, so there is need for an environmentally friendly fertilizer in which biofertilizers are a potent alternative to chemical fertilizers. Biofertilizers are the nutrient-rich microbes which increase the nutrient content of soil and enhance their availability to the crops. Plant growth-promoting microbes increase plant growth by several indirect or direct methods like regulation of plant growth regulators, siderophore production, mineral solubilization, etc. [9,19,20,45]. In a psychrotolerant bacterium *Pseudomonas vancouverensis*, some potent plant growth attributes have been detected like production of IAA, siderophores, and HCN as well as it is shows potent antibiosis against *Sclerotium rolfsii*, *Rhizoctonia solani*, *Pythium* sp., and *Fusarium oxysporum* [77]. Gulati et al. [78] characterized plant growth-promoting features like production of auxin, solubilization of phosphate, side-rophore production, ammonia generation, 1-aminocyclopropane-1-carboxylate deaminase activity in the *Acinetobacter rhizosphaerae* strain BIHB 723. Another study on psychrotolerant bacteria, associated with wheat plant, revealed great plant growth-promoting characteristics such as solubilization of minerals like potassium, phosphorus, and zinc, and production of phytohor-mones and siderophores. Isolates of bacteria also showed inhibition of patho-genic fungal genera such as *Fusarium graminearum*, *Rhizoctonia solani*, and *Macrophomina phaseolina* [38,79]. Microbial biopesticides are those microor-ganisms that are capable of promoting plant growth by controlling and inhibit-ing phytopathogenic agents through a wide array of mechanisms such as production of antibiotics, HCN, siderophores, production of hydrolytic enzymes, and induction of acquired and induced systemic resistance (ISR) [80–83]. Mycorrhizae play a pivotal role in increasing the uptake of nutrients and water, such as phosphorus from the soil, which is essential for plant growth and development and for high productivity. Similarly, mycorrhiza *Glomus ver-siformis* has been reported to enhance the inorganic phosphate transporter (Pi) to absorb phosphate ions from the soil for the host plant [84]. Potassium is a crucial macronutrient that regulates the activities of many enzymes, for exam-ple, amylases, which are also intricate in the coordination of the root-shoot ratio [85]. Some bacterial genera such as *Bacillus mucilaginous*, *Rhizobium* spp., and *Azotobacter chroococcum* have been reported for their capacity for potassium solubilization, which results in increased productivity of wheat, pepper, maize, cotton, chili, and sorghum [86]. Sulfur is a macronutrient which is needed in high concentration by plants. It is constituent of some amino acids, such as cys-teine and methionine, as well as it is the cofactor of many enzymes regulating plants, such as ascorbate peroxidase, superoxide dismutase, dehydroascorbate reductase, glutathione reductase, and monodehydroascorbate reductase. Con-version of organic sulfur into inorganic sulfur is performed by some sulfur-oxidizing microbes like *Alcaligenes*, *Xanthobacter*, *Bacillus*, and *Pseudomonas* [87]. As zinc is an essential element, its deficiency shows several abnormalities like chlorosis, light stress, and pathogenic attack to plants but is hazardous for the environment also. Thus, the application of microorganisms that have the capacity to solubilize zinc may be an alternative to Zn supply. Various strains of Zn solubilizing microbes have been identified and applied in biofertilizer production. These include *Rhizobium* spp., *Pseudomonas* spp., *Thiobacillus thioxidans*, *Azospirillum* spp., and *Bacillus aryabhattai* [88]. Biofertilizers increase plant growth and plant yield to increase food production and sustain-able agriculture. This statement is evident in research conducted by Dicko et al. [89] who reported improved maize yield after application of biofertilizer made from plant growth-promoting *Actinomycetes* sp. H7, O19, and AHB12. Further-more, the introduction of a consortium of bacteria, namely, *Bacillus lentus*,

Pseudomonas, and *Azospirillum brasilense*, was characterized as having plant growth-promoting attributes like increase in chlorophyll content in plants and expression of various antioxidant enzymes under stressful conditions [90]. Current studies reported that microbes such as *Azotobacter*, *Paenibacillus*, *Bacillus*, *Enterobacter*, and *Pseudomonas* are potent reducers of pesticide toxicity as well as involved to prevent pathogen proliferation [91]. VOCs production by bacteria has been involved in systemic resistance induction in plants and increases their tolerance to abiotic and biotic stress [92]. So the above description clearly show that the use of microbial biofertilizers are the key tool to modern agriculture which are renewable, cost-effective, and eco-friendly that provide sustainable agriculture as well alleviate abiotic and biotic stress in plants.

11 Conclusions

The beneficial microorganisms play many significant roles in many fields, especially in medicine, agriculture, and industry. They belong to groups like archaea, bacteria, actinomycetes, and fungi. They are not only involved in plant growth and development but also improve plant health by alleviating abiotic (acidic, alkaline, salinity, drought, temperatures, and pressure) as well as biotic stress (attack by pathogens). These functions are performed by several mechanisms such as N_2 fixation, mineral solubilization like potassium, phosphorus, and zinc, production of antagonistic compounds, siderophores, PGPRs like auxin and gibberellins. These microbes are also applicable as biofertilizers and biopesticides. So this article provides detailed information about the different functions and strategies used by plant PGPMs under difficult or stressful conditions to cure plant health and their growth and development. There is a need for further studies on the molecular pattern of plant microbe interaction and genetic expression of genes involved in this mechanism. New techniques like nanotechnology may be helpful in the field of biofertilizer formulation.

Acknowledgments

The authors are thankful to the University Grant Commission (UGC) under Startup Research Grant (UGC Faculty Research Promotion Scheme; FRPS), New Delhi, India for the financial assistance (No.F.30-476/2019 (BSR) FD Diary No. 5662). The authors are also grateful to their respective Universities for providing support during the work.

References

[1] P.N. Bhattacharyya, Diversity of Microorganisms in the Surface and Subsurface Soil of the Jia Bharali River Catchment Area of Brahmaputra Plains, 2012. http://hdl.handle.net/10603/114597.

[2] F.D. Andreote, T. Gumiere, A. Durrer, Exploring interactions of plant microbiomes, Sci. Agric. 71 (6) (2014) 528–539, https://doi.org/10.1590/0103-9016-2014-0195.

[3] P.A. Niklaus, J. Alphei, C. Kampichler, E. Kandeler, C. Körner, D. Tscherko, M. Wohlfender, Interactive effects of plant species diversity and elevated CO_2 on soil biota and nutrient cycling, Ecology 88 (12) (2007) 3153–3163, https://doi.org/10.1890/06-2100.1.

[4] V.K. Singh, M. Meena, A. Zehra, A. Tiwari, M.K. Dubey, R.S. Upadhyay, Fungal toxins and their impact on living systems, in: R.N. Kharwar, R.S. Upadhyay, N.K. Dubey, R. Raghuwanshi (Eds.), Microbial Diversity and Biotechnology in Food Security, Springer, New Delhi Dordrecht Heidelberg London New York, 2014, pp. 513–530, https://doi.org/10.1007/978-81-322-1801-2_47.

[5] M. Meena, V. Prasad, R.S. Upadhyay, Evaluation of biochemical changes in leaves of tomato infected with *Alternaria alternata* and its metabolites, Vegetos 30 (2017) 2, https://doi.org/10.5958/2229-4473.2017.00020.9.

[6] M. Meena, V. Prasad, R.S. Upadhyay, Evaluation of *Alternaria alternata* isolates for metabolite production isolated from different sites of Varanasi, India, J. Agric. Res. 2 (1) (2017) 00012.

[7] M. Meena, P. Swapnil, R.S. Upadhyay, Isolation, characterization and toxicological potential of tenuazonic acid, alternariol and alternariol monomethyl ether produced by *Alternaria* species phytopathogenic on plants, Sci. Rep. 7 (2017) 8777, https://doi.org/10.1038/s41598-017-09138-9.

[8] P. Verma, A.N. Yadav, S.K. Kazy, A.K. Saxena, A. Suman, Elucidating the diversity and plant growth promoting attributes of wheat (*Triticum aestivum*) associated acidotolerant bacteria from southern hills zone of India, 2013. http://krishi.icar.gov.in/jspui/handle/123456789/65733.

[9] A.N. Yadav, R. Kumar, S. Kumar, V. Kumar, T. Sugitha, B. Singh, V.S. Chauhan, H.S. Dhaliwal, A.K. Saxena, Beneficial microbiomes: biodiversity and potential biotechnological applications for sustainable agriculture and human health, J. Appl. Biol. Biotechnol. 5 (6) (2017) 45–57, https://doi.org/10.7324/JABB.2017.50607.

[10] M. Meena, A. Zehra, M.K. Dubey, M. Aamir, V.K. Gupta, R.S. Upadhyay, Comparative evaluation of biochemical changes in tomato (*Lycopersicon esculentum* Mill.) infected by *Alternaria alternata* and its toxic metabolites (TeA, AOH, and AME), Front. Plant Sci. 7 (2016) 1408, https://doi.org/10.3389/fpls.2016.01408.

[11] M. Meena, A. Zehra, M.K. Dubey, R.S. Upadhyay, Mannitol and proline accumulation in *Lycopersicum esculentum* during infection of *Alternaria alternata* and its toxins, Int. J. Biomed. Sci. Bioinformatics 3 (2) (2016) 64–68.

[12] M. Meena, P. Swapnil, K. Divyanshu, S. Kumar, Harish, Y.N. Tripathi, A. Zehra, A. Marwal, R.S. Upadhyay, PGPR-mediated induction of systemic resistance and physiochemical alterations in plants against the pathogens: current perspectives, J. Basic Microbiol. 60 (10) (2020) 828–861, https://doi.org/10.1002/jobm.202000370.

[13] X. Zhang, R. Zhang, J. Gao, X. Wang, F. Fan, X. Ma, Y. Deng, Thirty-one years of rice-rice-green manure rotations shape the rhizosphere microbial community and enrich beneficial bacteria, Soil Biol. Biochem. 104 (2017) 208–217, https://doi.org/10.1016/j.soilbio.2016.10.023.

[14] K. Farrar, D. Bryant, N. Cope-Selby, Understanding and engineering beneficial plant–microbe interactions: plant growth promotion in energy crops, Plant Biotechnol. J. 12 (9) (2014) 1193–1206, https://doi.org/10.1111/pbi.12279.

[15] P. Trivedi, J.E. Leach, S.G. Tringe, T. Sa, B.K. Singh, Plant–microbiome interactions: from community assembly to plant health, Nat. Rev. Microbiol. 18 (11) (2020) 607–621, https://doi.org/10.1038/s41579-020-0412-1.

[16] B.R. Glick, Plant growth-promoting bacteria: mechanisms and applications, Scientifica 2012 (2012), https://doi.org/10.6064/2012/963401.

[17] P. Kumari, M. Meena, P. Gupta, M.K. Dubey, G. Nath, R.S. Upadhyay, Plant growth promoting rhizobacteria and their biopriming for growth promotion in mung bean (*Vigna radiata* (L.) R. Wilczek), Biocatal. Agric. Biotechnol. 16 (2018) 163–171.

[18] P. Kumari, M. Meena, R.S. Upadhyay, Characterization of plant growth promoting rhizobacteria (PGPR) isolated from the rhizosphere of *Vigna radiata* (mung bean), Biocatal. Agric. Biotechnol. 16 (2018) 155–162.

[19] A. Suman, P. Verma, A.N. Yadav, R. Srinivasamurthy, A. Singh, R. Prasanna, Development of hydrogel based bio-inoculant formulations and their impact on plant biometric parameters of wheat (*Triticum aestivum* L.), Int. J. Curr. Microbiol. App. Sci. 5 (3) (2016) 890–901, https://doi.org/10.20546/ijcmas.2016.503.103.

[20] A. Suman, A.N. Yadav, P. Verma, Endophytic microbes in crops: diversity and beneficial impact for sustainable agriculture, in: Microbial Inoculants in Sustainable Agricultural Productivity, Springer, New Delhi, 2016, pp. 117–143, https://doi.org/10.1007/978-81-322-2647-5_7.

[21] J. Massoni, M. Bortfeld-Miller, L. Jardillier, G. Salazar, S. Sunagawa, J.A. Vorholt, Consistent host and organ occupancy of phyllosphere bacteria in a community of wild herbaceous plant species, ISME J. 14 (1) (2020) 245–258, https://doi.org/10.1038/s41396-019-0531-8.

[22] K.L. Rana, D. Kour, A.N. Yadav, V. Kumar, H.S. Dhaliwal, Endophytic microbes from wheat: diversity and biotechnological applications for sustainable agriculture, in: Proceeding of 57th Association of Microbiologist of India and International symposium on "Microbes and Biosphere: What's new What's next" (Vol. 453), 2016.

[23] S.M. Kang, A.L. Khan, Y.H. You, J.G. Kim, M. Kamran, I.J. Lee, Gibberellin production by newly isolated strain *Leifsonia soli* SE134 and its potential to promote plant growth, J. Microbiol. Biotechnol. 24 (1) (2014) 106–112, https://doi.org/10.4014/jmb.1304.04015.

[24] A.N. Yadav, Agriculturally important microbiomes: biodiversity and multifarious PGP attributes for amelioration of diverse abiotic stresses in crops for sustainable agriculture, Biomed. J. Sci. Tech. Res. 1 (4) (2017) 861–864, https://doi.org/10.26717/BJSTR.2017.01.000321.

[25] A.N. Yadav, S.G. Sachan, P. Verma, A.K. Saxena, Prospecting cold deserts of North Western Himalayas for microbial diversity and plant growth promoting attributes, J. Biosci. Bioeng. 119 (6) (2015) 683–693, https://doi.org/10.1016/j.jbiosc.2014.11.006.

[26] J.M. Barea, M.J. Pozo, R. Azcon, C. Azcon-Aguilar, Microbial co-operation in the rhizosphere, J. Exp. Bot. 56 (417) (2005) 1761–1778, https://doi.org/10.1093/jxb/eri197.

[27] K.V.B.R. Tilak, N. Ranganayaki, K.K. Pal, R. De, A.K. Saxena, C.S. Nautiyal, S. Mittal, A.K. Tripathi, B.N. Johri, Diversity of plant growth and soil health supporting bacteria, Curr. Sci. 89 (1) (2005) 136–150.

[28] T.G. Dobrovolskaya, K.A. Khusnetdinova, N.A. Manucharova, A.V. Golovchenko, Structure of epiphytic bacterial communities of weeds, Microbiology 86 (2) (2017) 257–263, https://doi.org/10.1134/S0026261717020072.

[29] P. Verma, A.N. Yadav, S.K. Kazy, A.K. Saxena, A. Suman, Evaluating the diversity and phylogeny of plant growth promoting bacteria associated with wheat (*Triticum aestivum*) growing in central zone of India, Int. J. Curr. Microbiol. App. Sci. 3 (5) (2014) 432–447.

[30] J. Lian, Z. Wang, S. Zhou, Response of endophytic bacterial communities in banana tissue culture plantlets to fusarium wilt pathogen infection, J. Gen. Appl. Microbiol. 54 (2) (2008) 83–92, https://doi.org/10.2323/jgam.54.83.

[31] M. Rajkumar, S. Sandhya, M.N.V. Prasad, H. Freitas, Perspectives of plant-associated microbes in heavy metal phytoremediation, Biotechnol. Adv. 30 (6) (2012) 1562–1574, https://doi.org/10.1016/j.biotechadv.2012.04.011.

[32] L. Epelde, A. Lanzen, F. Blanco, T. Urich, C. Garbisu, Adaptation of soil microbial community structure and function to chronic metal contamination at an abandoned Pb-Zn mine, FEMS Microbiol. Ecol. 91 (1) (2015) 1–11, https://doi.org/10.1093/femsec/fiu007.

[33] L.G. Li, Y. Xia, T. Zhang, Co-occurrence of antibiotic and metal resistance genes revealed in complete genome collection, ISME J. 11 (3) (2017) 651–662, https://doi.org/10.1038/ismej.2016.155.

[34] V. Shah, A. Daverey, Phytoremediation: a multidisciplinary approach to clean up heavy metal contaminated soil, Environ. Technol. Innov. 18 (2020), 100774, https://doi.org/10.1016/j.eti.2020.100774.

[35] B.R. Glick, Bacteria with ACC deaminase can promote plant growth and help to feed the world, Microbiol. Res. 169 (1) (2014) 30–39, https://doi.org/10.1016/j.micres.2013.09.009.

[36] Y. Ma, R.S. Oliveira, H. Freitas, C. Zhang, Biochemical and molecular mechanisms of plant-microbe-metal interactions: relevance for phytoremediation, Front. Plant Sci. 7 (2016) 918, https://doi.org/10.3389/fpls.2016.00918.

[37] E.H. Verbon, P.L. Trapet, I.A. Stringlis, S. Kruijs, P.A. Bakker, C.M. Pieterse, Iron and immunity, Annu. Rev. Phytopathol. 55 (2017) 355–375, https://doi.org/10.1146/annurev-phyto-080516-035537.

[38] P. Verma, A.N. Yadav, K.S. Khannam, S. Kumar, A.K. Saxena, A. Suman, Growth promotion and yield enhancement of wheat (*Triticum aestivum* L.) by application of potassium solubilizing psychrotolerant bacteria, in: Proceeding of 56th Annual Conference of Association of Microbiologists of India and International Symposium on "Emerging Discoveries in Microbiology", 2015, https://doi.org/10.13140/RG (Vol. 2, no. 1648.0080).

[39] P. Verma, A.N. Yadav, K.S. Khannam, N. Panjiar, S. Kumar, A.K. Saxena, A. Suman, Assessment of genetic diversity and plant growth promoting attributes of psychrotolerant bacteria allied with wheat (*Triticum aestivum*) from the northern hills zone of India, Ann. Microbiol. 65 (4) (2015) 1885–1899, https://doi.org/10.1007/s13213-014-1027-4.

[40] S.S. Gill, R. Gill, D.K. Trivedi, N.A. Anjum, K.K. Sharma, M.W. Ansari, A.A. Ansari, A.K. Johri, R. Prasad, E. Pereira, A. Varma, *Piriformospora indica*: potential and significance in plant stress tolerance, Front. Microbiol. 7 (2016) 332, https://doi.org/10.3389/fmicb.2016.00332.

[41] Y. Yuan, H. Feng, L. Wang, Z. Li, Y. Shi, L. Zhao, Z. Feng, H. Zhu, Potential of endophytic fungi isolated from cotton roots for biological control against verticillium wilt disease, PLoS One 12 (1) (2017), e0170557, https://doi.org/10.1371/journal.pone.0170557.

[42] E. Bakhshandeh, M. Gholamhosseini, Y. Yaghoubian, H. Pirdashti, Plant growth promoting microorganisms can improve germination, seedling growth and potassium uptake of soybean under drought and salt stress, Plant Growth Regul. 90 (1) (2020) 123–136, https://doi.org/10.1007/s10725-019-00556-5.

[43] K.S. Sattiraju, S. Kotiyal, A. Arora, M. Maheshwari, Plant growth-promoting microbes: contribution to stress management in plant hosts, Environ. Biotechnol. (2019) 199–236, https://doi.org/10.1007/978-981-10-7284-0_8.

[44] L.A. Msimbira, D.L. Smith, The roles of plant growth promoting microbes in enhancing plant tolerance to acidity and alkalinity stresses, Front. Sustain. Food Syst. 4 (2020) 106, https://doi.org/10.3389/fsufs.2020.00106.

[45] P. Verma, A.N. Yadav, V. Kumar, D.P. Singh, A.K. Saxena, Beneficial plant-microbes interactions: biodiversity of microbes from diverse extreme environments and its impact for crop improvement, in: Plant-Microbe Interactions in Agro-Ecological Perspectives, Springer, Singapore, 2017, pp. 543–580.

[46] K.L. Rana, D. Kour, P. Verma, A.N. Yadav, V. Kumar, D.H. Singh, Diversity and biotechnological applications of endophytic microbes associated with maize (*Zea mays* L.) growing in Indian Himalayan regions, in: Proceeding of National Conference on Advances in Food Science and Technology, 2017.

[47] A.N. Yadav, P. Verma, S.G. Sachan, R. Kaushik, A.K. Saxena, Psychrotrophic microbiomes: molecular diversity and beneficial role in plant growth promotion and soil health, in: Microorganisms for Green Revolution, Springer, Singapore, 2018, pp. 197–240.

[48] A.T.M.A. Choudhury, I.R. Kennedy, Prospects and potentials for systems of biological nitrogen fixation in sustainable rice production, Biol. Fertil. Soils 39 (4) (2004) 219–227, https://doi.org/10.1007/s00374-003-0706-2.

[49] V.T. Pham, H. Rediers, M.G. Ghequire, H.H. Nguyen, R. De Mot, J. Vanderleyden, S. Spaepen, The plant growth-promoting effect of the nitrogen-fixing endophyte *Pseudomonas stutzeri* A15, Arch. Microbiol. 199 (3) (2017) 513–517, https://doi.org/10.1007/s00203-016-1332-3.

[50] P. Verma, A.N. Yadav, K.S. Khannam, S. Kumar, A.K. Saxena, A. Suman, Molecular diversity and multifarious plant growth promoting attributes of *Bacilli* associated with wheat (*Triticum aestivum* L.) rhizosphere from six diverse agro-ecological zones of India, J. Basic Microbiol. 56 (1) (2016) 44–58, https://doi.org/10.1002/jobm.201500459.

[51] P.N. Bhattacharyya, M.P. Goswami, L.H. Bhattacharyya, Perspective of beneficial microbes in agriculture under changing climatic scenario: a review, J. Phytol. 8 (2016) 26–41, https://doi.org/10.19071/jp.2016.v8.3022.

[52] S. Tabatabaei, P. Ehsanzadeh, H. Etesami, H.A. Alikhani, B.R. Glick, Indole-3-acetic acid (IAA) producing *Pseudomonas* isolates inhibit seed germination and α-amylase activity in durum wheat (*Triticum turgidum* L.), Span. J. Agric. Res. 14 (1) (2016) 15 (ISSN 1695-971X, ISSN-e 2171-9292).

[53] A.K. Sharma, P.N. Bhattacharyya, D.J. Rajkhowa, D.K. Jha, Impact of global climate change on beneficial plant-microbe association, Ann. Biol. Res. 5 (3) (2014) 36–37.

[54] A. Kumar, B.R. Maurya, R. Raghuwanshi, V.S. Meena, M. Tofazzal Islam, Co-inoculation with *Enterobacter* and *Rhizobacteria* on yield and nutrient uptake by wheat (*Triticum aestivum* L.) in the alluvial soil under Indo-Gangetic plain of India, J. Plant Growth Regul. 36 (2017) 608–617, https://doi.org/10.1007/s00344-016-9663-5.

[55] M.C. Lefort, A. McKinnon, T.L. Nelson, T. Glare, Natural occurrence of the entomopathogenic fungi *Beauveria bassiana* as a vertically transmitted endophyte of *Pinus radiata* and its effect on above-and below-ground insect pests, N. Z. Plant Prot. 69 (2016) 68–77.

[56] M.L. Russo, A.C. Scorsetti, M.F. Vianna, N. Allegrucci, N.A. Ferreri, M.N. Cabello, S.A. Pelizza, Effects of endophytic *Beauveria bassiana* (Ascomycota: Hypocreales), J. King Saud Univ. Sci. 31 (4) (2019) 1077–1108, https://doi.org/10.1016/j.jksus.2018.11.009.

[57] C.K. Dash, B.S. Bamisile, R. Keppanan, M. Qasim, Y. Lin, S.U. Islam, L. Wang, Endophytic entomopathogenic fungi enhance the growth of *Phaseolus vulgaris* L. (Fabaceae) and negatively affect the development and reproduction of *Tetranychus urticae* Koch (Acari: Tetranychidae), Microb. Pathog. 125 (2018) 385–392, https://doi.org/10.1016/j.micpath.2018.09.044.

[58] A.S. Elnahal, M.T. El-Saadony, A.M. Saad, E.S.M. Desoky, A.M. El-Tahan, M.M. Rady, K.A. El-Tarably, The use of microbial inoculants for biological control, plant growth promotion, and sustainable agriculture: a review, Eur. J. Plant Pathol. 1-34 (2022), https://doi.org/10.1007/s10658-021-02393-7.

[59] Y.D. Jing, Z.L. He, X.E. Yang, Role of soil rhizobacteria in phytoremediation of heavy metal contaminated soils, J. Zhejiang Univ. Sci. B 8 (2007) 192–207, https://doi.org/10.1631/jzus.2007.B0192.

[60] T. Barupal, M. Meena, K. Sharma, Comparative analysis of bioformulations against *Curvularia lunata* (Wakker) Boedijn causing leaf spot disease of maize, Arch. Phytopathol. Plant Protect. 54 (5–6) (2020) 261–272, https://doi.org/10.1080/03235408.2020.1827657.

[61] P. Bharti, R. Tewari, Purification and structural characterization of a phthalate antibiotic from *Burkholderia gladioli* OR1 effective against multi-drug resistant *Staphylococcus aureus*, J. Microbiol. Biotechnol. Food Sci. 5 (3) (2015) 207–211, https://doi.org/10.15414/jmbfs.2015/16.5.3.207-211.

[62] J.S. Bale, J.C. Van Lenteren, F. Bigler, Biological control and sustainable food production, Philos. Trans. R. Soc. Lond. Ser. B Biol. Sci. 363 (1492) (2008) 761–776, https://doi.org/10.1098/rstb.2007.2182.

[63] O. Degani, S. Dor, *Trichoderma* biological control to protect sensitive maize hybrids against late wilt disease in the field, J. Fungi 7 (4) (2021) 315, https://doi.org/10.3390/jof7040315.

[64] Y. Shi, Q. Yang, Q. Zhao, S. Dhanasekaran, J. Ahima, X. Zhang, H. Zhang, *Aureobasidium pullulans* S-2 reduced the disease incidence of tomato by influencing the postharvest microbiome during storage, Postharvest Biol. Technol. 185 (2022), 111809, https://doi.org/10.1016/j.postharvbio.2021.111809.

[65] E.I. Ochiai, Bioinorganic Chemistry: An Introduction, Allyn and Bacon, Boston, 1977, pp. 218–262.

[66] L. Chen, S. Luo, X. Li, Y. Wan, J. Chen, C. Liu, Interaction of cd-hyperaccumulator *Solanum nigrum* L. and functional endophyte *Pseudomonas* sp. Lk9 on soil heavy metals uptake, Soil Biol. Biochem. 68 (2014) 300–308, https://doi.org/10.1016/j.soilbio.2013.10.021.

[67] Z. Kong, B.R. Glick, The role of plant growth-promoting bacteria in metal phytoremediation, Adv. Microb. Physiol. 71 (2017) 97–132, https://doi.org/10.1016/bs.ampbs.2017.04.001.

[68] W. Ju, L. Liu, L. Fang, Y. Cui, C. Duan, H. Wu, Impact of co-inoculation with plant-growth-promoting rhizobacteria and rhizobium on the biochemical responses of alfalfa-soil system in copper contaminated soil, Ecotoxicol. Environ. Saf. 167 (2019) 218–226, https://doi.org/10.1016/j.ecoenv.2018.10.016.

[69] H. Korir, N.W. Mungai, M. Thuita, Y. Hamba, C. Masso, Co-inoculation effect of rhizobia and plant growth promoting rhizobacteria on common bean growth in a low phosphorus soil, Front. Plant Sci. 8 (2017) 141, https://doi.org/10.3389/fpls.2017.00141.

[70] M. Vocciante, M. Grifoni, D. Fusini, G. Petruzzelli, E. Franchi, The role of plant growth-promoting rhizobacteria (PGPR) in mitigating plant's environmental stresses, Appl. Sci. 12 (3) (2022) 1231, https://doi.org/10.3390/app12031231.

[71] B.R. Glick, Z. Cheng, J. Czarny, J. Duan, Promotion of plant growth by ACC deaminase-producing soil bacteria, Eur. J. Plant Pathol. 119 (2007) 329–339, https://doi.org/10.1007/978-1-4020-6776-1_8.

[72] S. Chandwani, N. Amaresan, Role of ACC deaminase producing bacteria for abiotic stress management and sustainable agriculture production, Environ. Sci. Pollut. Res. 2022 (2022) 1–17, https://doi.org/10.1007/s11356-022-18745-7.

[73] S. Ali, W.C. Kim, Plant growth promotion under water: decrease of waterlogging-induced ACC and ethylene levels by ACC deaminase-producing bacteria, Front. Microbiol. 9 (2018) 1096, https://doi.org/10.3389/fmicb.2018.01096.

[74] Y.S. Moon, S. Ali, Possible mechanisms for the equilibrium of ACC and role of ACC deaminase-producing bacteria, Appl. Microbiol. Biotechnol. 106 (3) (2022) 877–887, https://doi.org/10.1007/s00253-022-11772-x.

[75] X. Li, Z. Yan, D. Gu, D. Li, Y. Tao, D. Zhang, L. Su, Y. Ao, Characterization of cadmium-resistant rhizobacteria and their promotion effects on *Brassica napus* growth and cadmium uptake, J. Basic Microbiol. 59 (6) (2019) 579–590, https://doi.org/10.1002/jobm.201800656.

[76] D. Blaha, C. Prigent-Combaret, M.S. Mirza, Y. Moënne-Loccoz, Phylogeny of the 1-amino-cyclopropane-1-carboxylic acid deaminase-encoding gene *acdS* in phytobeneficial and pathogenic *Proteobacteria* and relation with strain biogeography, FEMS Microbiol. Ecol. 56 (3) (2006) 455–470, https://doi.org/10.1111/j.1574-6941.2006.00082.x.

[77] P.K. Mishra, S. Mishra, G. Selvakumar, S.C. Bisht, J.K. Bisht, S. Kundu, H.S. Gupta, Characterisation of a psychrotolerant plant growth promoting *Pseudomonas* sp. strain PGERs17 (MTCC 9000) isolated from North Western Indian Himalayas, Ann. Microbiol. 58 (4) (2008) 561–568, https://doi.org/10.1007/BF03175558.

[78] A. Gulati, P. Vyas, P. Rahi, R.C. Kasana, Plant growth-promoting and rhizosphere-competent *Acinetobacter rhizosphaerae* strain BIHB 723 from the cold deserts of the Himalayas, Curr. Microbiol. 58 (2009) 371–377, https://doi.org/10.1007/s00284-008-9339-x.

[79] A. Zehra, N.A. Raytekar, M. Meena, P. Swapnil, Efficiency of microbial bio-agents as elicitors in plant defense mechanism under biotic stress: a review, Curr. Res. Microb. Sci. 2 (2021), 100054, https://doi.org/10.1016/j.crmicr.2021.100054.

[80] D. Chandler, G. Davidson, W.P. Grant, J. Greaves, G.M. Tatchell, Microbial biopesticides for integrated crop management: an assessment of environmental and regulatory sustainability, Trends Food Sci. Technol. 19 (5) (2008) 275–283, https://doi.org/10.1016/j.tifs.2007.12.009.

[81] E. Somers, J. Vanderleyden, M. Srinivasan, Rhizosphere bacterial signaling: a love parade beneath our feet, Crit. Rev. Microbiol. 30 (4) (2004) 205–240, https://doi.org/10.1080/10408410490468786.

[82] A. Zehra, M. Meena, M.K. Dubey, M. Aamir, R.S. Upadhyay, Synergistic effects of plant defense elicitors and *Trichoderma harzianum* on enhanced induction of antioxidant defense system in tomato against fusarium wilt disease, Bot. Stud. 58 (2017) 44, https://doi.org/10.1186/s40529-017-0198-2.

[83] A. Zehra, M. Meena, M.K. Dubey, M. Aamir, R.S. Upadhyay, Activation of defense response in tomato against fusarium wilt disease triggered by *Trichoderma harzianum* supplemented with exogenous chemical inducers (SA and MeJA), Braz. J. Bot. 21 (2017) 1–14, https://doi.org/10.1007/s40415-017-0382-3.

[84] M. Parihar, M. Chitara, P. Khati, A. Kumari, P.K. Mishra, A. Rakshit, K. Rana, V.S. Meena, A. K. Singh, M. Choudhary, J.K. Bisht, Arbuscular mycorrhizal fungi: abundance, interaction with plants and potential biological applications, in: Advances in Plant Microbiome and Sustainable Agriculture, Springer, Singapore, 2020, pp. 105–143, https://doi.org/10.1007/978-981-15-3208-5_5.

[85] D. Kour, K.L. Rana, A.N. Yadav, N. Yadav, M. Kumar, V. Kumar, P. Vyas, H.S. Dhaliwal, A. K. Saxena, Microbial biofertilizers: bioresources and eco-friendly technologies for agricultural and environmental sustainability, Biocatal. Agric. Biotechnol. 23 (2020), 101487, https://doi.org/10.1016/j.bcab.2019.101487.

[86] Y. Zhao, M. Zhang, W. Yang, H.J. Di, L. Ma, W. Liu, B. Li, Effects of microbial inoculants on phosphorus and potassium availability, bacterial community composition, and chili pepper growth in a calcareous soil: a greenhouse study, J. Soils Sediments 19 (10) (2019) 3597–3607,- https://doi.org/10.1007/s11368-019-02319-1.

[87] H. Etesami, S. Emami, H.A. Alikhani, Potassium solubilizing bacteria (KSB): mechanisms, promotion of plant growth, and future prospects: a review, J. Plant Nutr. Soil Sci. 17 (4) (2017) 897–911, https://doi.org/10.4067/S0718-95162017000400005.

[88] M. Ijaz, Q. Ali, S. Ashraf, M. Kamran, A. Rehman, Development of future bioformulations for sustainable agriculture, in: V. Kumar, R. Prasad, M. Kumar, D.K. Choudhary (Eds.), Microbiome in Plant Health and Disease, 1st, Springer Nature, Singapore, 2019, pp. 421–446. https://doi.org/10.1007/978-981-13-8495-0_19.

[89] A.H. Dicko, A.H. Babana, A. Kassogué, R. Fané, D. Nantoumé, D. Ouattara, K. Maiga, S. Dao, A Malian native plant growth promoting actinomycetes based biofertilizer improves maize growth and yield, Symbiosis 75 (3) (2018) 267–275, https://doi.org/10.1007/s13199-018-0555-2.

[90] G.P. Brahmaprakash, P.K. Sahu, G. Lavanya, S.S. Nair, V.K. Gangaraddi, A. Gupta, Microbial functions of the rhizosphere, in: Plant-Microbe Interactions in Agro-Ecological Perspectives, Springer, Singapore, 2017, pp. 177–210, https://doi.org/10.1007/978-981-10-5813-4_10.

[91] M. Shahid, A. Zaidi, A. Ehtram, M.S. Khan, *In vitro* investigation to explore the toxicity of different groups of pesticides for an agronomically important rhizosphere isolate *Azotobacter vinelandii*, Pestic. Biochem. Physiol. 157 (2019) 33–44, https://doi.org/10.1016/j. pestbp.2019.03.006.

[92] W. Raza, Q. Shen, Volatile organic compounds mediated plant-microbe interactions in soil, in: Molecular Aspects of Plant Beneficial Microbes in Agriculture, Academic Press, 2020, pp. 209–219, https://doi.org/10.1016/B978-0-12-818469-1.00018-3.

Chapter 15

Plant-microbe interactions to reduce salinity stress in plants for the improvement of the agricultural system

Yashika Maheshwari[a], Shalini Tailor[b], Avinash Marwal[b], and Anita Mishra[c]

[a]*Department of Biotechnology, School of Science, GSFC University, Vadodara, Gujarat, India,* [b]*Department of Biotechnology, Mohanlal Sukhadia University, Udaipur, Rajasthan, India,* [c]*Department of Science (Biotechnology), Biyani Girls College, University of Rajasthan, Jaipur, Rajasthan, India*

1 Introduction

The world is rich in plant biodiversity and to protect this enormous number of species, many researchers have studied the effects of plant hormones and factors responsible for encouraging and limiting agricultural crop productivity. Eventually, as the research carried further, they discovered many points that have an outcome on plant development in both positive and negative manner such as proper sunlight, water, fertile soil, nutrients, drought, salinity, high temperature, cold, high winds, flood, and pH. Researchers recognized that salinization is the major abiotic stress to the environment [1]. Stress is a limitation that restricts the metabolic processes as well as decreases the ability of plants by converting energy into a nonrenewable energy source [2]. Stress affects one billion hectares of land globally [3]. Salt stress is an undesirable outcome of redundant minerals like Na^+ along with Cl or either one of them on the plant [2], and accrual of sodium and chloride ions on the Earth's surface results in natural salinity [3]. An inappropriate practice of irrigation exerts influence on 50% of the total fertile area, which is the major reason for increase in salt concentrations which further cause plant impairment [3]. On the basis of the living world, the quality, and the relation of the plant growth in the salinity area, Szabolcs [4] coined two main types of soils [2], which are as follows: (1) brackish soils—sodium chloride and sodium sulfate are the salts with good solubility, and often this soil

Plant-Microbe Interaction—Recent Advances in Molecular and Biochemical Approaches
https://doi.org/10.1016/B978-0-323-91876-3.00002-6

finds space to accommodate ample amount of chloride and sulfate ions of calcium and magnesium ions, respectively. It could be observed that sodium chloride (NaCl) in large amount has harmful effect on most of the crop plants. (2) Alkali soils—one of the example of this type of soil is sodium carbonate which consists of sodium salts and can hydrolyze alkali [2]. Therefore, with each passing decade, some plants have developed tolerance to salinity to alleviate the effect of soil concentration. To make plants more tolerable against salinity stress, the detection of salinity in plants is required. For monitoring and keeping a record of the changes that can occur in salinity with additional deterioration, and to depict the outline is important so that the decisions, recovery, and treatment measures can take place within the given time limit [5]. Microbes are involved in the nutrition as well as the growth of plants; they also help to improve salinity tolerance in plants. Some microbes are in a symbiotic relationship with plant roots, thus helping the plant by enhancing its productivity, immunity, and developing the overall system of plants for salt tolerance against salty and dry land, this symbiotic relationship between plant and microbe is very beneficial to counteract the accrual of NO^{3-} and PO_4^{3-} in soils of cultivated land.

2 Types and causes of salinity

Salinity affects agriculture in many ways but first, we have to understand how many types of salinity are there and how it affects the soil. (1) Natural salinity: Natural salinity is also known as primary salinity; in this the assemblage of salts occurs through natural processes over a long period of time [2]. Alkaline soils emerged as a result of the natural processes of the Earth's physical structures, movement of water, and soil formation [6]. This is the result of two naturally occurring processes: in the first process weathering of rocks and minerals take place [2]. Weathering is the breakdown of rocks such as igneous rocks; volcanic rocks [6] release a variety of soluble salts largely consisting of chlorides of Ca^{2+}, Na^+, and Mg^{2+}, and SO_4^{2-} and CO_3^{2-} in trace amounts. NaCl produces ions of sodium and chloride when dissociates and interacts with the solvent, so NaCl is a soluble salt [2] and in the second process, the accumulation of ocean salts takes place during breeze and rainfall [2]. These oceanic soluble NaCl called "cyclic salts" are carried interior by breeze as well as deposited during rainfall [2]. (2) Anthropogenic salinity: also known as secondary salinity when salt-affected soil has been salinized by anthropogenic action and adjust the hydrological stability among irrigated water, which is then used for the agricultural purpose as a result salt concentration in water increases and that same quality of poor water is further used in irrigated land [2,6]. In nonfertile and partly fertile land, anthropic salinization occurs because of waterlogging introduced by improper irrigation practices [6]. Salinization is caused by many factors: the movement of salt in both top and surface layers of soil is caused by the deforestation and salinity of soil [6]. High levels of salts are found in sewage obtained from intensive agriculture and industrial wastewater [6]. In countries such as Japan and Netherlands, with intensive agricultural systems, the salinization caused by contamination with chemicals appears more frequently. In closed and semiclosed systems, if the chemicals are not removed it will result in the accumulation of salt [6].

3 Salinity impact on the agricultural system

Plants regularly face stress due to drought, cold, and high temperature conditions [6]. Stress conditions can delay growth and productivity of the crop plant. In severe conditions, it can cause the death of the crop plant. Salinity exerts an effect that is unfavorable for the growth of a plant, osmotic potential and nutritional imbalance [6]. It has been estimated that almost 2000 ha of fertile agricultural land convert to unfertile degraded land every day because of salinity conditions. The two possible reasons for the inhibition of plant growth because of the presence of salt in the soil are as follows: (1) Accumulation of salts decreases the uptake of water by plants, which causes a reduction in growth rate. This reduction in the growth rate of plant due to salt accumulation is known as osmotic or water scarcity due to salinity stress [2]. (2) Salt enter the transpiration stream in huge amounts, thus transported to the cells of leaves. This reduces the growth of plant [2]. This effect is known as the ion-excess effect of salinity [2]. Earth is a salty planet because water present on the Earth contains about 30 g of sodium chloride per liter water. Increased salinity affects plant morphology, functioning, homeostasis, and biomass [6]. Salinity stress affects different developmental processes in plant such as germination, water relations, photosynthetic efficiency, the overall growth rate, photosynthesis, and nutrient absorption. Salinity stress induces oxidative stress and reduces crop yield [2].

- **Seed germination**: The most crucial part of plant growth is seed germination that ultimately affects the yield. Salinity stress has adverse effects on the germination process in various species of plants including *Oryza sativa* and *Brassica* spp. [7–9]. When there is a reduction in osmotic potential of germination media, the uptake of water by seeds decreases [10]. Decrease in osmotic potential of media provided for germination also causes toxic effects, which can modify the activity of enzymes [11] and metabolism of protein [12], and also result in hormonal imbalance [13]. The effect of decreased osmotic potential on rates and percentage of germinated seed varies in many plant species [2]. In an investigation, Bordi [14] reported a decrease in germination percentage of *Brassica napus* when treated with 150 and 200 mM of NaCl concentration. There is an inverse relation between salinity and germination percentage because when salinity increases, the germination rate decreases [2]. Khodarahmpour et al. [15] have reported a significant decline in seed vigor (95%), germination rate (32%), root length (80%), shoot length (78%), and overall seedling length (78%) in the *Zea mays* seeds when exposed to 240 mM NaCl.

- **Plant growth**: Salt affects plant growth because (1) it reduces plant ability to imbibe water, which leads to slow plant growth. The salt present in the salt solution impairs leaf and root growth. (2) Salt enters transpiration stream, which ultimately increases the Na^+ and Cl concentrations in leaves. Excess salt concentration in the cytoplasm alters the activity of enzymes present in cytoplasm. Accumulation of salt ions in cell wall results in the dehydration of cell. The impacts that these ions have on plants are the physiological disorders, imbalance in stomatal regulation, and chlorotic toxicity (Cl ions affect chlorophyll synthesis) [2].
- **Photosynthetic efficiency**: Photosynthesis is an essential biochemical process in which plants transform the trapped solar energy into chemical energy. Accumulation of Na^+ and/or Cl ions in chloroplast reduces water potential, which decreases photosynthetic rate [2].
- **Water potential**: To know the plant water status, water potential is a required physiological parameter. Romero-Aranda et al. [16] stated that when salinity levels increase in rhizosphere, a decrease in leaf water potential is observed. A decrease in leaf water potential affects many physiological processes in a plant. At low or moderate soil salinity, plants can accumulate solutes and maintain the potential gradient required for the influx of water.
- **Nutrient imbalance**: Salinity stress can affect plant nutrient availability. Improper uptake, transport, and distribution of nutrients within the plant cause nutritional deficiencies. Different studies have shown that increase in salinity reduces the uptake, absorption, and accumulation of nutrients in plant. Salinity stress increases pH, which is responsible for micronutrient deficiency in plants. Salinity stress due to the interaction between Na^+ and NH_4^+ and between Cl and NO_3^- affects the uptake and assimilation of nitrogen. This reduces crop productivity. An increase in soil salinity also affects phosphorus availability in plants. Salinity affects phosphorus availability due to following reasons: (a) ionic strength effects tend to reduce the activity of PO_4^- ions, (b) sorption processes firmly guard the phosphate concentrations in soil, and (c) Ca-P minerals show low solubility. Qadir and Schubert [17] reported that an increase in salinity decreases phosphate concentration in agronomic crops. Tuna et al. [18] performed an investigation which shows that treatment with high concentration of NaCl leads to increase in sodium concentration in plant tissues and decrease in leaf Ca^{2+}, K^+, and N. The assimilation of nutrients (K and Ca) takes place due to the elevated levels of sodium chloride (NaCl) in the rhizosphere, which results in ion imbalances of magnesium (Mg), calcium (Ca), and potassium (K) [19].
- **Oxidative stress**: Salinity causes a decrease plant growth and development. Excessive accumulation of reactive oxygen species (ROS) due to salinity stress is responsible for inhibiting plant growth. Accumulation of reactive oxygen species (ROS) due to salinity-induced oxidative stress leads to lipid peroxidation, protein oxidation, enzymes inactivation, damage to DNA, and

damage to other vital components of plant cells. Carbon fixation is the process of fixing atmospheric carbon that is present in the form of carbon dioxide in nature. When salt stress increases in plant leaves, stomata present on surface of leaf close, which causes a reduction in carbon dioxide availability required for carbon fixation. This causes exposure of chloroplasts to excessive excitation energy, which increases the production of ROS such as hydroxyl radical (OH•), hydrogen peroxide (H_2O_2), superoxide (O_2•$-$), and singlet oxygen ($^1O_2$) [20–24]. Membrane damage caused by reactive oxygen species is a major reason for cellular toxicity induced by salinity in various crop plant species [25–29]. Regulation of ROS production is the most vital step to reduce cellular oxidative stress and toxicity [2].

- **Yield**: Many studies have reported that agricultural productivity decreases due to salinity stress. It always reduces the productivity of the soil. Salinity stress significantly decreases the number of seeds and pods per plant; it also causes a decrease in seed weight. Nahar and Hasanuzzaman [30] reported that the salinity causes an extreme reduction in the number of pods in *Vigna radiata* crop. It a reduction in the yield, decrease in productivity, induces senescence, physiologically less active green vegetation, and reduces photosynthetic activity [31]. Salinity stress reduces fertility rate, which may be due to inhibition of carbohydrate supply to the developing panicles [32]. Salinity stress reduces the viability of pollen, which causes unsuccessful germination of the seed set [33]. Semiz et al. [34] have done an investigation on *Foeniculum vulgare* to analyze the effect of salinity on plant growth parameters including plant height, fresh weight yield, and biomass. It has been shown that an increase in the level of salt ions in water supplied to plant can significantly affect many growth parameters (Fig. 1).

According to Ghassemi et al. [35] globally, the world's irrigated soil was estimated and different parameters were taken including total cropped land area, irrigated land area, and salt-affected irrigated land in which soil of 11 countries was taken into consideration. These countries are China, India, the Russia the United States, Pakistan, Iran, Thailand, Egypt, Australia, Argentina, and South Africa. Russia has the highest cropland area, which is around 233 Mha (million ha.). Egypt has the least cropland area with only 3 Mha. China has the highest irrigated area, around 45 Mha. South Africa has the least irrigated land area, which is only 1 Mha. Salt-affected irrigated land is highest in India, around 7 Mha. South Africa has the least salt-affected irrigated land, only 0.1 Mha.

4 Detection of salinity

Detection of salinity with economic and efficient tools, namely remote sensing (RS) data technology and geographic information system (GIS) technology [36], is a route to monitor and map soil-affected areas. RS data technology, for example, video image illustrations, microwave image illustrations, aerial

FIG. 1 Suppressed plant growth due to salinization. *(Source: Earth observing system (EOS).)*

image illustrations, infrared thermography, and visible and infrared multi-spectral images [37], has been utilized for spotting, scanning, and recording salt-affected area. According to Morshed et al. [36], Landsat imagery is a technique that is frequently utilized for finding different aspects of the surface as it provides a wide array of bands and permits image augmentation. Several authors have claimed that when information provided by satellite images and information from field data analysis are combined, salinity detection can be done more accurately [38,39]. In an investigation, Pearson correlation among 18 salinity indices and field data on soil salinity, which is expressed in decisie-mens/meter, have been performed to get information on soil salinity [39]. They established that salinity around an infrared (NIR) band shows a significant relation with the field EC. To predict soil salinity, they used NIR band and SI_2 in the form of the regression equation, which proved precise for satellite image-based salinity diagnosis techniques for spectral, temporal, and spatial discrepancies. Salinity levels in soil were predicted by Khan et al. [40], using principal component analysis (PCA) and soil salinity indices. Garcia et al. [41] evaluated soil salinity by using near-infrared (NIR) band, blue band, near-infrared (NIR)/red (band 4/band 3) ratio, and normalized difference vegetation index (NDVI) in the form of regression equation. Three regression models have been utilized by them, i.e., spatial autoregression (SAR), ordinary least squares (OLS), and spatial lag (SLAG) model, and the salinity forecasting model was used for evaluating and assessing high R2 and low P-value at low standard error [36] (Fig. 2).

Standard techniques used for finding salinity in the soil by using field data analysis are (i) exchangeable sodium percentage (ESP), (ii) total dissolved salts (TDS), (iii) electrical conductivity (EC), and (iv) sodium adsorption ratio (SAR) [36].

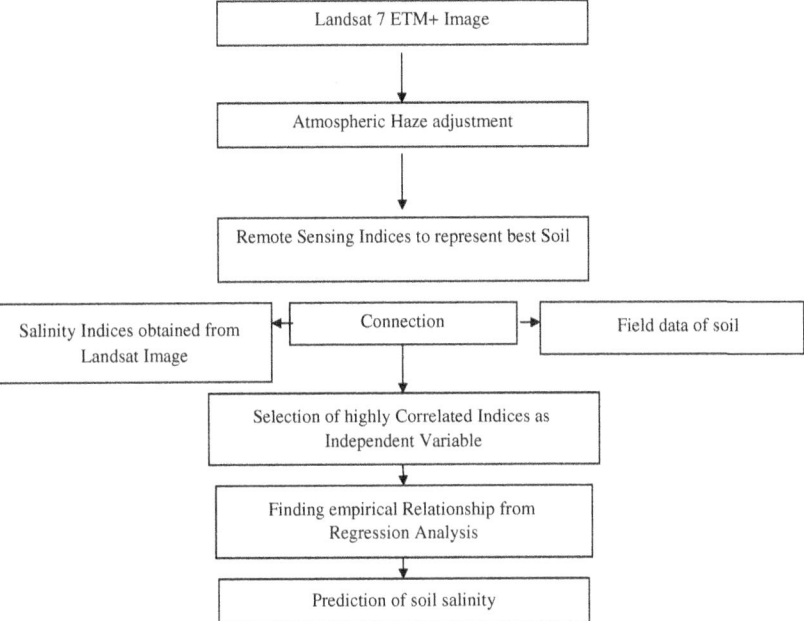

FIG. 2 Integrated approach for soil salinity detection [36].

- **Satellite image editing process**—Three techniques are used to get accurate data and information regarding the salinity quality of surface soil [36]. The satellite image of the Earth's surface will be recorded accurately only if the atmosphere is free from different kinds of particles [36]. Chavez [42] provided a method to get an accurate data and information from the satellite image. This is an improved dark-object subtraction approach for atmospheric scattering correction (haze value correction) of multispectral data. All the data and information that were documented in digital number (DN) were converted to reflection factor (reflectance). Huang et al. [43] established that noise from reflectance was 50% less than noise from DN value. For this, DN values were first converted to radiance by considering the gain and bias of different bands using the following expression: $L = (gain*DN)+bias$, where L is radiance [36]. Later, the radiance was converted to reflectance by taking into consideration the Earth-Sun distance and the Sun's elevation with the help of the following equation:

$$R = \pi * L * d2 / E * \theta se$$

where

R stands for reflectance
L stands for radiance

d stands for distance between Earth and Sun (which relies on the day when image is achieved)

θse stands for sun elevation, expressed in radian [36].

- **Selecting salinity indices and their correlations**—First, the salinity indices were selected, studied, and the interrelations between these indices were analyzed. Field data of soil salinity (regarding electrical conductivity) were enumerated to investigate the highly correlated or interrelated indices [36].
- **Analysis by regression model**—In this method, stepwise regression analysis was done to determine the best possible correlation between the field electrical conductivity (EC) values and salinity. A multiple regression equation was developed by using OLS, SAR, and SLAG as regression models. In multiple regression equations, highly correlated indices were taken as independent variables in order to find the salinity of soil. The GeoDa software was used to calculate the regression model. The anticipated EC values were found to deviate from the normal trend line. The above correlation studies show that the histogram of residuals was very close to a normal distribution. These correlation studies established that residuals show no correlation among themselves and showed spatial independence with each other [36].
- **Soil salinity**—The soil salinity map has been categorized into seven classes, with EC values ranging from nonsaline soil to heavy saline soil [36].

4.1 Visual diagnosis of saline and sodic soils

- **Flocculation and salt crust**: The process of aggregation of fine particles into clusters is known as flocculation and is favorable for soil aeration and penetration of roots. Salt crystals and stains of light gray or white color on surface soils are some of the physical indicators for salinity in soil. Saline soils tend to keep the clay in a flocculated condition due to excessive salts present in them.

5 Strategies to tolerate salinity in flora

Plants can develop and perform their metabolic processes due to the elevated concentration of NaCl salt substrate [2]. Primary minerals are the major source of salts in the Earth's outermost layer.

- **Fluctuation in the estimation of ion**: The drift of sodium and potassium ions was measured by the vibrating microelectrodes [44,45] and determined by an electrochemical potential difference among the two locations [46].
- **Assessment of soluble substances**: Roots, as well as leaves, were taken to assess the soil matter, then it is kept in salt-stress conditions and continuously provided water for 7 days, and then it was reaped. The Solarbio Science & Technology, Beijing calculated free amino acids with Solarbio Assay Kit [47].
- **Determining the concentration of chlorophyll**: First, the fresh leaves were taken and punched into discs and then soaked in C_3H_6O and H_2O. Then the

chloropropyl is extracted at room temperature within complete darkness until the leaves become completely white. Measurements were done in a spectrophotometer using acetone as blank [47]. Both chlorophyll A and B were calculated at different wavelengths and the equations were determined for evaluating concentrations [48].

6 Plant-bacteria interaction to reduce salinity

The rhizosphere region in several plants remains colonized by the plant growth-promoting rhizobacteria (PGPR). These PGPR are found to show various beneficial results such as enhanced plant growth and better defense system against the plant pathogens (bacteria, virus, fungus, nematodes, and more) [49,50]. These rhizobacteria are known to cause enhancement in the germination of seed along with the growth of the root and shoot. They also play a vital role in the absorption of the nutrients and provide tolerance to stress in plants. Hence, these bacteria are termed as plant-growth-promoting bacteria (PGPB). Although the main purpose of the PGPB is to improve the productivity of the crop plants even under salinity, a noninvasive method for checking its efficiency under high salt conditions is still lacking. A study conducted by the South Korean scientists showed improvement in the salinity resistance when *Brevibacterium linens* RS16 (a PGPR) was inoculated in the rice plant. They concluded that the volatile emission of foliage and the photosynthetic characteristic may serve as the noninvasive markers and might help improve salinity resistance. This finding led the scientists to concentrate on the emission of volatile organic compounds, which eventually act as a plant defense system. But this emission was found to be harmful for the climate as well as the environment.

Increased salinity in the farm field causes elevation in the oxidative stress, and eventually decreased photosynthesis that affects the productivity of the crops. Hence, it could be concluded that salt stress once developed beyond the tolerance level could impact photosynthesis through the volatile emissions thereby disrupting the physiology of the plants. A study was conducted on two different genotypes of rice plant, one with salt-sensitive gene and another with salt-tolerant gene in order to understand the actual amount of foliage and the volatile emission when the crop plant was injected with the halo-tolerant PGPB along with mixing of the soil with salt. It was observed that the plant injected with the halo-tolerant PGPB gene were found to show reduced salt stress with higher photosynthetic rate and reduced emission of the volatile organic compounds. Hence, the salt-stressed plants showed positive results in response to the PGPB gene in comparison to the control plants [51].

7 Plant-fungi interaction to reduce salinity

Fungi usually grow close to some specific plant species below ground. Beneath the earth, the fungi are present in a network of fine white filaments known as mycelia, which permeate through the ground and link to the adjacent plant root system. Countless mycelia build a concentrated web of filaments surrounding the exposed roots of a tree. With the help of branched mycelium, the fungi take water and nutrition in the form of phosphorus and nitrogen and deliver them directly into plant roots. Roots deliver water and nutrition to leaves. During photosynthesis dextrose is produced from water and sunlight. If plenty of dextrose is produced for the plant then the leftover is transferred beneath the Earth's surface. This process helps fungi to sustain because they are unable to build dextrose. This type of symbiotic relationship between plants and fungi referred to as mycorrhiza is advantageous. In this symbiotic relationship, the tree was benefitted by taking in water and fungi get a sufficient amount of dextrose to continue their living and development. Around 30 different varieties of mycorrhiza fungi are present in the tree's root system. Fungus filaments make a network among adjacent trees and provide nutrients to the weaker tree. It is possible that the same mycorrhiza coexists in a variety of plants. One of the most popular fungi involved in the interaction with the plant is arbuscular mycorrhiza. Arbuscular mycorrhiza when present in a symbiotic relationship with plants confers some benefits: (1) absorption of fewer mobility ions, (2) enhances the physical parameters of soil structure, (3) increases the flora, (4) ameliorates biogeochemical cycle, and (5) makes the plant resistant to biotic and abiotic stress [52–55].

8 Conclusions

Agricultural system is extremely affected by abiotic stress especially salinity because it exerts influence on plant metabolic processes [56]. Therefore, to detect salinity a remote sensing technique was developed. Thus, slowly and gradually plant develops a tolerance power and becomes resistant to salinity stress [2]. Ion-flux measurements, assessment of soluble substances, and determination of the chlorophyll content are the strategies adopted by plants to tolerate salinity. Salinity can be reduced by using various methods which involve the interaction between the plants and the microbes. Rhizobacteria and arbuscular mycorrhiza are the bacteria and fungi, respectively, that help in the growth of plants and diminish salinity thus helping in the overall development of plants and the whole agricultural system [57].

References

[1] A. Marwal, A. Sahu, R.K. Gaur, New insights in the functional genomics of plants responding to abiotic stress, in: R.K. Gaur, P. Sharma (Eds.), Molecular Approaches in Plant Abiotic Stress, CRC Press, Science Publishers, Taylor & Francis Group, 2014, pp. 158–180, https://doi.org/10.1201/b15538 (Chapter 10, ISBN: 13: 978-1-4665-8894-3).

[2] P. Parihar, S. Singh, R. Singh, V.P. Singh, S.M. Prasad, Effect of salinity stress on plants and its tolerance strategies, 22 (2014) 3739.

[3] D. Egamberdieva, B. Lugtenberg, Use of plant growth-promoting rhizobacteria to alleviate salinity stress in plants, in: M. Miransari (Ed.), Use of Microbes for the Alleviation of Soil Stresses, vol. 1, Springer, New York, 2014. https://doi.org/10.1007/978-1-4614-9466-9_4.

[4] I. Szabolcs, Salt Affected Soils in Europe, Martinus Nijhoff, The Hague, 1974, p. 63.

[5] B. Nwer, H. Zurqani, E. Rhoma, The use of remote sensing and geographic information system for soil salinity monitoring in Libya, GSTF Int. J. Geol. Sci. 1 (1) (2013) 1–5.

[6] H. Safdar, A. Amin, Y. Shafiq, A. Ali, R. Yasin, A. Shoukat, M.U. Hussan, M.I. Sarwar, A review: impact of salinity on plant growth, Nat. Sci. 17 (1) (2019) 34–40.

[7] M. Ibrar, M. Jabeen, J. Tabassum, F. Hussain, I. Ilahi, Salt tolerance potential of Brassica juncea Linn, J. Sci. Technol. Univ. Peshawar 27 (2003) 79–84.

[8] M. Ulfat, H. Athar, M. Ashraf, N.A. Akram, A. Jamil, Appraisal of physiological and biochemical selection criteria for evaluation of salt tolerance in canola (Brassica napus L.), Pak. J. Bot. 39 (2007) 1593–1608.

[9] S. Xu, B. Hu, Z. He, F. Ma, J. Feng, W. Shen, J. Yan, Enhancement of salinity tolerance during rice seed germination by presoaking with hemoglobin, Int. J. Mol. Sci. 12 (2011) 2488–2501.

[10] M.A. Khan, D.J. Weber, Ecophysiology of High Salinity Tolerant Plants (Tasks for Vegetation Science), first ed., Springer, Amsterdam, 2008.

[11] E. Gomes-Filho, C.R.F. Machado Lima, J.H. Costa, A.C. da Silva, M. da Guia Silva Lima, C.F. de Lacerda, J.T. Prisco, Cowpea ribonuclease: properties and effect of NaCl salinity on its activation during seed germination and seedling establishment, Plant Cell Rep. 27 (2008) 147–157.

[12] B.F. Dantas, R.L. De Sa, C.A. Aragao, Germination, initial growth and cotyledon protein content of bean cultivars under salinity stress, Rev. Bras. Sementes 29 (2007) 106–110.

[13] M.A. Khan, Y. Rizvi, Effect of salinity, temperature and growth regulators on the germination and early seedling growth of Atriplex griffithii var. Stocksii, Can. J. Bot. 72 (1994) 475–479.

[14] A. Bordi, The influence of salt stress on seed germination, growth and yield of canola cultivars, Not. Bot. Horti Agrobot. 38 (2010) 128–133.

[15] Z. Khodarahmpour, M. Ifar, M. Motamedi, Effects of NaCl salinity on maize (Zea mays L.) at germination and early seedling stage, Afr. J. Biotechnol. 11 (2012) 298–304.

[16] R. Romero Aranda, T. Soria, S. Cuartero, Tomato plant water uptake and plant water relationships under saline growth conditions, Plant Sci. 160 (2001) 265–272.

[17] M. Qadir, S. Schubert, Degradation processes and nutrient constraints in sodic soils, Land Degrad. Dev. 13 (2002) 275–294.

[18] L.A. Tuna, C. Kaya, M. Ashraf, H. Altunlu, I. Yokas, B. Yagmur, The effects of calcium sulphate on growth, membrane stability and nutrient uptake of tomato plants grown under salt stress, Environ. Exp. Bot. 59 (2007) 173–178.

[19] A.J. Keutgen, E. Pawelzik, Impacts of NaCl stress on plant growth and mineral nutrient assimilation in two cultivars of strawberry, Environ. Exp. Bot. 65 (2009) 170–176.

[20] P. Ahmad, C.A. Jaleel, M.A. Salem, G. Nabi, S. Sharma, Roles of enzymatic and non-enzymatic antioxidants in plants during abiotic stress, Crit. Rev. Biotechnol. 30 (2010) 161–175.

[21] P. Ahmad, G. Nabi, M. Ashraf, Cadmium induced oxidative damage in mustard Brassica juncea (L.) Czern. & Coss. plants can be alleviated by salicylic acid, S. Afr. J. Bot. 77 (2011) 36–44.

[22] P. Ahmad, S. Sharma, Salt stress and phytobiochemical responses of plants, Plant Soil Environ. 54 (2008) 89–99.

[23] A. Marwal, S.S. Verma, S. Trivedi, R. Prajapat, Reactive oxygen species: its effects on various diseases, J. Adv. Biotechnol. 3 (1) (2014) 122–134, https://doi.org/10.24297/jbt.v3i1.1684.

[24] A.K. Parida, A.B. Das, Salt tolerance and salinity effect on plants: a review, Ecotoxicol. Environ. Saf. 60 (2005) 324–349.

[25] P. Ahmad, C.A. Jaleel, S. Sharma, Antioxidative defence system, lipid peroxidation, proline-metabolizing enzymes and biochemical activity in two genotypes of Morus alba L. subjected to NaCl stress, Russ. J. Plant Physiol. 57 (2010) 509–517.

[26] P. Ahmad, C.A. Jaleel, M.M. Azooz, G. Nabi, Generation of ROS and non-enzymatic antioxidants during abiotic stress in plants, Bot. Res. Int. 2 (2009) 11–20.

[27] M.L. Dionisio-Sese, S. Tobita, Antioxidant responses of rice seedlings to salinity stress, Plant Sci. 135 (1998) 1–9.

[28] Y. Gueta-Dahan, Z. Yaniv, B.A. Zilinskas, G. Ben Hayyim, Salt and oxidative stress: similar and specific responses and their relation to salt tolerance in citrus, Planta 204 (1997) 460–469.

[29] V. Mittova, M. Guy, M. Tal, M. Volokita, Salinity upregulates the antioxidative system in root mitochondria and peroxisomes of the wild salt tolerant tomato species Lycopersicon pennellii, J. Exp. Bot. 55 (2004) 1105–1113.

[30] K. Nahar, M. Hasanuzzaman, Germination, growth, nodulation and yield performance of three mung bean varieties under different levels of salinity stress, Green Farming 2 (2009) 825–829.

[31] A. Wahid, R. Rao, E. Rasul, Identification of salt tolerance traits in sugarcane lines, Field Crop Res. 54 (1997) 9–17.

[32] P.S.S. Murty, K.S. Murty, Spikelet sterility in relation to nitrogen and carbohydrate contents in rice, Indian J. Plant Physiol. 25 (1982) 40–48.

[33] Z. Abdullah, M.A. Khan, T.J. Flowers, Causes of sterility in seed set of rice under salinity stress, J. Agron. Crop Sci. 167 (2001) 25–32.

[34] G.D. Semiz, A. Ünlukara, E. Yurtseven, D.L. Suarez, I. Telci, Salinity impact on yield, water use, mineral and essential oil content of fennel (Foeniculum vulgare Mill.), J. Agric. Sci. 18 (2012) 177–186.

[35] F. Ghassemi, A.J. Jakeman, H.A. Nix, Salinisation of Land and Water Resources: Human Causes, Extent, Management and Case Studies, UNSW Press/CAB International, Sydney, Australia/Wallingford, UK, 1995.

[36] M.M. Morshed, M.T. Islam, R. Jamil, Soil salinity detection from satellite image analysis: an integrated approach of salinity indices and field data, Environ. Monit. Assess. 188 (2) (2016), 119, https://doi.org/10.1007/s10661-015-5045-x.

[37] G.I. Metternicht, J.A. Zinck, Remote sensing of soil salinity: potentials and constraints, Remote Sens. Environ. 85 (1) (2003) 1–20.

[38] T.F.A. Bishop, A.B. McBratney, A comparison of prediction method for the creation of field-extent soil property maps, Geoderma 103 (2001) 149–160.

[39] M. Bouaziz, J. Matschullat, R. Gloaguen, Improved remote sensing detection of soil salinity from a semi-arid climate in Northeast Brazil, Compt. Rendus Geosci. 343 (2011) 795–803.

[40] N.M. Khan, V.V. Rastoskuev, E. Shalina, Y. Sato, Mapping salt affected soil using remote sensing indicators. A simple approach with the use of GisIdrissi, in: 22nd Asian Conference on Remote Sensing, 5–9 November 2001, Singapore, 2001.

[41] L. Garcia, A. Eldeiry, A. Elhaddad, Estimating soil salinity using remote sensing data, in: Proceedings of the 2005 Central Plains Irrigation Conference, 2005, pp. 1–10.

[42] P.S. Chavez Jr., An improved dark-object subtraction technique for atmospheric scattering correction of multispectral data, Remote Sens. Environ. 24 (1988) 459–479.

[43] C. Huang, L. Yang, C. Homer, B. Wylie, J. Vogelman, T. DeFelice, At Satellite Reflectance: A First Order Normalization of LANDSAT 7 ETM+ Images, 2002.

[44] T.A. Cuin, D. Parsons, S. Shabala, Wheat cultivars can be screened for NaCl salinity tolerance by measuring leaf chlorophyll content and shoot sap potassium, Funct. Plant Biol. 37 (7) (2010) 656–664, https://doi.org/10.1071/FP09229.

[45] H.H. Wu, L. Shabala, X.H. Liu, E. Azzarello, M. Zhou, C. Pandolfi, Z.H. Chen, J. Bose, S. Mancuso, S. Shabala, Linking salinity stress tolerance with tissue-specific Na+ sequestration in wheat roots, Front. Plant Sci. 6 (2015) 71, https://doi.org/10.3389/fpls.2015.00071.

[46] I.A. Newman, Ion transport in roots: measurement of fluxes using ion-selective microelectrodes to characterize transporter function, Plant Cell Environ. 24 (1) (2001) 1–14, https://doi.org/10.1046/j.1365-3040.2001.00661.x.

[47] Y. Yuan, C. Wu, L. Liu, Q. Ma, Q. Yang, B. Fang, Unravelling the distinctive growth mechanism of proso millet (*Panicum miliaceum* L.) under salt stress: from root-to-leaf adaptations to molecular response, GCB Bioenergy (2021), https://doi.org/10.1111/gcbb.12910.

[48] H.K. Lichtenthaler, C. Buschmann, Chlorophylls and carotenoids: measurement and characterization by UV-VIS spectroscopy, Curr. Protocol Food Anal. Chem. 1 (1) (2001) F4.3.1–F4.3.8, https://doi.org/10.1002/0471142913. faf0403s01.

[49] M. Meena, P. Swapnil, K. Divyanshu, S. Kumar, Y.N. Tripathi, A. Zehra, A. Marwal, R.S. Upadhyay, PGPR-mediated induction of systemic resistance and physiochemical alterations in plants against the pathogens: current perspectives, J. Basic Microbiol. 60 (10) (2020) 828–861.

[50] R. Prajapat, A. Marwal, P.N. Jha, *Erwinia carotovora* associated with potato: a critical appraisal with respect to Indian perspective, Int. J. Curr. Microbiol. App. Sci. 2 (10) (2013) 83–89.

[51] P. Chatterjee, A. Kanagendran, S. Samaddar, L. Pazouki, T.M. Sa, U. Niinemets, Inoculation of Brevibacterium linens RS16 in Oryza sativa genotypes enhanced salinity resistance: impacts on photosynthetic traits and foliar volatile emissions, Sci. Total Environ. 645 (2018) 721–732.

[52] S. Mahmood, N. Lakra, A. Marwal, N.M. Sudheep, K. Anwar, Crop genetic engineering: an approach to improve fungal resistance in plant system, in: D.P. Singh, H.B. Singh, R. Prabha (Eds.), Plant-Microbe Interactions in Agro-Ecological Perspectives. Volume 2: Microbial Interactions and Agro-Ecological Impacts, Springer Nature, Singapore, 2017, pp. 581–591, https://doi.org/10.1007/978-981-10-6593-4_23 (Chapter 23).

[53] A. Marwal, R. Kumar, R.K. Verma, M. Mishra, R.K. Gaur, S.M.P. Khurana, Genomics and molecular mechanisms of plant's response to abiotic and biotic stresses, in: S.M.P. Khurana, R.K. Gaur (Eds.), Plant Biotechnology: Progress in Genomic Era, 2019, pp. 131–146, https://doi.org/10.1007/978-981-13-8499-8_6 (Chapter 6).

[54] A. Porcel, J.R. Lozano, Salinity stress alleviation using arbuscular mycorrhizal fungi. A review, Agron. Sustain. Dev. 32 (2012) 181–200, https://doi.org/10.1007/s13593-011-0029-x.

[55] N.M. Sudheep, A. Marwal, N. Lakra, K. Anwar, S. Mahmood, Fascinating fungal endophytes role and possible beneficial applications: an overview, in: D.P. Singh, H.B. Singh, R. Prabha (Eds.), Plant-Microbe Interactions in Agro-Ecological Perspectives. Volume 1: Fundamental Mechanisms, Methods and Functions, 2017, pp. 255–273, https://doi.org/10.1007/978-981-10-5813-4_13 (Chapter 13).

[56] P. Swapnil, M. Meena, S.K. Singh, U.P. Dhuldhaj, A. Marwal, Vital roles of carotenoids in plants and humans to deteriorate stress with its structure, biosynthesis, metabolic engineering and functional aspects, Curr. Plant Biol. 26 (2021) 100203.

[57] S. Arora, J.C. Dagar, Research developments in saline, Agriculture (2019), https://doi.org/10.1007/978-981-13-5832-6_5.

Chapter 16

Metabolomic studies of medicinal plant-fungi interaction

Mahinder Partap[a,b,*], Abhishek Kumar[a,*], Pankaj Kumar[c], Shiv Shanker Pandey[a,b], and Ashish R. Warghat[a,b]

[a]*Biotechnology Division, CSIR-Institute of Himalayan Bioresource Technology, Palampur, Himachal Pradesh, India,* [b]*Academy of Scientific and Innovative Research (AcSIR), Ghaziabad, Uttar Pradesh, India,* [c]*Department of Biotechnology, Dr. Y.S. Parmar University of Horticulture and Forestry, Solan, Himachal Pradesh, India*

1 Literature search process

The review of literature has been searched using PubMed, Google Scholar, Web of Science, Sci-Finder, Science-Direct, and Scopus databases. The databases were searched with the keywords such as medicinal plants, plant-fungi interaction, metabolomics, secondary metabolites, analytical chemistry, or omics analysis. The data presented in this chapter were obtained from published research articles, review papers, books, and reports in English version only.

2 Introduction

The microsymbionts are closely associated with the host plants through roots or soil close to the origins [1,2]. The plant and microbial associations can be categorized as positive associations such as symbiotic or nonsymbiotic benefiting the associates, and negative associations including competition or parasitism [3]. During symbiotic association, dynamic variation in the genome, transcriptome, metabolic pathways, and signaling networks have been observed in the symbiotic allies [1]. The fungi can augment or modify the function, growth, development, reproduction, and metabolism of the host plants [4]. Primarily, during plant-fungi association, the elicitor molecules are secreted by the fungal partner (exogenous) or produced inside the plant cell (endogenous) by a specific physical or chemical stimulus. These are the key interacting molecules

---

*Authors contributed equally to this work and considered as joint first author.

Plant-Microbe Interaction—Recent Advances in Molecular and Biochemical Approaches
https://doi.org/10.1016/B978-0-323-91876-3.00003-8

responsible for successful plant-fungi association [5]. Elicitors are the molecules secreted in traces and are responsible for activating plant defense mechanisms [6]. Based on the nature, elicitors are categorized into two groups: biotic and abiotic elicitors. However, it is categorized as exogenous or endogenous based on its origin. The elicitors have been widely used as molecules that could produce or stimulate/induce the production of desired secondary metabolites in medicinal plants [6]. These metabolites are synthesized in a sequential reaction, where plants are exposed to abiotic and biotic stress as a part of the defense system [6]. Industries use such specialized metabolites as therapeutics, nutraceuticals, and dietary components [1]. In recent years, the interest of pharmaceutical industries increased toward applications of elicitors to enhance the production of natural compounds in medicinal plants [1].

The development of new methods and approaches for metabolomics studies helps to understand the molecular level of the plant-fungal interaction. Metabolomics is an efficient method for the qualitative and quantitative analysis of various metabolites [7]. The metabolomics methodology follows two principal techniques: (1) analytical and (2) computational. The analytical methods include accurate, simultaneous, comprehensive, and high-throughput estimation of metabolites. However, in the computational technique, the multivariate analysis and statistical processing of the analytical chemistry data are processed using numerical bioinformatics [8]. Therefore, metabolomics is a sensitive, reliable, and appropriate method for the complete analysis of active metabolites and metabolism in the living system. The practical application of metabolomics approaches in plant-fungi interaction studies has been widely described in numerous research and review articles [1,2]. In addition, metabolomics is also commonly employed in other important research areas such as plant-microbe interaction, agriculture, nutrition sciences, environmental sciences, natural product chemistry, and many more [1]. Therefore, this chapter focuse on metabolomics studies involving medicinal plants and fungi interaction with qualiquantitative metabolites analysis. This chapter also discusses the effect of fungal elicitors on the medicinal plant and their applications toward secondary metabolite production. The mechanism involved in medicinal plants and fungi interaction is also discussed.

3 Plant-fungi interactions

The interaction of plants with fungi is considered to be as old as the emergence of plants [1]. Positive plant and fungi associations provide stabilization to both partners. But negative associations may have resulted in host destabilization [9]. Nevertheless, most associations enhance plant growth, development, and metabolisms. The fungi actively act as symbiotic partners that improvise biochemical processes, soil-resource utilization, adaptation, and stress tolerance in medicinal plants. In return, the plants delivered a carbon source to the symbiotic fungi and contributed toward stable association [9–11]. In the review of literature, the symbiotic plant-fungal interactions can be classified into endophytic and mycorrhizal associations. However, endophytic fungi showed symptomless growth in the living plant tissues until senescence. At the senescence stage, the fungi may change to some extent into a pathogenic form [1,9,12]. With so many fungal endophytes, most plants are likely to embrace the benefits of stress tolerance to the various pathogen(s), or biotic and abiotic environmental/stress conditions. Mycorrhizae are different from endophytes based on nutrient transfer, absorption, and uptake during host interactions [1,2]. The evolutionary advantage of the plant-fungal symbiotic relationship likely allowed plants to change from aquatic to terrestrial [13].

4 Metabolomics approach

Metabolomics is an omics approach for the comprehensive and simultaneous profiling of phytochemicals in a sample or organism [7]. The metabolome analysis involves multiple sequential steps, including preparation and extraction of samples, quali-quantitative analysis, and data investigation (Fig. 1). This technique is highly effective for the precise quantification of primary metabolites (carbohydrates, amino acids, and organic acids) and secondary metabolites (phenolics, alkaloids, terpenes, and steroids) in plants [8]. The secondary metabolites are more stable than primary metabolites because of a lower turnover rate. Therefore, the quenching process is eliminated during sample preparation for secondary metabolite extraction. After quenching, different organic or inorganic solvents (ethanol, ethyl acetate, methanol, dichloromethane or hexane, etc.) have been used for the extraction procedure. The selection of appropriate solvents for extraction depends on the metabolites of interest and the sample source. Various techniques, including classical solvents, supercritical fluids, subcritical water, steam extraction, sonication, microwave-assisted, and high hydrostatic pressure, have been used for the extraction of phytochemicals. The internal standards have been added at the initiation stage of the extraction procedure to enable precise quantification and normalization against procedural/technical variability. After that, the qualitative and quantitative profiling of phytochemicals can be done using targeted and untargeted metabolomics approaches [7,8,14].

The untargeted metabolomics approach aims to simultaneously detect various classes of phytochemicals in a given sample at the same time. Liquid or gas chromatography (LC or GC), hyphenated mass spectrometry (MS), and nuclear magnetic resonance (NMR) are often used for this purpose [8]. In untargeted metabolomics, the analysis of metabolites depends on applied detection technologies, mass libraries, and bioinformatics tools. This analysis process can be executed using various commercial software such as ThermoFisher, Agilent technologies, Waters Corporation, and software developed by the National Institute of Standards and Technology. However, the targeted metabolomic approaches depend on analytical chemistry or quali-quantitative estimation

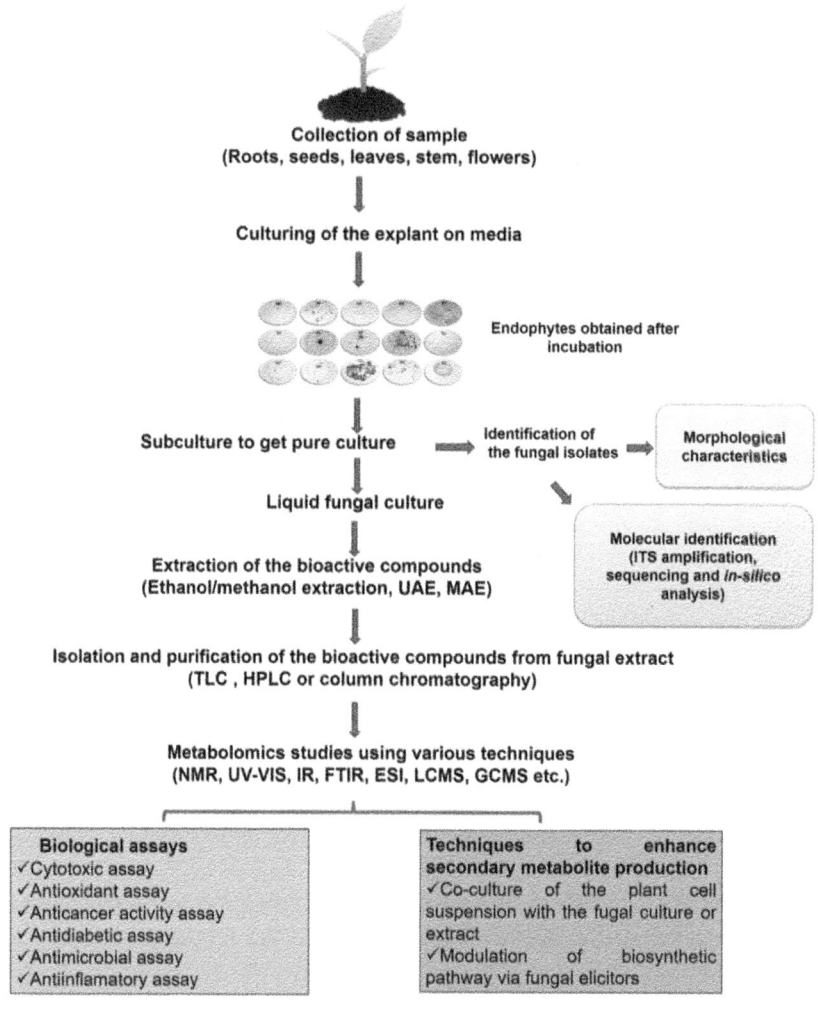

FIG. 1 Various steps and techniques involved in the isolation of the fungal endophytes and bioactive compounds analysis. *UAE*, ultrasonic-assisted extraction; *MAE*, microwave-assisted extraction; *TLC*, thin layer chromatography; *HPLC*, high-performance liquid chromatography; *NMR*, nuclear magnetic resonance; *UV-Vis*, ultraviolet visible spectroscopy; *IR*, infrared spectroscopy; *FTIR*, Fourier-transform infrared spectroscopy; *ESI*, electrospray ionization; *LCMS*, liquid chromatography mass spectrometry; *GCMS*, gas chromatography mass spectrometry. *(The images adopted from the Clipart (clipartkey.com).)*

of targeted metabolites in a given sample. This method commences with evaluating the performance and validation of the analytical method using retention time, UV/mass spectra, linear range, correlation coefficient, the limit of quantification, limit of detection, intraday, interday, and recovery parameters [15]. Authentic standards are required to study the performance of these parameters.

Further, the validation of the developed method is often achieved by spiking experiments (to obtain the recovery percentage) [15,16]. Following the development and validation of a precise analytical approach, the method can be used to quantify targeted metabolites in the given sample. Various metabolomics approaches have been reported for profiling and estimation of secondary metabolites in the medicinal plants during fungal associations (Table 1).

5 Fungi elicitors and their action in medicinal plants

In medicinal plants, the secondary metabolite accumulation is often very low or in traces. Their content depends on the plant source (including plant tissue type), plant developmental stage, physiological status, and environmental factors. Therefore, researchers have focused on the elicitation process as an alternative strategy to increase the production of secondary metabolites. Primarily fungal elicitors are being used to improve the production of specific metabolites [1,2]. Several studies have demonstrated the application of fungal elicitors to enhance secondary metabolites with metabolomics techniques in medicinal plants (Table 1). During plant-fungi association, the fungi secrete the elicitor molecules that induce/stimulate the synthesis of the secondary metabolites in the host plant and are also found to enhance the production of secondary metabolites [1,2]. First, the fungal elicitors bind to the receptor present on the plant cell wall leading to the stimulation of the downstream signaling cascade. Also, it activates the structural and regulatory genes involving transcription factors, specific gene expression, and biosynthetic pathways [1,63]. However, the first response to the binding of the elicitors to the receptor is the alteration of the ionic fluxes of ions (Cl^-, K^+, H^+, Ca^{2+}), and the cytoplasm becomes acidic [64]. As a result of the change in the pH of the cytosol, the physiological responses are affected within the cell. Changes in the optimum level of the Ca^{2+} concentration inside the cell and organelles (chloroplast and mitochondria) initiate a signaling cascade, which triggers salicylic acid, jasmonic acid, ethylene, and calcium-dependent protein kinase pathways. As a result, various defense and stress-responsive genes are activated, enhancing the secondary metabolite production inside the cell. The change in the pH of the cell also initiates NADPH oxidase and the production of the reactive oxygen species (ROS), which stimulates stress-responsive genes [65,66]. Further, it initiates the stress-responsive transcription factors (TFs), which bind to the DNA, resulting in the transcription and translation of the genes and ultimately enhancing the secondary metabolite production. In another response, mitogen-activated protein kinases (MAPKs) are activated, which leads to the phosphorylation of the TFs. The TFs further enhance the production of the JA, SA, and secondary metabolites, which help to protect the plant from pathogens [67]. The mechanism for enhancing the secondary metabolite by the fungal elicitors response is shown in Fig. 2.

TABLE 1 Metabolomics approach for secondary metabolites analysis during plant and fungi interaction.

Plant species	Fungi	Secondary metabolites	Metabolomics approaches	References
Ambrosia artemisiifolia	Protomyces gravidus	Thiarubrine A	HPLC-UV	Bhagwath and Hjortsø [17]
Andrographis paniculata	Glomus mosseae and Trichoderma harzianum	Andrographolide	Spectrophotometric	Arpana and Bagyaraj [18]
Artemisia annua	Penicillium chrysogenum	Artemisinin	HPLC	
	Aspergillus terreus (AFL, AFSt, AFR)	Alkaloids, coumarins and polyketides	LC-HRMS	Sayed et al. [19]
	Aspergillus terreus, A. favus, A. oryzae, Penicillium commune, P. chrysogenum, P. chrysogenum, Talaromyces piophilus, T. piophilus, Fusarium oxysporum, F. nematophilum, Pleosporaceae sp.	Physcion, emodin, katenarin, norjavanicin, dechlorogriseofulvin, benzyl benzoate, 4-hydroxy benzyl benzoate, benzyl anisate	LC-HRMS	Alhadrami et al. [20]
Atractylodes lancea	Gilmaniella sp.	Atractylone	GC	Wang et al. [21]
Cajanus cajan	Fusarium solani, Fusarium oxysporum, Hypocrea lixii, and Fusarium proliferatum	Cajaninstilbene acid and Cajanol	LC-MS/MS	Zhao et al. [22] and Zhao et al. [23]
Capsicum annuum	Alternaria alternata	Capsaicin	LC-ESI-MS/MS	Devari et al. [24]
Catharanthus roseus	Phytophthora megasperma, Alternaria carthami	Polyacetylenes	UV spectra	Tietjen et al. [25]

Plant	Endophyte	Compound	Method	Reference
Catharanthus roseus	Fusarium oxysporum, Talaromyces radicus, and Eutypella spp.	Vinblastin and vincristine	LC-ESI-MS, TLC, HPLC, ESI-MS, and NMR	Palem et al. [26] Kumar et al. [27] Pandey et al. [28]
Cephalotaxus hainanensis	Alternaria tenuissima	Homoharringtonine	HPLC, LC-MS/MS, and NMR	Hu et al. [29]
Cichorium intybus	Phytophthora parasitica	Coumarin	HPLC and NMR	Bais et al. [3]
Cinchona ledgeriana	Phomopsis, Diaporthe, Schizophyllum, Penicillium, Fomitopsis, and Arthrinium	Cinchona	HPLC	Maehara et al. [30]
Coleus blumei	Pythium aphanidermatum	Rosmarinic acid	HPLC	Szabo et al. [31]
Coleus forskohlii	Rhizoctonia bataticola	Forskolin	TLC	Mir et al. [32]
Daucus carota	Pythium aphanidermatum and Aspergillus flavus	p-Hydroxybenzoic acid and Anthocyanin	RP-HPLC, HPLC, and PC	Schnitzler et al. [33] and Rajendran et al. [34]
Digitalis lanata	Alternaria spp., Penicillium spp., and Aspergillus spp.	Digoxine	HPLC	Kaul et al. [35]
Dioscorea deltoidea	Rhizopus arrhizus and Alternaria tenuis	Diosgenin	GC	Rokem et al. [36] and Rojas et al. [37]
Euphorbia pekinensis	Fusarium sp.	Euphol	HPLC	Gao et al. [38]

Continued

TABLE 1 Metabolomics approach for secondary metabolites analysis during plant and fungi interaction—cont'd

Plant species	Fungi	Secondary metabolites	Metabolomics approaches	References
Forsythia suspensa	Colletotrichum gloeosporioides	Philliryn	TLC, HPLC, and HPLC-MS	Zhang et al. [15,16]
Fritillaria cirrhosa	Fusarium redolens	Peimisine; imperaline-3-β-D-glucoside	HPLC-ELSD-MS	Pan et al. [39]
Ginkgo biloba	Fusarium oxysporum SY0056	Glinkolide B	TLC, HPLC/ESI-MS and NMR	Cui et al. [40]
Glycine max	Phytophthora megasperma	Glyceollin		Hille et al. [41]
Gymnema sylvestre	Aspergillus niger	Gymnemic acid	HPLC	Chodisetti et al. [42]
Hyoscyamus muticus	Rhizoctonia solani	Sesquiterpenes		Singh [43]
Hypericum perforatum	Yeast	Hypericin	HPLC-UV	Kirakosyan et al. [44]
Macleaya cordata	Fusarium proliferatum BLH51	Sanguinarine	TLC and HPLC	Wang et al. [45]
Medicago sativa	Verticillium albo-atrum Aspergillus terreus	Phytoalexins phenolic metabolites	TLC LC-HRES-MS	Walton et al. [46] Sayed et al. [19]
Nerium indicum	Geomyces sp.	Vincamine	TLC, HPLC, and LC-MS	Na et al. [47]

Ocinum basilicum	Aspergillus niger	Rosmarinic acid	HPLC-MS	Bais et al. [3]
Passiflora incarnata	Alternaria alternata, Colletotrichum capsici, and Chryseobacterium taiwanense	Chrysin	UV-vis, FT-IR, LC-ESI-MS, and NMR	Seetharaman et al. [48]
Piper nigrum	Colletotrichum gloeosporioides	Piperine	LC-MS/MS	Chithra et al. [49]
Podocarpus gracilior	Aspergillus terreus	Taxol	HPLC, NMR and FTIR	El-Sayed et al. [50]
Rheum palmatum	Fusarium solani	Emodin and rhein	TLC HPLC, and LC-MS	You et al. [51]
Rhodiola rosea	Phialocephala fortinii	Salidroside; p-tyrosol	UPLC/Q-TOF-MS and NMR	Cui et al. [52]
Rhodiola sachalinensis	Aspergillus niger, Coriolus versicolor, and Ganoderma lucidum	Salidroside	HPLC	Zhou et al. [53]
Salvia miltiorrhiza	Phoma glomerata D14	Salvianolic acid	HPLC	Li et al. [54]
Solanum nigrum	Aspergillus flavus	Solamargine	TLC, LC-HRESIMS, and NMR	El-Hawary et al. [55]
Taverniera cuneifolia	Mucor hiemalis, Fusarium moniliforme, and Aspergillus niger	Glycyrrhizic acid	TLC	Awad et al. [56]
Taxus chinensis	Aspergillus niger Fusarium maire Fusarium mairei	Taxol Taxol Taxol	HPLC-UV HPLC HPSEC	Wang et al. [57,58]

Continued

TABLE 1 Metabolomics approach for secondary metabolites analysis during plant and fungi interaction —cont'd

Plant species	Fungi	Secondary metabolites	Metabolomics approaches	References
Taxus cuspidata				Li et al. [59]
Withania somnifera	Aspergillus terreus, Penicillium oxalicum, Sarocladium kiliense	Withanolide	HPLC	Pandey et al. [60] and Kushwaha et al. [61,62]

TLC, thin layer chromatography; HPLC, high-performance liquid chromatography; LC, liquid chromatography; UV, ultraviolet; NMR, nuclear magnetic resonance; UPLC/Q-TOF-MS, ultrahigh-performance liquid chromatography-quadrupole time-of-flight mass spectrometry; PC, paper chromatography; LC-MS/MS, liquid chromatography with tandem mass spectrometry; IR, infrared spectroscopy; FT-IR, Fourier-transform infrared spectroscopy; RP-HPLC, reverse-phase high-performance liquid chromatography; ESI-MS, electrospray ionization mass spectrometry; LC-HRESIMS, liquid chromatography-high-resolution electrospray ionization mass spectrometry; HPLC-ELSD, high-performance liquid chromatography evaporative light scattering detection; GC, gas chromatography; HPSEC, high-performance size exclusion chromatography.

(Data adopted from I. Chamkhi, T. Benali, T. Aanniz, N. El Menyiy, F.E. Guaouguaou, N. El Omari, … A. Bouyahya, Plant-microbial interaction: the mechanism and the application of microbial elicitor induced secondary metabolites biosynthesis in medicinal plants, Plant Physiol. Biochem. 167 (2021) 269–295 and O.C. Gómez, J.H.H. Luiz, Endophytic fungi isolated from medicinal plants: future prospects of bioactive natural products from Tabebuia/Handroanthus endophytes, Appl. Microbiol. Biotechnol. 102(21) (2018) 9105–9119.)

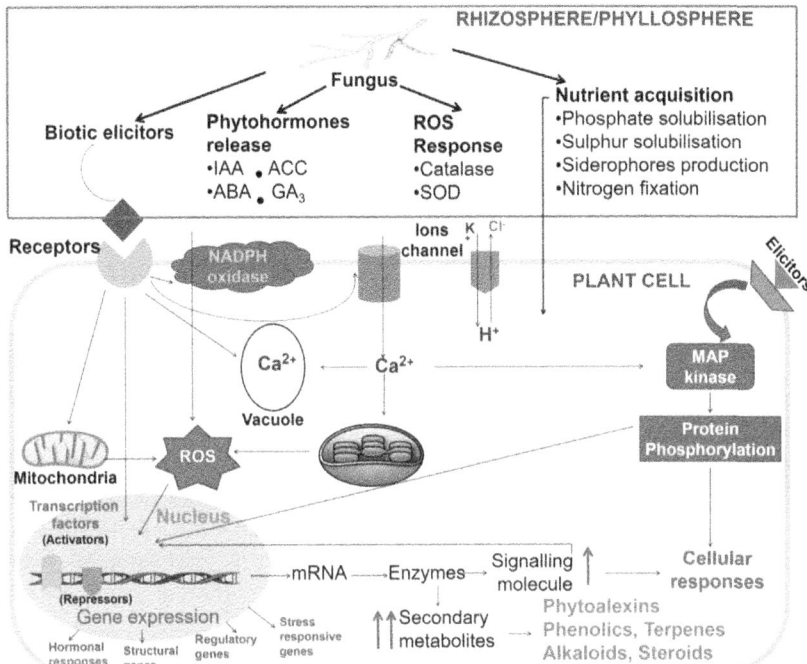

FIG. 2   Mechanism involved in the production and enhancement of secondary metabolites in plant-fungi interaction. *(The images adopted from the Clipart (clipartkey.com) and concept adopted from S.N.R. Shasmita, S.K. Rath, S. Behera, S.K. Naik, In vitro secondary metabolite production through fungal elicitation: an approach for sustainability, in: Fungal Nanobionics: Principles and Applications, Springer, Singapore, 2018, pp. 215–242.)*

Several reports on plant-fungi interaction associated with secondary metabolites are presented in Table 1. The β-glucan elicitor from *Phytophthora megasperma* helps to enhance the production of glyceollin in the soybean [68]. In the report of Walton et al. [46], elicitor (glycoprotein) from the fungi *Verticillium albo-atrum* has been found to increase the production of phytoalexins in *Medicago sativa* suspension culture [69]. Li et al. [70] reported the increase in diosgenin accumulation in *Dioscorea zingiberensis* suspension culture using the fungus *Fusarium oxysporum* as elicitor [36,37]. Similarly, in the *Artemisia annua* artemisinin, content increased by using an elicitor of fungi *Colletotrichum gloeosporioides* [57,58]. Stimulation of biosynthesis of metabolites in cell cultures of *Ambrosia artemisiifolia* was also achieved using thiarubrine, an elicitor from the *Protomyces graidus* [17]. A report by Schnitzler et al. [33] demonstrated that the fungus *Pythium aphanidermatum* enhanced the accumulation of *p*-hydroxybenzoic acid in *Daucus carota* [71]. Veit et al. [72] reported that *P. aphanidermatum* stimulated the synthesis of 4-hydroxybenzoic acid in the cell culture of *D. carota*.

6 Case studies on medicinal plant-fungi interaction

6.1 *Catharanthus roseus* (L.) G. Don (Madagascar periwinkle)

Catharanthus roseus is a widely studied medicinal plant and is considered as a model species for studying the plant-microbe interactions [73,74]. This plant produces anticancerous compounds such as bisindole alkaloids, vinblastine, and vincristine. Different fungal endophytes like *Curvularia* sp. and *Choanephora infundibulifera* [28], *F. oxysporum* [27], *Talaromyces radicus*, and *Eutypella* spp. [26] have been reported from *C. roseus*. Kumar et al. [27] reported *F. oxysporum* endophyte from *C. roseus*. These fungal endophytes help to produce the content of vincristine and vinblastine. These compounds were purified using the TLC and HPLC technique and characterized using ESI-MS, UV-Vis spectroscopy, MS/MS, and NMR techniques. However, Palem et al. [26] isolated 22 endophytic fungi from this medicinal plant, and only *Eutypella* sp.-CrP14 and *T. radicus*-CrP20 showed the strongest antiproliferative activity. While tryptophan decarboxylase (a key enzyme of the biosynthetic pathway of terpenoid indole alkaloid) was only amplified in *T. radicus*-CrP20, which yields vincristine and vinblastine. Pandey et al. [28] reported enhanced production of vindoline content (229%–403%) in *C. roseus*, when cultured with fungal endophytes, i.e., *Curvularia* sp. CATDLF5 and *C. infundibulifera* CATDLF6. These fungal elicitors may increase the expression of key regulatory genes involved in vindoline biosynthesis [28].

6.2 *Withania somnifera* L. (Indian ginseng or Ashwagandha)

Withania somnifera is widely distributed worldwide and is an important medicinal plant used in Ayurveda, Unani, Siddha, and the Chinese system. All parts of this plant have therapeutic importance due to bioactive compounds, including steroidal lactones withanolides and withaferin-A [75,76]. About 40 endophytes have been found in *W. somnifera*, and among them, 11 were fungal endophytes [60]. Endophytes enhanced photosynthesis and increased the production of withanolide content by upregulating the expression of key genes in *W. somnifera*. In this report, the metabolite content was measured using the HPLC technique. However, in another study of *W. somnifera*, 97%–100% enhancement in withanolide content was observed using HPLC analysis when the plant was inoculated with fungal endophytes 2aWF (*Aspergillus terreus*), 5aWF (*Penicillium oxalicum*), and 10aWF (*Sarocladium kiliense*) [61]. Further, in the cell suspension culture of *W. somnifera*, *A. terreus* endophyte elicited the withanolide (10.29 μg/g FCB) content [62].

6.3 *Artemisia annua* L. (sweet wormwood)

Artemisia annua L. is a well-known medicinal herb and is considered as the major source of artemisinin. About 11 fungal endophytes were isolated from the stem and leaf tissue of *A. annua* plant [20]. Further, LC-HRMS-based metabolomics and multivariate analysis have been used to screen the antimalarial compounds from these endophytes. The active metabolites including norjavanicin, physcion, emodin, benzyl benzoate, katenarin, dechlorogriseofulvin, benzyl anisate, and 4-hydroxy benzyl benzoate were obtained from these endophytes. Further in another study by Sayed et al. [19], three fungal endophytes AFL, AFSt, and AFR (*A. terreus*) were isolated from the leaves, stem, and roots of *A. annua*. Metabolomics studies using LC-MS technique and multivariate statistical analysis showed that all the fungal extracts possess antioxidant compounds. In later studies, phenolics, polyketides, and coumarins were bioactive metabolites of the three *A. terreus* fungal strains.

6.4   *Medicago sativa* L. (alfa alfa)

Medicago sativa, commonly known as "father of all foods," is an important medicinal plant used in traditional medicine systems for the treatment of digestive, central nervous system disorders, management of diabetes, asthma, inflammation, gallstones, kidney disorders, and microbial infections [77–79]. This plant produces phytoalexins in response to the elicitor released by the fungal phytopathogens. Coculture of *M. sativa* cell suspension culture with glycoprotein elicitor from the extract of endophytic fungi *V. albo-atrum*, enhanced the level of phytoalexins up to ~160%, as detected by radiolabeled-thin layer chromatography (R-TLC) [46]. In another study, endophytic fungi *A. terreus* was isolated from *M. sativa*, and its extract was analyzed for metabolomics study using liquid chromatography coupled with high-resolution mass spectrometry (LC-HRES-MS) analysis [19]. In this study, a total of 632 compounds have been identified. Among these, fungi extracted from the leaves have been the richest source of phenolic metabolites and found maximum antioxidant capacity. The ethyl acetate extracts from fungal culture were found to show antioxidant properties against the prooxidant xanthine oxidase [19].

6.5   *Taxus baccata* L. (yew trees)

Taxus baccata is an evergreen tree that belongs to the family Taxaceae and is well known for its medicinal important taxol compounds. Taxol, also known as paclitaxel (an anticancer drug), was synthesized by 18 different genera of fungi, mostly by the endophytic fungi living within *Taxus* sp. [80,81]. *Taxus cuspidata* cell suspension cultures, when cocultured with endophytic fungi (*Fusarium mairei*) have been reported to increase the paclitaxel amount by 2–6.8 fold [82]. The active compound of the fungal extract was identified by the high-performance size exclusion chromatography (HPSEC) and found to be an exopolysaccharide of ~79 kD. However, the taxol yield was more than two fold higher when the plant was cocultured with fungus compared to fungal culture extract [83]. Soliman et al. [82] reported that the endophyte (*Paraconiothyrium*) enhanced the production of paclitaxel (detected by the HPLC) in the *Taxus*. They also found the fungicidal activity of the paclitaxel against conifer wood fungal pathogens. Li et al. [59] also reported 25.63 mg/L of paclitaxel within 15 days of the coculture of the *F. maire* endophytic fungi with *Taxus chinensis* cell suspension culture, which was about 38-fold higher than the control experiment in the HPLC studies. However, *A. terreus* (an endophyte isolated from *Podocarpus gracilior*) enhanced the paclitaxel production by ∼2.4-folds in in vitro treated culture of *P. gracilior* [50]. They suggested that the physical interaction of the endophytic fungi isolated from *P. gracilior* with *A. terrus* must be essential for enhancing the biosynthesis of paclitaxel. They have confirmed the chemical identity of paclitaxel extracted from *A. terreus* using the NMR, HPLC, and FTIR analysis [50].

7 Conclusions

Through the coevolution of microbes and higher organisms, different fungi have established a close association with their host plants. As discussed in this chapter, different studies have reported on the interaction between plants and fungi with their stimulatory effect on secondary metabolite production. The published reports highlighted that fungal elicitors enhance the biosynthesis of specific secondary metabolites in host plants. Therefore, this approach is a promising and alternative discipline to increase the content of specialized compounds and induce the biosynthesis of metabolic pathways in plants. Furthermore, in the era of omics technologies, integrative approaches of metabolomics are applied successfully to elucidate the plant-fungi interaction and their mode of metabolism action. These approaches help to understand in deep at least three main aspects of medicinal plants: (1) qualitative and quantitative analysis of any metabolites present in the plant, (2) the metabolite biosynthesis, and their pathway elucidation, and (3) the mechanism and mode of action of these metabolites on diseases. Plant metabolomics provides a broad understanding of all types of metabolites involved in plant metabolism and during interaction with the fungi. This approach can be widely used in diverse area such as synthetic biology, medical science, plant cell, tissue culture, cocultivation, agriculture, and environmental sciences. The enhancement of metabolites in medicinal plants by fungal elicitor response hyphenated to metabolomics techniques could provide a promising platform for industrial applications.

Author's contribution

Mahinder Partap and Abhishek Kumar: Conceptualization, collecting data and interpretation, original figures and table preparation, original draft writing. **Pankaj Kumar**, **Shiv Shanker Pandey**, and **Ashish R. Warghat:** Conceptualization, supervision, writing, reviewing, and editing.

Acknowledgments

The authors are thankful to the Director, CSIR-IHBT, for providing all the necessary facilities. This study was financially supported by CSIR-IHBT, Palampur, Himachal Pradesh, Government of India, under the project Endophyte Network Project (MLP-0171), and National Medicinal Plant Board (NMPB, Ministry of Ayush, New Delhi) project (GAP-0274). Mahinder Partap was supported by the CSIR, New Delhi, through the Senior Research Fellowship and Academy of Scientific and Innovative Research (AcSIR), Ghaziabad, India, for PhD enrolment.

References

[1] I. Chamkhi, T. Benali, T. Aanniz, N. El Menyiy, F.E. Guaouguaou, N. El Omari, A. Bouyahya, Plant-microbial interaction: the mechanism and the application of microbial elicitor induced secondary metabolites biosynthesis in medicinal plants, Plant Physiol. Biochem. 167 (2021) 269–295.

[2] O.C. Gómez, J.H.H. Luiz, Endophytic fungi isolated from medicinal plants: future prospects of bioactive natural products from *Tabebuia/Handroanthus* endophytes, Appl. Microbiol. Biotechnol. 102 (21) (2018) 9105–9119.

[3] H.P. Bais, T.L. Weir, L.G. Perry, S. Gilroy, J.M. Vivanco, The role of root exudates in rhizosphere interactions with plants and other organisms, Annu. Rev. Plant Biol. 57 (2006) 233–266, https://doi.org/10.1146/annurev.arplant.57.032905.105159.

[4] M. Narayani, S. Srivastava, Elicitation: a stimulation of stress in in vitro plant cell/tissue cultures for enhancement of secondary metabolite production, Phytochem. Rev. 16 (2017) 1227–1252, https://doi.org/10.1007/s11101-017-9534-0.

[5] M.E. Maffei, G.-I. Arimura, A. Mithöfer, Natural elicitors, effectors and modulators of plant responses, Nat. Prod. Rep. 29 (2012) 1288, https://doi.org/10.1039/c2np20053h.

[6] M. Yuan, B.P.M. Ngou, P. Ding, X.F. Xin, PTI-ETI crosstalk: an integrative view of plant immunity, Curr. Opin. Plant Biol. 62 (2021), 102030, https://doi.org/10.1016/j.pbi.2021. 102030.

[7] K. Segers, S. Declerck, D. Mangelings, Y.V. Heyden, A.V. Eeckhaut, Analytical techniques for metabolomic studies: a review, Bioanalysis 11 (24) (2019) 2297–2318.

[8] A. Shafi, I. Zahoor, Metabolomics of medicinal and aromatic plants: goldmines of secondary metabolites for herbal medicine research, in: Medicinal and Aromatic Plants, Academic Press, 2021, pp. 261–287.

[9] S. Zeilinger, V.K. Gupta, T.E. Dahms, R.N. Silva, H.B. Singh, R.S. Upadhyay, S.C. Nayak, Friends or foes? Emerging insights from fungal interactions with plants, FEMS Microbiol. Rev. 40 (2) (2016) 182–207.

[10] T. Barupal, M. Meena, K. Sharma, Comparative analysis of bioformulations against *Curvularia lunata* (Wakker) Boedijn causing leaf spot disease of maize, Arch. Phytopathol. Plant Protect. 54 (5–6) (2020) 261–272, https://doi.org/10.1080/03235408.2020.1827657.

[11] M. Meena, P. Swapnil, A. Zehra, M.K. Dubey, M. Aamir, C.B. Patel, R.S. Upadhyay, Virulence factors and their associated genes in microbes, in: H.B. Singh, V.K. Gupta, S. Jogaiah (Eds.), New and Future Developments in Microbial Biotechnology and Bioengineering, Elsevier, 2019, pp. 181–208, https://doi.org/10.1016/B978-0-444-63503-7.00011-5.

[12] C.B. Patel, V.K. Singh, A.P. Singh, M. Meena, R.S. Upadhyay, Microbial genes involved in interaction with plants, in: H.B. Singh, V.K. Gupta, S. Jogaiah (Eds.), New and Future

Developments in Microbial Biotechnology and Bioengineering, Elsevier, Singapore, 2019, pp. 171–180, https://doi.org/10.1016/B978-0-444-63503-7.00010-3.

[13] E.T. Kiers, M.G.V.D. Heijden, Mutualistic stability in the arbuscular mycorrhizal symbiosis: exploring hypotheses of evolutionary cooperation, Ecology 87 (7) (2006) 1627–1636.

[14] M. Meena, A. Zehra, M.K. Dubey, M. Aamir, V.K. Gupta, R.S. Upadhyay, Comparative evaluation of biochemical changes in tomato (*Lycopersicon esculentum* Mill.) infected by *Alternaria alternata* and its toxic metabolites (TeA, AOH, and AME), Front. Plant Sci. 7 (2016) 1408, https://doi.org/10.3389/fpls.2016.01408.

[15] A. Zhang, H. Sun, P. Wang, Y. Han, X. Wang, Modern analytical techniques in metabolomics analysis, Analyst 137 (2) (2012) 293–300.

[16] Q. Zhang, X. Wei, J. Wang, Phillyrin produced by *Colletotrichum gloeosporioides*, an endophytic fungus isolated from *Forsythia suspensa*, Fitoterapia 83 (8) (2012) 1500–1505, https://doi.org/10.1016/j.fitote.2012.08.017.

[17] S.G. Bhagwath, M.A. Hjortsø, Statistical analysis of elicitation strategies for thiarubrine A production in hairy root cultures of *Ambrosia artemisiifolia*, J. Biotechnol. 80 (2000) 159–167, https://doi.org/10.1016/S0168-1656(00)00256-X.

[18] J. Arpana, D.J. Bagyaraj, Response of kalmegh to an arbuscular mycorrhizal fungus and a plant growth promoting rhizomicroorganism at two levels of phosphorus fertilizer, Am.-Euras. J. Agric. Environ. Sci. 2 (2007) 33–38.

[19] A.M. Sayed, N.H. Sherif, A.O. El-Gendy, Y.I. Shamikh, A.T. Ali, E.Z. Attia, M.M. El-Katatny, B.A. Khalifa, H.M. Hassan, U.R. Abdelmohsen, Metabolomic profiling and antioxidant potential of three fungal endophytes derived from *Artemisia annua* and *Medicago sativa*, Nat. Prod. Res. 36 (9) (2020) 1–5.

[20] H.A. Alhadrami, A.M. Sayed, A.O. El-Gendy, Y.I. Shamikh, Y. Gaber, W. Bakeer, N.H. Sheirf, E.Z. Attia, G.M. Shaban, B.A. Khalifa, C.J. Ngwa, A metabolomic approach to target antimalarial metabolites in the *Artemisia annua* fungal endophytes, Sci. Rep. 11 (1) (2021) 1.

[21] Y. Wang, C.C. Dai, J.L. Cao, D.S. Xu, Comparison of the effects of fungal endophyte *Gilmaniella* sp. and its elicitor on *Atractylodes lancea* plantlets, World J. Microbiol. Biotechnol. 28 (2012) 575–584, https://doi.org/10.1007/s11274-011-0850-z.

[22] J. Zhao, Y. Fu, M. Luo, Y. Zu, W. Wang, C. Zhao, C. Gu, Endophytic fungi from pigeon pea [*Cajanus cajan* (L.) Millsp.] produce antioxidant cajaninstilbene acid, J. Agric. Food Chem. 60 (17) (2012) 4314–4319, https://doi.org/10.1021/jf205097y.

[23] J. Zhao, C. Li, W. Wang, C. Zhao, M. Luo, F. Mu, Y. Fu, Y. Zu, M. Yao, *Hypocrea lixii*, novel endophytic fungi producing anticancer agent cajanol, isolated from pigeon pea (*Cajanus cajan* [L.] Millsp.), J. Appl. Microbiol. 115 (1) (2013) 102–113, https://doi.org/10.1111/jam.12195.

[24] S. Devari, S. Jaglan, M. Kumar, R. Deshidi, S. Guru, S. Bhushan, M. Kushwaha, A.P. Gupta, S. G. Gandhi, J.P. Sharma, S.C. Taneja, R.A. Vishwakarma, B.A. Shah, Capsaicin production by *Alternaria alternata*, an endophytic fungus from *Capsicum annum*; LC-ESI-MS/MS analysis, Phytochemistry 98 (2014) 183–189, https://doi.org/10.1016/j.phytochem.2013.12.001.

[25] K.G. Tietjen, D. Hunkler, U. Matern, Differential response of cultured parsley cells to elicitors from two non-pathogenic strains of fungi, Eur. J. Biochem. 131 (1983) 401–407, https://doi.org/10.1111/j.1432-1033.1983.tb07277.x.

[26] P.P.C. Palem, G.C. Kuriakose, C. Jayabaskaran, Correction: an endophytic fungus, *Talaromyces radicus*, isolated from *Catharanthus roseus*, produces vincristine and vinblastine, which induce apoptotic cell death, PLoS One 11 (4) (2016) 1–6, https://doi.org/10.1371/journal.pone.0153111.

[27] A. Kumar, D. Patil, P.R. Rajamohanan, A. Ahmad, Isolation, purification and characterization of vinblastine and vincristine from endophytic fungus *fusarium oxysporum* isolated from *Catharanthus roseus*, PLoS One 8 (9) (2013), e71805.

[28] S.S. Pandey, S. Singh, C.S. Babu, K. Shanker, N.K. Srivastava, A.K. Shukla, A. Kalra, Fungal endophytes of *Catharanthus roseus* enhance vindoline content by modulating structural and regulatory genes related to terpenoid indole alkaloid biosynthesis, Sci. Rep. 6 (1) (2016) 1–4.

[29] X. Hu, W. Li, M. Yuan, C. Li, S. Liu, C. Jiang, Y. Wu, K. Cai, Y. Liu, Homoharringtonine production by endophytic fungus isolated from *Cephalotaxus hainanensis* Li, World J. Microbiol. Biotechnol. 32 (7) (2016) 1–9, https://doi.org/10.1007/s11274-016-2073-9.

[30] S. Maehara, P. Simanjuntak, C. Kitamura, K. Ohashi, H. Shibuya, Cinchona alkaloids are also produced by an endophytic filamentous fungus living in *Cinchona* plant, Chem. Pharm. Bull. 59 (8) (2011) 1073–1074, https://doi.org/10.1248/cpb.59.1073.

[31] E. Szabo, A. Thelen, M. Petersen, Fungal elicitor preparations and methyl jasmonate enhance rosmarinic acid accumulation in suspension cultures of *Coleus blumei*, Plant Cell Rep. 18 (1999) 485–489, https://doi.org/10.1007/s002990050608.

[32] R.A. Mir, P.S. Kaushik, R.A. Chowdery, M. Anuradha, Elicitation of forskolin in cultures of *Rhizactonia bataticola*-a phytochemical synthesizing endophytic fungi, Int. J. Pharm. Pharm. Sci. 7 (10) (2015) 185–189 (ISSN: 0975-1491).

[33] J.P. Schnitzler, J. Madlung, A. Rose, H. Ulrich Seitz, Biosynthesis of phydroxybenzoic acid in elicitor-treated carrot cell cultures, Planta 188 (1992) 594–600, https://doi.org/10.1007/BF00197054.

[34] L. Rajendran, G. Suvarnalatha, G.A. Ravishankar, L.V. Venkataraman, Enhancement of anthocyanin production in callus cultures of *Daucus canota* L. under the influence of fungal elicitors, Appl. Microbiol. Biotechnol. 42 (1994) 227–231, https://doi.org/10.1007/BF00902721.

[35] S. Kaul, M. Ahmed, K. Zargar, P. Sharma, M.K. Dhar, Prospecting endophytic fungal assemblage of *Digitalis lanata* Ehrh. (foxglove) as a novel source of digoxin: a cardiac glycoside, 3 Biotech 3 (4) (2013) 335–340, https://doi.org/10.1007/s13205-012-0106-0.

[36] J.S. Rokem, J. Schwarzberg, I. Goldberg, Autoclaved fungal mycelia increase diosgenin production in cell suspension cultures of *Dioscorea deltoidea*, Plant Cell Rep. 3 (1984) 159–160, https://doi.org/10.1007/BF00270213.

[37] R. Rojas, J. Alba, I. Magaña-Plaza, F. Cruz, A.C. Ramos-Valdivia, Stimulated production of diosgenin in *Dioscorea galeottiana* cell suspension cultures by abiotic and biotic factors, Biotechnol. Lett. 21 (1999) 907–911, https://doi.org/10.1023/A:1005598623728.

[38] F. Gao, Y. Yong, C. Dai, Effects of endophytic fungal elicitor on two kinds of terpenoids production and physiological indexes in *Euphorbia pekinensis* suspension cells, J. Med. Plant Res. 5 (2011) 4418–4425.

[39] B.F. Pan, X. Su, B. Hu, N. Yang, Q. Chen, W. Wu, *Fusarium redolens* 6WBY3, an endophytic fungus isolated from *Fritillaria unibracteata* var. wabuensis, produces peimisine and imperialine-3β-d-glucoside, Fitoterapia 103 (2015) 213–221, https://doi.org/10.1016/j.fitote.2015.04.006.

[40] Y. Cui, D. Yi, X. Bai, B. Sun, Y. Zhao, Y. Zhang, Ginkgolide B produced endophytic fungus (*Fusarium oxysporum*) isolated from *Ginkgo biloba*, Fitoterapia 83 (5) (2012) 913–920, https://doi.org/10.1016/j.fitote.2012.04.009.

[41] A. Hille, C. Purwin, J. Ebel, Induction of enzymes of phytoalexin synthesis in cultured soybean cells by an elicitor from *Phytophthora megasperma* f. sp. glycinea, Plant Cell Rep. 1 (1982) 123–127, https://doi.org/10.1007/BF00272369.

[42] B. Chodisetti, K. Rao, S. Gandi, A. Giri, Improved gymnemic acid production in the suspension cultures of *Gymnema sylvestre* through biotic elicitation, Plant Biotechnol. Rep. 7 (2013) 519–525, https://doi.org/10.1007/s11816-013-0290-3.

[43] G. Singh, Fungal Elicitation of Plant Root Cultures-Application to Bioreactor Dosage, Pennsylvania State University, USA, 1995.

[44] A. Kirakosyan, H. Hayashi, K. Inoue, A. Charchoglyan, H. Vardapetyan, Stimulation of the production of hypericins by mannan in *Hypericum perforatum* shoot cultures, Phytochemistry 53 (2000) 345–348, https://doi.org/10.1016/S0031-9422(99)00496-3.

[45] X.J. Wang, X.W.C. Min, M. Ge, R. Zuo, An endophytic sanguinarine producing fungus from *Macleaya cordata*, *Fusarium proliferatum* BLH51, Curr. Microbiol. 68 (2014) 336–341,- https://doi.org/10.1007/s00284-013-0482-7.

[46] T.J. Walton, C.J. Cooke, R.P. Newton, C.J. Smith, Evidence that generation of inositol 1,4,5-trisphosphate and hydrolysis of phosphatidylinositol 4,5-biphosphate are rapid responses following addition of fungal elicitor which induces phytoalexin synthesis in lucerne (*Medicago sativa*) suspension culture cells, Cell. Signal. 5 (1993) 345–356, https://doi.org/10.1016/0898-6568(93)90026-I.

[47] R. Na, L. Jiajia, Y. Dongliang, P. Yingzi, H. Juan, L. Xiong, Z. Nana, Z. Jing, L. Yitian, Indentification of vincamine indole alkaloids producing endophytic fungi isolated from *Nerium indicum*, Apocynaceae, Microbiol. Res. 192 (2016) 114–121, https://doi.org/10.1016/j.micres.2016.06.008.

[48] P. Seetharaman, S. Gnanasekar, R. Chandrasekaran, G. Chandrakasan, M. Kadarkarai, S. Sivaperumal, Isolation and characterization of anticancer flavone chrysin (5,7-dihydroxy flavone)-producing endophytic fungi from *Passiflora incarnata* L. leaves, Ann. Microbiol. 67 (4) (2017) 321–331.

[49] S. Chithra, B. Jasim, C. Anisha, J. Mathew, E.K. Radhakrishnan, LCMS/MS based identification of piperine production by endophytic Mycosphaerella sp. PF13 from *Piper nigrum*, Appl. Biochem. Biotechnol. 173 (1) (2014) 30–35, https://doi.org/10.1007/s12010-014-0832-3.

[50] A.S. El-Sayed, S. Safan, N.Z. Mohamed, L. Shaban, G.S. Ali, M.Z. Sitohy, Induction of Taxol biosynthesis by *Aspergillus terreus*, endophyte of *Podocarpus gracilior* Pilger, upon intimate interaction with the plant endogenous microbes, Process Biochem. 1 (71) (2018) 31–40.

[51] X. You, S. Feng, S. Luo, D. Cong, Z. Yu, Z. Yang, J. Zhang, Studies on a rhein-producing endophytic fungus isolated from *Rheum palmatum* L, Fitoterapia 85 (1) (2013) 161–168,- https://doi.org/10.1016/j.fitote.2012.12.010.

[52] J. Cui, T. Guo, J. Chao, M. Wang, J. Wang, Potential of the endophytic fungus *Phialocephala fortinii* Rac56 found in *Rhodiola* plants to produce salidroside and p-tyrosol, Molecules 21 (4) (2016) 502, https://doi.org/10.3390/molecules21040502.

[53] X. Zhou, Y. Wu, X. Wang, B. Liu, H. Xu, Salidroside production by hairy roots of *Rhodiola sachalinensis* obtained after transformation with *Agrobacterium rhizogenes*, Biol. Pharm. Bull. 30 (2007) 439–442, https://doi.org/10.1248/bpb.30.439.

[54] X. Li, X. Zhai, Z. Shu, R. Dong, Q. Ming, L. Qin, C. Zheng, Phoma glomerata D14: an endophytic fungus from *Salvia miltiorrhiza* that produces salvianolic acid C, Curr. Microbiol. 73 (1) (2016) 31–37, https://doi.org/10.1007/s00284-016-1023-y.

[55] S.S. El-Hawary, R. Mohammed, S.F. Abouzid, W. Bakeer, R. Ebel, A.M. Sayed, M.E. Rateb, Solamargine production by a fungal endophyte of *Solanum nigrum*, J. Appl. Microbiol. 120 (4) (2016) 900–911, https://doi.org/10.1111/jam.13077.

[56] V. Awad, A. Kuvalekar, A. Harsulkar, Microbial elicitation in root cultures of *Taverniera cuneifolia* (Roth) Arn. for elevated glycyrrhizic acid production, Ind. Crop. Prod. 54 (2014) 13–16, https://doi.org/10.1016/j.indcrop.2013.12.036.

[57] C. Wang, J. Wu, X. Mei, Enhancement of Taxol production and excretion in *Taxus chinensis* cell culture by fungal elicitation and medium renewal, Appl. Microbiol. Biotechnol. 55 (2001) 404–410, https://doi.org/10.1007/s002530000567.

[58] J.W. Wang, Z. Zhang, R.X. Tan, Stimulation of artemisinin production in *Artemisia annua* hairy roots by the elicitor from the endophytic *Colletotrichum* sp, Biotechnol. Lett. 23 (2001) 857–860, https://doi.org/10.1023/A:1010535001943.

[59] Y.C. Li, W.Y. Tao, L. Cheng, Paclitaxel production using co-culture of *Taxus* suspension cells and paclitaxel-producing endophytic fungi in a co-bioreactor, Appl. Microbiol. Biotechnol. 83 (2) (2009) 233–239.

[60] S.S. Pandey, S. Singh, H. Pandey, M. Srivastava, T. Ray, S. Soni, A. Pandey, K. Shanker, C.S. Babu, S. Banerjee, M.M. Gupta, Endophytes of *Withania somnifera* modulate in planta content and the site of withanolide biosynthesis, Sci. Rep. 8 (1) (2018) 1–9.

[61] R.K. Kushwaha, S. Singh, S.S. Pandey, A. Kalra, C.S. Babu, Fungal endophytes attune withanolide biosynthesis in *Withania somnifera*, prime to enhanced withanolide A content in leaves and roots, World J. Microbiol. Biotechnol. 35 (2) (2019) 1–9.

[62] R.K. Kushwaha, S. Singh, S.S. Pandey, A. Kalra, C.S. Vivek Babu, Innate endophytic fungus, *Aspergillus terreus* as biotic elicitor of withanolide A in root cell suspension cultures of *Withania somnifera*, Mol. Biol. Rep. 46 (2) (2019) 1895–1908.

[63] M. Meena, P. Swapnil, K. Divyanshu, S. Kumar, Harish, Y.N. Tripathi, A. Zehra, A. Marwal, R.S. Upadhyay, PGPR-mediated induction of systemic resistance and physiochemical alterations in plants against the pathogens: current perspectives, J. Basic Microbiol. 60 (10) (2020) 828–861, https://doi.org/10.1002/jobm.202000370.

[64] Z.H. Siddiqui, A. Mujib, Mahmooduzzafar, J. Aslam, K.R. Hakeem, T. Parween, In vitro production of secondary metabolites using elicitor in *Catharanthus roseus*: a case study, in: K.R. Hakeem, P. Ahmad, M. Ozturk (Eds.), Crop Improvement, Springer, New York, 2013, pp. 401–419.

[65] M. Meena, P. Swapnil, R.S. Upadhyay, Isolation, characterization and toxicological potential of tenuazonic acid, alternariol and alternariol monomethyl ether produced by *Alternaria* species phytopathogenic on plants, Sci. Rep. 7 (2017) 8777, https://doi.org/10.1038/s41598-017-09138-9.

[66] M. Sagi, R. Fluhr, Production of reactive oxygen species by plant NADPH oxidases, Plant Physiol. 141 (2) (2006) 336–340.

[67] P. Paul, S.K. Singh, B. Patra, X. Sui, S. Pattanaik, A differentially regulated AP2/ERF transcription factor gene cluster acts downstream of a MAP kinase cascade to modulate terpenoid indole alkaloid biosynthesis in *Catharanthus roseus*, New Phytol. 213 (3) (2017) 1107–1123.

[68] J. Ebel, Oligoglucoside elicitor-mediated activation of plant defense, BioEssays 20 (1998) 569–576.

[69] C.E.C.C. Ejike, M. Gong, C.C. Udenigwe, Phytoalexins from the Poaceae: biosynthesis, function and prospects in food preservation, Food Res. Int. 52 (2013) 167–177, https://doi.org/10.1016/j.foodres.2013.03.012.

[70] P. Li, Z. Mao, J. Lou, Y. Li, Y. Mou, S. Lu, Y. Peng, L. Zhou, Enhancement of diosgenin production in *Dioscorea zingiberensis* cell cultures by oligosaccharides from its endophytic fungus *Fusarium oxysporum* Dzf17, Molecules 16 (2011) 10631–10644, https://doi.org/10.3390/molecules161210631.

[71] S.A. Heleno, I.C.F.R. Ferreira, A.P. Esteves, A. Ciric, J. Glamo clija, A. Martins, M. Sokovic, M.J.R.P. Queiroz, Antimicrobial and demelanizing activity of *Ganoderma lucidum* extract, p-hydroxybenzoic and cinnamic acids and their synthetic acetylated glucuronide methyl esters, Food Chem. Toxicol. 58 (2013) 95–100, https://doi.org/10.1016/j.fct.2013.04.025.

[72] S. Veit, J.M. Wörle, T. Nürnberger, W. Koch, H.U. Seitz, A novel protein elicitor (PaNie) from *Pythium aphanidermatum* induces multiple defense responses in carrot, arabidopsis, and tobacco, Plant Physiol. 127 (2001) 832–841, https://doi.org/10.1104/pp.010350.

[73] A.S. Ferreira Filho, et al., Endophytic *Methylobacterium extorquens* expresses a heterologous β-1, 4-endoglucanase A (EglA) in *Catharanthus roseus* seedlings, a model host plant for *Xylella fastidiosa*, World J. Microbiol. Biotechnol. 28 (2012) 1475–1481.

[74] A.K. Shukla, S.P.S. Khanuja, in: D. Barh (Ed.), OMICS Applications in Crop Science, CRC Press, Boca Raton, 2013, pp. 325–384 (Chapter 10).

[75] A. Kumar, M.K. Kaul, M.K. Bhan, P.K. Khanna, K.A. Suri, Morphological and chemical variation in 25 collections of the Indian medicinal plant, *Withania somnifera* (L.) Dunal (Solanaceae), Genet. Resour. Crop. Evol. 54 (2007) 655–660.

[76] V. Sharma, S. Sharma, Pracheta, R. Paliwal, *Withania somnifera*: a rejuvenating Ayurvedic medicinal herb for the treatment of various human ailments, Int. J. Pharmtech Res. 3 (2011) 187–192.

[77] M. Adams, F. Gmunder, M. Hamburger, Plants traditionally used in age related brain disorders - a survey of ethnobotanical literature, J. Ethnopharmacol. 113 (2007) 363–381.

[78] B.I. Cohen, E.H. Mosbach, N. Matoba, S.O. Suh, C.K. McSherry, The effect of alfalfa-corn diets on cholesterol metabolism and gallstones in prairie dogs, Lipids 25 (1990) 143–148.

[79] A.M. Gray, P.R. Flatt, Pancreatic and extra-pancreatic effects of the traditional antidiabetic plant, *Medicago sativa* (lucerne), Br. J. Nutr. 78 (1997) 325–334.

[80] A. Stierle, G. Strobel, D. Stierle, Taxol and taxane production by *Taxomyces andreanae*, an endophytic fungus of Pacific yew, Science 260 (5105) (1993) 214–216.

[81] X. Zhou, H. Zhu, L. Liu, J. Lin, K. Tang, A review: recent advances and future prospects of Taxol-producing endophytic fungi, Appl. Microbiol. Biotechnol. 86 (6) (2010) 1707–1717.

[82] S.S. Soliman, C.P. Trobacher, R. Tsao, J.S. Greenwood, M.N. Raizada, A fungal endophyte induces transcription of genes encoding a redundant fungicide pathway in its host plant, BMC Plant Biol. 13 (1) (2013) 1.

[83] Y.C. Li, W.Y. Tao, Interactions of taxol-producing endophytic fungus with its host (*Taxus* spp.) during taxol accumulation, Cell Biol. Int. 33 (1) (2009) 106–112.

Chapter 17

Sustainable agricultural approach to study interaction of plants and microbes

Parul Tyagi, Ayushi Singh, Pooja Saraswat, Ambika Chaturvedi, and Rajiv Ranjan
Department of Botany, Dayalbagh Educational Institute, Dayalbagh Agra, India

1 Introduction

Sustainable agricultural practice without polluting the environment is the major significant concern in the sector of agriculture [1]. Plant microbes will be involved in playing vital roles in achieving long-term agricultural sustainability for future generations. Plant growth stimulating bio-agents have recently attracted more attention due to their demand as bio-fertilizers and other beneficial impacts on crop development and agro-ecosystem fertility [2,3]. However, the role of sustainable development in agro-ecosystems and the mechanisms of plant-microbe interactions are still in their infancy. As a result, this provides an understanding of the functional and utility of plant-microbe association for sustainable agriculture and development of the environment. Phyto-microorganisms (bacteria, fungus, nematodes, and protozoans, among others) provide favorable conditions for crop yield and management of diseases.

Among the group of microorganisms found in the soil, i.e., bacteria, protozoa, fungi, actinomycetes, and algae, bacteria comprise the major proportion. The rhizosphere surrounding the plant roots contains a good bacterial amount as it is separated from most of the soil [4]. Regardless of the amount of soil bacteria, the plant can get affected by the bacteria in three ways. Bacterial association with the plant might serve to be useful, harmful, or impartial in reference to the plant [5–7]. Free-living bacteria, bacterial spp. of *Frankia* and *Rhizobia* forming a mutual association with plants, blue-green bacteria, and bacterial endophytes invading a few or all inner tissues of a plant are all examples of PGPB. Directly or indirectly, these bacteria can boost plant development as they can efficiently mitigate pathogenicity. Because they also function as biocontrol agents, they facilitate and enhance plant growth by moderating the levels of

Plant-Microbe Interaction—Recent Advances in Molecular and Biochemical Approaches
https://doi.org/10.1016/B978-0-323-91876-3.00011-7

plant hormones [8]. Plant requires a certain plant-microbe interaction for active growth. These interactions are more complex below the soil surface than those taking place above it, and understanding them is critical for plant health and development [9,10]. The literature suggests that the majority of the interactions, including those between rhizosphere microbes and those between plants and microbes, are complex and not completely known. Comprehending growth-promoting variations, microbial ecology, function and mechanism of action to stimulate plant growth is demanding for getting the most out for the increasing population. Fig. 1 and Table 1 highlight many studies of PGPR that help not only to understand their role ecologically, but also their biotechnological uses [11].

As a result, agricultural research intends to concentrate on other ways to boost food production. Plant-microbe interactions studied at the molecular level could serve to be a better choice for sustainable agriculture [25]. Plants with microorganisms (bacteria, fungus, and others) coexist in the phyllosphere and rhizosphere ecosystems. They can be found as endophytes, epiphytes, or even in the soil around the roots. These phyto-pathogens might be beneficial,

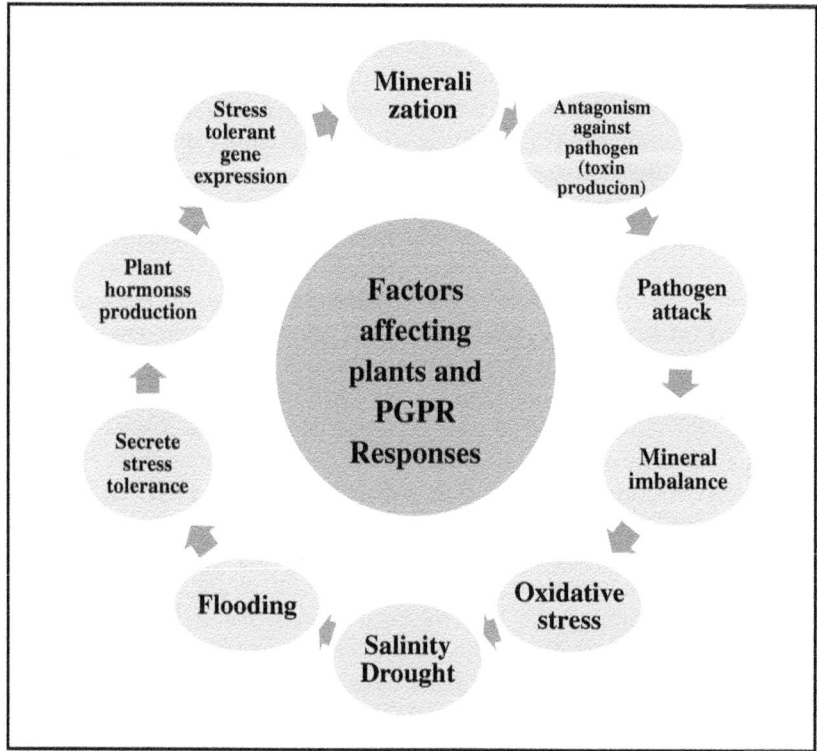

FIG. 1   Numerous types of PGPR response affecting factors.

TABLE 1 Regulation of PGPR in normal conditions.

Experimental crops	Phyto-pathogens	Experimental condition(s)	Proposed mechanism	Plant-microbes interactions	References
Sorghum bicolor (*Poaceae*)	*Azospirillum brasilense* SM	Axenic test	Production of indole-3-acetic (IAA)	Shoot length and seedling dry weight improved over control.	Malhotra and Srivastava [12]
Vigna radiate (*Fabaceae*)	*Acinetobacter baumannii* CR 1.8 and *Klebsiella pneumoniae* SN 1.1	Pot experiment	Production and solubilization of IAA and phosphorus	Adventitious root length increased over control.	Chaiharn and Lumyong [13]
Oryza sativa (*Gramineae*)	*Enterobacter cloacae* GS1	Hydroponic test	Production of IAA and solubilization of phosphorus	Compared to control, increased fresh weight, root, shoot, and nitrogen content.	Shankar et al. [14]
Zea mays (*Poaceae*)	*Acinetobacter baumannii* CR 1.8 and *Klebsiella pneumoniae* SN 1.1	Pot experiment	Production of IAA and solubilization of phosphorus	Slightly higher significance than the control.	Chaiharn and Lumyong [13]
Malus domestic (*Rosaceae*)	*Bacillus subtilis* OSU-142, *B. subtilis* M-3, *Burkholderia pseudomallei* OSU-7, and *Pseudomonas aeruginosa* BA-8	Field experiment	Production of IAA and cytokinin	Average shoot length increased over control.	Aslantas et al. [15]

Continued

TABLE 1 Regulation of PGPR in normal conditions—cont'd

Experimental crops	Phyto-pathogens	Experimental condition(s)	Proposed mechanism	Plant-microbes interactions	References
Musa paradisiaca (*Musaceae*)	*Azospirillum brasilense* Sp7, and *Bacillus sphaericus* UPMB1	Hydroponic test	Nitrogen fixation	Comparatively increased the bunch.	Baset et al. [16]
Beta vulgaris (*Amaranthaceae*)	*Acinetobacter johnsonii* 3-1	Pot experiment	Production of IAA and solubilization of phosphorus	Beet plant dry weight and yield increased over controls.	Shi et al. [17]
Cucumis melo (*Cucurbits*)	*Bacillus subtilis* Y-IVI	Pot experiment	Production of IAA and siderophore	Compared to the control, the shoot dry weight increased.	Zhao et al. [18]
Juglans regia (*Juglandaceae*)	*Pseudomonas chlororaphis* W24, *Pseudomonas fluorescens* W12, *Bacillus cereus* W9	Pot experiment	Solubilization of phosphorus	Walnut seedlings' plant height, shoot, and root dry weight all increased, while phosphorus and nitrogen uptake improved.	Yu et al. [19]
Arachis hypogaea (*Legumes*)	*Pseudomonas* spp. (PGPR1, PGPR2, PGPR4, and PGPR7)	Axenic, pot, and field experiment	Seed inoculation with PGPR containing ACC-deaminase	Improvements in the field-based pod, haulm, and nodule dry weight.	Dey et al. [20]
Nicotiana tabacum (*Nightshade*)	*Pantoea agglomerans* strain PVM	Axenic test	Production of IAA	Root length increased over control.	Apine and Jadhav [21]
Pennisetum glaucum (*Poaceae*)	*Pseudomonas* spp., *Citrobacter* spp., *Acinetobacter* spp., *Serratia* spp., and *Enterobacter* spp.	Pot experiment	Solubilization of phosphorus	Compared to control, significantly increased plant growth and biomass.	Santoyo et al. [22]

Plant	Microorganism	Experiment type	Mechanism	Effect	Reference
Triticum aestivum (Poaceae)	*Bacillus* spp. AW1, *Providencia* spp. AW5, and *Brevundimonas* spp. AW7	Pot experiment	Solubilization of nutrients and nitrogen fixation	Biometric and micronutrient enhancement of plants compared to controls.	Karamanos et al. [23]
Cucumis sativus (Cucurbitaceae)	*Ochrobactrum haematophilum* H10	Pot experiment	Production of IAA, solubilization of phosphorus, and deaminase of ACC	Comparatively increased cucumber leaf and root length.	Zhao et al. [18]
Arabidopsis thaliana (Brassicaceae)	*Burkholderia pyrrocinia* Bcc171, *Chromobacterium violaceum* CV01	Petri plate experiment	Production of volatile organic compounds	In LB media (1 drop) and MR-VP media (3 drops) showed growth-boosting effects compared to controls.	Cohen et al. [24]

or harmful, or inert on plant growth and health [26,27]. The mechanism underlying plant-microbe interaction is yet unknown, and nearby there are a slew of unanswered questions. These questions are more about the plant resistance, signaling pathways (of both microbes and plants), favorable and destructive interaction between microbes and plants, and so on. The investigated studies would aid in a better understanding about the overall mechanism of these interactions, as well as uncover microorganisms that might be useful to boost crop yields shortly [28]. In the agricultural field, the interaction of bacteria with plants acts as a catalyst for naturally increasing yield. Existing farming operations that rely primarily on the function of high-yielding agrochemicals frequently threaten the environment [1].

A climate shift throughout the globe, a decline in the agricultural land area, growing urbanization, and extensive practice of agrochemicals impose disastrous effects on both the environment and crop production, emphasizing the significance of environmentally friendly and sustainable agriculture development [29]. Exploiting microorganisms for enhancement of plant quality, nutrient enrichment, and crop productivity is a good strategy for climate-smart farming practices. Stress mitigation by priming, plant growth promotion (PGP), nutrient uptake, plant-mediated transfer of nutrients that are difficult to absorb, activation of plant defense mechanism, and mycorrhizal symbiosis are some of the benefits imposed by the interaction of microbes and plants [30]. Key proteins engaged in plant growth and providing tolerance against both biotic and abiotic stresses actively participate to maintain plant cellular activities by influencing biochemical and physiological pathways. According to a recent study, modern "omics" technologies have been developed as a crucial approach for discovering new genes that encode functional proteins that would be required in numerous developmental programs for crops [31]. We addressed the significance of interactions involving microbes and plants for crop development and management of stresses in this chapter, with a focus on the benefits of NGS and GWA mapping.

According to a study of the host plant and its connected microbiome (holobiont), plant-microbe interactions are coevolutionary [32]. Modern techniques such as omics approaches (e.g., metabolomics, metagenomics, transcriptomics, and proteomics) next-generation sequencing (NGS), and computational tools help researchers to investigate molecular features and components of plant-microbe interaction that affect plant traits or characteristics. Several recent researches investigated various features of plant microbiota, as well as the impact caused by the genotype of the host on various parts of microbiomes. Genetic data regarding plant-microbe interactions is readily accessible for varieties of plants and related microbes [33]. Understanding the genetic components of plant-microbe interactions will be critical for enhancing microbiome usage in agriculture. In the very same context, advanced technologies like CRISPR-based genome editing (GE), which can create accurate genetic modification, are an excellent platform for rapid learning that is fundamental to interacting with plants and viruses and facilitating genetic modification to increase plant and disease production resistance. In the current chapter, an important account on the current updates on plant interactions with bacteria is extended in terms of the structure, composition, and factors responsible for the formation of plant microbiome.

2 Role in agricultural sustainability

Despite the fact that little is known about the exact mechanism of microbial interaction with plants, increasing the usage of microorganisms in a targeted manner can help to ensure sustainability. Extensive study has shown that organic farming increases the prevalence of microorganisms such as fungal and bacterial load in the soil, which is generally referred to as plant probiotics [34]. In the agricultural industry, the use of beneficial microorganisms has gained traction against chemical-based and synthetic pesticides and fertilizers. The capability of helpful bacteria to colonize seeds when planted in soil, also to provide protection from pathogens, is seen when seeds are inoculated with beneficial microbes [35]. The seed inoculation approach using microbial consortiums has the benefit of delivering bacteria directly to the rhizosphere, where they can form a bond with plants.

Microorganism inoculation improves nutrient availability to plants while also assisting in successful carbon sequestration belowground. Inoculating seeds in leguminous plants causes a high presence of nodule-forming rhizobia colonized in the rhizosphere, responsible to fix nitrogen, and hence increase productivity [36]. *B. ambifaria* MCI 7 has an effective result on maize seedling growth when used as a seed treatment, but on the other hand, when directly inoculated into the soil, it has a harmful effect on the growth of the plant. The increasing expense and distribution in the difficulty associated with phosphorus-based fertilizers prompted the growth of microbial fertilizers that help plants get phosphorus from the soil [37]. "JumpStart" (Monsanto [38]), which contains the *P. bilaii* fungus, is one of the commercialized products for canola and wheat. In one study, it had a high yield (66%); nevertheless, it has been shown to have less beneficial characteristics in other investigations [23]. Just before sowing, the seeds are inoculated with the fungus to make the process easier. *Pseudomonas* species have been potentially known to promote plant growth and pathogen suppression; consequently, multiple methods for seed coating by Pseudomonas have been used with mixed results [39].

In a greenhouse setting, two variants of *P. syringae* were experimented on tomato plants, with the *P. syringae* pv. *syringae* variant 260-02 encouraging plant development and the *P. syringae* pv. tomato strain DC3000 exerting biocontrol of *B. cinerea* against *Cymbidium ringspot* virus. *P. syringae* can be advantageous in some situations, in addition to being a disease. This could be due to its unique volatile emission profiles and patterns of root colonization. Induced systemic resistance (resistance mechanism) was produced against the fungus *Colletotrichum graminicola* in a study when *P. putida* KT2440 was used as a root inoculant in corn plants, as evidenced by the considerably reduced leaf necrosis and low fungal burden in treated samples [40]. *Bacillus* species have emerged as interesting possibilities for developing stable bio-products against illnesses due to their ability to create heat- and drought-resistant endospores. At various phases of growth, inoculating tomato plants with *Pseudomonas* and *Bacillus* boosted yield, growth, and nutritional status [41]. In *Sulla coronaria*, coinoculation of *Pseudomonas* and *Rhizobium sullae* improved development and antioxidant levels while lowering cadmium buildup, whereas *Rhizobium* and *Pseudomonas* increased rice root and shoot dry weight as well as total yield [42]. Inoculating plants or seeds with microorganisms has been used in numerous trials to promote plant growth and development (both single and consortia).

Mycorrhiza is a symbiotic association that exists between root-colonizing fungi and plants. The mycorrhizal partnership begins with a signal exchange between the two partners. The host root of arbuscular mycorrhizal fungi secretes signaling molecules, known as "branching factors," which result in significant hyphal branching [43]. It has long been assumed that arbuscular mycorrhizal fungi produce "myc factors," which are molecular and cellular responses that allow arbuscular mycorrhizal fungi to colonize roots successfully. Until the discovery of "branching factors" in the root secretions of the legume *Lotus japonicus*, none of these signals had been isolated and chemically identified. 5-Deoxy-strigol, a strigolactone, was identified [44]. The association between mycorrhizae and plants has been extensively studied as a means of improving plant immunity.

Endophytic fungi are found in abundance in plant tissues, where they aid in plant health and play a vital role in plant-microbe communications [45–47]. Later in the ecological process, plants and endophytes collaborate in a reciprocal manner. *P. indica*, a beneficial endophyte isolated from the roots of plants growing in Rajasthan, India, is one of the beneficial endophytes [48]. It has been carefully researched and tested on various plants for its essential qualities. This fungus promotes yield and crop output while improving nutrient uptake and assisting plants in surviving under stressful situations such as salt and drought. It also possesses systemic resistance to diseases, heavy metals, and harmful chemicals [49]. Several other investigations have found higher biomass distribution and improved plant growth when plants are treated with these fungi [50]. In terms of agricultural, medicinal, decorative, and other plants, more than 150 host plant species have been researched and proven to be beneficially connected with *P. indica*. Early developmental gene expression was found in the roots colonized by *P. indica*, indicating that treated roots developed faster at the start than control roots [51].

The colonization of the outer root cortex was discovered after inoculating maize roots with *P. indica*, significantly enhancing the growth responses. In a study, the coinoculation of the endophytic fungus *Rhizophagus irregularis* (*Glomeromycotina* species) and *Serendipita* or *Piriformospora indica* reduces the lead (Pb) uptake in the shoots of *Ocimum basilicum* (sweet basil), whereas alone *S. indica* limits the copper (Cu) uptake in shoots [52]. Many countries produce useful *Trichoderma harzianum* goods; for example, the T-22 strain is used to stimulate the Tianum-P product in Poland. Numerous investigations have discovered that the *Trichoderma* species can create beneficial substances such as viriden, isonitryles, gliotoxines, peptaboils, and sesquiterpenes, as well as a variety of other essential compounds [53]. *Trichodermaatro viride* G79/11, according to a study, may produce the enzyme cellulase, making it a feasible candidate for the bio-preparation of antifungal drugs. *Talaromyces* is a fungus genus that belongs to the heat-resistant fungi (HRF) family, with the *Talaromyces flavus* strain being most familiar. In a glucose tartrate-rich solution at pH 5, a heat-resistant fungus can withstand temperatures ranging from [90°C (6 min) to 95°C (1 min)] [54]. Bioactive compounds generated by it have been identified as actofunicone, deoxyfunicone, and vermistatin. This strain has the potential to be used in disease biocontrol since these chemicals help them compete for nutrition and develop quicker [53]. Biosept 33 SL and Micosat F are two bio-products and bio-preparations used in the cultivation of organic fruits. Plant extracts (e.g., garlic-*Allium sativum*), animal-derived compounds (e.g., chitosan), and *Pythium oligandrum* inoculum are all active components. Agriculturalists value these bio-preparations since they are both safe and effective for plants and animals [55].

3 Microbial defense mechanisms

Microbes play a role in disease incidence as well as biocontrol. Phytotoxic chemicals produced by a few microbes can cause illness symptoms. *Pseudomonas syringae* is a pathogenic bacterium that has a wide range of hosts, including tomato, tobacco, olive, and green bean. *Erwinia amylovora* is a pathogenic bacterium that causes fire blight disease in fruit-bearing trees and ornamental plants. Due to the presence of *Xanthomonas campestris*, *R. solanacearum*, and *Xylella fastidiosa*, banana and potato crops are also susceptible to a number of illnesses [56]. Pathogen population size, a favorable habitat, and host vulnerability, as well as biotic elements involved in the overall development of plant-pathogen interactions, all influence the severity of plant disease. The host may develop resistance to pathogenic interventions as a result of the bacterial activities both above and below the ground, altering plant defense responses [57]. Pathogen invasions and disease, on the other hand, can be managed through a variety of biocontrol actions. Because the use of chemicals has raised many severe concerns about agricultural production, the adoption of a benign microbial population has grown in appeal as a cost-effective alternative. This can be aided by lytic enzymes, antibiotic synthesis, and the creation of pathogen-inhibiting siderophores and volatile chemicals [58,59].

Microorganisms use a variety of strategies to manage pathogenic microbes, including antagonism, competition for nutrients and habitats, and defense responses. Antibiotic bacteria prevent other germs from growing in their vicinity, thereby limiting pathogen growth. Furthermore, fast-growing bacteria can use nutrients for their own growth while depleting them for others, resulting in dangerous microorganisms growing slowly or not at all. A few bacteria assure the plant against diseases by controlling plant hormone levels and generating plant resistance. Consistent agricultural soil use can result in pathogenic pressure and the formation of disease-resistant soils containing disease-resistant bacteria [60]. In this experiment, scientists found that the three important bacterial taxa from the *Firmicutes* (Gram-positive), *Actinobacteria*, and *Acidobacteria* bacterial families were able to manage Fusarium wilt disease on a massive scale [61]. The importance of endosphere bacterial communities in suppressing the calamitous disease (*Gaeumannomyces graminis*) was discovered, and *Serratia* and *Enterobacter endophytes* were identified as the most promising challengers to *G. graminis*. When a plant is exposed to beneficial microorganisms, ISR protects the plant systemically through the participation of the phytohormones ET and JA [62–64]. Plant priming is well known during ISR, when dangerous microbe defense mechanisms are swiftly activated aboveground, and some species of PGPR have demonstrated plant priming characteristics [65]. MAMP-triggered immunity has been identified as a significant defense in SAR. Unlike ISR, it provides SA provides systemic plant protection, as discussed in Mechanism of Belowground Interactions in the Rhizosphere: Beyond Plant's Innate Immune Response [66].

4 Application development in the future

4.1 Microbiome (plant-microbe interaction) analysis techniques

Since people began to farm and were no longer nomads, plants have been an important element of our diet. Since then, the globe has been faced with the constant issue of feeding an ever-increasing population. Eutrophication is caused by excessive fertilizer use [27], and genetic engineering of plants is a costly and time-consuming process. Microbes' role in plant-microbe interactions has been extensively researched during the last decade. Extensive research suggests that utilizing beneficial bacteria is a superior long-term strategy for increasing crop productivity, which plays a vital role in the transmission and management of disease [67]. Plant-microbe interactions have so far been studied in three ways: symbiosis with mycorrhizae [68], rhizobacteria [69], and pathogenicity [70,71]. Plant stress (abiotic) research has made extensive use of the transcriptome, proteome, and metabolome, as well as bioinformatics. Plant protein profiles in response to abiotic stressors have been broadly studied using the proteome methods, which could lead to the evolution of new stress tolerance strategies [72]. Microbial metabolome is a procedure for investigating the collection of compounds found in microbial populations. Fig. 2 shows how Narasimhan et al. [73] employed a rhizosphere metabolomics-driven technique

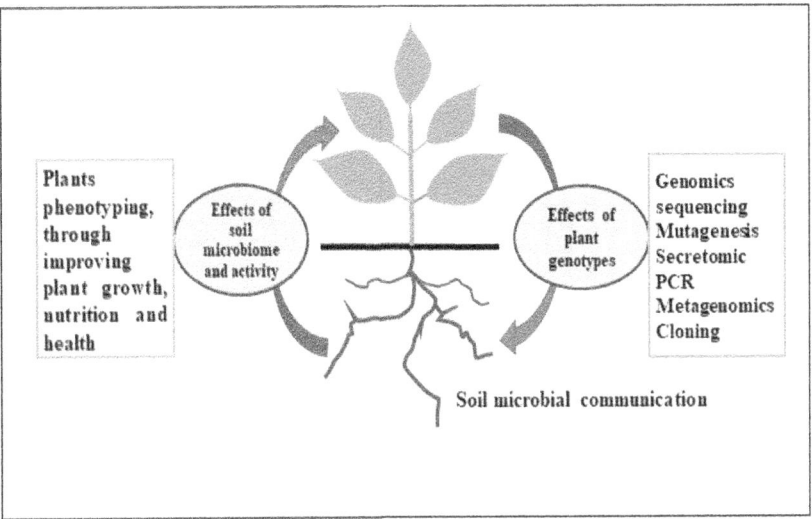

FIG. 2 Plant-microbe relations in the rhizosphere [36].

to study the plant and microbial interaction for the elimination of polychlorinated biphenyls.

Infections and their development have posed a severe warning to food safety, agricultural practices, and food species conservation, and understanding the emergence of new infectious agents and their relevance has become a key job. Previously, plant pathogen systems could only be used to study one gene or protein at a time; however, the genomic era heralded the start of in-depth research into plant-pathogen relationships [74]. Pathogen genomes have been sequenced, and analysis of these sequences has revealed pathogen evolution processes as well as previously unknown elements of pathogen biology. The genome scale renovation model (GSRM) is based on metabolic reconstructions on a genomic scale and is used to measure and evaluate metabolite absorptions under specific conditions [75]. GSRM has been utilized to successfully develop a diversity of species, including bacteria, fungi, plants, and animals. The microbial diversity of a sample can be determined using a variety of techniques. Table 2 shows that two next-generation sequencing methods, amplicon sequencing and metagenomics, are used to characterize the entire microbiome.

4.1.1 Sequencing of amplicons

The unique binding of universal primers to highly conserved locations throughout the genome of the microbe is the basis for these strategies. Amplicon sequencing is used to analyze microbial populations in microbial ecology investigations. D'Amore et al. describe the sequencing of subsequent polymerase chain reaction (PCR) products, including taxon-specific hypervariable areas [86]. The 16S rRNA (part of the 30S subunit) gene of prokaryotic ribosome such

TABLE 2 Biomolecules involved in microbial and root-based activities, both direct and indirect.

S. no.	Microorganisms name	Activity	Biomolecules	Functions	References
Microorganism-based direct activity					
1.	*Azotobacter, Bacillus, Clostridium,* and *Klebsiella*	Nitrogen fixation	EPS, EPS II, flavanols, nodulating factors, flavanones, lipochitooligosaccharides	Role of root cortical cells in nodule formation	Santi et al. [76]
2.	Mycorrhizae (ectomycorrhizae, and endomycorrhizae)	Mycorrhizal association	Sesquiterpene, Myc factor	Trigger mycorrhization	Dobbelaere and Okon [77]
3.	*Desulfuromonas* sp., *Klebsiella planticola, P. aeruginosa, Vibrio harveyi, Pseudomonas putida, Brevibacterium* sp., and *Bacillus* sp.	Metal uptake	Glutathione, chorismic acid, Caffeic acid, metallothioneins, and ferulic acid, p-, oxalic acid, etc.	Metallic bioavailability	Colangelo and Guerinot [78]
4.	*Ralstonia solanacearum*	Virulence factors	Extracellular polysaccharide, phytotoxins, effector proteins	Resistance response suppression is vital for virulence.	Sleator and Hill [79]
5.	*Azotobacter, Azospirillum, Bacillus, Klebsiella,* and *Pseudomonas, Enterobacter, Serratia.*	PGPR	LPS, antimicrobials, EPS, lipopeptides, etc.	It improves nutrient uptake activity and increases plant growth.	Kohler et al. [80]

Direct root-based activity

#	Organism	Activity	Metabolites	Function	Reference
1.	*Burkholderia* sp., *Rhizopus microspores, Geosiphon pyriforme, Gigaspora margarita, Nostoc punctiforme, Glomeribacter gigasporarum*	Bacterial and fungal symbionts	Flavonoids (glyceollin, coumestrol, daidzein, glyceollin and coumestrol, genistein), strigolactones, jasmonates, auxins, abscisic acid, ethylene, and gibberellin levels are increased.	Enhancing presymbiotic processes and AMF colonization of higher plant roots.	Wagg et al. [81]
2.	*Escherichia coli*	Uptake of carbon	Sugars like arabinose, ribose, hexose, and fructose.	Utilization and metabolism of carbon	Fellbaum et al. [82]
3.	*Ralstonia solanacearum*	In defense response and pathogenicity factors	Phytoalexins, glucosinolates naphthoquinones, indole, benzoxazinone, flavonoid, terpenoid, saponins, rosmarinic acid	Antipathogenic microorganisms	Hosseinzadeh et al. [83]

Indirect microorganism-based activity

#	Organism	Activity	Metabolites	Function	Reference
1.	*Staphylococcus aureus, Bacillus cereus, Pseudomonas aeruginosa,* and *Vibrio cholera*	Quorum sensing (Gram-negative and positive bacteria)	Peptide molecules, P-coumarate, quinolone, N-acyl homoserine	Swarming, biofilm, and antibiotic production	Miller and Bassler [84]

Indirect root-based activity

#	Organism	Activity	Metabolites	Function	Reference
1.	*Klebsiella pneumoniae* and *Pseudomonas aeruginosa*	Defense	Phospholipases, phosphatases, MAP kinases: methyl jasmonate jasmonate, Lipoxygenase, etc.	Defense reactions activation	Garmendia et al. [85]

as bacteria is the most commonly used amplicon for microbiome research [87]. For the bacterial (prokaryotic ribosome) 16S rRNA, several primer combinations for amplifying various HVRs and producing PCR products of varying lengths for sequencing platforms have been recommended (such as Pacific Biosciences vs Illumina). Metagenomic loci, as well as variable sequences of 16S rRNA (for prokaryotic ribosome), 18S rRNA (eukaryotic), and internal transcript spacer (ITS) segments (for fungi), contain information about the phylogeny of microorganisms, which can be used to infer and determine their taxonomy.

It should be highlighted, however, that the quality and completeness of the reference databases employed determine the accuracy of taxonomy identification using marker genes. Because of the large and well-curated database as well as the increasing sequence diversity, the ITS region of fungi was chosen over the 18S rRNA (eukaryotic) gene in a study [88]. However, it is debatable if varying-length ITS fragments cause advantageous PCR amplification of shorter-length ITS sequences, resulting in a skewed estimate of relative fungal taxon profusion. As a result, non-ITS targets of fungi can be used in fungal microbiome sequencing investigations [89]. It may be a challenge to distinguish between normal genetic changes and infrequent technical sequencing faults (less than 0.1% on the Illumina platform). Following amplicon-based sequencing, the microbiome is evaluated by clustering operational taxonomical units (OTUs) based on arbitrary decisive sequence equality thresholds (e.g., 97%). Similar but slightly different sequences are assigned to similar species by selecting OTUs, signifying that they have a biological origin. Amplicon sequence variants have higher specificity and sensitivity than OTU-based approaches, as well as a declined risk of false OTU set identification due to wrongly grouped sequences, but they may inflate microbial diversity [90].

4.1.2 Metagenomics

Instead of 16S rRNA gene segments or other focused amplicons, metagenomics utilizes the entire genome shotgun technique to access and sequence the complete DNA sequence of a microbiological sample. Following that, readings from bacteria (gram-negative and -positive), viruses, archaebacteria, bacteriophages, and fungi, as well as extrachromosomal fragments, plasmids, and host DNA from various eukaryotes, were obtained. Unlike 16S rRNA gene analysis, this method necessitates a larger amount of data to obtain the level of sequencing required to distinguish and categorize uncommon and rare microbiome members. Metagenomic raw reads are trimmed through KneadData, RAST, QIIME, and other quality control tools for robust data processing [91]. Web-based tools are becoming more widely available, and they can help compare and map readings in reference databases by providing the metrics needed. KEGG orthologs and clusters of orthologous genes are two databases where the annotated functions can be found.

Metagenomics-based research has enhanced researchers' ability to define microorganisms not only at the species level, but also at the strain level. In contrast, owing to the deep sequence conservation of the amplicons assembled at these taxonomic parameters, 16S rRNA-based NGS techniques have limited characterization determination [92]. To improve the sequencing response, a microbial genome must be redeveloped from a mixture of small fragments of DNA collected from various bacteria using an innovative bioinformatics approach. This is especially important when it comes to identifying and characterizing bacteria at the gram-strain level, where assembly algorithms can overcome challenges like intergenomic repetitive sequences and accurately detect small genetic changes [93]. Finally, metagenomics allows for functional annotation of gene sequences, providing a more comprehensive explanation of microbial characterization than focused amplicon sequencing surveys. Gene prediction and gene annotation are the two fundamental aspects of functional annotation. Bioinformatics techniques are used in gene prediction to identify sequences that may encode proteins. The sequences are then compared to a protein family database and annotated accordingly [94]. This data is also used to identify new functional gene sequences. It is critical to remember that gene prediction in metagenomics does not imply that the genes will be expressed in the initial sample. Even though amplicon sequencing and metagenomics are both next-generation sequencing methods, they may have some testing and analysis limitations [95].

4.1.3 Analysis of genome-wide association studies of plant-microbe interactions

Detecting single nucleotide polymorphisms (SNPs) and analyzing their relationship with important attributes using GWAS has become common practice in the genomic era due to the high quality of comparative genomics analysis. The GWAS method allows for the statistical identification of possible SNPs in people with a shared evolutionary history. In wild and crop plants, GWAS is now an effective technique for finding genomic areas linked with natural variability in disease resistance [96]. GWAS allows for phenotypic correlation in genetically diverse groups by using preexisting cumulative recombination events in wild populations. The reliance of GWAS on a context genome, in contrast to this, makes it difficult to recognize sequences that have considerably diverged from the reference, such as resistance genes. This obstacle was overcome by trait-dependent subsequences (k-mers) genetics with resistance gene enrichment sequencing (AgRenSeq), which enabled the finding and cloning of resistance genes from a diverse panel of plants. AgRenSeq's capacity to quickly clone agronomically essential resistance genes could be employed in breeding programs as specialized genetic markers or for resistance engineering [97]. GWAS has been utilized in a number of plant-microbe interaction studies to examine how a plant genotype affects interactions with a single microbial taxon in pairs. GWAS was used to examine 340 *japonica indica* accessions, and 16 blast counteraction loci were discovered, 2 of which were strongly connected with rice blast resistance in *japonica* and 1 locus in *indica* [98]. Knowledge of the coevolutionary factors that can lead to plant species accepting adaptive dynamics in plant communities might vastly enhance our understanding and prediction of emerging diseases (ED) [96].

Despite this, plant pathogen-based GWASs are used to find the genes responsible for a wide range of traits, including those that are influenced by the microbial ecosystem. *Fusarium graminearum*, a common fungal pathogen of wheat, barley, and maize, has shown intraspecific variation in traits such as aggressiveness in recent pathogen GWAS research. The restriction site-associated DNA sequencing (RADseq) method was used to analyze 213 fungal pathogen isolates from 13 German field communities of *F. graminearum*, and it was discovered that the high gene flow between these field populations would allow this pathogen to adapt quickly to changes in its environmental conditions, such as the yield of resistant crops and fungicide applications. In 2013, Dalman applied the GWA mapping procedure to categorize the genetic segments that support virulence in *Heterobasidion annosum*, a fungal necrotrophic disease that causes significant damage to forest conifers. Based on 23 haploid whole-genome sequenced *H. annosum* isolates collected in various geographic European countries, GWA mapping on virulence was performed under controlled conditions on *Pinus sylvestris* (scots pine) and Norway spruce using 33,018 nonsingleton SNPs; 12 SNPs are strongly associated with virulence in both host species [97]. In a study on *Phaeosphaeria nodorum* [99], 191 isolates were reported to be affected on 2 wheat lines to determine virulence, and over 3000 SNPs were genotyped across the genome, as well as genetic markers for putative genes. The discovery of SNPs in *SnToxA* and *SnTox3*, two previously cloned effector genes, demonstrated GWA mapping's ability to map virulence components in *P. nodorum* [99].

For 20 newly sequenced *Puccinia graminis tritici* isolates from Australia, a polygenic structure corresponding to 302 genes was recently revealed by utilizing a hybrid approach of GWA mapping and comparative genomics, including at least one SNP related to leaf rust virulence on wheat [100]. Using 306,474 SNPs, a polygenic structure corresponding to 302 genes with at least one SNP related to leaf rust virulence on wheat was found. More studies are needed in the future to show how GWAS may be used to find novel virulence factors. Because of advancements in NGS technology, DNA sequencing has become a more tempting choice for genotyping SNP arrays, allowing GWAS to go beyond common variants and maintain the possibility of detecting uncommon alleles and structural differences. Sorghum [101], *foxtail millet* [102], soybean [103], and maize have all benefited from genome sequencing-based GWAS [104]. Combining GWAS and gene-based association analysis with haplotype analysis to identify candidate genes for a number of illnesses is another effective technique. Furthermore, GWAS research and related experimental validation should be carried out to examine plant resistance and susceptibility pathways, which would lead to novel disease-fighting strategies.

4.1.4 Stress management through plant-microbe interaction

Abiotic and biotic stresses are always present in the agricultural environment, affecting crop yield, so il health, and fertility. It is possible for both abiotic and biotic stresses to be generated by natural or human factors. An example of an abiotic variable is dryness, while an example of a biotic component is bacteria, fungi, or nematodes. As a result of these environmental stressors, plant physiological and metabolic processes and gene regulation are adversely affected [105]. Some plants, through acclimatization and adaptation, can change gene expression and adapt to these conditions, but others cannot. Plant-associated microbial communities, such as mycorrhizal fungi (in higher plants) and plant growth-promoting bacteria (PGPB), are one of the better options since they aid plant growth and development in a variety of biotic and abiotic environments [106]. PGPB has been labeled a cost-effective and ecologically friendly approach for disease prevention since it activates the cellular component and accumulates secondary metabolites. Plant growth and metabolism are considered to be aided by PGPR, which help plants to recover from stress. Two PGPR that promote soil productivity and plant development are *Pseudomonas reactans* and *Chryseobacter iumhumi*. PGPR have a competitive edge over fungi for iron absorption due to the formation of siderophores. Bacteria can take up the iron-siderophore combination because such siderophores have a strong affinity for iron. PGPR use this method to prevent pathogen development by restricting iron availability, thereby protecting the plant from illness [107]. Two PGPR were examined in a study for their ability to stimulate certain genes in a rice host plant to tolerate the impact of a fungal disease (*Magnoporthe grisea*) [108].

In both developed and developing countries, recent advances in plant biotechnology, such as structural and functional genomics, can provide critical technologies for agronomic improvement and stress management. Many disease resistance genes have been identified, mapped, and transferred into tomato recognition using a variety of molecular markers. PGPR genes, which are recognized for their capacity to enhance food availability, reduce the pathogenic fungus, endure oxidative stress, quorum sensing (in bacteria), and the ability to break down aromatic and toxic compounds, as well as other abiotic stress, were recently discovered using NGS technology. Plant microbe interactions have been studied using gene-editing tools such as transcription activator-like effector nucleases (TALENs) and CRISPR-Cas to create transformed plants [109]. Microbes such as *Rhizobium* (Gram-negative), *Bacillus subtilis* (Gram-positive), *Pseudomonas fluorescens* (Gram-negative), *Methylobacterium* (fastidious and Gram-negative), *Variovorax paradoxus* (Gram-negative), *Enterobacter* (Gram-negative), and others have been discovered to provide host plants with resistance to stress such as biotech, drought, and salinity conditions [110]. Drought stress lowers agricultural productivity through lowering water content, cell size, membrane integrity, and reactive oxygen species generation (ROS). It also promotes increased leaf senescence. Microbes have developed, adapted, or constructed tolerance mechanisms to help them survive in low-water-potential environments. They can either develop thick walls (biofilm) or go inactive, accumulating osmolytes and producing exopolysaccharides (EPS). In arid settings, PGPR may also produce plant hormones (IAA and cytokinins) that promote plant growth and division [111]. Some PGPR bacteria strains generate antioxidants and cytokinin, resulting in the buildup of abscisic acid (ABA) and the elimination of reactive oxygen species. *Azospirillum brasilense*, for example, increases the quantity of ABA in the drought response of the plant, allowing it to tolerate drought stress [24]. As a result, in the postgenomic era, genetic modifications can be utilized to increase the attributes of PGPB strains and as a low-cost, long-term, and environmentally friendly approach for plant managing stress.

5 Using beneficial microbes to improve crop quality

Through the use of chemical fertilizers and pesticides, the green revolution, which began in the 1970s, dramatically increased agricultural productivity and produce. However, there have been reports over the years about the chemicals' potential risks to the soil, environment, and human health. Only about half of the nitrogen fertilizers are digested by plants; the remainder is lost to evaporation, drainage, or leaching. This results in extremely high levels of NO_3^- and NH_4^+ in groundwater, posing a risk to human health [112]. This situation has emphasized the potential use of microorganisms for agricultural development, which has been garnering support for decades, because traditional organic farming alone will not be sufficient to create crops with improved yields and disease resistance. Fungi such as PGPB, PGPR, and vesicular-arbuscular mycorrhizae (AMF) are examples of effective microbes (EMs) or plant growth-promoting microorganisms (PGPM) [113]. In research, various microorganisms have been related to crop improvement mechanisms, and we will go over a number of them below. Microorganisms called bio-fertility inoculants are introduced into the soil to boost plant development by increasing nutrient uptake and solubilization.

Insufficient supplies of the nutrients phosphorus and nitrogen are found in the soil. Microorganisms with improved nutrition acquisition capacities are being investigated as possible nutrient acquisition solutions [114]. Pesticide-tolerant PGPR *Bacillus* sp.1 (*FOW1*) and *Lysinibacillus sphaericus* (Gram-positive) (*FOW7*) have been proven to increase agricultural yields by improving soil aeration, soil water holding capacity, and plant development while also performing bioremediation. Six strains of nitrogen-fixing endophytic bacteria were studied to see if they could help *Picea glauca* trees grow quicker, and it was determined that they could boost plant biomass and seedling length while also boosting nitrogen fixation [115]. Fungi have been found to have a higher rate of phosphorus solubilization than bacteria. *Penicillium bilaiae* is a commercially available fungus that utilizes citric and oxalic acid to dissolve phosphate. Biocontrol organisms, on the other hand, are pathogenic organism antagonists that have been widely researched and used in the field [114]. Together with CIAT 899, *Rhizobium tropici* (CIAT 899) and endophytic nodule generating bacteria (*Pseudomonas* spp.), UFLA 02-286 (*Bacillus* spp.), and UFLA 04-227 (*Burkholderia fungorum*) formed consortiums that successfully controlled damping-off disease. As shown in Fig. 3, recent research has used the antagonistic action of the endophytic bacteria *Bacillus velezensis OEE1* against *Verticillium dahlia*, the cause of *verticillium* wilt in olive trees [116].

Many researchers are interested in the metagenomic method to study the whole genomes of all microorganisms both culturable and nonculturable available in various niches because it serves as a source of vast information of all the helpful microbes that can be used for PGP and as biocontrol agents [117]. Rather than focusing primarily on rhizosphere microbial consortia, researchers discovered that the microbiomes of lichens (algae and fungi), alpine mosses (*Diphasiastrum alpinum*), and *Primula vulgaris* can help economically productive plants such as maize and *Beta vulgaris* with progress and stress tolerance [118]. The functional study of metagenome from the rhizosphere, phyllosphere, and endosphere has provided good information about the microorganisms that are related to various plants and ecosystems. These microbiomes have also been connected to processes like nutrition acquisition, fixing nitrogen, phosphorus utilization, iron mobilization, and stress tolerance.

The phenol-adapted plant effluent generated a novel *Bradyrhizobiaceae* genome with unique characteristics such as nitrogen fixation, nitrate absorption,

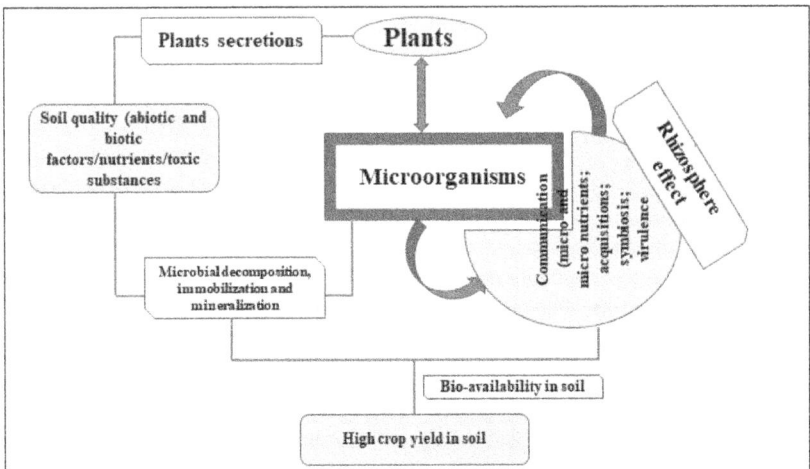

FIG. 3 Interactions between plants and microbes and their importance to sustainable agriculture.

and conversion to nitrite, sulfate, and aromatic chemical utilization, iron uptake, and so on [119]. DNA cloning, Sanger sequencing, denaturing gradient gel electrophoresis (DGGE), terminal restriction fragment length polymorphism (TRFLP), fluorescence in situ hybridization (FISH), and stable isotope probe (SIP) are some of the molecular biology techniques that reveal fascinating insights about culturable and nonculturable microbiome, as well as aid in the characterization of microbial functions [120]. The use of nonculturable microbial consortia for field activities is becoming more common, but it comes with its own set of issues, such as pathogenicity caused by unknown microorganisms in the consortium, food contamination that might be damaging to human health, and bacteria's failure to survive. The creation of novel media and cultural methods to culture nonculturable microbiome members is now being investigated. These strategies will aid in the discovery of both beneficial and harmful microorganisms connected with the consortium, as well as the manufacture of better agricultural formulas [121].

A single or consortia microbial inoculum offers considerably more advantages than disadvantages. They can, among other things, assist in restoring soil fertility, boosting nutrient availability, defending against biotic and abiotic stresses, increasing soil microbial activity, breaking down toxic compounds, fostering beneficial microbial colonization, and recycling organic materials. The need for microbial inoculants grows by around 12% each year, owing to the rising cost of chemical fertilizers and society's desire for an environmentally friendly technology. PGPR such as *Azotobacter* spp. (Gram-negative), *Bacillus* spp. (Gram-positive), *Azospirillum* spp. (Gram-negative), *Serratia* spp. (Gram-negative), *Rhizobium* spp. (Gram-negative), and others are now being commercially produced on a huge scale, despite the fact that different nations have their own guidelines for the use of microbe-based bio-fertilizers and bio-pesticides in agricultural practices [122]. The most difficult challenges for field trials are consistency, dependability, and shelf life of microbial inoculum. Gram-positive bacteria that do not produce spores have a longer shelf life than Gram-negative bacteria that do. In contrast, superinoculants have been found in studies to have all of the fundamental characteristics of a microbial inoculant [123].

However, some PGPR that are potentially harmful to humans, such as pathogenic microorganisms such as *Pseudomonas* spp. and *Burkholderia cepacia*, have been studied. Despite their PGP activity, some species can be toxic to humans; thus, they should be adequately treated before being commercially produced. More study is needed before PGPR microorganisms can be employed in sustainable agriculture. The biosafety of PGPR-based bio-fertilizers is being re-evaluated in many European and other nations, including the United States [124]. Plant-microbe interactions have been shown to be affected by climate change in studies; nevertheless, additional research is needed to fully understand the capabilities of PGPR before they are embraced by management rules, bio-fertilizer companies, and farmers. Farmers may be able to embrace and use cost-effective microbial consortia technology in the future if provisions are made. Farmers should be able to acquire bio-fertilizers/bio-pesticides with a longer shelf life and greater stability from government-regulated outlets at sub-sidized prices, with the option of replacing an old batch of inoculum with a new batch. Farmers can receive training from agricultural-based community admin-istrative organizations, which will emphasize the benefits, proper handling and usage, and basic principles.

6   Interactions between the mycorrhizosphere and agriculture are important for long-term sustainability

6.1   Nutrient supply that is long-term

Arbuscular mycorrhiza has been indicated as having the potential to promote phosphorus nutrition, nitrogen uptake, and disease resistance in its host plants. Other microorganisms, such as nitrogen-fixing bacteria or phosphorus-solubilizing bacteria, may interact synergistically with arbuscular mycorrhiza fungi to aid plant development and growth [125]. In sustainable agricultural sys-tems with reduced nutrient inputs, the mycorrhizal symbiosis becomes even more significant. AM mycelium (perhaps in collaboration with bacteria or other fungus) might play a critical role in nitrogen mobilization from agricultural wastes in these settings. The AM symbiotic *Glomus* fungus in the soil increased plant litter digestion and nitrogen absorption (15 N-13C tagged *Lolium perenne* leaves), according to Hodge et al. [126]. In the presence of organic compounds, the fungal symbiont's hyphal development was also boosted. Bacteria linked to the arbuscular mycorrhizal fungus can also help with soil nutrient mobilization. Diazotrophic bacteria give nitrogen fixation to both plants and fungi in bacterial-arbuscular mycorrhizal-legume tripartite symbiotic relationships; there are several examples of this. As previously stated, nitrogen-fixing bacteria nodulation of legumes and the formation of arbuscular mycorrhizal typically occur at the same time and in concert. Minerdi et al. [127] discovered genes of nitrogen fixation in endosymbiotic *Burkholderia* bacteria in arbuscular mycorrhizal hyphae, implying that atmospheric nitrogen fixation could boost mycorrhizal plant nitrogen supply.

6.2   Biocontrol

In agricultural cropping systems, microbial inoculants can be used to replace pesticides, allowing for a decrease in pesticide use that would otherwise damage human well-being. The biocontrol organisms may have an effect on AM fungi, or be impacted by them, in a similar fashion to the interactions outlined above. Biocontrol chemicals used to combat pathogenic fungi may have unfavorable impacts on "nontarget" AM fungi. Competition for colonization sites or nutri-ents, as well as the synthesis of fungistatic chemicals, may be involved in antag-onistic interactions that result in biocontrol. Despite the vast knowledge on biocontrol, few researches have looked specifically at interactions with the arbuscular mycorrhizal fungus. Some of the bacteria's beneficial effects on arbuscular mycorrhizal fungal root colonization may be due to antagonistic effects on competing pathogens as well as direct synergistic effects on mycorrhizal colonization [25].

Some of the favorable effects of the bacteria (*Pseudomonas* spp.) on arbuscular-mycorrhizal fungal colonization of roots might be related to antagonistic effects on competing pathogens as well as direct synergistic effects on mycorrhizal colonization [25]. *P. chlororaphis* was discovered to be an efficient colonizer of tomato roots and effective against the root pathogen *Fusarium oxysporum* in one investigation. Among other antifungal chemicals, the bacterial strain generated phenazine-1-carboxamide (PCN), hydrogen cyanide, chitinases, and proteases [22].

It was discovered that knocking down the phenazine-biosynthetic operon resulted in much decreased biocontrol activity, which demonstrate that the chemical was an essential antifungal component for disease suppression of tomato roots. The density of the hyphal network within tomato roots was considerably reduced by 70%–80% when biocontrol bacteria were introduced to identical fungal and bacterial strains that were labeled with green and red fluorescent protein, respectively [128]. The fluorescent effects on arbuscular mycorrhizal fungal hyphae (group of hypha), on the other hand, were not studied. Bacteria's capacity to colonize root surfaces quickly and interact closely with illnesses, as well as produce antifungal compounds, may promote pathogen suppression. For the past 30 years, scientists have been studying AM fungi as pathogen-defeating agents. Although numerous researches have been published, the fundamental mechanisms remain unknown. Some basic mechanisms that have been hypothesized are as follows: improved plant nutrition and competition for photo-synthates, as well as arbuscular mycorrhizal-induced root infection suppression and promotion of saprotrophs (also called saprophytes or saprobes) and PGP microbes [129]. Other processes that have been described by several studies include arbuscular mycorrhizal fungus-induced structural or morphological changes in root arrangement, as well as arbuscular mycorrhizal fungi-induced local elicitation of plant defense mechanisms, which are often inconsistent across studies. Due to the difficulties of acquiring considerable amounts of pure culture arbuscular mycorrhizal inoculum, few researches have looked at the practical use of AM fungi as inoculants to increase plant resistance against root-rotting diseases. The addition of growth-promoting bacteria to the arbuscular mycorrhizal fungus could help in inoculum production [130]. A variety of AM fungi have been shown in numerous studies to have biocontrol properties against root diseases. Its currently unknown whether arbuscular mycorrhizal fungi could be utilized as biocontrol agents in the real world, or if they could act as vectors for bacteria having biocontrol properties.

7 Conclusions

With the world's population growing at an alarming rate, crop production must expand to meet global food demands while also improving agricultural sustainability. Plant-associated microbes have a big impact on the health and performance of their hosts. Attempts to use beneficial bacteria in the field, on the other hand, have failed to reliably boost crops. Plant-symbiotic microbial community interactions, ecological repercussions of plant-associated microorganisms, and plant-microbial metabolic dynamics are all poorly understood at the moment [131]. Despite the fact that genomic approaches have greatly improved our understanding of plant-bacterial interactions, they are still unable to explain plant-microbe interactions effectively. Metagenomics and amplicon sequencing have increased our awareness of plant-bacterial interactions when combined with other omics technologies, databases (PHI-based) [132], and metabolomics. Plant genotypes and environmental conditions influence the ecological communities in which plants and microorganisms dwell. Physiological and immunological responses, as well as host-specific microbial populations, are all affected by genotype differences. Sugars, amino acids, organic acids, nucleotides, flavonoids, antimicrobial compounds, and enzymes found in plant root exudates aid in the formation of specialized communities, the attraction of plant growth-promoting colonization, and the fight against pathogen infections.

Despite the excess of studies on plant-microbe interactions, the molecular pathways driving gene functions and signal transduction during positive and detrimental interactions are scarce. Using next-generation sequencing technology and numerous "omics" technologies, plant-microbe genetic interactions will emerge as a powerful tool for learning more about biological phenomena and improving plant health, food quality, and stress management [133]. There will be many hurdles in this field of research shortly that must be overcome to gain a comprehensive understanding of plant-pathogen interactions. Identification of these interactions during immune responses, control of novel emergent plant diseases, and the creation of plant resistant crops are some of the issues. Understanding the mechanisms of plant-microbe interaction in the postgenomic era may be able to help solve these problems, resulting in more sustainable agriculture. Understanding plant-microbe and microbe-microbe interactions will be important in the future as a regulating microbiome for disease prevention and higher gross plant productivity. Beneficial plant-associated bacteria may also operate as an antidote to illnesses in the microbial ecosystem, helping to stabilize the environment, increase biodiversity, reduce pathogen outbreaks, and boost plant productivity. In the future, a well-studied plant-microbe partnership might help boost agricultural productivity at a cheap cost, perhaps resulting in a new "Green Revolution."

References

[1] J.S. Singh, V.C. Pandey, D.P. Singh, Efficient soil microorganisms: a new dimension for sustainable agriculture and environmental development, Agric. Ecosyst. Environ. 140 (3–4) (2011) 339–353.

[2] P. Tyagi, R. Ranjan, Comparative study of the pharmacological, phytochemical and biotechnological aspects of *Tribulus terrestris* Linn. and *Pedalium murex* Linn: an overview, Acta Ecol. Sin. (2021), https://doi.org/10.1016/j.chnaes.2021.07.008.

[3] S. Singh, K. Srivastava, S. Sharma, A.K. Sharma, Mycorrhizal inoculum production, in: Mycorrhizal Fungi: Use in Sustainable Agriculture and Land Restoration, Springer, Berlin, Heidelberg, 2014, pp. 67–79.

[4] A. Raza, F. Ashraf, X. Zou, X. Zhang, H. Tosif, Plant adaptation and tolerance to environmental stresses: mechanisms and perspectives, in: Plant Ecophysiology and Adaptation under Climate Change: Mechanisms and Perspectives, Springer, Singapore, 2020, pp. 117–145, https://doi.org/10.1007/978-981-15-2156-0_5.

[5] P. Kumari, M. Meena, P. Gupta, M.K. Dubey, G. Nath, R.S. Upadhyay, Plant growth promoting rhizobacteria and their biopriming for growth promotion in mung bean (*Vigna radiata* (L.) R. Wilczek), Biocatal. Agric. Biotechnol. 16 (2018) 163–171.

[6] P. Kumari, M. Meena, R.S. Upadhyay, Characterization of plant growth promoting rhizobacteria (PGPR) isolated from the rhizosphere of *Vigna radiata* (mung bean), Biocatal. Agric. Biotechnol. 16 (2018) 155–162.

[7] J.M. Lynch, J.M. Whipps, Substrate flow in the rhizosphere, Plant Soil 129 (1) (1990) 1–10.

[8] B.R. Glick, Plant growth-promoting bacteria: mechanisms and applications, Scientifica (2012) 1–15.

[9] J.M. Barea, D. Werner, C. Azcon-Guilar, R. Azcon, Interactions of arbuscular mycorrhiza and nitrogen-fixing symbiosis in sustainable agriculture, in: Nitrogen Fixation in Agriculture, Forestry, Ecology, and the Environment, Springer, Dordrecht, 2005, pp. 199–222.

[10] M. Meena, P. Swapnil, K. Divyanshu, S. Kumar, Harish, Y.N. Tripathi, A. Zehra, A. Marwal, R.S. Upadhyay, PGPR-mediated induction of systemic resistance and physiochemical alterations in plants against the pathogens: current perspectives, J. Basic Microbiol. 60 (10) (2020) 828–861, https://doi.org/10.1002/jobm.202000370.

[11] G. Berg, N. Roskot, A. Steidle, L. Eberl, A. Zock, K. Smalla, Plant-dependent genotypic and phenotypic diversity of antagonistic rhizobacteria isolated from different *Verticillium* host plants, Appl. Environ. Microbiol. 68 (7) (2002) 3328–3338.

[12] M. Malhotra, S. Srivastava, Stress-responsive indole-3-acetic acid biosynthesis by *Azospirillum brasilense* SM and its ability to modulate plant growth, Eur. J. Soil Biol. 45 (1) (2009) 73–80.

[13] M. Chaiharn, S. Lumyong, Screening and optimization of indole-3-acetic acid production and phosphate solubilization from rhizobacteria aimed at improving plant growth, Curr. Microbiol. 62 (1) (2011) 173–181.

[14] M. Shankar, P. Ponraj, D. Ilakkiam, P. Gunasekaran, Root colonization of a rice growth promoting strain of *Enterobacter cloacae*, J. Basic Microbiol. 51 (5) (2011) 523–530.

[15] R. Aslantas, R. Cakmakçi, F. Şahin, Effect of plant growth promoting rhizobacteria on young apple tree growth and fruit yield under orchard conditions, Sci. Hortic. 111 (4) (2007) 371–377.

[16] M. Baset, Z.H. Shamsuddin, Z. Wahab, M. Marziah, Effect of plant growth promoting rhizobacterial (PGPR) inoculation on growth and nitrogen incorporation of tissue-cultured *Musa* plantlets under nitrogen-free hydroponics condition, Aust. J. Crop. Sci. 4 (2) (2010) 85–90.

[17] Y. Shi, K. Lou, C. Li, Growth promotion effects of the endophyte *Acinetobacter johnsonii* strain 3-1 on sugar beet, Symbiosis 54 (3) (2011) 159–166.

[18] Q. Zhao, Q. Shen, W. Ran, T. Xiao, D. Xu, Y. Xu, Inoculation of soil by *Bacillus subtilis* Y-IVI improves plant growth and colonization of the rhizosphere and interior tissues of muskmelon (*Cucumis melo* L.), Biol. Fertil. Soils 47 (5) (2011) 507–514.

[19] X. Yu, X. Liu, T.H. Zhu, G.H. Liu, C. Mao, Isolation and characterization of phosphate-solubilizing bacteria from walnut and their effect on growth and phosphorus mobilization, Biol. Fertil. Soils 47 (4) (2011) 437–446.

[20] R.K. Dey, K.K. Pal, D.M. Bhatt, S.M. Chauhan, Growth promotion and yield enhancement of peanut (*Arachis hypogaea* L.) by application of plant growth-promoting rhizobacteria, Microbiol. Res. 159 (4) (2004) 371–394.

[21] O.A. Apine, J.P. Jadhav, Optimization of medium for indole-3-acetic acid production using *Pantoea agglomerans* strain PVM, J. Appl. Microbiol. 110 (5) (2011) 1235–1244.

[22] G. Santoyo, M.D. Orozco-Mosqueda, M. Govindappa, Mechanisms of biocontrol and plant growth-promoting activity in soil bacterial species of *Bacillus* and *Pseudomonas*: a review, Biocontrol Sci. Tech. 22 (8) (2012) 855–872.

[23] R.E. Karamanos, N.A. Flore, J.T. Harapiak, Re-visiting use of *penicillium bilaii* with phosphorus fertilization of hard red spring wheat, Can. J. Plant Sci. 90 (2010) 265–277, https://doi.org/10.4141/CJPS09123.

[24] A.C. Cohen, R. Bottini, M. Pontin, F.J. Berli, D. Moreno, H. Boccanlandro, C.N. Travaglia, P.N. Piccoli, *Azospirillum brasilense* ameliorates the response of *Arabidopsis thaliana* to drought mainly via enhancement of ABA levels, Physiol. Plant. 153 (1) (2015) 79–90.

[25] J.F. Johansson, L.R. Paul, R.D. Finlay, Microbial interactions in the mycorrhizosphere and their significance for sustainable agriculture, FEMS Microbiol. Ecol. 48 (1) (2004) 1–3.

[26] G. Berg, A. Krechel, M. Ditz, R.A. Sikora, A. Ulrich, J. Hallmann, Endophytic and ectophytic potato-associated bacterial communities differ in structure and antagonistic function against plant pathogenic fungi, FEMS Microbiol. Ecol. 51 (2) (2005) 215–229.

[27] V.H. Smith, G.D. Tilman, J.C. Nekola, Eutrophication: impacts of excess nutrient inputs on freshwater, marine, and terrestrial ecosystems, Environ. Pollut. 100 (1–3) (1999) 179–196.

[28] K. Farrar, D. Bryant, N. Cope-Selby, Understanding and engineering beneficial plant-microbe interactions: plant growth promotion in energy crops, Plant Biotechnol. J. 12 (9) (2014) 1193–1206.

[29] B.R. Glick, Bacteria with ACC deaminase can promote plant growth and help to feed the world, Microbiol. Res. 169 (1) (2014) 30–39.

[30] B. Lugtenberg, F. Kamilova, Plant-growth-promoting rhizobacteria, Annu. Rev. Microbiol. 63 (2009) 541–556.

[31] B.A. Olukolu, W.F. Tracy, R. Wisser, B. De Vries, P.J. Balint-Kurti, A genome-wide association study for partial resistance to maize common rust, Phytopathology 106 (7) (2016) 745–751.

[32] P.N. Bhattacharyya, P.M. Goswami, L.H. Bhattacharyya, Perspective of beneficial microbes in agriculture under changing climatic scenario: a review, J. Phytol. 8 (2016) 26–41.

[33] S. Withers, E. Gongora-Castillo, D. Gent, A. Thomas, P.S. Ojiambo, L.M. Quesada-Ocampo, Using next-generation sequencing to develop molecular diagnostics for *Pseudoperonospora cubensis*, the cucurbit downy mildew pathogen, Phytopathology 106 (10) (2016) 1105–1116.

[34] G. Yadav, K. Vishwakarma, S. Sharma, V. Kumar, N. Upadhyay, N. Kumar, R.K. Verma, R. Mishra, D.K. Tripathi, R.G. Upadhyay, Emerging significance of rhizospheric probiotics and its impact on plant health: current perspective towards sustainable agriculture, in: Probiotics and Plant Health, Springer, Singapore, 2017, pp. 233–251.

[35] M. Ahmad, L. Pataczek, T.H. Hilger, Z.A. Zahir, A. Hussain, F. Rasche, R. Schafleitner, S.O. Solberg, Perspectives of microbial inoculation for sustainable development and environmental management, Front. Microbiol. 9 (2018) 2992, https://doi.org/10.3389/fmicb.2018.02992.

[36] K. Vishwakarma, S. Sharma, N. Kumar, N. Upadhyay, S. Devi, A. Tiwari, Contribution of microbial inoculants to soil carbon sequestration and sustainable agriculture, in: Microbial Inoculants in Sustainable Agricultural Productivity, Springer, New Delhi, 2016, pp. 101–113.

[37] A.E. Richardson, R.J. Simpson, Update on microbial phosphorus. Soil microorganisms mediating phosphorus availability, Plant Physiol. 156 (989) (2011), e996.

[38] Monsanto BioAg, 2016. https://www.efeedlink.com/contents/12-28-2016/c5141bdd-e8a4-4f05-9502-279a7add15ff-0003.html.

[39] M. O'Callaghan, J. Swaminathan, J. Lottmann, D.A. Wright, T.A. Jackson, Seed coating with biocontrol strain *Pseudomonas fluorescens* F113, N. Z. Plant Prot. 59 (2006) 80–85.

[40] C. Planchamp, G. Glauser, B. Mauch-Mani, Root inoculation with *Pseudomonas putida* KT2440 induces transcriptional and metabolic changes and systemic resistance in maize plants, Front. Plant Sci. 5 (2015) 719.

[41] Y. He, H.A. Pantigoso, Z. Wu, J.M. Vivanco, Co-inoculation of *Bacillus* sp. and *Pseudomonasputida* at different development stages acts as a bio-stimulant to promote growth, yield and nutrient uptake of tomato, J. Appl. Microbiol. 127 (1) (2019) 196–207.

[42] M. Chiboub, S.H. Jebara, G. Abid, M. Jebara, Co-inoculation effects of *Rhizobium sullae* and *Pseudomonas* sp. on growth, antioxidant status, and expression pattern of genes associated with heavy metal tolerance and accumulation of cadmium in *Sulla coronaria*, J. Plant Growth Regul. 3 (2019) 1–3.

[43] M. Akhtar, J. Panwar, Arbuscular mycorrhizal fungi and opportunistic fungi: efficient root symbionts for the management of plant parasitic nematodes, Adv. Sci. Eng. Med. 3 (3) (2011) 165–175, https://doi.org/10.1166/asem.2011.1109.

[44] K. Akiyama, H. Hayashi, Strigolactones: chemical signals for fungal symbionts and parasitic weeds in plant roots, Ann. Bot. 97 (6) (2006) 925–931, https://doi.org/10.1093/aob/mcl063.

[45] J. Goutam, R. Singh, R.S. Vijayaraman, M. Meena, Endophytic fungi: carrier of potential antioxidants, in: P. Gehlot, J. Singh (Eds.), Fungi and Their Role in Sustainable Development: Current Perspectives, Springer, Singapore, 2018, pp. 539–551, https://doi.org/10.1007/978-981-13-0393-7_29.

[46] M. Meena, A. Zehra, P. Swapnil, Harish, A. Marwal, G. Yadav, P. Sonigra, Endophytic nanotechnology: an approach to study scope and potential applications, Front. Chem. 9 (2021) 613343, https://doi.org/10.3389/fchem.2021.613343.

[47] G. Yadav, M. Meena, Bioprospecting of endophytes in medicinal plants of Thar Desert: an attractive resource for biopharmaceuticals, Biotechnol. Rep. 30 (2021), e00629, https://doi.org/10.1016/j.btre.2021.e00629.

[48] A. Varma, M. Bakshi, B. Lou, A. Hartmann, R. Oelmüller, Functions of a novel plant growth-promoting mycorrhizal fungus: *Piriformospora indica*, Agric. Res. 1 (2) (2012) 117–131.

[49] M. Meena, P. Swapnil, A. Zehra, M.K. Dubey, M. Aamir, C.B. Patel, R.S. Upadhyay, Virulence factors and their associated genes in microbes, in: H.B. Singh, V.K. Gupta, S. Jogaiah (Eds.), New and Future Developments in Microbial Biotechnology and Bioengineering, Elsevier, 2019, pp. 181–208, https://doi.org/10.1016/B978-0-444-63503-7.00011-5.

[50] S.S. Gill, R. Gill, D.K. Trivedi, N.A. Anjum, K.K. Sharma, M.W. Ansari, A.A. Ansari, A.K. Johri, R. Prasad, E. Pereira, A. Varma, *Piriformospora indica*: potential and significance in plant stress tolerance, Front. Microbiol. 7 (2016) 332.

[51] F. Waller, B. Achatz, H. Baltruschat, J. Fodor, K. Becker, M. Fischer, T. Heier, R. Huckelhoven, C. Neumann, D. von Wettstein, P. Franken, The endophytic fungus

Piriformospora indica reprograms barley to salt-stress tolerance, disease resistance, and higher yield, Proc. Natl. Acad. Sci. 102 (38) (2005) 13386–13391.

[52] M. Sabra, A. Aboulnasr, P. Franken, E. Perreca, L.P. Wright, I. Camehl, Beneficial root endophytic fungi increase growth and quality parameters of sweet basil in heavy metal contaminated soil, Front. Plant Sci. 9 (2018) 1726.

[53] M. Pylak, K. Oszust, M. Frac, Review report on the role of bio-products, bio-preparations, biostimulants and microbial inoculants in organic production of fruit, Rev. Environ. Sci. Biotechnol. 18 (3) (2019) 597–616.

[54] J. Panek, M. Frąc, Development of a qPCR assay for the detection of heat-resistant *Talaromyces flavus*, Int. J. Food Microbiol. 270 (2018) 44–51.

[55] B. Marjanska-Cichon, A. Sapieha-Waszkiewicz, Efficacy of garlic extracts used to control of grey mould in strawberry, Prog. Plant Prot. 51 (2011) 413–420.

[56] J. Mansfield, S. Genin, S. Magori, V. Citovsky, M. Sriariyanum, P. Ronald, M.A. Dow, V. Verdier, S.V. Beer, M.A. Machado, I.A. Toth, Top 10 plant pathogenic bacteria in molecular plant pathology, Mol. Plant Pathol. 13 (6) (2012) 614–629.

[57] M. De Vrieze, F. Germanier, N. Vuille, L. Weisskopf, Combining different potato-associated *Pseudomonas* strains for improved biocontrol of *Phytophthora infestans*, Front. Microbiol. 29 (9) (2018) 2573.

[58] M. Meena, P. Swapnil, A. Zehra, M.K. Dubey, R.S. Upadhyay, Antagonistic assessment of *Trichoderma* spp. by producing volatile and non-volatile compounds against different fungal pathogens, Arch. Phytopathol. Plant Protect. 50 (13–14) (2017) 629–648, https://doi.org/10.1080/03235408.2017.1357360.

[59] R.K. Verma, M. Sachan, K. Vishwakarma, N. Upadhyay, R.K. Mishra, D.K. Tripathi, S. Sharma, Role of PGPR in sustainable agriculture: molecular approach toward disease suppression and growth promotion, in: Role of Rhizospheric Microbes in Soil, Springer, Singapore, 2018, pp. 259–290.

[60] P. Duran, G. Tortella, S. Viscardi, P.J. Barra, V.J. Carrión, M.D. Mora, M.J. Pozo, Microbial community composition in take-all suppressive soils, Front. Microbiol. 9 (2018) 2198.

[61] P. Trivedi, M. Delgado-Baquerizo, C. Trivedi, K. Hamonts, I.C. Anderson, B.K. Singh, Keystone microbial taxa regulate the invasion of a fungal pathogen in agro-ecosystems, Soil Biol. Biochem. 111 (2017) 10–14.

[62] C.M. Pieterse, C. Zamioudis, R.L. Berendsen, D.M. Weller, S.C. Van Wees, P.A. Bakker, Induced systemic resistance by beneficial microbes, Annu. Rev. Phytopathol. 4 (2014) 52.

[63] A. Zehra, M. Meena, M.K. Dubey, M. Aamir, R.S. Upadhyay, Synergistic effects of plant defense elicitors and *Trichoderma harzianum* on enhanced induction of antioxidant defense system in tomato against Fusarium wilt disease, Bot. Stud. 58 (2017) 44, https://doi.org/10.1186/s40529-017-0198-2.

[64] A. Zehra, M. Meena, M.K. Dubey, M. Aamir, R.S. Upadhyay, Activation of defense response in tomato against fusarium wilt disease triggered by *Trichoderma harzianum* supplemented with exogenous chemical inducers (SA and MeJA), Braz. J. Bot. 21 (2017) 1–14, https://doi.org/10.1007/s40415-017-0382-3.

[65] A. Martinez-Medina, V. Flors, M. Heil, B. Mauch-Mani, C.M. Pieterse, M.J. Pozo, J. Ton, N. M. van Dam, U. Conrath, Recognizing plant defense priming, Trends Plant Sci. 21 (10) (2016) 818–822.

[66] Z.Q. Fu, X. Dong, Systemic acquired resistance: turning local infection into global defense, Annu. Rev. Plant Biol. 64 (2013) 839–863.

[67] A. Reid, Microbes helping to improve crop productivity, Microbe 6 (10) (2011) 435.

[68] R.F. Denison, E.T. Kiers, Life histories of symbiotic rhizobia and mycorrhizal fungi, Curr. Biol. 21 (18) (2011) R775–R785.

[69] P. Vejan, R. Abdullah, T. Khadiran, S. Ismail, B.A. Nasrulhaq, Role of plant growth promoting rhizobacteria in agricultural sustainability—a review, Molecules 21 (5) (2016) 573.

[70] E.M. De Souza, C.E. Granada, R.A. Sperotto, Plant pathogens affecting the establishment of plant-symbiont interaction, Front. Plant Sci. 21 (2016) 7–15.

[71] A. Zehra, N.A. Raytekar, M. Meena, P. Swapnil, Efficiency of microbial bio-agents as elicitors in plant defense mechanism under biotic stress: a review, Curr. Res. Microb. Sci. 2 (2021), 100054, https://doi.org/10.1016/j.crmicr.2021.100054.

[72] B. Gupta, A. Sengupta, J. Saha, K. Gupta, Plant abiotic stress: 'omics' approach, J. Plant Biochem. Physiol. 1 (3) (2013).

[73] K. Narasimhan, C. Basheer, V.B. Bajic, S. Swarup, Enhancement of plant-microbe interactions using a rhizosphere metabolomics-driven approach and its application in the removal of polychlorinated biphenyls, Plant Physiol. 132 (1) (2003) 146–153.

[74] D.J. Schneider, A. Collmer, Studying plant-pathogen interactions in the genomics era: beyond molecular Koch's postulates to systems biology, Annu. Rev. Phytopathol. 48 (2010) 457–479.

[75] M. Durot, P.Y. Bourguignon, V. Schachter, Genome-scale models of bacterial metabolism: reconstruction and applications, FEMS Microbiol. Rev. 33 (1) (2008) 164–190.

[76] C. Santi, D. Bogusz, C. Franche, Biological nitrogen fixation in non-legume plants, Ann. Bot. 111 (5) (2013) 743–767.

[77] S. Dobbelaere, Y. Okon, The plant growth-promoting effect and plant responses, in: Associative and Endophytic Nitrogen-Fixing Bacteria and Cyanobacterial Associations, Springer, Dordrecht, 2007, pp. 145–170.

[78] E.P. Colangelo, M.L. Guerinot, Put the metal to the petal: metal uptake and transport throughout plants, Curr. Opin. Plant Biol. 9 (3) (2006) 322–330.

[79] R.D. Sleator, C. Hill, Bacterial osmoadaptation: the role of osmolytes in bacterial stress and virulence, FEMS Microbiol. Rev. 26 (1) (2002) 49–71.

[80] J. Kohler, J.A. Hernández, F. Caravaca, A. Roldan, Induction of antioxidant enzymes is involved in the greater effectiveness of a PGPR versus AM fungi with respect to increasing the tolerance of lettuce to severe salt stress, Environ. Exp. Bot. 65 (2–3) (2009) 245–252.

[81] C. Wagg, J. Jansa, B. Schmid, M.G. van der Heijden, Belowground biodiversity effects of plant symbionts support aboveground productivity, Ecol. Lett. 14 (10) (2011) 1001–1009.

[82] C.R. Fellbaum, E.W. Gachomo, Y. Beesetty, S. Choudhari, G.D. Strahan, P.E. Pfeffer, E.T. Kiers, H. Bücking, Carbon availability triggers fungal nitrogen uptake and transport in arbuscular mycorrhizal symbiosis, Proc. Natl. Acad. Sci. 109 (7) (2012) 2666–2671.

[83] S. Hosseinzadeh, M. Shams-Bakhsh, E. Hosseinzadeh, Effects of sub-bactericidal concentration of plant essential oils on pathogenicity factors of *Ralstonia solanacearum*, Arch. Phytopathol. Plant Protect. 46 (6) (2013) 643–655.

[84] M.B. Miller, B.L. Bassler, Quorum sensing in bacteria, Annu. Rev. Microbiol. 55 (1) (2001) 165–199.

[85] I. Garmendia, J. Aguirreolea, N. Goicoechea, Defence-related enzymes in pepper roots during interactions with arbuscular mycorrhizal fungi and/or *Verticillium dahliae*, BioControl 51 (3) (2006) 293–310.

[86] R. D'Amore, U.Z. Ijaz, M. Schirmer, J.G. Kenny, R. Gregory, A.C. Darby, M. Shakya, M. Podar, C. Quince, N. Hall, A comprehensive benchmarking study of protocols and sequencing platforms for 16S rRNA community profiling, BMC Genomics 17 (1) (2016) 1–20.

[87] S. Kittelmann, H. Seedorf, W.A. Walters, J.C. Clemente, R. Knight, J.I. Gordon, P.H. Janssen, Simultaneous amplicon sequencing to explore co-occurrence patterns of bacterial, archaeal and eukaryotic microorganisms in rumen microbial communities, PLoS ONE 8 (2) (2013), e47879.

[88] C.L. Schoch, K.A. Seifert, S. Huhndorf, V. Robert, J.L. Spouge, C.A. Levesque, W. Chen, Fungal Barcoding Consortium, Nuclear ribosomal internal transcribed spacer (ITS) region as a universal DNA barcode marker for Fungi, Proc. Natl. Acad. Sci. 109 (16) (2012) 6241–6246.

[89] F. De Filippis, M. Laiola, G. Blaiotta, D. Ercolini, Different amplicon targets for sequencing-based studies of fungal diversity, Appl. Environ. Microbiol. 83 (17) (2017), e00905-17.

[90] E. Kopylova, J.A. Navas-Molina, C. Mercier, Z.Z. Xu, F. Mahé, Y. He, H.W. Zhou, T. Rognes, J.G. Caporaso, R. Knight, Open-source sequence clustering methods improve the state of the art, MSystems 1 (1) (2016), e00003-15.

[91] S. Nayfach, K.S. Pollard, Toward accurate and quantitative comparative meta-genomics, Cell 166 (5) (2016) 1103–1116.

[92] K.T. Konstantinidis, J.M. Tiedje, Prokaryotic taxonomy and phylogeny in the genomic era: advancements and challenges ahead, Curr. Opin. Microbiol. 10 (5) (2007) 504–509.

[93] J.S. Ghurye, V. Cepeda-Espinoza, M. Pop, Focus: microbiome: metagenomic assembly: overview, challenges and applications, Yale J. Biol. Med. 89 (3) (2016) 353.

[94] T.J. Sharpton, An introduction to the analysis of shotgun metagenomic data, Front. Plant Sci. 5 (2014) 209.

[95] S.A. Boers, R. Jansen, J.P. Hays, Understanding and overcoming the pitfalls and biases of next-generation sequencing (NGS) methods for use in the routine clinical microbiological diagnostic laboratory, Eur. J. Clin. Microbiol. Infect. Dis. 38 (6) (2019) 1059–1070.

[96] L. Lambrechts, Dissecting the genetic architecture of host-pathogen specificity, PLoS Pathog. 6 (8) (2010), e1001019.

[97] K. Dalman, K. Himmelstrand, A. Olson, M. Lind, M. Brandström-Durling, J. Stenlid, A genome-wide association study identifies genomic regions for virulence in the non-model organism *Heterobasidion annosum* s.s, PLoS ONE 8 (1) (2013), e53525.

[98] L.M. Raboin, E. Ballini, D. Tharreau, A. Ramanantsoanirina, J. Frouin, B. Courtois, N. Ahmadi, Association mapping of resistance to rice blast in upland field conditions, Rice 9 (1) (2016) 1–2.

[99] Y. Gao, Z. Liu, J.D. Faris, J. Richards, R.S. Brueggeman, X. Li, R.P. Oliver, B.A. McDonald, T.L. Friesen, Validation of genome-wide association studies as a tool to identify virulence factors in *Parastagonospora nodorum*, Phytopathology 106 (10) (2016) 1177–1185.

[100] J.Q. Wu, S. Sakthikumar, C. Dong, P. Zhang, C.A. Cuomo, R.F. Park, Comparative genomics integrated with association analysis identifies candidate effector genes corresponding to Lr20 in phenotype-paired *Puccinia triticina* isolates from Australia, Front. Plant Sci. 8 (2017) 148.

[101] G.P. Morris, P. Ramu, S.P. Deshpande, C.T. Hash, T. Shah, H.D. Upadhyaya, O. Riera-Lizarazu, P.J. Brown, C.B. Acharya, S.E. Mitchell, J. Harriman, Population genomic and genome-wide association studies of agroclimatic traits in sorghum, Proc. Natl. Acad. Sci. 110 (2) (2013) 453–458.

[102] G. Jia, X. Huang, H. Zhi, Y. Zhao, Q. Zhao, W. Li, Y. Chai, L. Yang, K. Liu, H. Lu, C. Zhu, A haplotype map of genomic variations and genome-wide association studies of agronomic traits in foxtail millet (*Setaria italica*), Nat. Genet. 45 (8) (2013) 957–961.

[103] Z. Zhou, Y. Jiang, Z. Wang, Z. Gou, J. Lyu, W. Li, Y. Yu, L. Shu, Y. Zhao, Y. Ma, C. Fang, Resequencing 302 wild and cultivated accessions identifies genes related to domestication and improvement in soybean, Nat. Biotechnol. 33 (4) (2015) 408–414.

[104] W. Wen, D. Li, X. Li, Y. Gao, W. Li, H. Li, J. Liu, H. Liu, W. Chen, J. Luo, J. Yan, Metabolome-based genome-wide association study of maize kernel leads to novel biochemical insights, Nat. Commun. 5 (1) (2014) 1–10.

[105] V. Ramegowda, M. Senthil-Kumar, The interactive effects of simultaneous biotic and abiotic stresses on plants: mechanistic understanding from drought and pathogen combination, J. Plant Physiol. 176 (2015) 47–54.

[106] N. Tank, M. Saraf, Salinity-resistant plant growth promoting rhizobacteria ameliorates sodium chloride stress on tomato plants, J. Plant Interact. 5 (1) (2010) 51–58.

[107] R. Penyalver, P. Oger, M.M. Lopez, S.K. Farrand, Iron-binding compounds from *Agrobacterium* spp.: biological control strain *Agrobacterium rhizogenes* K84 produces a hydroxamate siderophore, Appl. Environ. Microbiol. 67 (2) (2001) 654–664.

[108] Y. Jha, B. Dehury, S.P. Kumar, A. Chaurasia, U.B. Singh, M.K. Yadav, U.B. Angadi, R. Ranjan, M. Tripathy, R.B. Subramanian, S. Kumar, Delineation of molecular interactions of plant growth promoting bacteria induced β-1, 3-glucanases and guanosine triphosphate ligand for antifungal response in rice: a molecular dynamics approach, Mol. Biol. Rep. (2021) 1–11.

[109] V. Kumar, M. Baweja, P.K. Singh, P. Shukla, Recent developments in systems biology and metabolic engineering of plant-microbe interactions, Front. Plant Sci. 7 (2016) 1421.

[110] A. Kumar, J.P. Verma, Does plant-microbe interaction confer stress tolerance in plants: a review? Microbiol. Res. 207 (2018) 41–52.

[111] M. Grover, S.Z. Ali, V. Sandhya, A. Rasul, B. Venkateswarlu, Role of microorganisms in adaptation of agriculture crops to abiotic stresses, World J. Microbiol. Biotechnol. 27 (5) (2011) 1231–1240.

[112] S. Savci, An agricultural pollutant: chemical fertilizer, Int. J. Environ. Sci. Dev. 3 (1) (2012) 73.

[113] K. Naik, S. Mishra, H. Srichandan, P.K. Singh, P.K. Sarangi, Plant growth promoting microbes: potential link to sustainable agriculture and environment, Biocatal. Agric. Biotechnol. 21 (2019), 101326.

[114] J.J. Parnell, R. Berka, H.A. Young, J.M. Sturino, Y. Kang, D.M. Barnhart, M.V. DiLeo, From the lab to the farm: an industrial perspective of plant beneficial microorganisms, Front. Plant Sci. 7 (2016) 1110.

[115] A. Puri, K.P. Padda, C.P. Chanway, Can naturally-occurring endophytic nitrogen-fixing bacteria of hybrid white spruce sustain boreal forest tree growth on extremely nutrient-poor soils? Soil Biol. Biochem. 140 (2020), 107642.

[116] M.C. Azabou, Y. Gharbi, I. Medhioub, K. Ennouri, H. Barham, S. Tounsi, M.A. Triki, The endophytic strain *Bacillus velezensis* OEE1: an efficient biocontrol agent against *Verticillium* wilt of olive and a potential plant growth promoting bacteria, Biol. Control 142 (2020), 104168.

[117] C.A. Muller, M.M. Obermeier, G. Berg, Bio-prospecting plant-associated microbiomes, J. Biotechnol. 235 (2016) 171–180.

[118] C. Zachow, H. Müller, R. Tilcher, C. Donat, G. Berg, Catch the best: novel screening strategy to select stress protecting agents for crop plants, Agronomy 3 (4) (2013) 794–815.

[119] H. Tikariha, H.J. Purohit, Assembling a genome for novel nitrogen-fixing bacteria with capabilities for utilization of aromatic hydrocarbons, Genomics 111 (6) (2019) 1824–1830.

[120] D.C. Hao, P.G. Xiao, Rhizosphere microbiota and microbiome of medicinal plants: from molecular biology to omics approaches, Chin. Herb. Med. 9 (3) (2017) 199–217.

[121] M.S. Sarhan, M.A. Hamza, H.H. Youssef, S. Patz, M. Becker, H. ElSawey, R. Nemr, H.S. Daanaa, E.F. Mourad, A.T. Morsi, M.R. Abdelfadeel, Culturomics of the plant prokaryotic microbiome and the dawn of plant-based culture media-a review, J. Adv. Res. 19 (2019) 15–27.

[122] J.A. Parray, S. Jan, A.N. Kamili, R.A. Qadri, D. Egamberdieva, P. Ahmad, Current perspectives on plant growth-promoting rhizobacteria, J. Plant Growth Regul. 35 (3) (2016) 877–902.

[123] M. Schoebitz, M.D. Lopez, A. Roldan, Bio-encapsulation of microbial inoculants for better soil–plant fertilization. A review, Agron. Sustain. Dev. 33 (2013) 751–765.

[124] A. Kumar, A. Munder, R. Aravind, S.J. Eapen, B. Tummler, J.M. Raaijmakers, Friend or foe: genetic and functional characterization of plant endophytic *Pseudomonas aeruginosa*, Environ. Microbiol. 15 (3) (2013) 764–779.

[125] G. Puppi, R. Azcon, G. Hoflich, Management of positive interactions of arbuscular mycorrhizal fungi with essential groups of soil microorganisms, in: Impact of Arbuscular Mycorrhizas on Sustainable Agriculture and Natural Ecosystems, 1994, pp. 201–215.

[126] A. Hodge, C.D. Campbell, A.H. Fitter, An arbuscular mycorrhizal fungus accelerates decomposition and acquires nitrogen directly from organic material, Nature 413 (6853) (2001) 297–299.

[127] D. Minerdi, R. Fani, R. Gallo, A. Boarino, P. Bonfante, Nitrogen fixation genes in an endosymbiotic *Burkholderia* strain, Appl. Environ. Microbiol. 67 (2) (2001) 725–732.

[128] E. Siasou, D. Standing, K. Killham, D. Johnson, Mycorrhizal fungi increase bio-control potential of *Pseudomonas fluorescens*, Soil Biol. Biochem. 41 (6) (2009) 1341–1343.

[129] A. Iavicoli, E. Boutet, A. Buchala, J.P. Metraux, Induced systemic resistance in *Arabidopsis thaliana* in response to root inoculation with *Pseudomonas fluorescens* CHA0, Mol. Plant-Microbe Interact. 16 (10) (2003) 851–858.

[130] S. Compant, B. Duffy, J. Nowak, C. Clément, E.A. Barka, Use of plant growth-promoting bacteria for bio-control of plant diseases: principles, mechanisms of action, and future prospects, Appl. Environ. Microbiol. 71 (9) (2005) 4951–4959.

[131] C.L. Bender, The post-genomic era: new approaches for studying bacterial diseases of plants, Australas. Plant Pathol. 34 (4) (2005) 471–474.

[132] R. Winnenburg, M. Urban, A. Beacham, T.K. Baldwin, S. Holland, M. Lindeberg, H. Hansen, C. Rawlings, K.E. Hammond-Kosack, J. Kohler, PHI-base update: additions to the pathogen-host interaction database, Nucleic Acids Res. 36 (suppl_1) (2007), D572-6.

[133] B.F. Quirino, E.S. Candido, P.F. Campos, O.L. Franco, R.H. Kruger, Proteomic approaches to study plant-pathogen interactions, Phytochemistry 71 (4) (2010) 351–362.

Chapter 18

Plant-microbe interactions: Role in sustainable agriculture and food security in a changing climate

Diksha Tokas[a], Siril Singh[a,b], Rajni Yadav[a], and Anand Narain Singh[a]
[a]*Soil Ecosystem and Restoration Ecology Lab, Department of Botany, Panjab University Chandigarh, Chandigarh, India,* [b]*Department of Environment Studies, Panjab University Chandigarh, Chandigarh, India*

1 Introduction

Global climate change, population growth, rapid industrialization, diminishing agroecosystems, and use of agrochemicals create a challenge to crop productivity and food security worldwide. There is a strong need to find an eco-friendly and sustainable solution to overcome the challenges in the agriculture sector such as abiotic and biotic stresses. Microbiomes in association with the plants have enormous ability to provide economical and sustainable solutions that will bring in innovative approaches for improving agricultural practices, thus increasing crop productivity. Plants stay in close alliance with a wide range of microorganisms. Plant-microbe interactions can be both symbiotic and hostile, and the comprehension of these interactions is equivalently crucial for the betterment of agricultural production [1,2]. An important scheme analogous to climate-smart agricultural practices is to find the role of microorganisms in improved plant nutrient quality and subsequently crop production.

Most of the "microbiome" components associated with the plants are only proficient to colonize and persevere on the surfaces of the plant tissue or in the soil. However, some can also introduce themselves as endophytes. All plants likely carry endophytes, which play a crucial role in plant fitness and development. The endophytes and the plant with which they are associated work in great coordination and they sustain an exceptional ecosystem. The interactions of plants and microbes can be beneficial and antagonistic: beneficial include improvement of plant health, disease suppression, increased yield, improvement

Plant-Microbe Interaction—Recent Advances in Molecular and Biochemical Approaches
https://doi.org/10.1016/B978-0-323-91876-3.00008-7
363

of soil structure, production of plant hormones, promotion of nutrient mineralization and absorption, and antagonistic interactions include pathogens, pests, and diseases. The instrumental plant-microbe interactions include adaptations to different environmental variables, uptake of nutrients by the plants, mycorrhizal associations, and biotic and abiotic stress tolerance [3,4]. At present, the farming activities are heavily dependent on the excessive use of high-production agrochemicals that are often responsible for various environmental hazards. The crop production across the world has been highly affected by the adverse global climate change, shrinking agricultural lands, use of harmful chemicals, and expeditious urbanization. This devastating effect on crop production has caused concerns for food security, thus requiring the need for eco-friendly and sustainable developments in agriculture. The association of the microorganisms with plants is observed to improve yield spontaneously. The beneficial plant-microbe interactions can be of enormous significance to the agroecosystems as the pathways of interactions can be harnessed and studied and can also be used for the production of biofertilizers and biopesticides, etc. This can also reduce the dependency on synthetic fertilizers that are not only expensive but also cause severe harm to the agricultural soils. A paramount approach related to climate-friendly agricultural practices is to explore the role of microorganisms in improved crop yield and plant nutrient quality.

2 Types of plant-microbe interactions

Several bacterial and fungal species thrive in the rhizosphere. These microbes interact within themselves and with the plants in their premises. The kind of interactions they have with the plants can be detrimental as well as beneficial. These interactions occur both below- and aboveground; however, the belowground plant-microbe interactions are much more complex as compared to the ones above the surface of the soil [5].

The interactions between plants and microbe communities are very intricate. These interactions, on the one hand, help in ameliorating the growth of the plant under normal environmental conditions, and on the other hand, these also indirectly conserve the plant from unfavorable environmental conditions by promoting plant growth.

Plant roots have associate microbes such as rhizobia and fungi that provide nutrients to plants in exchange for carbon that is required for their growth. Many reports have shown the effect of several bacterial strains on plant growth under adverse environmental conditions like temperature, salinity, pathogen, drought, and heavy metals [6].

It is also observed that microorganisms cause fatalistic effects on plant growth as they interact negatively. Those impacts happen to plants because of the pathogenic nature of microorganisms as they release some harmful compounds into the plant. Whether beneficial or antagonistic, the kind of interaction is determined by the type of microbial species and the kind of action mechanisms adopted by the microbe. For example, the production of cyanide by a few strains of bacteria may inhibit plant growth, while phytohormones produced by some bacteria may lead to plant growth enhancement.

2.1   Pathogenic

Fungal strains mostly reside as pathogens and give rise to specific disease in plants. The study on relationships of phytopathogenic fungi and plants has become a very crucial and fascinating subject of plant sciences. These pathogens can be classified as biotrophic, hemibiotrophic, or necrotrophic. Biotrophic fungi obtain their nutrients from living tissues of plants via haustoria, and necrotrophic fungi get their nutrients from dead host tissue after killing it via enzymes and toxins. On the other hand, hemibiotrophic fungi have both phases in their life cycle, i.e., a biotrophic stage followed by a necrotrophic one [7]. Due to their diverse rank, they are capable of colonizing plants successfully. The plant physiology can be adversely affected by pathogenic fungi.

The study of plant pathogenic fungi is very crucial from the economy's point of view due to the kind of detrimental effects they cause to the growth and production of most of the economically important crops. Forests, grasslands, and crops are being deprived of their worth due to the harmful effects of pathogenic fungi. Fungal strains have inconsistency among themselves concerning the severity pathogenicity. Dean et al. [8] reported a list of 10 pathogenic strains on the basis of their severity that include *Blumeria graminis, Fusarium graminearum, Mycosphaerella graminicola, Magnaporthe oryzae, Melampsora lini, Ustilago maydis, Colletotrichum* spp., *Fusarium oxysporum, Puccinia* spp., and *Botrytis cinerea*. Due to plant diseases caused by fungi, the annual crop loss has been estimated around 15% [7].

The magnification of plant growth is a familiar facet of rhizospheric bacteria. However, the harmful effects of these bacteria on plant growth and development have also been studied. This antagonistic influence might be due to the production of some compounds that are detrimental to plants or the excess production of some growth regulators. A few bacterial strains release cyanide that negatively affects plant growth and development. Microbial volatiles are organic molecules produced by all bacteria as a byproduct of metabolism. These chemicals have a larger role in plant-microbe interactions than nonvolatile compounds. The bacteria's volatile chemicals may have an inhibiting or stimulating impact.

2.2   Symbiotic

Fungi and plants have diversified associations which range from symbiotic to pathogenic associations. Mutualism is chiefly based on organic material produced by fungal decay for accessibility of nutrients otherwise unavailable to plants. In the rhizosphere, fungi and plants communicate at the molecular level as plants release amino acids, sugars, and organic compounds that activate the fungi to colonize the plant roots.

Mycorrhizae are the most common symbiotic association that is distinguished by unique morphological growth (fungi roots). Mycorrhizae fungi, which live on or in plant roots, are common in forest trees and are connected with more than 90% of plant species [9]. Mycorrhizae thrive in a variety of crops, including fruits, vegetables, grains, and ornamental plants. Fungi aid plants by expanding the root absorption surface area and preventing pathogens, which increases nutrients (nitrogen and phosphate) and water intake. Endo-mycorrhizae roots are comparable in size, shape, and color to typical plant roots.

In contrast, hyphae expand into the cortical cells of the feeder root and give birth to arbuscules (exclusive feeding hyphae) and, in rare situations, vesicles, i.e., food storage hyphal expansions. In most situations, endomycorrhizae have both arbuscules and vesicles, which are referred to as VAM (vesicular arbuscular mycorrhizae). In exchange, fungi consume sugar, which plants produce through photosynthesis. Because most plants are unable to metabolize plant sugars (mannitol and trehalose), ectomycorrhizae metabolize them.

Protease enzyme, which is responsible for protein breakdown in leaf litter, is also produced by ectomycorrhizal fungi. The endomycorrhizal fungus uses extra radicle hyphae to absorb nutrients from the soil and transport them to the plant via branching arbuscules.

Plant development is aided by favorable plant-microbe interactions. Plant growth-promoting rhizobacteria (PGPR) are recognized as one of the most important microbial communities in the rhizosphere due to their capacity to stimulate plant development. Solubilization of nutrients, hormone secretion, siderophores synthesis, and nitrogen fixation are all extremely important growth-promoting properties. These PGPR also shield the plant against the damaging effects of plant diseases. This can be accomplished by decreasing the pathogen's access to certain nutrients or by destroying its cell wall. The former is performed by creating siderophores, which bind to iron and render it inaccessible to the pathogen [10]. They can also mitigate the negative effects of pathogens by increasing plant resistance against diseases by a mechanism known as induced systemic resistance (ISR) [11] (Fig. 1).

3 Types of plant microbiomes

3.1 Based on the location concerning plant

3.1.1 Rhizosphere

The rhizosphere is a hub of microbial activity, and the microorganisms that live there react with the many compounds generated by plant roots. Thus, microbes and the compounds they produce interact with plant roots in a variety of ways, including positive, negative, and neutral interactions. These interactions can

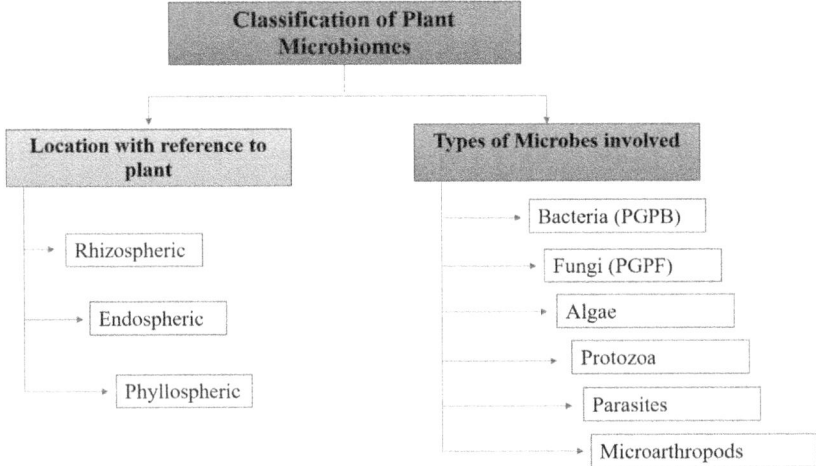

FIG. 1   Classification of plant microbiomes.

influence plant development, modify nutrient dynamics, and alter the plant's susceptibility to certain diseases, abiotic stress, and heavy metal resistance [12].

Bacteria are the most copious microbes in the rhizosphere, and hence they are destined to influence the plant effectively. The ratio of the microbial population to the rhizosphere (R) in the bulk soil (S), i.e., the R/S value, is greater than 20 for bacteria, whereas it is 10 and 2–3 for fungi and actinomycetes, respectively [13]. Because of the reduced oxygen level caused by root respiration, the typical percentage of aerobic bacteria is considerably lower in the rhizosphere. By receiving signals from the host, the rhizosphere is designed to attract a diverse range of bacterial species that are beneficial to plants. The beneficial bacteria are collectively known as plant growth-promoting rhizobacteria (PGPRs). *Agrobacterium, Flavobacterium, Alcaligenes, Arthrobacter, Azospirillum, Azotobacter Pseudomonas, Acinetobacter, Bacillus, Cellulomonas, Micrococcus, Mycobacterium*, and *Rhizobium* are the most common bacterial genera found in the rhizosphere. PGPRs are classified as biopesticides, biofertilizers, phytostimulators, and elicitors of tolerance to biotic and abiotic challenges depending on their modes of action [10].

The other rhizobacteria (*Enterobacter, Burkholderia, Rhizobium*) help plants combat stress due to reactive oxygen species. *Bacillus subtilis* and *Achromobacter piechaudii* improve salt tolerance in coastal plants, whereas *Paenibacillus polymyxa* and *Rhizobium tropici* have been explored for drought tolerance in Arabidopsis, tomato, and common bean. The rhizosphere is home to both symbiotic and harmful fungi. Approximately 105–106 organisms can be found in 1 g of the rhizosphere Arbuscular mycorrhizal fungi (AMF), belonging to Glomerales, are ancient fungi that establish the most primitive type of association with plant roots.

3.1.2 Endospheric

A large number of endophytes and microbial communities reside in the root system of different crop plants. Endophytes are the accessory microbes (such as bacteria, fungi, and viruses) that live in the endosphere of plants all over their life cycle or during a phase of their life cycle without causing any apparent changes to the host plants. The endophytic microbes are those microorganisms that live in the plant tissue as endosymbionts without any intention to evolve any apparent diseases in the host plant. Endophytic fungi affect plant health by ameliorating the plant's nutrient status; eventually, the plant growth and development are highly enhanced, restricting the entry of a pathogen. The microbial communities outside the host plants substantially influence the plant roots.

The endophytes sheltering in plants are divided into three groups: (1) obligate endophytes, microbes that are unable to live outside the host plant; (2) facultative endophytes, free-living and begin colonization of the roots when needed; and (3) passive endophytes, come into existence as a result of some specific events only such as open wounds [14]. Insect attacks are reduced by the presence of endophytic fungi in plants. Webber in 1981 reported on the protection of elm trees by the endophyte *Phomopsis oblonga*, which preserved the trees from the destructive beetle *Physocnemum brevilineum*. According to the findings, poisonous chemicals generated by *P. oblonga* were responsible for repelling the insects [15]. Entrance of endophytic bacteria into host plants happens spontaneously during plant development or through wounds. They spread from parent to offspring or among individuals.

3.1.3 Phyllospheric

The term phyllosphere refers to "the parts of a plant above the ground, usually surface of leaves, regarded as a habitat for microorganisms." This is a place where ordinarily a variety of microorganisms (bacteria and fungi) colonize. The phyllosphere is the environment in which bacteria invade and form associations with plants, most prominently epiphytes. The phyllosphere's microbial communities are extraordinarily complex, containing both uncultured and cultured bacteria. There are external factors that affect the diversity of microbes thriving in the phyllosphere like light, temperature, nutrient availability, water, and UV light. The phyllosphere has specific microenvironments that are according to the leaf physiology and arrangement of leaf epidermal cells and these highly affect the abundance of microorganisms on the surface. The cuticle layer outside has aliphatic compounds in it and it allows permeability and moisture, which provide better attachment of microorganisms. The permeability of water is the deciding factor for the growth and survival of epiphytes on the phyllosphere. The leaf surfaces that have high water content are heavily colonized by bacteria.

The phyllosphere is composed of diverse microbial communities that include bacteria, fungi, algae, and protozoans. Among the manifold communities of microbes, bacteria are the leading community on leaves. Molecular studies have shown that alpha-, beta-, and gamma-proteobacteria are the chief bacterial inhabitants of the phyllosphere. Acidobacteria, cyanobacteria, and actinobacteria also occur frequently in the phyllosphere. Yeasts like *Sporobolomyces*, *Rhodotorula*, and *Cryptococcus* are frequently found on the leaf's surface. Methylotrophic bacteria residing in the phyllosphere include genera such as *Methylophilus*, *Methylobacterium*, *Methylocella*, *Methylibium*, *Hyphomicrobium*, and *Methylocystis* [16].

The fungi mainly associated with the phyllospheric region include genera like *Fusarium oxysporum*, *Aspergillus niger*, *Penicillium aurantiogriseum*, *Alternaria alternata*, *Talaromyces funiculosus*, *Aspergillus flavus*, *Trichoderma aureoviride* [17]. The phyllosphere microflora affects the ecological links of the plants. The phyllosphere usually has fungi, bacteria, algae, lichens, and viruses that actively participate in the growth, resistance, infection, and adaptation of the host plant [18]. Phyllospheric microorganisms play a significant role in leaf functions, apical growth and flowering, seed mass and development of fruit.

3.2 Classification

3.2.1 Bacteria (PGPB)

Microbes present in the rhizosphere carry out several functions toward the growth and development of the host plant. Rhizobacteria that are mainly associated with plant growth promotion and disease resistance are classified as PGPB (plant growth-promoting bacteria), a class of microorganisms that improve plant growth and increase yield via a variety of plant growth-promoting substances, which serve as bioprotectants or biofertilizers. The different genera of bacteria like *Bacillus*, *Pseudomonas*, *Klebsiella*, *Enterobacter*, *Variovorax*, *Burkholderia*, *Serratia*, *Azospirillum*, and *Azotobacter* are included under PGPR, out of which *Bacillus* and *Pseudomonas* sp. are mostly reported [19]. PGPR can be divided into two groups, intracellular PGPR (iPGPR) and extracellular PGPR (ePGPR), based on their location. PGPR present in the rhizospheric soil are called ePGPR and the ones which are present on the inner side of roots are known as iPGPR [20]. The ePGPR mainly observed are *Flavobacterium*, *Agrobacterium*, *Serratia*, *Caulobacter*, *Azospirillum*, *Burkholderia*, *Pseudomonas*, *Arthrobacter*, *Bacillus*, *Erwinia*, *Micrococcus*, *Chromobacterium*, *Azotobacter*, etc. the best known iPGPR are *Allorhizobium*, *Bradyrhizobium*, *Azorhizobium*, *Rhizobium*, and *Mesorhizobium* [10]. Bacteria inhabiting the rhizosphere of plants have an important role and they have plant growth-promoting traits, which help in improving nutrient cycling and reducing the use of chemicals [21]. Numerous rhizospheric bacteria are used in organic farming as biofertilizers for sustainable agriculture. It is well reported that PGPR can be used as biofertilizers as well as coherent soil-inhabiting bacteria for sustainable agriculture.

PGPR produce a lot of special metabolites, which help inhibit the growth of pathogenic bacteria by inducing tolerance and resistance to the plants against various pathogens and stress conditions [22]. PGPR stimulate plant growth by any of the two mechanisms, direct or indirect. Direct mechanisms include activities such as nutrient solubilization (P, K, and Zn), nitrogen fixation, production of plant growth regulators, and production of organic acids [23].

3.2.2 Fungi (PGPF)

Fungi that live in several habitats in a plant system (roots, leaves, stem, rhizosphere, and phyllosphere) are used to promote plant growth by activating various important pathways during plant development or disease resistance during pathogenesis or resisting harsh situations. Interactions between fungi and the plants they are associated with within the phyllosphere and rhizosphere help in the promotion of plant growth and induction of resistance systemically (ISR) on attacking pathogens are called plant growth-promoting fungi (PGPF). A large number of heterogeneous classes of fungi from different habitats can augment plant growth. The chief fungi genera that are reported to have PGPF traits include *Penicillium, Aspergillus, Fusarium, Phoma, Piriformospora, Trichoderma*, and many more [24]. The interactions of PGPF with the host plant positively affect the plant belowground and aboveground parts. PGPF also effectively improve root hair growth, flowering, seed germination and yield, photosynthetic efficiency, and seed composition [25]. PGPF are also helpful in controlling several pathogens by providing induced systemic resistance. They improve the host plant's abilities to increase nutrient uptake and hormone production, which further causes gene expression by activating different plant signaling pathways [26]. PGPF have received substantial attention as biofertilizers because of their immense benefits to plants in different ways.

3.2.3 Algae

Algae have a lesser proportion in the soil as compared to fungi. Algae can be unicellular (*Chlamydomonas*) or filamentous (*Ulothrix* and *Spirogyra*). Algae are phototrophic organisms as they contain chlorophyll and synthesize their food. The main genera of fungi residing in the soil include *Chlamydomonas, Chlorella, Chlorococcum, Chlorochitrum, Oedogonium*, and *Protosiphon*. This microbial group constitutes an integral part of the soil biota, which is chiefly composed of cyanobacteria. Mainly, cyanobacteria play a significant role in maintaining soil fertility, hence increase rice yield as a natural biofertilizer [27]. Cyanobacteria are a group of microorganisms that help in balancing atmospheric nitrogen. Due to its wide adaption to environmental changes and different soils, blue-green algae (cyanobacteria) make it cosmopolitan. The most efficient nitrogen-fixing fungi like *Anabaena variabilis, Nostoc linckia, Calothrix* sp., *Aulosira fertilisima, Tolypothrix* sp., and *Scytonema* sp. have been reported from many agricultural habitats and are being utilized for the manufacture of rice.

3.2.4 Protozoa

Protozoa are heterotrophic, unicellular, eukaryotic organisms that include four types of organizations: ciliates, amoebae, flagellates, and parasitic sporozoans. They are mainly functional with the plant's rhizospheric regions and form a ubiquitous group in the region. The number of protozoa in the soil varies with soil fertility, being many times higher in highly fertile soils. It has been reported that the presence of protozoa and their ratio in the rhizosphere can significantly promote plant growth [28]. Protozoa stimulate microbial decomposition and the release of organic matter and supply plants with sufficient nitrogen that is otherwise not wholly accessible. The increased availability of nitrogen benefits the fungi associated with the plants and transfers it to the plant roots via hyphae. This, in turn, increases the photosynthetic ability of the plants and overall plant growth and development.

3.2.5 Parasites

Nematodes are very complex eukaryotic invertebrate worms that are primarily free-living but they parasitize plants. Among the different nematodes residing in the soil, the genera *Heterorhabditis* and *Steinernema* have been reported as potent microbial controllers that inhibit the growth of various pathogenic insects and pests in the rhizosphere. The secondary metabolites that are released from the plant roots in the form of exudates play a crucial role in attracting the nematodes [29]. Reports have shown that insect herbivory at the roots can lead to the secretion of volatile substances that attract nematodes like *Heterorhabditis megidis* in many plants. Thus, this was adopted as an essential plant defense mechanism against insects [30].

3.2.6 Microarthropods

Soil microarthropods are an important component of terrestrial ecosystems due to their main regulators of crucial processes like plant litter decomposition and mineralization. Most microarthropods that are often thought to be saprophagic can be omnivorous [31]. Soil microarthropods, mostly mites and collembolans, are among the lesser-known faunal variety found in nearly all agricultural soils. They also play a role in the complex food webs of soils. The existence of various microbial groups in the soil, as well as the physical and chemical qualities of the soil, has a significant impact on the distribution of diverse arthropods in the soil.

4 Affinities of microbes with plants

4.1 Beneficial effects

4.1.1 Nutrient uptake facilitation

Microorganisms play an essential role in the cycling of N in the ecosystems, and the main contributor of the N pool is organic matter present in the soil. The main proportion of N pools present in the organic matter is not available to the plants due to its complex decomposition process. The decomposition of organic matter is chiefly done by soil microbes mostly through bacteria, actinomycetes, and fungi, leading to the release of nutrients in the soil. The microbial biomass plays a significant role in the soil N cycle [32]. The functional diversity of bacteria play an essential role in plant litter decomposition and recycling nutrients. They constitute a good source and sink of nutrients and occupy an important rank in the ecosystem and soil food chain.

Soil microorganisms also affect the availability of many other nutrients in addition to nitrogen. The availability of phosphorus in the soil is affected by phosphate-solubilizing bacteria (PSB) and phosphate-solubilizing fungi (PSF). The P solubilization potential of PSB is 1%–50%, whereas that of PSF is only 0.1%–0.5% [33]. The strains of bacteria like *Bacilli*, *Pseudomonas*, and endosymbiotic rhizobia with a few fungal strains like *Aspergillus*, *Arthrobotrys oligospora*, and *Penicillium* have been reported as efficient phosphate solubilizers [34].

Potassium (K) is the third essential plant nutrient that plays an important role in photosynthesis, protein synthesis, and enzyme activation. The potassium levels have decreased in soils of India due to a lack of replenishment of K after harvest. The cost of potassium fertilizers is also very high in India. The primary agent that is responsible for the capacity to solubilize potassium by microorganisms is the low-molecular-weight organic acids, viz., oxalic acid, citric acid, tartaric acid, succinic acids, etc. Along with this, the production of coumaric, ferulic, syringic, and malic acid by K-solubilizing bacteria is responsible for K solubilization [35]. The chemical fertilizers in the soil can be used for the replenishment of soil potassium reserve but it hurts the environment. From this perspective, potassium-solubilizing bacteria could be used as a wise alternative for replenishing potassium in the soil in an eco-friendly manner.

Sulfur is available in inorganic and organic forms. The accessibility of sulfur in soil is mainly affected by the availability of microorganisms residing in the soil. They are the only factor responsible for the generation of the sulfur pool by the oxidative transformation of organic sulfur [36]. The photoautotrophic and chemolithotrophic bacteria oxidize sulfur and generate sulfates that the plants utilize, and they get energy from the process. The microbes aiding the process include chemolithotrophs (*T. thiooxidans and T. ferrooxidans*), photoautotrophs (including green and purple S bacteria), and heterotrophs (Fig. 2).

Iron is an excellent element for the growth of plant as it is the cofactor of many enzymes important for plant metabolism [37]. The iron deficiency in the plant may cause trouble in many metabolic processes that may ultimately alter plant growth and development. Iron oxidation by soil bacteria occurs mostly in acidic environments under aerobic circumstances, whereas chelation is favored in neutral environments. Iron reduction and iron sulfide precipitation occur mostly in anaerobic environments. Iron oxidation can be mediated for energy generation by *Sulfobacillus acidophilus* and *Thiobacillus ferrooxidans*

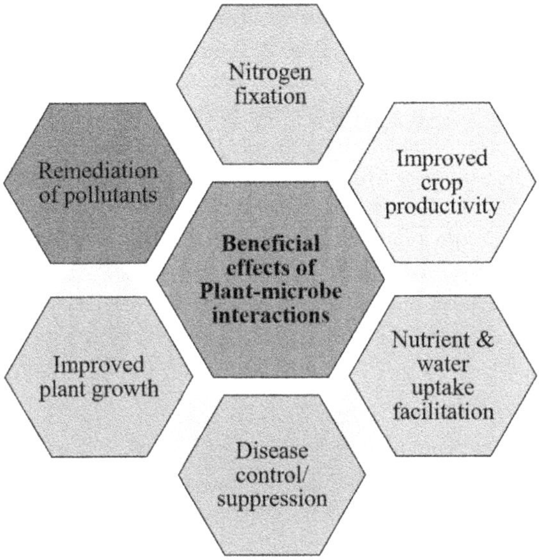

FIG. 2 Beneficial effects of plant-microbe interactions.

in acidic and aerobic conditions. Microbes such as *Crenothrix*, *Leptothrix*, *Sphaerotilus*, *Gallionella*, *Metallogenium*, and others may convert Fe (II) to Fe (III) in neutral soil without producing energy.

Zinc influences enzymatic activities like hydrogenase and carbonic anhydrase in plants and thus affects plant metabolism. It also contributes to the stability of ribosomal fractions. Because of their ability to solubilize zinc, a few microbial species, including *Bacillus* sp., *Aspergillus* sp., and *Pseudomonas* sp., are regarded as significant for the soil. Organic acid release by microorganisms enhances Zn availability in the soil. Vaid et al. [38] discovered that inoculating rice with zinc-solubilizing bacteria from the genera *Burkholderia* and *Acinetobacter* resulted in increased growth and production.

4.1.2 Mycorrhizal associations

Mycorrhizas are the association of fungi with the roots of higher plants that increase the uptake of water and nutrients. The arbuscular mycorrhizal (AM) association is an important mutualistic interaction that results in a significant beneficial impact worldwide and in over 65% of the plants. The mycorrhizal fungi form an extensive hyphal network in the soil in association with the roots and act as an artificial network for increasing the availability of water and nutrients to the plants. The mycorrhizal fungi improve the nutrient uptake by the host plant and are capable of distributing significant quantities of essential macroelements (N, P, K, and S) as well as trace elements (Cu, Zn). Many mycorrhizal fungi also provide help to the plants under stress conditions in addition to benefits of nutrients by the development of fitness apposing abiotic stresses (e.g., drought, heavy metals, and salinity) as well as biotic (pathogens) stresses. The association of fungi with the host plant also affects their relationship with the ecosystem. The mycorrhizal mycelia also influence the qualitative and quantitative alterations in the microbial community in the rhizosphere. The presence of mycorrhiza in the soil also affects the process of decomposition by altering the microbial behavior. They also support their host by affecting the physiology and morphology of the plant under various situations (stress, diseases, and others). Thus, under various stress situations, they produce growth-regulating chemicals, increase the photosynthetic rate, and improve the osmotic adjustment, all of which have a negative influence on pests and soilborne diseases [39].

4.1.3 Nitrogen fixation

Nitrogen is an essential element for plant development. The majority of this element is in gaseous form (N_2) that is not available to the plants. To fulfill the needs of plants in the agroecosystem, dependence on chemical fertilizers has been increasing; however, the detrimental effects of these are more than their benefits. Free-living bacteria like *Bacillus*, *Azotobacter*, *Acetobacter*, *Klebsiella*, *Corynebacterium*, *Clostridium*, *Diazotrophicus*, *Arthrobacter*, and *Pseudomonas*; and symbiotic *Azospirillum* are very useful in the process of nitrogen fixation. One of the many benefits of diazotrophic bacteria is to provide nitrogen to plants in exchange for the carbon that is released by the plant roots. This makes it crucial for these diazotrophs to live near the plants either in the rhizosphere or as endophytes in the plants. Nitrogen fixation has a potential role in enhancing soil fertility and productivity. Many nitrogen-fixing bacteria that inhabit the plant rhizosphere, particularly the plant roots: *Herbaspirillum seropedicae*, *Azotobacter diazotrophicus*, and *Azoarcus*, improve the yield of wheat, barley, rice, and sugarcane [40].

The complex enzyme (nitrogenase) that is for the process of nitrogen fixation is composed of two parts of metalloenzymes, mainly (1) dinitrogenase reductase (an iron protein) and (2) dinitrogenase (metal cofactor). Nitrogen reductase releases electrons having a high reducing ability, and dinitrogenase uses these electrons for reducing N_2 to NH_3. The ability to fix nitrogen varies according to the bacterial strain and the plant species. Most of the biological nitrogen fixations are regulated by the molybdenum nitrogenase enzyme that is present in all diazotrophs. The application of biological nitrogen-fixing organisms to agricultural soils can be a tool for the fulfillment of the nitrogen requirement along with benefits like disease suppression.

4.1.4 Promotion of plant growth

The microbes associated with the plants (either in the rhizosphere, phyllosphere, or endosphere) affect the plant growth and development at different parameters. They not only help the plant with nutrient mobilization but also significantly affect the plants by altering the hormonal balance. Plant growth and development are mainly affected by the interactions between plant roots and the surrounding microbial communities residing in the soil. The rhizosphere harbors both types of microorganisms that have negative as well as positive effects on plant growth. Although most of the microorganisms are beneficial for the plants, some pathogens affect plant health by releasing toxins. The major processes for promoting plant growth include the enhancement of nutrient availability (biofertilization), suppression of parasitic and nonparasitic pathogens (biocontrol), and production of plant hormones/and or plant growth-promoting substances (phytostimulation).

Biological nitrogen fixation that is done by nonsymbiotic bacteria, including *Azospirillum, Gluconacetobacter, Burkholderia*, and *Pseudomonas* species, may be employed in biofertilization of nonleguminous plants, including wheat, maize, rice, and sugarcane. Agricultural soils have considerable phosphorus available in the rhizosphere but that is exhausted frequently. A large number of soil microorganisms such as *Pseudomonas*, Actinomycetes, *Rhizobium*, and *Bacillus* are effective at solubilizing phosphorus and making it available for plants. Mycorrhizal associations can also increase plant growth by improvement of plant establishment, improved soil structure and nutrient uptake, and stress tolerance.

4.1.5 Disease control or suppression

Plant-associated microbes play a very crucial role in the suppression of disease and controlling the different pathogens attacking the plants. There are several mechanisms in which the microbes help in the process of disease control that includes both indirect as well as direct mechanisms. Direct mechanisms include the release of toxins. Antibiotics are an effective means of self-protection methods that include bacteria like *Pseudomonas, Bacillus* as well as fungi like *Gliocladium, Chaetomium, Trichoderma*, and *Amelomyces*. These microbes are thus very effective soil conditioners. Multifunctional microorganisms like *Trichoderma harzianum* solubilize important plant nutrients as well as help in suppressing plant pathogens. The production of hydrogen cyanide inhibits the growth of microbes and also suppresses pathogens that cause diseases like bacterial canker in tobacco and tomato plants, black rot, and root-knot disease [41].

Many reports signify that a lot of fungi and bacteria produce siderophores that act as chelating sources in conditions of iron deficiency such as *Pseudomonas, Rhizobium, Serratia*, and *Azospirillum*, and these can consume the surrounding iron, making it unavailable for pathogenic fungi. Also, the plant's defense system, known as induced systemic resistance (ISR), can be considered as a defense mechanism. The release of volatile organic compounds by plant growth-promoting bacteria and fungi may induce ISR, causing increased expression of defense-related genes in the plants [42].

4.1.6 Increased crop yield

The challenge of food security around the world has urged the need to develop techniques that help in crop production. However, intensive farming practices and intense use of agrochemicals without giving much attention to the deteriorating soil health has raised the concern. There is a need to develop strategies to use improved resources for farming practices that include water and nutrient use efficiency. This may increase agricultural benefits as well as benefit the environment by reducing GHG emissions and leaching losses. It is believed that only increased soil health can elevate productivity by 10%–15%. In agroecosystems, plant-microbe interactions are the key factors governing the fertility of the soil. Microbes utilize the carbon released by the plants and, in turn, provide nutrients to them. Several microbes are known to enhance the availability of nutrients to the plants that include nitrogen-fixing bacteria, mycorrhizal fungi, etc. that provide nutrients to the plants by converting them into available forms. Arbuscular mycorrhiza is known to enhance phosphate availability to the plants due to their vast hyphal network associated with the plant roots.

There is evidence that inoculated entophytes (*Klebsiellas pneumonniae* 342) had provided up to 44% N_2 in inoculated wheat. The identification of the signals that are responsible for the communication of plants with the microbes can be genetically used to enhance plant-microbe signaling. Considering the different benefits of microbes for the plants, some seeds can be developed by harnessing the use of biotechnology and that can be utilized in the fields.

4.1.7 Remediation of pollutants

Plant-microbe interactions are nowadays extensively studied for decontamination and remediation of soils. Microbes that are capable of breaking harmful chemicals like herbicides, pesticides, and other organic compounds used in farming practice can be an effective means to dispose of toxic compounds on the field. The microbial communities can sequester heavy metals, and hence they can be used for the remediation of heavy metal contaminated sites [43].

The endophytic microbes that have the efficiency to sequester heavy metals can be isolated from the plants growing in contaminated soil. The endophytes involved in the process can improve growth and photosynthetic ability of the plants under stress conditions. Many endophytic microbes isolated from the inside of legumes could detox heavy metals and can be used as a promising tool for the remediation of contaminated soils. A cadmium accumulator plant *Solanum nigrum* has endophytic colonies of the bacterium *Serratia* that has tolerance against heavy metals in addition to its ability to promote plant growth and phosphate solubilization potential [44].

4.2 Detrimental effects

Out of all the plant-microbe interactions, fungi are the main threat to the plants. Most of the pathogenic fungi are host-specific. The pathogenic microbes that are present in the agricultural soils can have detrimental effects on the crops that may include—disease, reduced soil fertility, soil deterioration, and poor crop health. This consequently reduces the food quality and promotes disease spread. Once spread, these pathogenic fungi become very difficult to control. The pathogenic organisms mainly include bacteria, fungi, and viruses. The effect of different microbes is different for the plants. Some may decompose the roots, enhance the leaching of nutrients from the soil, reduce nutrient uptake efficiency, and decrease plant growth. These also affect the metabolic activities of the plant by altering its physiology. The impacts may include impaired photosynthetic ability, reduced water uptake, necrosis, etc., which may ultimately lead to the death of the plant if not managed and treated properly at the initial stages.

Symptoms of pathogenic invasion may also include bacteria and fungal leaf spots, cankers, leaf distortion, and gummosis. Pathogenic microbes include *Fusarium, Puccinia, Ustilago, Rhizoctonia, Phytophthora, Alternaria*, etc. *Phytophthora* is considered very harmful for the plants as it has more than 100 species that can be pathogenic to the plants and it can damage ornamental and horticultural crops. The spores of these fungi can survive in plant debris for many years and can infect all the plant parts, ultimately leading to the death of the plant. *Fusarium* is another fungal strain that has a wide distribution in the soil. Some species of *Fusarium* have devastating effects on the crops. They can cause diseases like canker, wilt, and root rot. *Pythium* is also a pathogen that can cause common crop diseases like seed decay, reduced root growth, etc.

The loss of crops due to pathogens or weeds leads to the reduced production of food and cash crops. Crop losses may be qualitative or quantitative. Qualitative losses include reduced content of valuable ingredients and reduced market value due to the contamination of the harvested products with pests. Quantitative losses may be due to reduced crop productivity due to invasion of pathogens.

5 Role of plant-microbe interactions during stress conditions

Abiotic and biotic stresses highly affect crop productivity all around the world. Due to increasing environmental fluctuations, the different abiotic factors like temperature, salinity, drought, floods are posing a threat to crops. There is an urgent need to combat these stress conditions, and for the same purpose, microbes can be considered as an alternative to chemicals by exploiting their ability to tolerate environmental stress. This may open emerging doors for sustainable agriculture at no risk of environmental problems (Fig. 3).

5.1 Abiotic stresses

5.1.1 Salinity and mineral toxicity

Salinity stress is the stress due to excess ions like Na^+ (sodium), K^+ (potassium), Ca^{2+} (calcium), and Cl^- (chloride) in the soil. In modern agricultural systems, this is the most common biotic stress. This may result in altered microbial

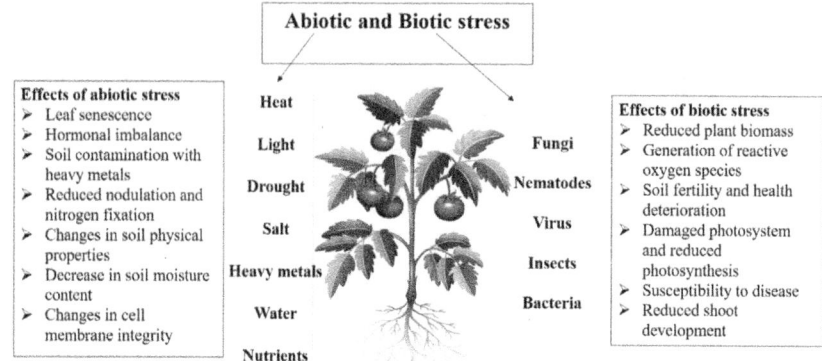

FIG. 3 Biotic and abiotic stresses that affect the growth and developmental features of plants.

activities, ultimately leading to reduced development of plants. Salinity stress causes low water potential in the soil and this makes it difficult for the plant to uptake nutrients and water from the soil, resulting in stress. Salts have harmful effects on all the plant parameters that include productivity, germination of seeds, uptake of nutrients, and disturbed physiological and ecological balance. This also affects the process of nodulation reducing nitrogen fixation and ultimately crop yield. These effects are mainly due to ion toxicity and osmotic effects.

Fungi are very sensitive to increased salt concentrations than other microbes. Stress due to salinity arises when the concentration of salt exceeds the tolerance limits of microbes. Initially, the salinity has bad impacts on the metabolism of soil microbes that reduces soil productivity, then it leads to the destruction of all the vegetation and organisms residing in the soil that ultimately leads to the transformation of fertile soils into barren lands. It is estimated that almost 20% of the irrigated land, producing one-third of total food, is affected by salt stress [45].

Soil salinity has a remarkable effect on the germination of seeds which is very crucial for crop production. The most efficient solution is to use bacterial inoculants that are salt-tolerant and produce auxins, gibberellins under such conditions. Plants tolerate salt stress by the accumulation of proline and proteins in the leaves. This is an adaptation mechanism as these proteins bind to the membrane and regulate the water permeability in the cells influencing water movement among tissues and organs (Table 1).

5.1.2 Extreme temperatures (cold, frost, and heat)

The increasing global warming has substantially caused differences in temperature that lead to stress conditions in the plants. The temperature stress may cause damage in the cell membranes, cell division, photosynthetic efficiency of the plants, and affected water potential. Temperature can affect the different

TABLE 1 Salt stress tolerance in plants under the application of microbes.

S. no.	Microorganism	Plant	Effect	Reference
1.	Azospirillum	Lettuce	Plant biomass increased	Fasciglione et al. [46]
2.	Rhizobium, Pseudomonas	Mung bean	Improved growth, nodulation, and yield	Ahmad et al. [47]
3.	Hartmannibacter diazotrophicus	Barley	Increased root and shoot dry weight	Suarez et al. [48]
4.	Brachybacterium saurashtrense	Groundnut	Increased nitrogen content, high concentration of auxin in root and shoot	Shukla et al. [49]
5.	Pseudomonas putida	Maize and Mustard	ACC deaminase activity and differential gene expression	Cheng et al. [50]
6.	Acinetobacter, Pseudomonas	Oats and Barley	Promotion of plant growth decreased ethylene production	Chang et al. [51]
7.	Pseudomonas, Bacillus	Rice	Reduced toxicity of reactive oxygen species, and reduced enzymatic activities	Jha and Subramanian [52]
8.	Azospirillum	Wheat	Accumulation of organic solutes increased plant biomass	Bacilio et al. [53]
9.	Bacillus amyloliquefaciens	Rice	Transcription modulation in different genes	Nautiyal et al. [54]
10.	Pseudomonas, Enterobacter	Maize	Reduced triple response and increased nutrient uptake	Nadeem et al. [55]

cell organelles differently. The high-temperature stress may increase the membrane fluidity while the low temperature can make it rigid. This also leads to the changes in hormone concentrations like jasmonic acid increases many times during stress conditions. Microbes that are associated with the plants and are tolerant to stress conditions can help the plant to survive such conditions of stress. The microbes that live in close association with the plants have special enzymatic machinery that helps them to regulate their metabolism by the changing temperature, thus maintaining membrane integrity. Heat and cold shock proteins play an important role under such stress conditions. These molecular chaperones provide a defense against heat stress. The bacteria associated with heat tolerance have been isolated from wheat that is observed to enhance plant growth and development at stress conditions and these include *Bacillus, Methylobacterium, Arthrobacter*, and several others [56]. Endophytes increase the adaptation of plants with low temperatures that reduce cellular damage, increase photosynthetic activity, and accumulate various metabolites related to cold stress such as phenolic compounds, proline, and starch. Heat shock proteins (HSP20, HSP 60, HSP70, HSP 90, HSP100) and reactive oxygen species (ROS)-scavenging enzymes (ascorbate peroxidase and catalase) are major proteins that play an important role during stress. But most of the crops are unable to tolerate environmental stress, and hence there is a need to develop a strategy that may help them overcome such situations.

Based on the growth of microbes in temperature fluctuations, the microbes are divided into two groups, psychrophilic and psychrotrophic microorganisms. The psychrophilic microbes grow at or below 15°C while psychrotrophic microbes grow at or above 15°C. These both can be utilized to understand gene expression under stress conditions and can be used for developing crop varieties that can combat stress.

5.1.3 Drought (water stress)

This is a major concern for agriculture worldwide as water being the main basis for crops. Limited water availability to the plants can affect them at different scales like reduced cell size, production of reactive oxygen species, and hence reduced crop productivity. The plants undergo significant physiological and molecular changes under stress conditions like damaged photosynthetic apparatus, photosynthesis inhibition, chlorophyll degradation, etc. This also causes accumulation of radicles leading to changes in membrane function, protein transformation, and finally cell death. Drought also disrupts root-microbe associations that play a major role in plant nutrient acquisition. Drought stress-tolerant microbes may help the plant to overcome such conditions. These microbes have various mechanisms to cope with the detrimental effects of drought. They help the plant to grow and develop under water scarcity by various direct and indirect mechanisms like induced systemic resistance, production of phytohormones, etc. The produced phytohormones help the plant to grow under such conditions. Furthermore, PGPR can produce plant hormones that accelerate plant growth and division under stress circumstances, such as auxin, which controls cell division, vascular tissue differentiation, adventitious and lateral root differentiation, and shoot development during drought stress [57]. ABA is also an important plant hormone that regulates the physiology of the plant during stress conditions. Cho et al. [58] reported that root colonization of plants with rhizobacteria *Pseudomonas chlororaphis* induces tolerance to drought stress.

5.1.4   Flood

Waterlogging is a serious threat to agriculture and is prevalent in many agroecosystems across the world. It causes changes in plant morphology as well as physiology by altering different processes. The activities of ACC synthase in the submerged roots and ACC oxidase in the shoots were found to be enhanced [59]. Grichko and Glick [60] found that 55-day-old tomato plants treated with ACC deaminase-producing strains of Pseudomonas and Enterobacter showed substantial resistance to flooding. There is a need to investigate the potential of rhizobacterial species in reducing stress caused by waterlogging.

5.2   Biotic stress

Biotic stress is mainly referred to as the stress caused by the biotic pathogens that include fungi, bacteria, viruses, insects, pests, weeds that destruct the plant by altering its metabolic processes. It is of high concern these days as the pathogens have developed resistance against most of the fungicides and herbicides and are very harmful to the crops. Hence, these have a direct effect on the economy of the world. The harmful effects of these pathogens include hormonal imbalances in the plant, nutrient imbalance, and physiological disorders. Therefore, there is a need to manage biotic pathogens by using eco-friendly tools with the help of biotechnology. The naturally occurring bacteria and fungi colonize root hair and enhance plant growth and development. Plant growth-promoting microorganisms have been considered as an environmentally friendly and cost-effective method of disease management. They promote pathogen protection by activating cellular components such as cellular burst, cell wall reinforcement, and secondary metabolite accumulation. JA, ethylene, and salicylic acid (SA) are defense-related hormones that play a key role in signal transduction and defense mechanisms [61]. They protect plants from pathogens by decreasing susceptibility to disease and increasing growth attributes. Under abiotic stress conditions, biological controls of soilborne illnesses, which substitute chemical agents, greatly contribute to crop yield. The interaction of microorganisms with plants produces many elicitors, which cause physiological and biochemical changes in plants. These modifications cause the plant to be resistant to disease for several months. An essential mechanism for biotic stress tolerance is the production of reactive oxygen species (ROS) and oxidative burst.

    Trichoderma can be used as a biocontrol agent that is resistant to many fungal and bacterial diseases of plants [62]. It is an efficient producer of many antimicrobial compounds, hydrolytic enzymes, and also causes plant defense induction. Some *Trichoderma* spp., known as biocontrol agents, are

T. harzianum, T. viride, and *T. hamatum.* Most of the biocontrol agents are applied to the soil where they may protect the plant roots and help in plant growth. The foliar application may also help in controlling foliar pathogens. *Trichoderma* is mainly used for getting rid of disease caused by a variety of fungi such as *Sclerotinia sclerotiorum, Pythium, Botrytis cinerea, Rhizoctonia,* etc. [63]. The advantages of *Trichoderma* to agriculture include release of plant growth stimulators, help in nutrient cycling, mechanism of plant defense activation, etc. The strains of *Trichoderma* that are also commercially available can be applied to the crops for protection from pathogens. Microbe-activated defensive response mechanisms involve two distinct pathways: induced systematic resistance (ISR) and systemic acquired resistance (SAR).

5.2.1 Induced systemic resistance

The infection caused by the microbes can induce the plant to develop resistance to further attack and this is called induced systemic resistance. This is induced by phytopathogens. The one accompanied by plant growth-promoting microbes is induced by the production of allopathic compounds such as siderophores, antibiotics that act efficiently against pathogens and inhibit their growth [64]. Through generated systemic resistance, *Pseudomonas* and *Bacillus* strains may control plant disease in a variety of crops. *Paenibacillus* P16 was shown to be an efficient biological control agent (BCA) for black rot (*Xanthomonas campestris*) disease in cabbage and has the potential to produce systemic resistance.

5.2.2 Systemic acquired resistance

The systemic acquired resistance in plants develops as a fully active defense mechanism in response to the primary invasion by plant pathogens. The host plant detects the pathogen's nature based on molecular patterns and subsequently detoxifies its effects by modifying gene expression, hormone synthesis, and metabolite production [65,66]. According to Banerjee et al. [67], *Arthrobacter* sp. and *Bacillus* sp. isolated from the tomato rhizosphere have plant growth-promoting qualities such as phosphate solubilization, IAA synthesis, and biocontrol capabilities. Fungicides generated by fungi are inhibited by several bacterial species. Siderophores, phytohormones, hydrogen cyanide, and ammonia are also produced under stress conditions (Table 2).

6 Molecular pathways associated with the microbes (MAMPs)

Plants have different mechanisms that help them to protect themselves from pathogens, and for the same purpose, plants use an innate immune system. Activation of the innate immune system includes local defense responses that include hypersensitive responses that are characterized by the death of the cells at the site of infection. The activation of the immune system of plants can also lead to systemic acquired resistance against a broad spectrum of potential

TABLE 2 Biotic stress tolerance in plants mediated by microbe association.

S. no.	Plant	Disease	Biocontrol microbe	Reference
1.	Rice	Bacterial leaf blight (*Xanthomonas oryzae*)	*Bacillus* sp.	Udayashankar et al. [68]
2.	Cabbage	Black rot (*Xanthomonas campestris*)	*Paenibacillus* sp.	Ghazalibiglar et al. [69]
3.	Pepper	Gray leaf spot disease (*Stemphylium lycopersici*)	*Brevibacterium iodinum*	Son et al. [70]
4.	Cucumber	Cucumber mosaic cucumovirus (CMV)	*Bacillus subtilis*	El-Borollosy and Oraby [71]
5.	Ginseng	Root disease (*Phytophthora cactorum*)	*Bacillus amyloliquefaciens*	Lee et al. [72]
6.	Tomato	Wilt disease (*Verticillium dahliae*)	*Pseudomonas* sp., *Bacillus amyloliquefaciens*	Vitullo et al. [73]
7.	Arabidopsis	*Pseudomonas syringae*	*Bacillus cereus*	Niu et al. [74]
8.	Maize	*Bipolaris maydis*	*Bacillus subtilis*	Ding et al. [75]
9.	Grapevine	*Botrytis cinerea*	*Bacillus subtilis*	Farace et al. [76]
10.	Tobacco	*Tobacco mosaic virus (TMV)*, *Ralstonia solanacearum*	*Pytophthora parastica*	Chang et al. [51]

pathogens [77]. The pattern of the defense responses varies with the organisms and the host plants. The recognition of the pathogen by the host plant precedes induced defense response. The host directly recognizes the foreign molecules associated with the microbes or the alterations in the host itself that are caused by the microbes. Some recognition events conform to the conceptually simple model in which a host receptor interacts directly with a molecule of the microbe. Interactions between microbe-associated molecular patterns (MAMPs) and

MAMP-receptors are examples of this. They also include interactions between some effectors and the proteins that serve as their cognate resistance (R). Other types of recognition occur in a roundabout way. These are consistent with the so-called guard hypothesis, in which R-proteins identify pathogen-derived effectors indirectly through host disruption.

An MAMP is a structural element found inside a potential pathogen's molecule. One distinguishing feature of these components is that they do not exist in the host. MAMPs are recognized directly by receptors encoded by the host. MAMPs are frequently discovered in highly conserved compounds that are required for the survival of a diverse range of species. As a result, many MAMP-receptors identify MAMPs that the virus finds difficult to shed or change. MAMPs are necessary structures for bacteria and are thus conserved among pathogens, nonpathogenic microorganisms, and saprophytic microorganisms. MAMPs are identified by pattern recognition receptors (PRRs) on the surface of plant cells; this initial stage of defense induction is known as MAMP-triggered immunity (MTI). MAMP-induced defensive responses include reactive oxygen species (ROS, commonly known as the oxidative burst), reactive nitrogen species such as nitric oxide (NO), changes in the plant cell wall, induction of antimicrobial compounds, and synthesis of pathogenesis-related (PR) proteins. ROS and NO can both operate as signaling molecules and have direct antibacterial effects [78].

7 Plant-microbe interactions as an approach to sustainable agriculture and food security

The beneficial plant-microbe interactions can be a promising solution for sustainable agriculture. These interactions have played a vital role in the development of biocontrol, biopesticides, biofungicides, biofertilizers, and bioremediation agents. The various interactions between plants and bacteria may be both detrimental and helpful to the plant. Plant growth-promoting bacteria (PGPB) are bacteria that are advantageous to plants and are extremely promising plant growth enhancers. These PGPB are thought to have started developing mutually beneficial partnerships with plants some 80–100 million years ago. It has also been stated that certain fungi began forming beneficial relationships with plants as long as 450 million years ago. Scientists have attempted to use PGPB and plant-associated fungus in agriculture, horticulture, and other areas in recent years as a result of their results. The successful implementation of microbes in agriculture requires a deep understanding of the mechanisms that they use to promote plant health and defense mechanisms against pathogens. PGPB may facilitate plant growth directly or indirectly. Direct mechanisms include the accumulation of nutrients from the environment, modulation of levels of hormones, and indirect mechanisms include the protection of plants from the deleterious effects of the pathogens. As a result, it is acceptable to conclude that using PGPB as an inherent part of current agronomic practices is a

TABLE 3 Role of plant-microbe interactions in agriculture.

S. no.	Microorganisms	Associated plants	Mechanisms	References
1.	Pseudomonas grimontii	Vicia sativa	Biological control	Mokrani et al. [79]
2.	Serratia marcescens strain B2	Oryza sativa	Biological control	Someya et al. [80]
3.	Trichoderma harzianum	Macrophomina Phaseolina infected Arachis hypogaea	Biological control	Sreedevi et al. [81]
4.	Serratia sp. SY5	Echinochloa crus-galli	Biofertilization	Koo and Cho [82]
5.	Bacillus subtilis	Capsicum annum	Biological control	Huang et al. [83]
6.	Bacillus amyloliquefaciens	Phaseolus vulgaris	Biofertilization	Mokrani et al. [84]
7.	Bacillus subtilis, Arthrobacter	Triticum aestivum	Biofertilization	Upadhyay et al. [85]
8.	Enterobacter sakazakaii	Zea mays	Biofertilization	Babalola et al. [86]

technology. Several PGPB strains have already been marketed and are being used effectively in agriculture in a variety of nations. Furthermore, as the world's population continues to rise, the demand for higher food production increases, necessitating the development of strategies to help regulate the situation without the use of dangerous chemicals and in harmony with the environment (Table 3).

8 Conclusions and recommendations

There is plenty of literature available on the plant-microbe interactions in plant growth and development, altered plant physiology, and defense mechanism against pathogens but a thorough knowledge of the process is lacking which may help in the development of crop varieties resistant to pathogens. The understanding of the genetic basis of these interactions is important to understand their benefits to plant health, disease control, improved food quality, and tolerance to stress. Plant-microbe interactions are responsible for several transformations in the rhizosphere like nutrient cycling, carbon sequestration, and ecosystem functioning. The kind and quantity of microorganisms in the soil impact a plant's capacity to receive nitrogen and other nutrients. Plants can influence these ecological changes by depositing secondary metabolites into the rhizosphere, which attract or inhibit microorganisms. Various nitrogen-fixing microorganisms are present in the rhizosphere of agricultural plants, but the contribution of fixed nitrogen to plant nutrition is controversial. Different bacterial and fungal groups are known to play very beneficial roles to the plants. The microbes residing in the rhizosphere are mainly associated with the roots and nutrient dynamics; the endospheric microbes are mainly associated with the resistance to the pathogens and phyllospheric microbes are mainly associated with the growth and development enhancement of the plants. Shortly, mankind may face many challenges due to global warming and significant environmental changes as well as the changing interactions of pathogens with the plants. These problems primarily include identifying essential variables engaged in such interactions during plant immune responses, detecting and effectively managing novel emerging and reemerging plant diseases, and developing pathogen-resistant crops. This requires the dire need to understand the different mechanisms that involve the study of microbes and their interactions that involve signaling and other molecular pathways. Then only we will be able to develop the right and effective means to tackle the problem of food security using eco-friendly and sustainable means.

References

[1] P. Kumari, M. Meena, P. Gupta, M.K. Dubey, G. Nath, R.S. Upadhyay, Plant growth promoting rhizobacteria and their biopriming for growth promotion in mung bean (*Vigna radiata* (L.) R. Wilczek), Biocatal. Agric. Biotechnol. 16 (2018) 163–171.

[2] P. Kumari, M. Meena, R.S. Upadhyay, Characterization of plant growth promoting rhizobacteria (PGPR) isolated from the rhizosphere of *Vigna radiata* (mung bean), Biocatal. Agric. Biotechnol. 16 (2018) 155–162.

[3] M. Meena, P. Swapnil, K. Divyanshu, S. Kumar, Harish, Y.N. Tripathi, A. Zehra, A. Marwal, R.S. Upadhyay, PGPR-mediated induction of systemic resistance and physiochemical alterations in plants against the pathogens: current perspectives, J. Basic Microbiol. 60 (10) (2020) 828–861, https://doi.org/10.1002/jobm.202000370.

[4] A. Zehra, N.A. Raytekar, M. Meena, P. Swapnil, Efficiency of microbial bio-agents as elicitors in plant defense mechanism under biotic stress: a review, Curr. Res. Microb. Sci. 2 (2021), 100054, https://doi.org/10.1016/j.crmicr.2021.100054.

[5] H.P. Bais, S.W. Park, T.L. Weir, R.M. Callaway, J.M. Vivanco, How plants communicate using the underground information superhighway, Trends Plant Sci. 9 (1) (2004) 26–32.

[6] A.A. Belimov, N. Hontzeas, V.I. Safronova, S.V. Demchinskaya, G. Piluzza, S. Bullitta, B.R. Glick, Cadmium-tolerant plant growth-promoting bacteria associated with the roots of Indian mustard (*Brassica juncea* L. Czern.), Soil Biol. Biochem. 37 (2) (2005) 241–250.

[7] L. Lo Presti, D. Lanver, G. Schweizer, S. Tanaka, L. Liang, M. Tollot, R. Kahmann, Fungal effectors and plant susceptibility, Annu. Rev. Plant Biol. 66 (2015) 513–545.

[8] R. Dean, J.A. Van Kan, Z.A. Pretorius, K.E. Hammond-Kosack, A. Di Pietro, P.D. Spanu, G.D. Foster, The top 10 fungal pathogens in molecular plant pathology, Mol. Plant Pathol. 13 (4) (2012) 414–430.

[9] C.S. Delavaux, P. Weigelt, W. Dawson, J. Duchicela, F. Essl, M. van Kleunen, J.D. Bever, Mycorrhizal fungi influence global plant biogeography, Nat. Ecol. Evol. 3 (3) (2019) 424–429.

[10] P.N. Bhattacharyya, D.K. Jha, Plant growth-promoting rhizobacteria (PGPR): emergence in agriculture, World J. Microbiol. Biotechnol. 28 (4) (2012) 1327–1350.

[11] D. Saravanakumar, C. Vijayakumar, N. Kumar, R. Samiyappan, PGPR-induced defense responses in the tea plant against blister blight disease, Crop Prot. 26 (4) (2007) 556–565.

[12] J.A.W. Morgan, G.D. Bending, P.J. White, Biological costs and benefits to plant–microbe interactions in the rhizosphere, J. Exp. Bot. 56 (417) (2005) 1729–1739.

[13] D.J. Bagyaraj, G. Rangaswami, Microorganisms in Soil. Agricultural Microbiology, second ed., Prentice Hall of India Private Limited, New Delhi, 2005, pp. 1–254.

[14] P.R. Hardoim, L.S. van Overbeek, J.D. van Elsas, Properties of bacterial endophytes and their proposed role in plant growth, Trends Microbiol. 16 (10) (2008) 463–471.

[15] J.L. Azevedo, W. Maccheroni Jr., J.O. Pereira, W.L. De Araújo, Endophytic microorganisms: a review on insect control and recent advances on tropical plants, Electron. J. Biotechnol. 3 (1) (2000) 15–16.

[16] M. Mizuno, H. Yurimoto, H. Iguchi, A. Tani, Y. Sakai, Dominant colonization and inheritance of *Methylobacterium* sp. strain OR01 on perilla plants, Biosci. Biotechnol. Biochem. 77 (7) (2013) 1533–1538.

[17] F.A. Ripa, W.D. Cao, S. Tong, J.G. Sun, Assessment of plant growth promoting and abiotic stress tolerance properties of wheat endophytic fungi, Biomed. Res. Int. 2019 (2019) 1–12.

[18] A.P. Walker, M.L. McCormack, J. Messier, I.H. Myers-Smith, S.D. Wullschleger, Trait covariance: the functional warp of plant diversity? New Phytol. 216 (4) (2017) 976–980.

[19] P. Verma, A.N. Yadav, S.K. Kazy, A.K. Saxena, A. Suman, Evaluating the diversity and phylogeny of plant growth promoting bacteria associated with wheat (*Triticum aestivum*) growing in central zone of India, Int. J. Curr. Microbiol. App. Sci. 3 (5) (2014) 432–447.

[20] M.D.V.B. Figueiredo, A. Bonifacio, A.C. Rodrigues, F.F. de Araujo, Plant growth-promoting rhizobacteria: key mechanisms of action, in: Microbial-Mediated Induced Systemic Resistance in Plants, Springer, Singapore, 2016, pp. 23–37.

[21] R. Cakmakci, M.F. Dönmez, Ü. Erdoğan, The effect of plant growth promoting rhizobacteria on barley seedling growth, nutrient uptake, some soil properties, and bacterial counts, Turk. J. Agric. For. 31 (3) (2007) 189–199.

[22] M. Meena, P. Swapnil, A. Zehra, M. Aamir, M.K. Dubey, R.S. Upadhyay, Beneficial microbes for disease suppression and plant growth promotion, in: D. Singh, H. Singh, R. Prabha (Eds.), Plant-Microbe Interactions in Agro-Ecological Perspectives, Springer, Singapore, 2017, pp. 395–432, https://doi.org/10.1007/978-981-10-6593-4_16.

[23] R. Hayat, S. Ali, U. Amara, R. Khalid, I. Ahmed, Soil beneficial bacteria and their role in plant growth promotion: a review, Ann. Microbiol. 60 (4) (2010) 579–598.

[24] M.M. Elsharkawy, M.B. Shivanna, M.S. Meera, M. Hyakumachi, Mechanism of induced systemic resistance against anthracnose disease in cucumber by plant growth-promoting fungi, Acta Agric. Scand. Sect. B Soil Plant Sci. 65 (4) (2015) 287–299.

[25] M. Murali, K.N. Amruthesh, Plant growth-promoting Fungus *Penicillium oxalicum* enhances plant growth and induces resistance in pearl millet against downy mildew disease, J. Phytopathol. 163 (9) (2015) 743–754.

[26] M.M. Hossain, F. Sultana, S. Islam, D. Singh, H. Singh, R. Prabha, Plant growth-promoting fungi (PGPF): phytostimulation and induced systemic resistance, in: Plant-Microbe Interactions in Agro-Ecological Perspectives, 2017, pp. 135–191.

[27] T. Song, L. Mårtensson, T. Eriksson, W. Zheng, U. Rasmussen, Biodiversity and seasonal variation of the cyanobacterial assemblage in a rice paddy field in Fujian, China, FEMS Microbiol. Ecol. 54 (1) (2005) 131–140.

[28] K.B. Zwart, P.J. Kuikman, J.A. Van Veen, Rhizosphere protozoa: their significance in nutrient dynamics, in: J.F. Darbyshire (Ed.), Soil Protozoa, CAB International, Wallingford, 1994, pp. 93–121.

[29] I. Hiltpold, G. Jaffuel, T.C. Turlings, The dual effects of root-cap exudates on nematodes: from quiescence in plant-parasitic nematodes to frenzy in entomopathogenic nematodes, J. Exp. Bot. 66 (2) (2015) 603–611.

[30] S. Rasmann, T.G. Köllner, J. Degenhardt, I. Hiltpold, S. Toepfer, U. Kuhlmann, T.C. Turlings, Recruitment of entomopathogenic nematodes by insect-damaged maize roots, Nature 434 (7034) (2005) 732–737.

[31] M. Bonkowski, W. Cheng, B.S. Griffiths, J. Alphei, S. Scheu, Microbial-faunal interactions in the rhizosphere and effects on plant growth, Eur. J. Soil Biol. 36 (3) (2000) 135–147.

[32] J. Aislabie, J.R. Deslippe, J. Dymond, Soil Microbes and Their Contribution to Soil Services. Ecosystem Services in New Zealand-Conditions and Trends, vol. 1(12), Manaaki Whenua Press, Lincoln, New Zealand, 2013, pp. 143–161.

[33] Y.P. Chen, P.D. Rekha, A.B. Arun, F.T. Shen, W.A. Lai, C.C. Young, Phosphate solubilizing bacteria from subtropical soil and their tricalcium phosphate solubilizing abilities, Appl. Soil Ecol. 34 (1) (2006) 33–41.

[34] R. Duponnois, M. Kisa, C. Plenchette, Phosphate-solubilizing potential of the nematophagous fungus *Arthrobotrys oligospora*, J. Plant Nutr. Soil Sci. 169 (2) (2006) 280–282.

[35] T.C. Setiawati, L. Mutmainnah, Solubilization of potassium containing mineral by microorganisms from sugarcane rhizosphere, Agric. Agric. Sci. Proc. 9 (2016) 108–117.

[36] R. Vidyalakshmi, R. Paranthaman, R. Bhakyaraj, Sulphur oxidizing bacteria and pulse nutrition—a review, World J. Agric. Sci. 5 (3) (2009) 270–278.

[37] W. Radzki, F.G. Mañero, E. Algar, J.L. García, A. García-Villaraco, B.R. Solano, Bacterial siderophores efficiently provide iron to iron-starved tomato plants in hydroponics culture, Antonie Van Leeuwenhoek 104 (3) (2013) 321–330.

[38] S.K. Vaid, B. Kumar, A. Sharma, A.K. Shukla, P.C. Srivastava, Effect of Zn solubilizing bacteria on growth promotion and Zn nutrition of rice, J. Soil Sci. Plant Nutr. 14 (4) (2014) 889–910.

[39] C. Plenchette, C. Clermont-Dauphin, J.M. Meynard, J.A. Fortin, Managing arbuscular mycorrhizal fungi in cropping systems, Can. J. Plant Sci. 85 (1) (2005) 31–40.

[40] J. Döbereiner, Biological nitrogen fixation in the tropics: social and economic contributions, Soil Biol. Biochem. 29 (5–6) (1997) 771–774.

[41] C. Voisard, C. Keel, D. Haas, G. Dèfago, Cyanide production by *Pseudomonas fluorescens* helps suppress black root rot of tobacco under gnotobiotic conditions, EMBO J. 8 (2) (1989) 351–358.

[42] H.A. Naznin, D. Kiyohara, M. Kimura, M. Miyazawa, M. Shimizu, M. Hyakumachi, Systemic resistance induced by volatile organic compounds emitted by plant growth-promoting fungi in *Arabidopsis thaliana*, PLoS ONE 9 (1) (2014), e86882.

[43] P.H. Kao, C.C. Huang, Z.Y. Hseu, Response of microbial activities to heavy metals in a neutral loamy soil treated with biosolid, Chemosphere 64 (1) (2006) 63–70.

[44] S. Luo, Y. Wan, X. Xiao, H. Guo, L. Chen, Q. Xi, J. Chen, Isolation and characterization of endophytic bacterium LRE07 from cadmium hyperaccumulator Solanum nigrum L. and its potential for remediation, Appl. Microbiol. Biotechnol. 89 (5) (2011) 1637–1644.

[45] P. Shrivastava, R. Kumar, Soil salinity: a serious environmental issue and plant growth promoting bacteria as one of the tools for its alleviation, Saudi J. Biol. Sci. 22 (2) (2015) 123–131.

[46] G. Fasciglione, E.M. Casanovas, V. Quillehauquy, A.K. Yommi, M.G. Goni, S.I. Roura, C.A. Barassi, Azospirillum inoculation effects on growth, product quality and storage life of lettuce plants grown under salt stress, Sci. Hortic. 195 (2015) 154–162.

[47] M. Ahmad, Z.A. Zahir, H.N. Asghar, M. Asghar, Inducing salt tolerance in mung bean through coinoculation with rhizobia and plant-growth-promoting rhizobacteria containing 1-aminocyclopropane-1-carboxylate deaminase, Can. J. Microbiol. 57 (7) (2011) 578–589.

[48] C. Suarez, M. Cardinale, S. Ratering, D. Steffens, S. Jung, A.M.Z. Montoya, S. Schnell, Plant growth-promoting effects of *Hartmannibacter diazotrophicus* on summer barley (*Hordeum vulgare* L.) under salt stress, Appl. Soil Ecol. 95 (2015) 23–30.

[49] P.S. Shukla, P.K. Agarwal, B. Jha, Improved salinity tolerance of *Arachis hypogaea* (L.) by the interaction of halotolerant plant-growth-promoting rhizobacteria, J. Plant Growth Regul. 31 (2) (2012) 195–206.

[50] Z. Cheng, E. Park, B.R. Glick, 1-Aminocyclopropane-1-carboxylate deaminase from Pseudomonas putida UW4 facilitates the growth of canola in the presence of salt, Can. J. Microbiol. 53 (7) (2007) 912–918.

[51] P. Chang, K.E. Gerhardt, X.D. Huang, X.M. Yu, B.R. Glick, P.D. Gerwing, B.M. Greenberg, Plant growth-promoting bacteria facilitate the growth of barley and oats in salt-impacted soil: implications for phytoremediation of saline soils, Int. J. Phytoremediation 16 (11) (2014) 1133–1147.

[52] Y. Jha, R.B. Subramanian, PGPR regulate caspase-like activity, programmed cell death, and antioxidant enzyme activity in paddy under salinity, Physiol. Mol. Biol. Plants 20 (2) (2014) 201–207.

[53] M. Bacilio, H. Rodriguez, M. Moreno, J.P. Hernandez, Y. Bashan, Mitigation of salt stress in wheat seedlings by a gfp-tagged *Azospirillum lipoferum*, Biol. Fertil. Soils 40 (3) (2004) 188–193.

[54] C.S. Nautiyal, S. Srivastava, P.S. Chauhan, K. Seem, A. Mishra, S.K. Sopory, Plant growth-promoting bacteria *Bacillus amyloliquefaciens* NBRISN13 modulates gene expression profile of leaf and rhizosphere community in rice during salt stress, Plant Physiol. Biochem. 66 (2013) 1–9.

[55] S.M. Nadeem, Z.A. Zahir, M. Naveed, M. Arshad, Rhizobacteria containing ACC-deaminase confer salt tolerance in maize grown on salt-affected fields, Can. J. Microbiol. 55 (11) (2009) 1302–1309.

[56] K.S. Verma, S. Ul Haq, S. Kachhwaha, S.L. Kothari, RAPD and ISSR marker assessment of genetic diversity in *Citrullus colocynthis* (L.) Schrad: a unique source of germplasm highly adapted to drought and high-temperature stress, 3 Biotech 7 (5) (2017) 1–24.

[57] D. Goswami, J.N. Thakker, P.C. Dhandhukia, Portraying mechanics of plant growth promoting rhizobacteria (PGPR): a review, Cogent Food Agric. 2 (1) (2016) 1–19.

[58] S.M. Cho, B.R. Kang, S.H. Han, A.J. Anderson, J.Y. Park, Y.H. Lee, Y.C. Kim, 2R, 3R-butanediol, a bacterial volatile produced by *Pseudomonas chlororaphis* O6, is involved in induction of systemic tolerance to drought in *Arabidopsis thaliana*, Mol. Plant-Microbe Interact. 21 (8) (2008) 1067–1075.

[59] Q. Chao, M. Rothenberg, R. Solano, G. Roman, W. Terzaghi, J.R. Ecker, Activation of the ethylene gas response pathway in Arabidopsis by the nuclear protein ETHYLENE-INSENSITIVE3 and related proteins, Cell 89 (7) (1997) 1133–1144.

[60] V.P. Grichko, B.R. Glick, Flooding tolerance of transgenic tomato plants expressing the bacterial enzyme ACC deaminase controlled by the 35S, rolD or PRB-1b promoter, Plant Physiol. Biochem. 39 (1) (2001) 19–25.

[61] A. Verhage, S.C. van Wees, C.M. Pieterse, Plant immunity: it's the hormones talking, but what do they say? Plant Physiol. 154 (2) (2010) 536–540.

[62] A. Zehra, M. Meena, M.K. Dubey, M. Aamir, R.S. Upadhyay, Synergistic effects of plant defense elicitors and *Trichoderma harzianum* on enhanced induction of antioxidant defense system in tomato against Fusarium wilt disease, Bot. Stud. 58 (2017) 44, https://doi.org/10.1186/s40529-017-0198-2.

[63] A.J. Almeida, J.A. Carmona, C. Cunha, A. Carvalho, C.A. Rappleye, W.E. Goldman, F. Rodrigues, Towards a molecular genetic system for the pathogenic fungus *Paracoccidioides brasiliensis*, Fungal Genet. Biol. 44 (12) (2007) 1387–1398.

[64] S. Jain, A. Vaishnav, A. Kasotia, S. Kumari, R.K. Gaur, D.K. Choudhary, Bacteria-induced systemic resistance and growth promotion in *Glycine max* L. Merrill upon challenge inoculation with fusarium oxysporum, Proc. Natl. Acad. Sci. India Sect. B: Biol. Sci. 83 (4) (2013) 561–567.

[65] C.B. Patel, V.K. Singh, A.P. Singh, M. Meena, R.S. Upadhyay, Microbial genes involved in interaction with plants, in: H.B. Singh, V.K. Gupta, S. Jogaiah (Eds.), New and Future Developments in Microbial Biotechnology and Bioengineering, Elsevier, Singapore, 2019, pp. 171–180, https://doi.org/10.1016/B978-0-444-63503-7.00010-3.

[66] R. Sunkar, Y.F. Li, G. Jagadeeswaran, Functions of microRNAs in plant stress responses, Trends Plant Sci. 17 (4) (2012) 196–203.

[67] S. Banerjee, R. Palit, C. Sengupta, D. Standing, Stress induced phosphate solubilization by *Arthrobacter* sp. and *Bacillus* sp. isolated from tomato rhizosphere, Aust. J. Crop. Sci. 4 (6) (2010) 378–383.

[68] A.C. Udayashankar, S.C. Nayaka, M.S. Reddy, C. Srinivas, Plant growth-promoting rhizobacteria mediate induced systemic resistance in rice against bacterial leaf blight caused by *Xanthomonas oryzae* pv. oryzae, Biol. Control 59 (2) (2011) 114–122.

[69] H. Ghazalibiglar, J.G. Hampton, E.V.Z. de Jong, A. Holyoake, Is induced systemic resistance the mechanism for control of black rot in *Brassica oleracea* by a *Paenibacillus* sp.? Biol. Control 92 (2016) 195–201.

[70] J.S. Son, J. Sumayo, Y.J. Hwang, B.S. Kim, S.Y. Ghim, Screening of plant growth-promoting rhizobacteria as elicitor of systemic resistance against gray leaf spot disease in pepper, Appl. Soil Ecol. 73 (2014) 1–8.

[71] A.M. El-Borollosy, M.M. Oraby, Induced systemic resistance against cucumber mosaic cucumovirus and promotion of cucumber growth by some plant growth-promoting rhizobacteria, Ann. Agric. Sci. 57 (2) (2012) 91–97.

[72] B.D. Lee, S. Dutta, H. Ryu, S.J. Yoo, D.S. Suh, K. Park, Induction of systemic resistance in *Panax ginseng* against *Phytophthora cactorum* by native *Bacillus amyloliquefaciens* HK34, J. Ginseng Res. 39 (3) (2015) 213–220.

[73] D. Vitullo, A. Di Pietro, A. Romano, V. Lanzotti, G. Lima, Role of new bacterial surfactins in the antifungal interaction between *Bacillus amyloliquefaciens* and *Fusarium oxysporum*, Plant Pathol. 61 (4) (2012) 689–699.

[74] D. Niu, X. Wang, Y. Wang, X. Song, J. Wang, J. Guo, H. Zhao, *Bacillus cereus* AR156 activates PAMP-triggered immunity and induces a systemic acquired resistance through a NPR1-and SA-dependent signaling pathway, Biochem. Biophys. Res. Commun. 469 (1) (2016) 120–125.

[75] T. Ding, B. Su, X. Chen, S. Xie, S. Gu, Q. Wang, H. Jiang, An endophytic bacterial strain isolated from *Eucommia ulmoides* inhibits southern corn leaf blight, Front. Microbiol. 8 (2017) 903.

[76] G. Farace, O. Fernandez, L. Jacquens, F. Coutte, F. Krier, P. Jacques, S. Dorey, Cyclic lipopeptides from *Bacillus subtilis* activate distinct patterns of defence responses in grapevine, Mol. Plant Pathol. 16 (2) (2015) 177–187.

[77] J. Ryals, S. Uknes, E. Ward, Systemic acquired resistance, Plant Physiol. 104 (4) (1994) 1109.

[78] M. Meena, P. Swapnil, A. Zehra, M.K. Dubey, M. Aamir, C.B. Patel, R.S. Upadhyay, in: H.B. Singh, V.K. Gupta, S. Jogaiah (Eds.), Virulence factors and their associated genes in microbes, Elsevier, New and Future Developments in Microbial Biotechnology and Bioengineering, 2019, pp. 181–208, https://doi.org/10.1016/B978-0-444-63503-7.00011-5.

[79] S. Mokrani, A. Rai, L. Belabid, A. Cherif, H. Cherif, M. Mahjoubi, E. Nabti, Pseudomonas diversity in western Algeria: role in the stimulation of bean germination and common bean blight biocontrol, Eur. J. Plant Pathol. 153 (2) (2019) 397–415.

[80] N. Someya, M. Nakajima, K. Watanabe, T. Hibi, K. Akutsu, Potential of *Serratia marcescens* strain B2 for biological control of rice sheath blight, Biocontrol Sci. Tech. 15 (1) (2005) 105–109.

[81] B. Sreedevi, M. Charitha Devi, D.V.R. Saigopal, Isolation and screening of effective *Trichoderma* spp. against the root rot pathogen *Macrophomina phaseolina*, J. Agric. Technol. 7 (3) (2011) 623–635.

[82] S.Y. Koo, K.S. Cho, Isolation and characterization of a plant growth-promoting rhizobacterium, Serratia sp. SY5, J. Microbiol. Biotechnol. 19 (11) (2009) 1431–1438.

[83] Y. Huang, Z. Wu, Y. He, B.C. Ye, C. Li, Rhizospheric *Bacillus subtilis* exhibits biocontrol effect against Rhizoctonia solani in pepper (*Capsicum annuum*), Biomed. Res. Int. 2017 (2017) 1–9.

[84] S. Mokrani, L. Belabid, B. Bedjaoui, E. Nabti, Growth stimulation of *Phaseolus vulgaris* L plantules by strain *Bacillus amyloliquefaciens* Hla producer of beneficial agricultural enzymes, JOJ Hortic. Arboric. 2 (2) (2018) 1–7.

[85] S.K. Upadhyay, J.S. Singh, A.K. Saxena, D.P. Singh, Impact of PGPR inoculation on growth and antioxidant status of wheat under saline conditions, Plant Biol. 14 (4) (2012) 605–611.

[86] O.O. Babalola, E.O. Osir, A.I. Sanni, G.D. Odhiambo, W.D. Bulimo, Amplification of 1-amino-cyclopropane-1-carboxylic (ACC) deaminase from plant growth promoting rhizobacteria in Striga-infested soil, Afr. J. Biotechnol. 2 (6) (2003) 157–160.

Chapter 19

Microbial interventions for improving agricultural performance under salt stress

Anisha Shashidharan[a] and Lhea Blue[b]
[a]Department of Botany, St. Albert's College (Autonomous), Ernakulam, Kerala, India,
[b]Avila College of Education, Ernakulam, Kerala, India

1 Introduction

Microorganisms have been on the Earth since the first signs of life started to appear and they have been phenomenal in the evolution of life forms during all geological periods. Since they are present everywhere, their influence is felt in all parts of the Earth. Although very small in size, they are significant players in all types of ecosystems. Plants being sessile organisms are exposed to all kinds of stress, both biotic and abiotic, without the option for escaping. Hence, they are forced to develop mechanisms to deal with the various types of stress. Those plants that fail to develop such mechanisms eventually perish when they get exposed to stress. Drought and salinity are the two major abiotic stresses to which plants are usually exposed to. Accumulation of large amount of salt in soil and water makes the land unfit for cultivation due to loss of fertility [1]. The threat from salinity has continued to increase in the last few years. Salinity affects all plants but the impact on crop plants has to be taken with great seriousness since it is a potential threat to food production and the availability of food to everyone. The menace of soil salinity is more disturbing in the arid and semiarid parts of the world [2]. Around the globe, 20% of total cultivated land and 33% of irrigated land are affected by salinity and the rate of salinization is also increasing because of the change in climate patterns [3]. In the coming years, more areas of land will become unsuitable for agriculture due to salinity and hence solutions to this issue need to be worked out as an emergency requirement.

Plants maintain strong coordination with the other organisms around them for mutual good. This is also necessary to regulate their metabolic activities in accordance with the surroundings so as to efficiently regulate and support their own growth and development. Microorganisms are a significant group of

Plant-Microbe Interaction—Recent Advances in Molecular and Biochemical Approaches
https://doi.org/10.1016/B978-0-323-91876-3.00016-6

organisms that interact with plants quite frequently. Various organic compounds such as sugars, vitamins, and organic acids are produced by plants which serve as signals for various microbial species. At the same time, microbes produce small molecules, phytohormones, or volatile compounds, which may be significant players in processes like growth, development, or immunity of the plant in a direct or indirect manner [4,5]. Healthy and properly regulated interactions between plants and microbes have prime importance for improving plant growth along with maintenance of proper soil conditions. The region of soil surrounding the plant roots or the rhizosphere is mostly colonized by microbes [6]. The narrow region of soil that surrounds the plant roots is called rhizosphere. This region is influenced by secretions from root and can harbor about 1011 cells of microbes per gram of root [7]. The rhizosphere may contain more than 30,000 prokaryotic species [8]. The microbes in the rhizosphere are involved in various interactions with the plant and can have varied influence on the host. Therefore, soil microbes have the potentiality to modify the rhizosphere and thus can ameliorate its harmful effects on plant growth and development under conditions of stress.

2 Relationship between microbes and plants

Microorganisms have been found to improve the growth of diverse crops grown under various stress conditions. Under natural conditions, most nutrients such as N, P, and S occur in bound form in organic molecules and thus are least bioavailable for plants. Soil microbes are equipped with metabolic pathways for depolymerizing and mineralizing the organic forms of N, P, and S. Thus, plants depend on the soil microbes such as bacteria and fungi for getting these nutrients [9]. The utilization of microbes for improving the availability of nutrients to plants is an important strategy to improve agricultural production.

Though soil may be considered as just a source of nutrients for the plants, it actually represents a complex ecosystem possessing organisms such as bacteria, fungi, animals, and protists [10–12]. Plants exhibit all sorts of possible ecological interactions (neutral, competitive, commensal, exploitative, and mutualistic) with these soil-dwelling organisms. Due to the omnipresent nature of microbes (present below the ground, above the ground as well as within the plants), their associations with plants are highly diverse [13,14].

Microbial interactions with plants may be both endophytic and epiphytic, and also with the nearby environment and soil in the proximity of plant roots. Microorganisms associated with plants may be of three types: rhizosphere microorganisms, residing in the vicinity of roots; rhizoplane microorganisms, residing on the surface of root; and endophytic microorganisms, living in the interior of tissues without causing harm to the host [15]. The epiphytic microorganisms which live on the surface of plants are the ones isolated from the surface of roots and leaves [16]. The root and its surroundings harbor numerous microbes and the region is busy with plant-microbe interactions. The microbial wealth of this region is significant with regard to the nutrition of plant, abiotic stress tolerance, and protection against pathogens [17–19]. Moreover, the plant-microbe interactions in the rhizosphere are important. The beneficial bacteria and fungi of this region can have a positive impact on plant in the form of better response to stress and increased yield [20].

Both microbes and plants provide a range of services to each other and hence are beneficial to each other. The microbes improve the nutrition of plants and also provide protection to some extent against plant diseases. The interaction between plants and microbes is a complex, dynamic, and continuous process. The millions of years long association of plants with microbes has led to the formation of an assemblage of host and nonhost species, forming a discrete ecological unit referred to as "holobiont" [21].

2.1 Beneficial and harmful interactions

The beneficial interactions between plants and microbes may include mechanisms like nutrient transfer, whereby microorganisms associate with roots and serve the plants with mineral nutrients, fixed nitrogen, direct stimulation of growth through phytohormones, and mitigation of stresses. However, the interactions may be harmful to plants as the invading microbes may be saprophytic and cause necrotrophy in the plants they colonize. In other cases of interaction, several plants produce allelochemicals, which are antagonistic to the growth of microorganisms in their vicinity.

Beneficial microbes in the soil are helpful to the host plants in satisfying their nutritional requirements, overcoming the abiotic stresses (if any), enhancing their growth and fitness, and sustaining plant productivity. Plant growth-promoting bacteria (PGPB), rhizobia (PGPR), and arbuscular mycorrhizal fungi (AMF) form the major plant growth-promoting microorganisms. Plant growth-promoting rhizobia (PGPR) can exist as symbiotic bacteria or rhizobacteria. As symbiotic bacteria, they live inside plants and exchange metabolites with them directly, while rhizobacteria live outside plant cells [22,23].

Plant growth-promoting rhizobacteria (PGPR) adopt various mechanisms to affect plant growth and development, which may be direct or indirect. The direct mechanisms of PGPRs include fixation of nitrogen, sequestration of iron by releasing siderophores, release of various plant growth-promoting hormones, etc. [24,25]. These actions of the PGPR are beneficial to plants under normal conditions of growth and can also be put to use judiciously for those growing under stress conditions. Salinity stress is a major growing concern among agriculturists worldwide and any strategy to diminish the impact of salt stress will be received with great enthusiasm, especially when it is based on natural agents without the use of any chemical or synthetic components. In the recent times, many studies have been conducted for the isolation and characterization of PGPRs from crops growing under salt stress conditions [26–30]. The PGPRs investigated in the above studies include *Xanthobacter autotrophicus*,

TABLE 1 The mode of action of the major PGPR genera [31].

Name of the PGPR	Mode of action
Bacillus	Cycling of nutrients; increased production of IAA; mineralization of insoluble phosphate.
Serratia	Formation of nodules; eases access to nutrients; enhances yield and production.
Pseudomonas	Increase in cell division; maintains hormonal balance; increases uptake of water and nutrients from the soil; elongation of roots; overall growth and development is enhanced.
Rhizobium	Increased colonization of roots facilitated by the release of molecules like flavonoids; enhanced growth of plants.

Enterobacter aerogenes, Bacillus brevis, Pseudomonas sp., *Rhizobium cicer, Alcaligenes* sp., *Ochrobactrum* sp., etc. The mode of action of some of the major PGPRs has been shown in Table 1.

The beneficial microbial communities have been grouped into three categories namely endophytic, phyllospheric, and rhizospheric. Endophytic microbes are beneficial plant growth microbes that enter into the internal tissues of plants such as root, stem, flower, fruits, or seeds. A large number of microbes have been identified as PGP endophytic microbes. They are *Azoarcus, Klebsiella, Enterobacter, Pseudomonas*, etc. Phyllospheric microbes are plant microbes found on the surface of plants and such microbes have the ability to tolerate more abiotic stress like high temperature (35–50°C) and UV radiation. Some examples of phyllospheric microbes are *Arthrobacter, Bacillus, Agrobacterium, Methylobacterium*, etc. The most crucial and predominant association between plants and microbes is the interaction of soil microbes with plant root ecosystems known as rhizospheric microbiomes. This interaction is due to the release of root exudates and microbial activities. A large number of species of diverse genera are rhizospheric microbes such as *Erwinia, Enterobacter, Methylobacterium, Burkholderia, Arthrobacter, Pseudomonas, Serratia, Rhizobium, Paenibacillus, Flavobacterium, Bacillus, Azospirillum, Alcaligenes*, and *Acinetobacter* that have been characterized for plant growth promotion [32–35]. Beneficial microbes have also been classified based on the goal of their application: biofertilizers (such as rhizobia, which have been applied commercially for over a century), phytostimulators (such as auxin-producing, root-elongating *Azospirillum*), rhizoremediators (pollutant degraders which use root exudate as their carbon source), and biopesticides [36].

However, plant-microbial interactions can be negative or harmful also. This happens when the net effect of all soil organisms including pathogens, symbiotic mutualists, and decomposers results in a reduced plant performance. The visible poor performance of plants can be due to the impact of pathogenic microbes that interact with the plant or due to a decrease in the number or activity of beneficial microorganisms [37].

A higher proportion of negatively interacting microbes in the rhizosphere of susceptible host plants than the beneficial ones threatens the survival of the host as well as leads to a deterioration of the diversity of native rhizosphere microflora. There may be changes in the structure of microbial community surrounding the host in the soil as well as change in the number of native microorganisms. These may altogether be disadvantageous to the host plant and its associated microbes with regard to nutrient availability or plant- and microbial-related metabolic activities.

3   Salt stress in plants

Plants require small amount of minerals and salts as a part of nutrition for maintaining healthy growth and development. However, excess salt in soil or water may lead to adverse consequences. Though plants may be exposed to various types of stress conditions, both abiotic and biotic, salinity forms one of the major abiotic stresses that badly affect the performance of crop plants all over the world [38]. Moreover, in the recent times, the problem of excess salinity in soil as well as irrigation water has aggravated significantly. The major reasons behind the emergence of salinity as a significant abiotic stress factor are increasing human population that is putting great pressure on the agriculture system, pollution, excess use of fertilizers, changing climate patterns, salts from irrigation water, etc. The increasing population and urbanization has led to the reallocation of fresh water from agriculture to domestic and industrial uses forcing the use of alternative sources for agriculture, which generally involves wastewater which is usually high in terms of salinity [39]. Also, water used for irrigation leaves salt behind on a regular basis making lands under irrigation more prone to salinity than drylands [40].

Salinity, be it in soil or water, is a major threat to agricultural productivity worldwide and needs to be addressed at all possible levels to keep the impact on crop production at bay. Salinity adversely affects plants in a number of ways eventually affecting the production. Not just plants, but all organisms depending directly or indirectly on soil for survival are affected by high salinity, which deteriorates the physical and chemical properties of soil. In plants, continuous exposure to high salinity leads to development of ionic and osmotic stress, which hampers the various biochemical procedures [41].

Some plants are able to maintain low rate of Na^+ and Cl^- transport to leaves, and also ensure accumulation of these ions in the vacuoles rather than in cytoplasm or cell walls making them salt tolerant. In this manner, they avoid the toxicity of the excess salt. However, salt-sensitive plants don't have this potential [42]. Since salinity is a growing problem in agriculture, it needs to be addressed wisely in order to ensure food security for the future. The problem of salinity cannot be eliminated completely but it can be controlled by putting a check on the entry of salt into soil or irrigation water and also by making plants capable of tolerating excess salt by appropriate strategies.

Salinity can have different negative impacts on the plants depending on the species (stress tolerance potential) and the duration for which it is exposed to salinity stress. Initially, the plant experiences osmotic stress due to decrease in the water potential of soil caused by the high concentration of salts. Salinity leads to various physiological, molecular, and biological changes in the affected plants [43]. More specifically, salinity reduces photosynthetic pigments (e.g., carotenoids and chlorophylls), protein synthesis, respiration, lipid metabolism, energy transformation, etc. It also causes alterations in their morphological and anatomical features. There is reduction in yield due to the ionic imbalance caused by salinity by the accumulation of Na and Cl in tissue and the subsequent inhibition of the uptake of minerals [39,44,45]. As the secondary impact of salinity, the plant develops oxidative stress and there is damage to biomolecules such as membrane proteins, lipids, nucleic acids, and dysfunctional metabolism [46]. The damage is caused by the overproduction of reactive oxygen species (ROS), like hydroxyl radical ($\cdot$OH), singlet oxygen ($^1O_2$), superoxide anion ($O_2^{\cdot-}$), and hydrogen peroxide (H_2O_2) [47]. Leguminous plants undergo decreased nodulation, nitrogenase activity, and also reduced biological nitrogen fixation under the influence of salinity [47].

Plants have developed a number of mechanisms in order to deal with the problem of high salinity. The major responses include compartmentalization of ions, synthesis of compatible solutes, and detoxification of ROS by the production of antioxidative enzymes and compounds [48–50]. Superoxide dismutase (SOD), catalase (CAT) and phenol peroxidase (POX) are the enzymes that protect the plants from ROS. Antioxidant compounds, such as ascorbate, glutathione, tocopherol, and carotenoids, also neutralize ROS. These compounds stabilize the subcellular structures and free radical scavenging and impart protection from dehydration caused by the osmotic stress [40,51]. Hence, by increasing the levels of ROS-scavenging compounds, plants are able to develop tolerance to various stressors, including salinity [47].

Since, salinity is gradually turning out to be a colossal problem, the innate mechanisms of plants may not be sufficient to effectively deal with this stress without compromising on the productivity of the plant. Hence, novel strategies are needed to overcome the complexities caused by salinity. Moreover, the world nations need to ensure food security for the coming generations amidst these increasingly stressful agricultural conditions. So, we are in a situation where on one hand, most of the crop plants are challenged by salinity conditions for maintaining their existing production rates and on the other hand there is an increasing global population which asks for much higher crop production rates in order to satisfy the coming generations. Many strategies have been proposed to cope up with the salinity stress and one of the most promising is the use of microbial inoculants, which are capable of alleviating salt stress, enhance the growth of plants, and control plant diseases [52].

4 Role of microbes in the alleviation of salt stress in plants

The relationship between plants and microbes has been studied quite frequently for both positive and negative impacts. The utilization of microbial associations to overcome salt stress is a concept that has developed in the last few years. The phytomicrobiome, which includes a plant together with its associated microbial community, functions as a holobiont. The features of the host plant are also influenced by the phytomicrobiome, which in turn facilitates its adaptation to the habitat. Members of the phytomicrobiome, which include plant growth-promoting rhizobacteria (PGPR), arbuscular mycorrhizal fungi (AMF), and other facultative endosymbionts are inoculated as microbial consortia for helping plants alleviate salt stress [40].

Generally, the zone called rhizosphere contains greater number of microbes than the other parts of soil [52]. The free-living beneficial bacteria living in the rhizosphere are known as plant growth-promoting rhizobacteria (PGPR) [53]. PGPR may be endophytic or live outside the plants. The externally living PGPR get into different types of associations with the plant roots. Most of the PGPR colonize the root surface and live in spaces between root hairs and rhizodermal layers [23]. A number of compounds are secreted by the plant roots, many of which act as signals of different sorts for the microbes in the rhizosphere. Such secretions include phenols, flavonoids, and organic acids. These compounds act as chemical signals for various microbial phenomena like bacterial secretion of exopolysaccharides, chemotaxis, biofilm formation, and quorum sensing during rhizosphere colonization [54–56].

Microbial forms may also be sensitive to excess salts but there are salt-tolerant microbial forms that can survive the stressed conditions and also help plants survive the salt stress. However, bacteria associated with roots have been found to be more tolerant to salt stress than other soil bacteria, since salt stress is higher in the rhizosphere region because water in this region is taken up by the root which leads to an increase in the ionic strength as well as osmolality [57]. Many PGPR strains have shown high tolerance to salt stress, as high as 3% NaCl, which enables them withstand the conditions in saline soils [52].

Even under normal soil conditions, the PGPR support plant growth and development through a number of mechanisms like enhanced nutrient assimilation (biofertilizers) by biological nitrogen fixation, iron acquisition or phosphorous solubilization, control of pathogens by antagonism and competition (biocontrol agents), degradation of organic pollutants and reduction of metal toxicity of contaminated soils (bioremediation), and facilitating phytoremediation [40] (Fig. 1). In an extended role, PGPR have been known to have an impact on the regulation of abiotic stress via direct and indirect mechanisms that induce systemic tolerance [58]. Many PGPR have been investigated for their role in

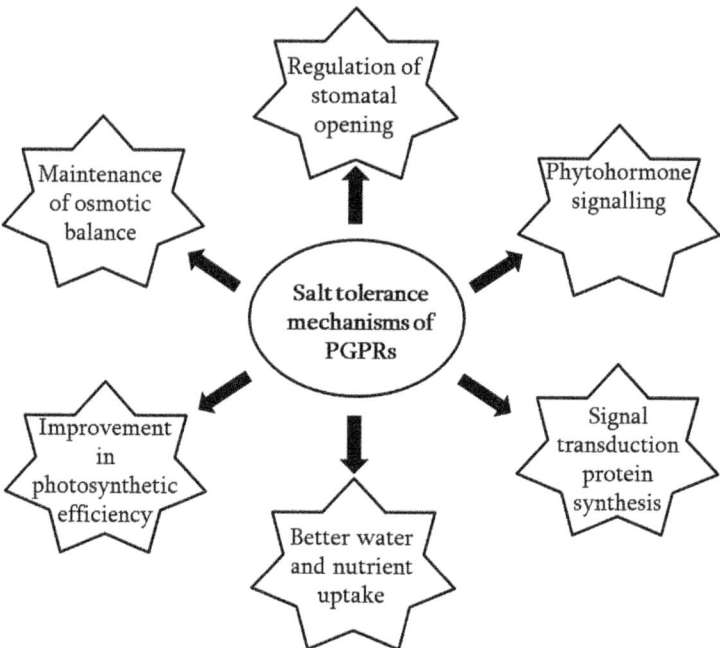

FIG. 1 Strategies of PGPR for dealing with salt stress.

improving ion homeostasis, plant-water relations, and photosynthetic efficiency of plants under salt stress. However, the mechanisms involved are intricate and often not understood properly. Even though the exact mechanisms and pathways involved in imparting salt tolerance is complicated and not understood well, various high-throughput techniques have shown that the major mechanisms involved in the action of PGPR are related to stomatal conductance, water and nutrient uptake, phytohormonal status, ion transport, antioxidant enzymes, signal transduction proteins, and carbohydrate metabolism [40].

5   Mechanism of action of microbes

The PGPR impart salt tolerance to plants by various mechanisms which are a significant area of research in crop plants.

5.1   Osmotic balance

The first direct impact of salinity on plants is the development of osmotic stress. PGPR under saline conditions help in maintaining osmotic balance. They regulate water potential and stomatal opening by affecting hydraulic conductivity and transpiration rate. In addition to this, PGPR stimulate the accumulation of osmolytes and phytohormone signaling, which is helpful to plants in withstanding the initial osmotic stress due to salinization. Some bacteria like *Bacillus subtilis*, *B. amyloliquefaciens* have shown these effects in their salt-tolerant strains [59,60]. Under stress conditions, microbes concentrate large quantities of osmoprotectants in their cytoplasm [61]. Not only this, most likely the PGPRs also exhibit quicker biosynthesis of osmolytes including proline, trehalose, and glycine betaines than their associated host plants. The compatible solutes absorbed through plant roots aid in maintaining osmotic balance and preventing cellular oxidative damage under saline conditions.

5.2 Ion homeostasis

PGPR limit the uptake of salt by the plant by complicated mechanisms like trapping cations in the exopolysaccharide matrix, changes in the root structure with extensive rhizosheaths, and regulation of the expression of ion affinity transporters. All sorts of mineral nutrient exchanges are increased by PGPR and this reduces the nutrient imbalance caused by the high influx of Na^+ and Cl^- ions [40].

K^+ is an inorganic solute and plays an important role in the maintenance of ionic strength, osmotic pressure, and osmotic adjustment in plants cultivated under nonsaline as well as salt stress conditions [49]. Generally, the K^+ ions uptake decreases, while the entry of Na^+ increases in untreated plants under salt stress. However, under the influence of inoculation with *Bradyrhizobium* or coinoculation with *Bradyrhizobium* and *Bacillus*, cowpea plants showed unchanged K^+ levels when compared to plants without salt stress [47]. Changes in the expression patterns of the genes involved in ionic homeostasis were also observed as an indication of salt tolerance [62].

5.3 Phytohormone signaling

Plant growth is regulated by a number of hormones or growth regulators, which are produced by the plant themselves. PGPRs also produce exogenous hormones, which impart enhanced tolerance to salinity [40]. Rhizobacteria produce auxins mainly by the utilization of tryptophan present in the plant root exudates. However, other mechanisms also exist. The tryptophan is converted to indole-3-acetic acid (IAA), which is taken up by the plants through roots. Thus, an auxin signaling pathway is initiated ultimately leading to cell division and growth [63].

Another hormone in plants is ethylene, which imparts tolerance to stress but at the same time growth is suspended. There are PGPR which produce 1-aminocyclopropane-1-carboxylase (ACC) deaminase, which converts ACC, the precursor of ethylene to ammonia and α-ketobutyrate. Thus, ethylene production is hindered but the plant develops better salt tolerance [64]. Cytokinins are the other group of plant hormones, which are produced by PGPR and help in developing tolerance to salinity [39].

5.4 Others

Normally, salt stress leads to a decrease in the amount of chlorophyll in the plants. However, in the presence of PGPR, not only chlorophyll but also phenol and polyphenol contents in the leaves increased [65]. The maintenance of chlorophyll ensures the normal plant metabolism. Not only have this, PGPR been reported to have induced the accumulation of signaling molecules like salicylic acid and jasmonate. These molecules are significant in the signaling mechanisms in plants under stress [66]. The association with PGPR equips the plant for a better signaling mechanism so that necessary and appropriate response can be initiated.

6 Impact on agricultural production

The role played by PGPR in alleviating the effect of salt stress is indeed remarkable. A number of microbe and crop combinations have been tried to find out the contribution made by microbes and the results are promising. The strategy to use microbial association with plants to overcome the impact of stress has got many advantages. It helps in ensuring normal produce from crop plants, no issues of pollution are involved, and the agricultural product is totally safe for use. With all these benefits, the use of PGPRs as a remedy for dealing with the issue of increasing salinity in soil and irrigation water appears to be the most promising strategy.

7 Conclusions

Salinity is a serious problem in agriculture and it is expected to aggravate in the coming years. There would not be many options for better soil or irrigation water. The future is indeed challenging because the human population keeps on increasing significantly, while the area of land available for agriculture keeps on reducing. Hence, effective strategies to ensure maximum yields from the available land masses are mandatory. In this circumstance, the use of PGPR to help plants to grow even under salt stress conditions shall turn out to be a successful approach but more trials to understand the microbes with salt tolerance potential and the preferred plants. The widespread use of this approach will definitely provide a solution to the issue of food security.

References

[1] A. Vaishnav, A.K. Shukla, A. Sharma, R. Kumar, D.K. Choudhary, Endophytic bacteria in plant salt stress tolerance: current and future prospects, J. Plant Growth Regul. 38 (2) (2019) 650–668.

[2] S. El hasini, I.O. Halima, M.E. Azzouzi, A. Douaik, K. Azim, A. Zouahri, Organic and inorganic remediation of soils affected by salinity in the Sebkha of Sed El Mesjoune—Marrakech (Morocco), Soil Tillage Res. 193 (2019) 153–160.

[3] R. Mukhopadhyay, B. Sarkar, H.S. Jat, P.C. Sharma, N.S. Bolan, Soil salinity under climate change: challenges for sustainable agriculture and food security, J. Environ. Manag. 280 (2021), 111736.

[4] M. Meena, P. Swapnil, A. Zehra, M.K. Dubey, R.S. Upadhyay, Antagonistic assessment of *Trichoderma* spp. by producing volatile and non-volatile compounds against different fungal pathogens, Arch. Phytopathol. Plant Protect. 50 (13–14) (2017) 629–648.

[5] R. Ortíz-Castro, H.A. Contreras-Cornejo, L. Macías-Rodríguez, J. López-Bucio, The role of microbial signals in plant growth and development, Plant Signal. Behav. 4 (8) (2009) 701–712.

[6] G. Berg, L. Eberl, A. Hartmann, The rhizosphere as a reservoir for opportunistic human pathogenic bacteria, Environ. Microbiol. 7 (11) (2005) 1673–1685.

[7] D. Egamberdieva, F. Kamilova, S. Validov, L. Gafurova, Z. Kucharova, B. Lugtenberg, High incidence of plant growth-stimulating bacteria associated with the rhizosphere of wheat grown on salinated soil in Uzbekistan, Environ. Microbiol. 10 (1) (2008) 1–9.

[8] R. Mendes, M. Kruijt, I. De Bruijn, E. Dekkers, M. van der Voort, J.H. Schneider, J.M. Raaijmakers, Deciphering the rhizosphere microbiome for disease-suppressive bacteria, Science 332 (6033) (2011) 1097–1100.

[9] R. Jacoby, M. Peukert, A. Succurro, A. Koprivova, S. Kopriva, The role of soil microorganisms in plant mineral nutrition—current knowledge and future directions, Front. Plant Sci. 8 (2017) 1617.

[10] M. Bonkowski, C. Villenave, B. Griffiths, Rhizosphere fauna: the functional and structural diversity of intimate interactions of soil fauna with plant roots, Plant Soil 321 (1) (2009) 213–233.

[11] D.B. Müller, C. Vogel, Y. Bai, J.A. Vorholt, The plant microbiota: systems-level insights and perspectives, Annu. Rev. Genet. 50 (2016) 211–234.

[12] C.B. Patel, V.K. Singh, A.P. Singh, M. Meena, R.S. Upadhyay, Microbial genes involved in interaction with plants, in: H.B. Singh, V.K. Gupta, S. Jogaiah (Eds.), New and Future Developments in Microbial Biotechnology and Bioengineering, Elsevier, Singapore, 2019, pp. 171–180, https://doi.org/10.1016/B978-0-444-63503-7.00010-3.

[13] D. Bulgarelli, K. Schlaeppi, S. Spaepen, E.V.L. Van Themaat, P. Schulze-Lefert, Structure and functions of the bacterial microbiota of plants, Annu. Rev. Plant Biol. 64 (2013) 807–838.

[14] J.A. Vorholt, Microbial life in the phyllosphere, Nat. Rev. Microbiol. 10 (12) (2012) 828–840.

[15] G. Yadav, M. Meena, Bioprospecting of endophytes in medicinal plants of Thar Desert: an attractive resource for biopharmaceuticals, Biotechnol. Rep. 30 (2021), e00629, https://doi.org/10.1016/j.btre.2021.e00629.

[16] A.V. Sturz, B.R. Christie, J. Nowak, Bacterial endophytes: potential role in developing sustainable systems of crop production, Crit. Rev. Plant Sci. 19 (1) (2000) 1–30.

[17] M. Meena, A. Zehra, P. Swapnil, Harish, A. Marwal, G. Yadav, P. Sonigra, Endophytic nanotechnology: an approach to study scope and potential applications, Front. Chem. 9 (2021), 613343, https://doi.org/10.3389/fchem.2021.613343.

[18] B. Mitter, G. Brader, M. Afzal, S. Compant, M. Naveed, F. Trognitz, A. Sessitsch, Advances in elucidating beneficial interactions between plants, soil, and bacteria, in: Advances in Agronomy, vol. 121, Academic Press, 2013, pp. 381–445.

[19] S.T. Ramírez-Puebla, L.E. Servín-Garciadueñas, B. Jiménez-Marín, L.M. Bolaños, M. Rosenblueth, J. Martínez, E. Martínez-Romero, Gut and root microbiota commonalities, Appl. Environ. Microbiol. 79 (1) (2013) 2–9.

[20] C. Dimkpa, T. Weinand, F. Asch, Plant–rhizobacteria interactions alleviate abiotic stress conditions, Plant Cell Environ. 32 (12) (2009) 1682–1694.

[21] A. Dolatabadian, Plant–microbe interaction, Biology 10 (1) (2021) 15.

[22] J. Goutam, R. Singh, R.S. Vijayaraman, M. Meena, Endophytic fungi: carrier of potential anti-oxidants, in: P. Gehlot, J. Singh (Eds.), Fungi and Their Role in Sustainable Development: Current Perspectives, Springer, Singapore, 2018, pp. 539–551, https://doi.org/10.1007/978-981-13-0393-7_29.

[23] E.J. Gray, D.L. Smith, Intracellular and extracellular PGPR: commonalities and distinctions in the plant–bacterium signaling processes, Soil Biol. Biochem. 37 (3) (2005) 395–412.

[24] S. Mayak, T. Tirosh, B.R. Glick, Plant growth-promoting bacteria confer resistance in tomato plants to salt stress, Plant Physiol. Biochem. 42 (6) (2004) 565–572.

[25] M. Meena, P. Swapnil, K. Divyanshu, S. Kumar, Harish, Y.N. Tripathi, A. Zehra, A. Marwal, R.S. Upadhyay, PGPR-mediated induction of systemic resistance and physiochemical alterations in plants against the pathogens: current perspectives, J. Basic Microbiol. 60 (10) (2020) 828–861.

[26] S.A. Abd El-Azeem, M.W. Elwan, J.K. Sung, Y.S. Ok, Alleviation of salt stress in eggplant (*Solanum melongena* L.) by plant-growth-promoting rhizobacteria, Commun. Soil Sci. Plant Anal. 43 (9) (2012) 1303–1315.

[27] F. Azarmi-Atajan, M.H. Sayyari-Zohan, Alleviation of salt stress in lettuce (*Lactuca sativa* L.) by plant growth-promoting rhizobacteria, J. Hortic. Postharvest Res. 3 (Special Issue-Abiotic and Biotic Stresses) (2020) 67–78.

[28] H.B. Bal, L. Nayak, S. Das, T.K. Adhya, Isolation of ACC deaminase producing PGPR from rice rhizosphere and evaluating their plant growth promoting activity under salt stress, Plant Soil 366 (1) (2013) 93–105.

[29] S. Sultana, S.C. Paul, S. Parveen, S. Alam, N. Rahman, B. Jannat, M.M. Karim, Isolation and identification of salt-tolerant plant-growth-promoting rhizobacteria and their application for rice cultivation under salt stress, Can. J. Microbiol. 66 (2) (2020) 144–160.

[30] H. Yilmaz, H. Kulaz, The effects of plant growth promoting rhizobacteria on antioxidant activity in chickpea (*Cicer arietinum* L.) under salt stress, Legum. Res. 42 (1) (2019) 72–76.

[31] N.K. Arora, S. Tewari, R. Singh, Multifaceted plant-associated microbes and their mechanisms diminish the concept of direct and indirect PGPRs, in: Plant Microbe Symbiosis: Fundamentals and Advances, Springer, New Delhi, 2013, pp. 411–449.

[32] E. Puglisi, S. Pascazio, N. Suciu, I. Cattani, G. Fait, R. Spaccini, M. Trevisan, Rhizosphere microbial diversity as influenced by humic substance amendments and chemical composition of rhizodeposits, J. Geochem. Explor. 129 (2013) 82–94.

[33] A. Singh, R. Kumari, A.N. Yadav, S. Mishra, A. Sachan, S.G. Sachan, Tiny microbes, big yields: microorganisms for enhancing food crop production for sustainable development, in: New and Future Developments in Microbial Biotechnology and Bioengineering, Elsevier, 2020, pp. 1–15.

[34] A.N. Yadav, R. Kumar, S. Kumar, V. Kumar, T.C.K. Sugitha, B. Singh, A.K. Saxena, Beneficial microbiomes: biodiversity and potential biotechnological applications for sustainable agriculture and human health, J. Appl. Biol. Biotechnol. 5 (6) (2017) 4–7.

[35] A.N. Yadav, V. Kumar, H.S. Dhaliwal, R. Prasad, A.K. Saxena, Microbiome in crops: diversity, distribution, and potential role in crop improvement, in: Crop Improvement Through Microbial Biotechnology, Elsevier, 2018, pp. 305–332.

[36] B.J. Lugtenberg, G.V. Bloemberg, Microbe–plant interactions: principles and mechanisms, Antonie Van Leeuwenhoek 81 (1) (2002) 373–383.

[37] J.D. Bever, T.G. Platt, E.R. Morton, Microbial population and community dynamics on plant roots and their feedbacks on plant communities, Annu. Rev. Microbiol. 66 (2012) 265–283.

[38] Y. Ma, M.C. Dias, H. Freitas, Drought and salinity stress responses and microbe-induced tolerance in plants, Front. Plant Sci. 11 (2020) 591911, https://doi.org/10.3389/fpls.2020.591911.

[39] E. Yildirim, M. Turan, M.F. Donmez, Mitigation of salt stress in radish (*Raphanus Sativus* l.) by plant growth promoting rhizobacteria, Rom. Biotechnol. Lett. 13 (5) (2008) 3933–3943.

[40] G. Ilangumaran, D.L. Smith, Plant growth promoting rhizobacteria in amelioration of salinity stress: a systems biology perspective, Front. Plant Sci. 8 (2017) 1768, https://doi.org/10.3389/fpls.2017.01768.

[41] F. Orhan, Alleviation of salt stress by halotolerant and halophilic plant growth-promoting bacteria in wheat (*Triticum aestivum*), Braz. J. Microbiol. 47 (3) (2016) 621–627, https://doi.org/10.1016/j.bjm.2016.04.001.

[42] R. Munns, Comparative physiology of salt and water stress, Plant Cell Environ. 25 (2) (2002) 239–250.

[43] A. Gupta, A. Bano, S. Rai, M. Kumar, J. Ali, S. Sharma, N. Pathak, ACC deaminase producing plant growth promoting rhizobacteria enhance salinity stress tolerance in *Pisum sativum*, 3 Biotech 11 (12) (2021) 1–17.

[44] A.K. Parida, A.B. Das, Salt tolerance and salinity effects on plants: a review, Ecotoxicol. Environ. Saf. 60 (2005) 324–349, https://doi.org/10.1016/j.ecoenv.2004.06.010.

[45] N. Tuteja, Mechanisms of high salinity tolerance in plants, in: Methods in Enzymology, Academic Press Inc, New York, 2007, pp. 419–438.

[46] H. Hossein, M.P. Rezvani, Effect of Water and Salinity Stress in Seed Germination on Isabgol (*Plantago ovata*), vol. 4, 2006, pp. 15–22.

[47] A.D.A. Santos, J.A.G.D. Silveira, A. Bonifacio, A.C. Rodrigues, M.D.V.B. Figueiredo, Antioxidant response of cowpea co-inoculated with plant growth-promoting bacteria under salt stress, Braz. J. Microbiol. 49 (2018) 513–521.

[48] P. Sharma, A.B. Jha, R.S. Dubey, M. Pessarakli, Reactive oxygen species, oxidative damage, and antioxidative defense mechanism in plants under stressful conditions, J. Bot. 2012 (2012), 217037.

[49] P. Shrivastava, R. Kumar, Soil salinity: a serious environmental issue and plant growth promoting bacteria as one of the tools for its alleviation, Saudi J. Biol. Sci. 22 (2) (2014) 123–131.

[50] A. Zehra, M. Meena, M.K. Dubey, M. Aamir, R.S. Upadhyay, Synergistic effects of plant defense elicitors and *Trichoderma harzianum* on enhanced induction of antioxidant defense system in tomato against Fusarium wilt disease, Bot. Stud. 58 (2017) 44, https://doi.org/10.1186/s40529-017-0198-2.

[51] A. Zehra, N.A. Raytekar, M. Meena, P. Swapnil, Efficiency of microbial bio-agents as elicitors in plant defense mechanism under biotic stress: a review, Curr. Res. Microb. Sci. 2 (2021), 100054.

[52] D. Egamberdieva, B. Lugtenberg, Use of plant growth-promoting rhizobacteria to alleviate salinity stress in plants, in: Use of Microbes for the Alleviation of Soil Stresses, vol. 1, Springer, New York, NY, 2014, pp. 73–96.

[53] B. Lugtenberg, F. Kamilova, Plant-growth-promoting-rhizobacteria, Annu. Rev. Microbiol. 63 (2009) 541–556.

[54] D.V. Badri, T.L. Weir, D. van der Lelie, J.M. Vivanco, Rhizosphere chemical dialogues: plant–microbe interactions, Curr. Opin. Biotechnol. 20 (2009) 642–650, https://doi.org/10.1016/j.copbio.2009.09.014.

[55] W.D. Bauer, U. Mathesius, Plant responses to bacterial quorum sensing signals, Curr. Opin. Biotechnol. 7 (2004) 429–433, https://doi.org/10.1016/j.pbi.2004.05.008.

[56] N. Narula, E. Kothe, R.K. Behl, Role of root exudates in plant-microbe interactions, J. Appl. Bot. Food Qual. 82 (2009) 122–130.

[57] A.K. Tripathi, B.M. Mishra, P. Tripathi, Salinity stress responses in the plant growth promoting rhizobacteria, *Azospirillum* sp, J. Biosci. 23 (1998) 463–471.

[58] J. Yang, J.W. Kloepper, C.M. Ryu, Rhizosphere bacteria help plants tolerate abiotic stress, Trends Plant Sci. 14 (2009) 1–4, https://doi.org/10.1016/j.tplants.2008.10.004.

[59] A. Marulanda, R. Azcon, F. Chaumont, J.M. Ruiz-Lozano, R. Aroca, Regulation of plasma membrane aquaporins by inoculation with a *Bacillus megaterium* strain in maize (*Zea mays* L.) plants under unstressed and salt-stressed conditions, Planta 232 (2010) 533–543, https://doi.org/10.1007/s00425-010-1196-8.

[60] C.S. Nautiyal, S. Srivastava, P.S. Chauhan, K. Seem, A. Mishra, S.K. Sopory, Plant growth-promoting bacteria *Bacillus amyloliquefaciens* NBRISN13 modulates gene expression profile of leaf and rhizosphere community in rice, Plant Physiol. Biochem. 66 (2013) 1–9.

[61] B. Kempf, E. Bremer, Uptake and synthesis of compatible solutes as microbial stress responses to high-osmolality environments, Arch. Microbiol. 170 (5) (1998) 319–330.

[62] I. Pinedo, T. Ledger, M. Greve, M.J. Poupin, *Burkholderia phytofirmans* PsJN induces long-term metabolic and transcriptional changes involved in *Arabidopsis thaliana* salt tolerance, Front. Plant Sci. 6 (2015) 466, https://doi.org/10.3389/fpls.2015.00466.

[63] S. Spaepen, J. Vanderleyden, Auxin and plant-microbe interactions, Cold Spring Harb. Perspect. Biol. 3 (2011), a001438, https://doi.org/10.1101/cshperspect.a001438.

[64] B.R. Glick, Z. Cheng, J. Czarny, J. Duan, Promotion of plant growth by ACC deaminase-producing soil bacteria, Eur. J. Plant Pathol. 119 (2007) 329–339, https://doi.org/10.1007/s10658-007-9162-4.

[65] D. Rojas-Tapias, A. Moreno-Galván, S. Pardo-Díaz, M. Obando, D. Rivera, R. Bonilla, Effect of inoculation with plant growth-promoting bacteria (PGPB) on amelioration of saline stress in maize (*Zea mays*), Appl. Soil Ecol. 61 (2012) 264–272.

[66] S. van Wees, M. Luijendijk, I. Smoorenburg, L.C. Van Loon, C.M. Pieterse, Rhizobacteria-mediated induced systemic resistance (ISR) in *Arabidopsis* is not associated with a direct effect on expression of known defense-related genes but stimulates the expression of the jasmonate-inducible gene Atvsp upon challenge, Plant Mol. Biol. 41 (4) (1999) 537–549.

Index

Note: Page numbers followed by *f* indicate figures and *t* indicate tables.

CPI Antony Rowe
Eastbourne, UK
April 21, 2023